Light Scattering Reviews 10

Light Scattering and Radiative Transfer

Alexander A. Kokhanovsky (Editor)

Light Scattering Reviews 10

Light Scattering and Radiative Transfer

Springer

Published in association with
Praxis Publishing
Chichester, UK

PRAXIS

Editor
Dr. Alexander A. Kokhanovsky
EUMETSAT
Darmstadt, Germany

ISBN 978-3-662-51571-6 ISBN 978-3-662-46762-6 (eBook)
DOI 10.1007/978-3-662-46762-6

Springer Heidelberg New York Dordrecht London

Softcover reprint of the hardcover 1st edition 2016

Cover design: Jim Wilkie
Project copy editor: Christine Cressy
Author-generated LaTex, processed by EDV-Beratung Herweg, Germany

Printed on acid-free paper

Springer-Verlag GmbH Berlin Heidelberg is part of Springer Science+Business Media (www.springer.com)

Contents

List of Contributors

Baize, Rosemary R.
NASA Langley Research Center
Mail Stop 420
Hampton
VA 23681
USA
rosemary.r.baize@nasa.gov

Berdnik, Vladimir V.
Kazan Federal University
18 Kremlyovskaya Str.
Kazan
420008
Russia
berdnik@pochta.ru

Bi, Lei
Department of Atmospheric Sciences
Texas A&M University
College Station
TX 77843
USA
bilei@tamu.edu

Harmel, Tristan
Université Pierre et Marie Curie
Laboratoire Océanographie de Villefranche
06230 Villefranche sur Mer
France
harmel@obs-vlfr.fr

Hioki, Souichiro
Department of Atmospheric Sciences
Texas A&M University
College Station
TX 77843
USA
s.hioki@tamu.edu

Horvath, Helmuth
University of Vienna
Faulty of Physics
Aerosol and Environmental Physics
Boltzmanngasse 5
1090 Vienna
Austria
Helmuth.Horvath@univie.ac.at

Hu, Yongxiang
NASA Langley Research Center
Mail Stop 420
Hampton
VA 23681
USA
yongxiang.hu-1@nasa.gov

Huang, Xin
Department of Atmospheric Sciences
Texas A&M University
College Station
TX 77843
USA
huangxin.tamu@gmail.com

Kattawar, George W.
Department of Physics & Astronomy
Texas A&M University
College Station
TX 77843
USA
kattawar@physics.tamu.edu

Lin, Bing
NASA Langley Research Center
Mail Stop 420
Hampton
VA 23681
USA
bing.lin-1@nasa.gov

Loiko, Valery A.
B.I. Stepanov Institute of Physics of the National Academy of Sciences of Belarus
Nezaleznasti Avenue 68
Minsk
Belarus
loiko@dragon.bas-net.by

Lukashin, Constantine
Mail Stop 420
NASA Langley Research Center
Hampton
VA 23681
USA
constantine.lukashin-1@nasa.gov

Panchenko, Mikhail Vasilievich
V.E. Zuev Institute of Atmospheric Optics
Tomsk
Russia
pmv@iao.ru

Savenkov, Sergey N.
Department of Radiophysics
Kyiv Taras Shevchenko University
Kyiv
Ukraine
sns@univ.kiev.ua

Sun, Wenbo
Mail Stop 420
NASA Langley Research Center
Hampton
VA 23681
USA
wenbo.sun-1@nasa.gov

Videen, Gorden
US Army Research Laboratory
2800 Powder Mill Road
Adelphi
MD 20783
USA
gorden.videen@us.army.mil

Xie, Yu
Department of Atmospheric Sciences
Texas A&M University
College Station
TX 77843
USA
xieyupku@gmail.com

Yang, Ping
Department of Atmospheric Sciences
Texas A&M University
College Station
TX 77843
USA
pyang@tamu.edu

You, Yu
Department of Physics & Astronomy
Texas A&M University
College Station
TX 77843
USA
youyu3@gmail.com

Zhuravleva, Tatyana B.
V.E. Zuev Institute of Atmospheric Optics
Tomsk
Russia
ztb@iao.ru

Preface

This volume of *Light Scattering Reviews* is aimed at the discussion of modern problems in polarized light radiative transfer, atmospheric optics, and remote sensing.

The first paper, prepared by Kattawar et al., is devoted to a thorough review of polarized light scattering in the atmosphere and ocean. Presented is a brief review of the history of the study of light-polarization properties. The utilization of light polarization can be traced to the use of sunstones by Vikings in sailing navigation. Many famous scientists, including Thomas Young, Etinne-Louis Malus, and Dominique Francois Arago, substantially contributed to enhancing the knowledge of light-polarization properties. The concepts and terminology introduced by pioneers remain in use today. In this chapter, the authors revisit the basic formalisms of the polarization characteristics of light within the framework of the Stokes parameters and Mueller matrix. In addition, the neutral points in the atmosphere and ocean and polarotaxis for marine animals are discussed. To illustrate the applications of the polarization of radiation, it is shown that the polarimetric observations made by a passive space-borne polarimeter, the POLDER-3 instrument aboard the PARASOL (Polarization and Anisotropy of Reflectances for Atmospheric Sciences, coupled with observations from a lidar) satellite platform can be used to effectively infer the microphysical properties of atmospheric ice crystals and dust particles, particularly the degree of surface roughness of ice crystals and the aspect ratios of dust particles.

Harmel discusses recent developments in the use of light polarization for the studies of marine environment from space. Radiometric measurements by satellite sensors of the light field backscattered from the atmosphere–ocean system represent a powerful tool to monitor marine ecosystems, carbon cycle, or water quality on the global scale. Estimations from space of the chlorophyll-a concentration and the subsequent estimate of the marine primary production, for instance, are currently being based on multispectral measurements of the water-leaving radiance regardless of its state of polarization. On the other hand, new investigations have been focused on the exploitation of polarization of light in the water column and exiting the sea surface to improve our capacities of observing and monitoring coastal and oceanic environments. This chapter attempts to give a brief overview of the recent developments on the use of polarization for marine environment monitoring including assessment of aerosol and atmospheric correction, sea state and associated winds, oceanic and coastal water content, and, potentially, estimation of the ocean carbon stock. First, a short historical review of the successive discoveries punctuated our understanding of light polarization in the marine environment is given. After a description of the transfer of light in the atmosphere–ocean system, impacts of the water constituents (i.e., suspended particles, absorbing material) on polarization are summarized. Recent illustrations of the use and exploitation of light polarization for studying marine environment from laboratory to satellite applications are given. Through the chapter, benefits of polarimetric measurements for monitoring ocean, coastal, or lake environments are discussed in view of the future launch of polarimetric Earth-observing satellite missions.

Polarimetric investigations of terrestrial surfaces are discussed by Savenkov. To develop polarimetric methods for scattering scenes identification and classification, one needs to understand the relation between polarimetric and physical properties of the scatterers. Polarization properties of the scattered light contain extensive information on morphological and functional properties of the scatterers. Since polarization of scattered light depends on the morphological and physical parameters of scatterers (i.e., density, size, distribution, shape, refractive index, etc.) forming the studied object, this information can be utilized for making the scatterers identification techniques. Many constituents of a scattering object also exhibit polarization properties such as birefringence, dichroism, and depolarization which might serve to discriminate between surface and volume scattering as well. The importance of the optical and radar matrix polarimetry is that it contains all the information that one can obtain from the scattering scene. The polarimetric information has many useful applications in such diverse fields as interaction with various optical systems, cloud diagnostics, remote sensing of the ocean, atmosphere, and planetary surfaces, and biological tissue optics. The goal of the chapter is to explain the basics of polarimetry, outline its current state of the art, and review numerous important applications to characterize the terrestrial scattering both in optical range and in radar polarimetry.

Sun et al. present the results of polarized radiative-transfer modeling in the ocean–atmosphere-coupled system. Reflected solar radiation from Earth's ocean–atmosphere system is polarized. If a non-polarimetric radiometric sensor is sensitive to polarization, it will be a source of measurement errors in satellite remote sensing. To correct the errors due to this effect, the polarization state of the reflected solar light must be known with sufficient accuracy. In this chapter, recent studies of the polarized solar radiation for the ocean–atmosphere system with the adding-doubling radiative-transfer model are reviewed. The comparison of the modeled polarized solar radiation quantities with the PARASOL satellite measurements and other radiative-transfer model results, the sensitivities of reflected solar radiation's polarization to various ocean-surface and atmospheric conditions, and a novel super-thin cloud detection method based on polarization measurements are the focus of this chapter.

Panchenko and Zuravleva stress the importance of the aerosol vertical structure for radiative-transfer studies. In this chapter, the authors present a brief overview of the studies devoted to the problems of retrieving vertical profiles of microphysical and optical characteristics of tropospheric aerosols and their subsequent application in radiation calculations. The focus is on the descriptions of the approaches, which have been used to solve these tasks in the course of (i) specific comprehensive experiments and (ii) long-term observations (optical and microphysical approaches, method of microphysical extrapolation). Results obtained during the specific comprehensive experiments (TARFOX, ACE-2, SAFARI, SHADE) were performed by different research groups with the use of instruments installed on board aircraft, ships, and satellites, as well as ground-based equipment. This ensured the implementation of aerosol and radiative closure tests to examine the magnitude of the uncertainties associated with the various techniques used to estimate the vertical structure of certain types of aerosols. Multi-year aircraft observations, which are held in a pre-determined area of the planet, are the basis for models of vertical profiles of climatically significant aerosol parameters specific for this region. The authors discuss the approaches which made it possible (i) to reveal the seasonal differences in the vertical profiles of the optical properties of dry matter aerosol in the visible range over the territory of rural Oklahoma (US) and (ii) to create the empirical model of the vertical profiles of the extinction coefficient, albedo, and scattering-phase function developed for the territory of Western Siberia (Russia).

Horwath discusses the aerosol absorption effects in the terrestrial atmosphere. Light absorption in the atmosphere is due to gases and particles. Most of them show considerable spatial and temporal variations. In the visible, broadband light absorption is due to O_3 and NO_2. Oxygen has some absorption lines. Ozone has a strong absorption in the ultraviolet (UV), shielding the short-wave UV radiation of the Sun. Also, SO_2 is absorbing in the near UV, allowing tracing of plumes. In the infrared (IR) water vapor, CO_2, CH_4, N_2O, halogenated hydrocarbons are strongly absorbing, leading to the well known greenhouse effect. Particles containing elemental (black) carbon, hematite, or other colored substances absorb, but also scatter, light. For particle sizes below 30 nm, the scattering is small compared to absorption. Coating absorbing particles or mixing with a transparent substance enhances the light absorption – that is, the mixed particle absorbs more light than the sum of the absorption of its constituents. The most important light-absorbing substance in atmospheric particles is black (graphitic) carbon, formed during combustion, and thus omnipresent. Since graphite is a conductor, the electrons can absorb any portion of energy, therefore absorbing in the visible, UV, and IR. A slight dependence of the absorption coefficient on wavelength mainly is an effect of the particle size. The nearly black mineral hematite (α-Fe_2O_3) is strongly absorbing up to wavelengths of approximately 600 nm due to a band gap of 1.9–2.2 eV. About 10% of the organic carbon in the atmosphere has a brown appearance and thus is called brown carbon. Humic-like substances (HULIS), particles formed in biomass fires, plant debris and degradation products, soil humics, and atmospheric reaction products can have this property. Again, the light absorption of these particles is higher in the short-wavelength range. The light absorption by pure gases (having negligible light scattering) can be measured by well-established standard technologies. When aerosol particles interact with light, in addition to absorption, the scattering of light by the particles is unavoidable. This complicates the absorption measurement.

Berdnik and Loiko discuss the application of neural networks for optical particle characterization. The review of neural networks to retrieve size and refractive index of homogeneous particles by angular dependence of the single light scattering is presented. Operating principles and training methods of the Multilayer Perceptron Neural Networks (MLP NN), Radial Basis Function Neural Networks (RBF NN), High-Order Neural Networks (HO NN), and Sequences of Neural Networks are discussed and evaluated for the problem considered. The basic part of the article relates to the homogeneous spherical particle characteristics retrieval. Some data on retrieval of characteristics of the absorbing particles and sizing of soft spheroidal particles are discussed as well. The problem on the retrieval of sizes of an individual optically soft particle taken from binary mixtures of either oblate and prolate spheroids or cylinders and oblate spheroids is considered. The multilevel neural networks method with a linear activation function and the method of the discrimination functions are used.

This volume of Light Scattering Reviews is dedicated to the memory of Georgii V. Rozenberg (1914–82), who made outstanding contributions in various branches of atmospheric optics, remote sensing, and radiative transfer.

Darmstadt, Germany
December 2014

Alexander Kokhanovsky

Part I

Polarimetry

1 Polarization of light in the atmosphere and ocean

George W. Kattawar, Ping Yang, Yu You, Lei Bi, Yu Xie, Xin Huang, and Souichiro Hioki

1.1 Introduction

In this chapter, we will present a brief history of polarization, a short description of the Stokes parameter–Mueller matrix representation of polarized light, and techniques for using polarization for remote sensing of both the atmosphere and the ocean. For a collection of the diverse applications of polarization, the reader is referred to the excellent books by Gehrels (1974) and Mishchenko et al. (2010).

The use of polarization for navigation, sustenance, and recognition by both terrestrial and marine organisms has been in effect for several hundred million years. The *raison d'être* is that both skylight and underwater light can be strongly polarized, and nature has found a way for many organisms to utilize this property. Extensive research on skylight polarization has been undertaken as atmospheric observations are quite a natural part of our daily routine, and blue skylight is a source of polarized light. Perhaps because man is not a resident of the sea, research on the polarization properties of the ocean and the hydrosols contained therein has, unfortunately, been very limited.

1.2 Brief history of polarization

A more complete history of polarization can be found in the references (Brosseau, 1998) and at *www.polarization.com/*.

Whether the Vikings found their way to America by using skylight polarization as a navigational compass (Roslund and Beckman, 1994; Dougherty, 2013) by looking at the sky through a crystal, oftentimes referred to as a sunstone, is certainly speculative. The Vikings' amazing sailing achievements were made before the magnetic compass reached Europe from China, and the mechanics of steering a true course on their long voyages out of land sight, especially in the frequent bad weather and low visibility at high latitudes, are still in question. Six and a half centuries elapsed before other polarization properties were reported. Before the end of the 17th century, Erasmus Bartholinus had discovered the birefringence (double refraction) of calcite, $CaCO_3$, from the observation of double images viewed through the crystals, and wrote a 60-page memoir discussing his results (Bartholinus, 1670). He was the first to use the terms 'ordinary ray' and 'extraordinary ray,' which remain in use today. Roughly three years later, Christiaan Huygens explained the double-refraction property of calcite by extending the geometric construction method he used to explain refraction. Huygens (using what we now call the Huygens Principle) realized that, if the speed of light varied with direction, the spherical wavefronts would metamorphose into ellipsoids which would explain the double-refraction property of the crystals. Following the turn of the nineteenth century, rapid discoveries of the polarization properties of various materials and natural media were made. In 1801, the brilliant Thomas Young, who many historians believe deciphered the Egyptian Hieroglyphs before Champollion, illustrated that polarization arose from the transverse nature of light. In 1808, Etienne-Louis Malus, after surviving the French revolution, discovered that polarized light was not restricted only to certain crystals but could be generated by the reflection of sunlight from ordinary surfaces (Malus, 1808). From his experiments, Malus was able to deduce what we now call Malus's Law relating the quantity of transmitted light to the relative position of a polarizing filter when viewing polarized light. These first findings seemed to create a great deal of research activity dealing with polarization. In 1812, Sir David Brewster (probably best known for his discovery of the kaleidoscope), an ex-minister in the church of Scotland, repeated many of Malus's experiments and was able to conclude: 'The index of refraction is the tangent of the angle of polarization' and 'when a ray is polarized by reflection, the reflected ray forms a right angle with the refracted ray.' As history has it, the angle is called the 'Brewster angle' with no credit being given to Malus – an example of Stigler's Law of Eponymy, which states that 'No scientific discovery is named after its original discoverer.' Around 1811, Dominique Francois Arago, while observing interference colors by placing a thin sheet of mica between a glass reflector and a calcite prism, noticed the colors did not disappear when he removed the reflector leaving only blue sky as the background. From this observation, he immediately deduced that blue sky must indeed be polarized! Arago also saw circular polarization when the mica was replaced by a quartz crystal, which we know to be birefringent, and presented his findings to the Paris Academy. Unfortunately, a year later, a former colleague named Jean-Baptiste Biot presented two much more detailed papers and stole the limelight from Arago, and now Biot is sometimes credited with the discovery of circular polarization. One of Arago's greatest accomplishments was his persistence in convincing the brilliant mathematical physicist Augustin Fresnel to keep pursuing optics. Fresnel once stated, 'I have decided to remain a modest engineer ... and even abandon physics ... I now see it as a stupid plan troubling oneself to acquire a small bit of glory.' In 1816, Fresnel developed his theory of diffraction without mentioning polarization but, later that year and in the next year, Fresnel theoretically explained the existence of polarization as well as reflection, refraction, and Brewster's Law. Around 1821, Fresnel discovered that, if light waves were indeed transverse, he could explain many earlier experiments, and went on to show that two light beams polarized at right angles to one another do not interfere (note that many researchers are not familiar with Fresnel and Arago's four laws of interference, which are not usually covered in most modern optics books, and the reader is referred to an article by Collett (1971)). More optical phenomena, such as polarized rotation of quartz and various fluids, linear polarization of reflected light, polarization of comet tails, neutral points (points with no polarization) in the sky, rainbows, and 22° halos, were either discovered or explained in rapid sequence. For the next century and a half, improvements were made in the construction of efficient linear polarizers, and more observations of the polarization effects of natural phenomena were reported. Furthermore, the Lorenz–Mie theory for the scattering of light by small spheres (Lorenz, 1890; Mie, 1908; van de Hulst, 1981) has been a powerful tool to facilitate the theoretical explanation of the polarization characteristics associated with atmospheric scattering phenomena, such as rainbows, blue sky polarization, and glory (a bright, halo-like optical pattern due to the backscattering of light by a cloud of uniformly sized water droplets). Note, a quite extensive review of the history of the Lorenz–Mie theory can be found in Logan (1965).

1.3 Stokes–Mueller formulation

William B. Herapath, an English physician and surgeon, discovered linear dichroism in crystals in what is now known as herapathite. After this, Sir George Stokes introduced the four measurable quantities completely describing the state of polarization of a light beam, which we now call the Stokes parameter or Stokes vector (note that vector is not used in the mathematical sense). Once Edwin Land was able to make the first sheet polarizers by orienting crystalline needles of herapathite in a sheet of plastic, the use of polarization techniques began to blossom. People began to focus on how the state of polarization of a light beam changed when it interacted with optical elements such as polarizers, quarter-wave plates, etc. Soleillet (1929) was one of the first pioneers to suggest how one might characterize an optical device by relating the output Stokes vector to the input Stokes vector; however, his elegant paper received virtually no attention. Soleillet's method was rediscovered by Hans Mueller, who used these matrices in a series of lectures given at MIT from 1945 to 1948, and now the 4×4 matrix transformations bear the name *Mueller matrices* – another example of Stigler's Law of Eponymy at work. Mueller did not publish the work in the open literature, but the theory was later published by N.G. Parke, an MIT student, as part of his Ph.D. dissertation. Another polarization pioneer was Robert C. Jones, who was interested in the optical properties of birefringence and dichroism, and developed a method of transformations which used 2×2 matrices. The method had several limitations, of which one of the most important was being only specific to polarized light beams.

The polarized nature of the underwater light field was discovered over half a century ago, and has been found to aid in target identification and to be detected and utilized by a variety of marine animals. Waterman (1954) and Waterman and Westell (1956) probably made the earliest measurements of the polarized underwater light field and studied the sensitivity of the polarization state to the solar and viewing configuration. Substantially polarized light fields (75% to 80%) in the clear waters near Bermuda were observed by Ivanoff et al. (1961). The laboratory measurements made by Timofeeva (1974) showed a value of the degree of linear polarization (DoLP) as high as 40% in turbid waters. The more recent work on ocean-water polarization includes the large RaDyO (Radiation in a Dynamic Ocean) project (Dickey et al., 2011). The work of Sabbah and Shashar (2007) is the most comprehensive combination of both measurement and theory brought to the fore. As shown by Chami (2007), underwater polarization can also be used by remote sensing to study the optical signature of inorganic particles in coastal waters.

Multispectral underwater polarization properties have drawn a great deal of attention in the experimental biology community, particularly in the study of the impact of light polarization state on marine animals. Evidence has shown that some marine animals are able to detect both the DoLP and the angle of linear polarization (AoLP) specified by the orientation of the electric field vector; therefore, both are important. The impact of atmospheric conditions and water compositions on underwater polarized light fields was discussed by Tonizzo et al. (2009), who reported a hyperspectral and multangular analysis of polarized light in coastal waters.

An alternative approach to study underwater polarization is model simulation, which has been conducted since numerical models that solve the vector radiative-transfer equation became available in the 1970s. Such models include the Multi-Component Approximation (MCA) (Tynes et al., 2001) and the Monte Carlo (Kattawar et al., 1973) methods. With appropriate input parameters including solar irradiance, atmospheric conditions, aerosol loading, and water inherent optical properties (IOPs) in numerical simulations, close agreement has been achieved between theoretical and measured radiances and polarization states. In particular, comparisons between simulated and measured underwater polarized light fields (Adams et al., 2002; Tonizzo et al., 2009; You et al., 2011) demonstrate reasonable consistency in the cases of both the DoLP and the AoLP.

1.4 Stokes parameters

Polarization parameters of light are generally referred to specific planes. Precision is required when defining the planes to which the Stokes vector is referred when considering the polarization of light in radiative-transfer calculations. For a scattered beam in a single-scattering process, a natural selection is the scattering plane defined by the propagation directions of the incident beam and the scattered beam. The radiance of beams can be resolved into two orthogonal components referring to the vibrations of the electric field parallel and perpendicular to the scattering plane. In radiative-transfer calculations, the polarization parameters of a pencil of light are usually defined with respect to meridional planes. If we consider Fig. 1.1, and denote the radiance associated with the incident beam by I_{inc} and the counterpart of the scattered beam by I_{sca}, the directions of the incident beam and the scattered beam determine the scattering plane (the scattering angle is denoted as Θ), and the directions of two beams and the z-axis define two meridional planes. In addition to the reference plane, a proper treatment of the Stokes vector formulation requires careful consideration of relevant definitions due to various freedoms of choice.

Let us consider the meridional plane defined by the directions of the incident beam and the z-axis. Two orthogonal unit vectors $\hat{e}_1$ and $\hat{e}_2$ in the meridional plane

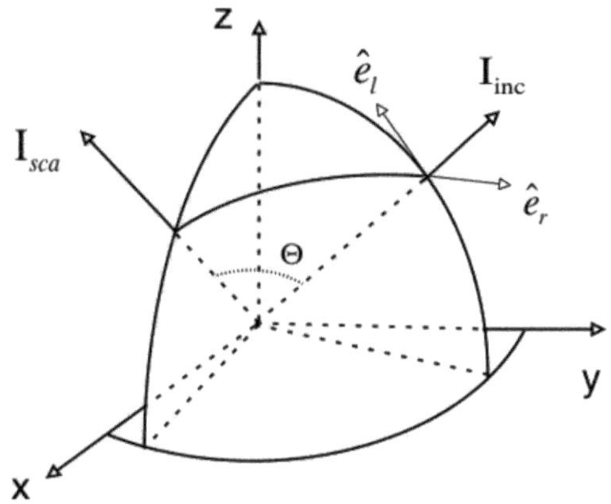

Fig. 1.1. Reference planes used for describing the Stokes vector.

and perpendicular to the plane are defined in such a way that $\hat{e}_r \times \hat{e}_l$ is aligned with the direction of propagation (note this is one of the freedoms of choice; switching $\hat{e}_l$ and $\hat{e}_r$ will result in differences in the foregoing definitions; most texts in the literature adopted this convention but, in Mishchenko et al. (2000), $\hat{e}_\varphi$ and $\hat{e}_\theta$ are used to represent the directions perpendicular and parallel to the meridional plane). Based on the transverse nature of local electromagnetic waves, the electric field vector in conjunction with the incident beam can be resolved into two components:

$$\overline{E} = E_l \hat{e}_l + E_r \hat{e}_r \,, \tag{1.1}$$

where E_l and E_r are complex oscillatory functions with respect to the propagation distance and time. The Stokes vector with four real components can be defined as

$$\begin{bmatrix} I \\ Q \\ U \\ V \end{bmatrix} = \begin{bmatrix} |E_l|^2 + |E_r|^2 \\ |E_l|^2 - |E_r|^2 \\ E_l E_r^* + E_r E_l^* \\ i(E_l E_r^* - E_r E_l^*) \end{bmatrix} , \tag{1.2}$$

where $|\cdot|$ denotes the modulus, the asterisk denotes complex conjugation, and $i = (-1)^{1/2}$. It can be proven that the following relation holds:

$$I^2 = Q^2 + U^2 + V^2 \,. \tag{1.3}$$

The first component I is commonly referred to as the radiance (in most texts, I represents irradiance in single scattering) and the other three parameters have the same units as I. The definition of Eq. (1.2) is consistent with the literature (van de Hulst, 1981; Bohren and Huffman, 1983). Liou (2002) included a minus sign into the definition of the V component. The freedom of sign choice in the definition of the V component is also mixed with the convention adopted in representing the time dependence of harmonic waves. The wave is propagating along the z-axis, in van de Hulst (1981) and Liou (2002) it is

$$E_l = a_l \, e^{-i\varepsilon_l} \, e^{-ikz+i\omega t}, \quad E_r = a_r \, e^{-i\varepsilon_r} \, e^{-ikz+i\omega t} \,, \tag{1.4}$$

while, in Bohren and Huffman (1983), it takes the form

$$E_l = a_l\, e^{-i\varepsilon_l}\, e^{ikz-i\omega t}, \quad E_r = a_r\, e^{-i\varepsilon_r}\, e^{ikz-i\omega t}, \tag{1.5}$$

where t is the time, a_l and a_r are real numbers that represent amplitudes, ε_1 and ε_2 are the phases of the wave at the spatial origin of space at time $t = 0$, k is the modified wave number, and ω is the circular frequency. Both Eqs (1.4) and (1.5) lead to

$$\begin{bmatrix} I \\ Q \\ U \\ V \end{bmatrix} = \begin{bmatrix} a_l^2 + a_r^2 \\ a_l^2 - a_r^2 \\ 2a_r a_l \cos\delta \\ 2a_r a_l \sin\delta \end{bmatrix}, \tag{1.6}$$

and $\delta = \varepsilon_l - \varepsilon_r$. However, if the simple waves

$$E_l = a_l\, e^{i\varepsilon_l}\, e^{ikz-\omega t}, \quad E_r = a_r\, e^{i\varepsilon_r}\, e^{ikz-i\omega t}, \tag{1.7}$$

are employed, the V component will have an opposite sign. Note that the adoption of a different time factor will lead to a complex refractive index with an opposite sign for the imaginary part, and the sign difference of the V component will result in sign differences of the first three Mueller matrix elements in the fourth row or column. Moreover, the sign difference of the V component will cause confusion in the definition of left-handed and right-handed circular light. Note that the I, Q, and U components in Eq. (1.6) are immaterial to the choice of the time dependence, but the sign of the V component depends on both the sign choice in Eq. (1.2) and the representation of harmonic waves in Eq. (1.5).

Consider the most general state of polarization, namely elliptical polarization, where the axes of the ellipse may not be in the $\hat{e}_l$ and $\hat{e}_r$ directions. Let $\hat{p}$ and $\hat{q}$ be orthogonal unit vectors that align with the semi-major and semi-minor axes of the polarization ellipse (see Fig. 1.2). A geometric description of the simple wave can be written as

$$\mathrm{Re}\,(\vec{E}) = \hat{p}a_p \cos(kz - \omega t + \alpha) + \hat{q}a_q \sin(kz - \omega t + \alpha)\,. \tag{1.8}$$

Note that the two components along the $\hat{p}$ and $\hat{q}$ directions satisfy the ellipse equation with the semi-axis of $|a_p|$ and $|a_q|$, and α is an arbitrary phase angle. Because the radiance should be independent of the choice of the representation, the following equation must be satisfied:

$$a_p^2 + a_q^2 = a_l^2 + a_r^2\,. \tag{1.9}$$

For convenience, an auxiliary angle $\beta \in [0, \pi/2]$ is introduced, such that

$$a_p = a\cos\beta, \quad a_q = a\sin\beta, \quad a = \sqrt{a_r^2 + a_l^2}\,. \tag{1.10}$$

With the projection of the electric field given by Eq. (1.8) to the $\hat{e}_l$ and $\hat{e}_r$ directions, we have

$$a\cos\beta\cos(kz - \omega t + \alpha)\cos\chi - a\sin\beta\sin(kz - \omega t + \alpha)\sin\chi$$
$$= a_l\cos(kz - \omega t + \varepsilon_1)\,, \tag{1.11}$$
$$a\cos\beta\cos(kz - \omega t + \alpha)\sin\chi + a\sin\beta\sin(kz - \omega t + \alpha)\cos\chi$$
$$= a_r\cos(kz - \omega t + \varepsilon_2)\,. \tag{1.12}$$

In Eqs (1.11) and (1.12), χ is the angle between $\hat{p}$ and $\hat{e}_l$. Because of the arbitrary position and time, the coefficients of $\cos(kz - \omega t)$ and $\sin(kz - \omega t)$ on both sides of Eqs (1.11) and (1.12) must be the same. Then, we have

$$a\cos\beta\cos\alpha\cos\chi - a\sin\beta\sin\alpha\sin\chi = a_l\cos\varepsilon_1\,, \tag{1.13}$$
$$a\cos\beta\sin\alpha\cos\chi + a\sin\beta\cos\alpha\sin\chi = a_l\sin\varepsilon_1\,, \tag{1.14}$$
$$a\cos\beta\cos\alpha\sin\chi + a\sin\beta\sin\alpha\cos\chi = a_r\cos\varepsilon_2\,, \tag{1.15}$$
$$-a\cos\beta\sin\alpha\sin\chi + a\sin\beta\cos\alpha\cos\chi = -a_r\sin\varepsilon_2\,. \tag{1.16}$$

Based on Eqs (1.13)–(1.16), the Stokes vector can be written with respect to the amplitude and the orientation angle as depicted in Fig. 1.2. On squaring and adding Eqs (1.13) and (1.14), we obtain the intensity along the $\hat{e}_l$ direction

$$a_l^2 = a^2\left(\cos^2\beta\cos^2\chi + \sin^2\beta\sin^2\chi\right)\,. \tag{1.17}$$

Similarly, from Eqs (1.15) and (1.16), we have

$$a_r^2 = a^2\left(\cos^2\beta\sin^2\chi + \sin^2\beta\cos^2\chi\right)\,. \tag{1.18}$$

Adding the product of Eq. (1.13) and (1.15), and that of (1.14) and (1.16) yields

$$2a_r a_l\cos\delta = a^2\cos 2\beta\sin 2\chi\,, \tag{1.19}$$

where $\delta = \varepsilon_1 - \varepsilon_2$. Adding the product of Eq. (1.14) and (1.15), and that of (1.13) and (1.16) yields

$$2a_r a_l\sin\delta = a^2\sin 2\beta\,. \tag{1.20}$$

Based on Eqs (1.13) and (1.14), we have

$$\frac{\text{Eq.[13]}\times\cos\alpha + \text{Eq.[14]}\times\sin\alpha}{\text{Eq.[13]}\times\sin\alpha - \text{Eq.[14]}\times\cos\alpha} \Rightarrow \tan(\alpha - \varepsilon_1) = -\tan\beta\tan\chi\,. \tag{1.21}$$

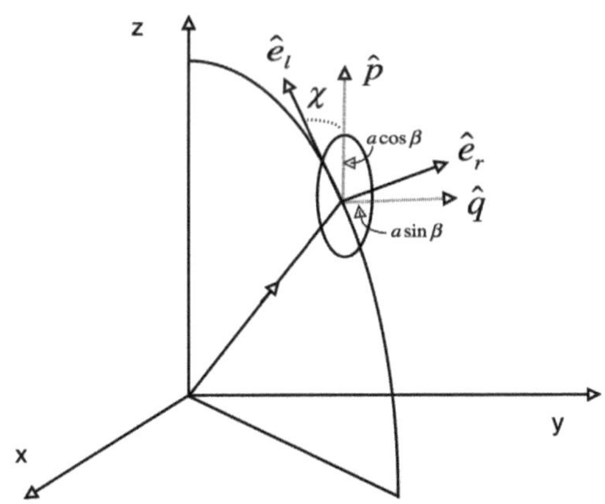

Fig. 1.2. Geometric description of an elliptical wave using the representation.

Similarly, we have

$$\frac{\text{Eq.}[15] \times \cos\alpha - \text{Eq.}[16] \times \sin\alpha}{\text{Eq.}[15] \times \sin\alpha + \text{Eq.}[16] \times \cos\alpha} \Rightarrow \tan(\alpha - \varepsilon_2) = \tan\beta \cot\chi\,, \tag{1.22}$$

therefore

$$\tan\delta = \tan\left[(\alpha - \varepsilon_2) - (\alpha - \varepsilon_1)\right] = -\frac{\tan 2\beta}{\sin 2\chi}\,. \tag{1.23}$$

Note that the sine and cosine terms can be formally switched to represent the geometry of the polarization. In that case, Eqs (1.17), (1.18), (1.19), (1.20), and (1.23) will not be changed, but Eqs (1.21) and (1.22) will be given by

$$\tan(\alpha + \varepsilon_1) = -\cot\beta\cot\chi\,, \tag{1.24}$$
$$\tan(\alpha + \varepsilon_2) = \cot\beta\tan\chi\,. \tag{1.25}$$

Based on Eqs (1.6), (1.17), (1.18), (1.19), and (1.20), the Stokes parameters in the geometric notation now become (van de Hulst, 1981)

$$\begin{bmatrix} I \\ Q \\ U \\ V \end{bmatrix} = \begin{bmatrix} a^2 \\ a^2\cos 2\beta\cos 2\chi \\ a^2\cos 2\beta\sin 2\chi \\ a^2\sin 2\beta \end{bmatrix}. \tag{1.26}$$

Note that Q, U, and V are spherical coordinates, which can be represented in a Poincare sphere (Born and Wolf, 2001).

Given the Stokes vector, it is straightforward from Eq. (1.26) to obtain the orientation of the polarization ellipse, $\tan 2\chi = U/Q$ and the ellipticity $\tan 2\beta = V/(Q^2+U^2)^{1/2}$. Other quantities extracted from the Stokes vector which are used quite often when discussing polarization states, namely the degree of polarization, $(Q^2+U^2+V^2)^{1/2}/I$, the DoLP, $(Q^2+U^2)^{1/2}/I$, and the degree of circular polarization, $|V|/I$.

The Stokes parameters can be defined with respect to different planes of reference, such as the scattering plane or the meridional plane as seen Fig. 1.1. For convenience, in Fig. 1.3, we consider two reference planes with an intersection angle of ϕ. The relationship between the two sets of Stokes parameters is easily shown to be

$$\begin{bmatrix} I' \\ Q' \\ U' \\ V' \end{bmatrix} = \mathbf{R}(\phi)\begin{bmatrix} I \\ Q \\ U \\ V \end{bmatrix} = \begin{bmatrix} 1 & 0 & 0 & 0 \\ 0 & \cos 2\phi & \sin 2\phi & 0 \\ 0 & -\sin 2\phi & \cos 2\phi & 0 \\ 0 & 0 & 0 & 1 \end{bmatrix}\begin{bmatrix} I \\ Q \\ U \\ V \end{bmatrix}, \tag{1.27}$$

where ϕ is defined as the clockwise rotation angle of the $\hat{e}_l$-axis to the $\hat{e}_l'$-axis, and the reader is facing the propagation direction of the beam. The rotation matrix warrants I, Q^2+U^2 and V to be invariant. As expected, the degree of polarization, the DoLP, the degree of circular polarization, and the ellipticity are independent of the plane of reference. The rotation matrix $\mathbf{R}(\phi)$ defined in Eq. (1.27) has some other important properties, namely:

$$\begin{aligned} \mathbf{R}(-\phi) &= \mathbf{R}(\pi - \phi) \\ \mathbf{R}(\phi_1)\mathbf{R}(\phi_2) &= \mathbf{R}(\phi_1 + \phi_2) \\ \mathbf{R}^{-1}(\phi) &= \mathbf{R}(-\phi) \end{aligned} \tag{1.28}$$

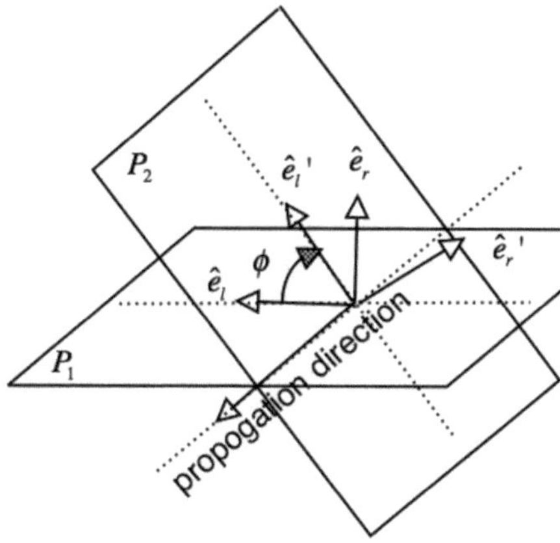

Fig. 1.3. A clockwise rotation by an angle ϕ of the two axes for the Stokes vector representation.

To measure the Stokes parameters associated with a light beam, the measurement set-up schematically illustrated in Fig. 1.4 can be used. The compensator, also called a wave plate or retarder, introduces a certain amount of retardation in the phase of one of the two components of the electric field decomposed with respect to the $\hat{e}_l$ and $\hat{e}_r$ directions. In the following discussion, we assume the phase difference between the $\hat{e}_l$ and $\hat{e}_r$ components of the electric field vector to be ε. The polarizer constrains the measurement of the electric field vector component to vibrating in only one direction and deviating from the $\hat{e}_r$-axis by an angle φ, which is called the direction of the transmission axis of the linear polarizer.

After algebraic manipulation (the details of which can be found in Chandrasekhar (1950)), the measured radiance is a function of ε and φ in the form of

$$I(\varepsilon, \varphi) = I_{i,l} \cos^2 \varphi + I_{i,r} \sin^2 \varphi + \frac{1}{2}(U_i \cos \varepsilon - V_i \sin \varepsilon) \sin 2\varphi \,, \tag{1.29}$$

where the subscript 'i' indicates the Stokes parameters associated with the incident beam. With Eq. (1.29), we consider the following four scenarios:

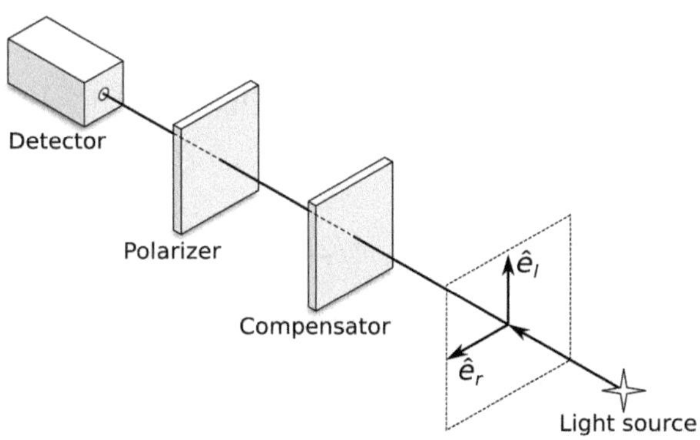

Fig. 1.4. Schematic diagram for measuring the Stokes parameters.

1. **No polarization** (i.e., the incident light is natural light), $I_{i,l} = I_{i,r} = I_i/2$, $U_i = 0$, and $V_i = 0$. Thus, we obtain

$$I(\varepsilon, \varphi) = I_{i,l} \cos^2 \varphi + I_{i,r} \sin^2 \varphi = \frac{I_i}{2}(\cos^2 \varphi + \sin^2 \varphi) = I_i/2 \tag{1.30}$$

or

$$I_i = 2I(\varepsilon, \varphi) \,. \tag{1.31}$$

2. **Horizontal and vertical polarization configurations without a compensator** (i.e., $\varepsilon = 0$, $\varphi = 0°$, and $\varphi = 90°$) for a polarized incident beam:

$$I(\varphi = 0°,\ \varepsilon = 0) = I_{i,l} \,, \tag{1.32}$$
$$I(\varphi = 90°,\ \varepsilon = 0) = I_{i,r} \,. \tag{1.33}$$

Using the above two equations, we obtain the first two Stokes parameters in the form of

$$I_i = I_{i,l} + I_{i,r} = I(\varphi = 0°,\ \varepsilon = 0) + I(\varphi = 90°,\ \varepsilon = 0) \,, \tag{1.34}$$
$$Q_i = I_{i,l} - I_{i,r} = I(\varphi = 0°,\ \varepsilon = 0) - I(\varphi = 90°,\ \varepsilon = 0) \,. \tag{1.35}$$

3. **+45° and −45° polarization configurations without a compensator** (i.e., $\varepsilon = 0$, $\varphi = 45°$, and $\varphi = -45°$) for a polarized incident beam:

$$I(\varphi = 45°,\ \varepsilon = 0) = \frac{1}{2}(I_{i,l} + I_{i,r}) + \frac{1}{2}U_i \,, \tag{1.36}$$
$$I(\varphi = -45°,\ \varepsilon = 0) = \frac{1}{2}(I_{i,l} + I_{i,r}) - \frac{1}{2}U_i \,. \tag{1.37}$$

Thus, we obtain

$$U_i = I(\varphi = 45°,\ \varepsilon = 0) - I(\varphi = -45°,\ \varepsilon = 0) \,. \tag{1.38}$$

4. **Measurement of circular polarization using +45° and −45° polarization configurations with a quarter-wave plate** ($\varepsilon = \pi/2$, $\varphi = 45°$, and $\varphi = -45°$)

$$I(\varphi = 45°,\ \varepsilon = \pi/2) = \frac{1}{2}(I_{i,l} + I_{i,r}) - \frac{1}{2}V_i \,, \tag{1.39}$$
$$I(\varphi = -45°,\ \varepsilon = \pi/2) = \frac{1}{2}(I_{i,l} + I_{i,r}) + \frac{1}{2}V_i \,, \tag{1.40}$$

Thus, we obtain

$$V_i = I(\varphi = -45°,\ \varepsilon = \pi/2) - I(\varphi = 45°,\ \varepsilon = \pi/2) \,. \tag{1.41}$$

Note that the foregoing measurements in scenarios (2)–(4) provide the Stokes parameters regardless of the polarization state of the incident beam. It should be pointed out that a thorough review of the polarized light, particularly light scattered by isotropic opalescent media, can be found in Perrin (1942).

1.5 Mueller matrices

We have shown a method to calculate the complete Stokes vector, and our next task is to determine the nature of the 4 × 4 matrix which transforms an incident Stokes vector into another Stokes vector due to the interaction between light and any object. The resulting matrix is called the Mueller matrix (Mueller, 1948) but, as pointed out earlier, was first discovered by Soleillet (1929). The matrix gives essentially all the optical information possible about an elastic scattering system where some interaction has occurred, and thus characterizes the optical properties of the involved scattering system. Many excellent references on the theory of Mueller matrices exist, but we refer the reader to the book by Shurcliff (1962).

As an example, we consider an ideal linear polarizer – that is, the compensator in Fig. 1.4 is removed (i.e., $\varepsilon = 0$) and Eq. (1.29) reduces to

$$I(\varphi) = I_{i,l}\cos^2\varphi + I_{i,r}\sin^2\varphi + \frac{1}{2}U_i\sin 2\varphi\,. \tag{1.42}$$

Using the relation $I_{i,l}\cos^2\varphi + I_{i,r}\sin^2\varphi = (I_i + Q_i\cos 2\varphi)/2$, the preceding equation can be further simplified in the form

$$I(\varphi) = \frac{1}{2}(I_i + Q_i\cos 2\varphi + U_i\sin 2\varphi)\,, \tag{1.43}$$

which is the first row of the following Mueller matrix associated with a linear polarizer (Bohren and Huffman, 1983), given by

$$\begin{bmatrix} I \\ Q \\ U \\ V \end{bmatrix} = \frac{1}{2}\begin{bmatrix} 1 & \cos 2\varphi & \sin 2\varphi & 0 \\ \cos 2\varphi & \cos^2 2\varphi & \cos 2\varphi\sin 2\varphi & 0 \\ \sin 2\varphi & \cos 2\varphi\sin 2\varphi & \sin^2 2\varphi & 0 \\ 0 & 0 & 0 & 0 \end{bmatrix}\begin{bmatrix} I_i \\ Q_i \\ U_i \\ V_i \end{bmatrix}. \tag{1.44}$$

Obviously, the linear polarizer is not able to generate the V component. Similarly, the Mueller matrix can characterize the optical properties of any optical element.

The Mueller matrices of the optical elements can be obtained in a straightforward manner because their optical properties are already known. For air–ocean interfaces, molecules, and particles involved in the atmospheric and oceanic radiative-transfer simulations, the Mueller matrices must be obtained from either physical principles or Maxwell's equations. Consider the dielectric surface between the atmosphere and the ocean; the Mueller matrices associated with the reflection and the transmission of light either from air to medium or medium to air have to be obtained from physical principles. For example, in the case of the air-to-medium reflection based on the Fresnel's reflection, the reflection Mueller matrix from air into a medium can be obtained as

$$R_{AM} = \begin{bmatrix} \alpha+\eta & \alpha-\eta & 0 & 0 \\ \alpha-\eta & \alpha+\eta & 0 & 0 \\ 0 & 0 & \gamma_{\mathrm{Re}} & 0 \\ 0 & 0 & 0 & \gamma_{\mathrm{Re}} \end{bmatrix}, \tag{1.45}$$

where

$$\alpha = \frac{1}{2}\frac{\tan^2(\theta_i - \theta_t)}{\tan^2(\theta_i + \theta_t)}, \tag{1.46}$$

$$\eta = \frac{1}{2}\frac{\sin^2(\theta_i - \theta_t)}{\sin^2(\theta_i + \theta_t)}, \tag{1.47}$$

$$\gamma_{\mathrm{Re}} = \frac{\tan(\theta_i - \theta_t)\sin(\theta_i - \theta_t)}{\tan(\theta_i + \theta_t)\sin(\theta_i + \theta_t)}. \tag{1.48}$$

In Eqs (1.46) and (1.47), α and η are reflectivities associated with polarization in the parallel and perpendicular directions. The Mueller matrices associated with the transmissivity and the scenarios from the medium to air can be similarly obtained (Kattawar and Adams, 1989).

To obtain the Mueller matrices for various particles, one must turn to Maxwell's equations. Based on the far-field approximation, the amplitude-scattering matrix is usually defined to transform the incident field to the scattered field:

$$\begin{bmatrix} E_{\parallel} \\ E_{\perp} \end{bmatrix} = \frac{e^{ikr}}{-ikr}\begin{bmatrix} S_2 & S_3 \\ S_4 & S_1 \end{bmatrix}\begin{bmatrix} E_{\parallel,i} \\ E_{\perp,i} \end{bmatrix}. \tag{1.49}$$

The scattered field by molecules, for particles much smaller than the incident wavelength, specifically, when $x \ll 1$ (Shifrin, 1951), can be easily obtained from Maxwell's equations, in which case the amplitude-scattering matrix is given by

$$\begin{bmatrix} S_2 & S_3 \\ S_4 & S_1 \end{bmatrix} \propto \begin{bmatrix} \cos\Theta & 0 \\ 0 & 1 \end{bmatrix}. \tag{1.50}$$

The Mueller matrix for Rayleigh scattering is given by

$$\frac{1}{2}\begin{bmatrix} \mu^2+1 & \mu^2-1 & 0 & 0 \\ \mu^2-1 & \mu^2+1 & 0 & 0 \\ 0 & 0 & \mu & 0 \\ 0 & 0 & 0 & \mu \end{bmatrix}, \tag{1.51}$$

where $\mu = \cos\Theta$ and Θ is the scattering angle. Several interesting features of the preceding matrix can be noted. In particular, complete linear polarization is noted at a scattering angle of 90°; ellipticity does not exist in the multiple-scattering-induced radiation if the source is unpolarized; and the relevant depolarization factor is zero.

The Mueller matrix for homogeneous spheres can be readily obtained from the Lorenz–Mie theory. However, the calculation is not trivial in mathematical and computational physics for arbitrarily shaped nonspherical particles. A number of numerical methods have been developed to obtain the Mueller matrix for particles within a limited size parameter range $(0, X_m)$ including the finite-difference-time-domain method (Yee, 1966; Yang and Liou, 1996a), the discrete-dipole-approximation method (Purcell and Pennypacker, 1973; Draine and Flatau, 1994; Yurkin et al., 2007), the separation of variables method (Asano,1979; Farafonov, 1983), and the T-matrix method (Waterman, 1971; Mishchenko and Travis, 1998; Mackowski and Mishchenko, 1996; Doicu et al., 2006; Kahnert, 2013; Bi et al., 2013, and references therein). X_m depends on the particle shape, the refractive index, the computational resources, and the selected computational method. For large-sized parameters beyond the capabilities of numerical methods, a common approach is based on geometric-and-physical optics approximations. A review of the physical-geometric optics approximations for a solution of the optical properties of ice crystals and aerosols can be found in Bi and Yang (2013).

Before one can theoretically calculate the polarization of the submarine radiance fields (see Kattawar et al., 1973, 1988; Kattawar and Adams, 1989), the Mueller matrix for ocean water must be obtained. Beardsley (1968) performed the first measurement of the Mueller matrix for ocean-water samples. He noticed a great deal of symmetry in the matrices measured from different sources, and the normalized matrix element values were similar to those for a normalized matrix derived for Rayleigh scattering (presented earlier). Later measurements made by Soviet scientists (Kadyshevich, 1977; Kadyshevich et al., 1971, 1976) suggested much larger variations in the Mueller matrix as a function of depth and location. The apparent discrepancy was resolved by Voss and Fry (1984), who made Mueller matrix measurements in the Atlantic, Pacific, and Gulf of Mexico, and showed that the Mueller matrix has little variation between sites and confirmed Beardsley's Rayleigh-like appearance of the matrices. Shown in Fig. 1.5 is a depiction of P_{12}/P_{11}, P_{22}/P_{11}, and P_{33}/P_{11} for scattering angles from 10° to 160°. Note that P_{12}/P_{11} and P_{33}/P_{11} for the ocean-water average values are close to those for a Rayleigh-scattering simulation. Kokhanovsky (2003) presented the parameterization of the Mueller matrix of oceanic water measured by Voss and Fry (1984), which could be used for theoretical polarized radiative-transfer studies involving oceanic waters. One of the most important hydrosol observations was that the particles could not be spherical. The explanation is the normalized Mueller matrix element S_{22} was not unity and, for any spherical polydispersion, S_{22} must be unity. The zero values in the upper and lower 2 × 2 submatrices indicate little, if any, optical activity in the samples or preferred orientation by the particulates; however, this is not to suggest that highly

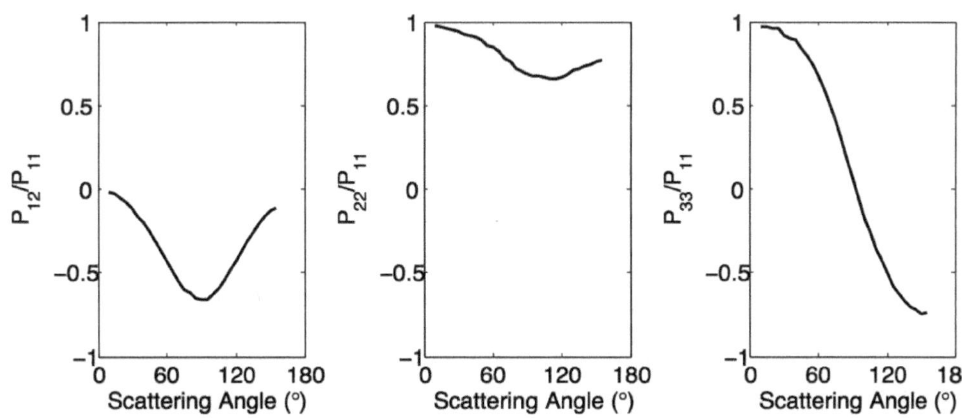

Fig. 1.5. Sample Mueller matrix elements (averaged values) regenerated from Voss and Fry (1984, Table VI). P_{21} is equal to P_{12}, and P_{44} is equal to P_{33}. All other elements associated with polarization are zero.

concentrated phytoplankton samples would not demonstrate optical activity. As is well known, amino acids and sugars have a 'handedness' or chirality and, therefore, when associated with the cells' pigments will induce a certain degree of optical activity (Houssier and Sauer, 1970). Also, Pospergelis (1969) measured the ratio of V/I for the Stokes vector of light reflected from green leaves and found the ratio to be of the order of a few percent but, for the low concentrations observed in the open oceans, measurement accuracies of one part in 10^4 are required.

1.6 Neutral points in the atmosphere and ocean

We have previously shown that, although incident sunlight is unpolarized, one of the important characteristics of the light field in the natural atmosphere and ocean is the polarization state, including the degree and plane of polarization. The polarization state of light contains valuable information about the optical properties of the atmosphere and ocean. On the other hand, in the angular distribution of the polarized light field, directions may exist where the light is actually unpolarized (or neutral) with a vanishing degree of polarization. These points in an angular distribution plot are called neutral points. The positions of neutral points are of great interest, as they are found to be sensitive indicators of various atmospheric optical properties, including atmospheric turbidity, air pollution, etc. (Bellver, 1987).

Measurements of neutral points were first made in a simple Rayleigh atmosphere. Early measurements included those by Arago (Barral, 1858), Babinet (1840), and Brewster (1842), who independently observed three different neutral points. All three neutral points are on the principal plane – the plane containing both the zenith and the Sun. The Arago point is normally 20° to 30° above the antisolar point, the Babinet point is approximately the same angular distance above the Sun, and the Brewster point is about 15° to 16° below the Sun. For a review of the observations of these three neutral points, readers are referred to Coulson (1988). The three neutral points have been well known for over a century and have been observed in separate measurements (e.g., Bellver, 1987; Coulson, 1983; Gal et al., 2001). As suggested by theory, a fourth neutral point should be in the principal plane that lies below the antisolar point, but was not observed until over a century later by Horvath et al. (2002) using measurements from a balloon-borne polarimeter.

From a theoretical perspective, a neutral point appears when a positive/negative polarization is exactly canceled by a negative/positive polarization. At the time the first three neutral points were observed, the origin of the opposite polarization state was not clear. Lord Rayleigh (1871) showed that Rayleigh scattering of light by atmospheric molecules introduced a large polarization at 90° to the incident beam, while the forward- and backward-scattered light (i.e., the scattered light at the solar and antisolar points) is unpolarized if only a single scattering is considered. This implied that two neutral points should be in these two directions, which differed from the observations. Later, Chandrasekhar (1951) using radiative-transfer theory showed that the opposite polarization state could be explained by multiple scattering of light in a Rayleigh atmosphere. According to radiative-transfer calculations, the neutral points at the solar point splits into two neutral points in the principal plane, which move away from the solar point as multiple-scattering effects increase, and which applies to the neutral point at the antisolar direction as well. Therefore, when multiple scattering dominates, a total of four neutral points occur in a Rayleigh-scattering atmosphere, which is consistent with observations (see, e.g., Chandrasekhar 1951).

Observations of atmospheric neutral points outside of the principal plane were reported by Cornu (1884) and Soret (1888). During Cornu's measurement, high-altitude air pollution was present, which introduced aerosol particles in the atmosphere. Soret's measurements were made from lake shores or over a sea, such that polarized light reflected by the air–sea surface came into play in determining the skylight polarization pattern. For a brief review of the measurements, readers are referred to Fraser (1968). Neutral points were also observed in airborne measurements. Kawata and Yamazaki (1998) reported airborne POLarization and Directionality of Earth Reflectances (POLDER) measurements of upwelling polarized light over an ocean surface at an aircraft altitude of 4,700 m and at wavelengths of 0.45 μm and 0.85 μm. Although the authors did not specifically discuss neutral points, at both wavelengths, two neutral points can be seen outside of the principal plane.

Neutral points have also been observed in the subsurface light field by Voss et al. (2011) and predicted by Chowdhary et al. (2006). The observation showed that, at a shorter wavelength (436 nm), two neutral points were outside the principal plane in the upwelling polarized light field, while, at a longer wavelength (526 nm), two neutral points were on the principal plane. The points are the well-known Babinet and Brewster neutral points.

To understand how the position of neutral points is related to various parameters, numerical simulations are helpful. One of the earliest simulations was made by Fraser (1968), who used model simulation for a Rayleigh atmosphere and a smooth water surface to explain the appearance of neutral points out of the principal plane. The polarized light reflected by the air–sea surface contributes to the skylight polarization and moves the neutral points out of the principal plane.

Kawata and Yamazaki (1998) also compared their POLDER measurements with simulation results from the adding–doubling algorithm for an atmosphere–surface system (i.e., no oceanic scattering was included). The model simulated polarization patterns were generally consistent with the measurements, but the model predicted neutral points on the principal plane. The discrepancy could be attributed to the fact that the in-water scattering was not included.

In a simulation effort, Takashima and Masuda (1985) investigated the upwelling polarized light field at the top of the atmosphere (TOA) in a coupled atmosphere–ocean system, where the in-water scattering and the interaction between the atmosphere and ocean were properly considered. The position of neutral points in the principal plane was specifically discussed. The authors suggested that oceanic conditions have little effect on the positions of TOA neutral points. They further reported that, in hazy conditions, two neutral points exist in the wavelength region from 0.4 μm to 0.8 μm, while, in clear conditions, the neutral points appear only at wavelengths shorter than 0.75 μm and disappear at longer wavelengths. The study did not show any results with neutral points outside of the principal plane.

Adams and Kattawar (1997) and Adams and Gray (2011) presented Monte Carlo simulations of positions of neutral points in a coupled atmosphere–ocean system in the upwelling and downwelling light fields. The authors suggested that the effects of highly polarized underwater scattering on the positions of neutral points above the interface were significant. The model predicted neutral points could move a few degrees when the underwater scattering was considered. Furthermore, the polarization introduced by the air–sea interface was found to play an important role in determining the underwater polarization. As a result, the polarization pattern and the positions of neutral points just beneath the interface could differ substantially from those just above the interface. Furthermore, neutral points off the principal plane could be found in the light field beneath the surface when the solar zenith angle was large enough (greater than 75° in the cases examined). Adams and Gray (2011) pointed out that, in the simulated case, as many as eight separate neutral points could exist under certain conditions, some on the principal plane and some outside.

Based on available observations and simulations, the positions of in-air neutral points can be explained. Multiple Rayleigh scattering by atmospheric molecules alters the polarization pattern and causes the neutral point in the solar and antisolar directions to split and form four separate neutral points on the principal plane. The atmospheric optical thickness is greater at shorter wavelengths, thus the neutral points are further away from the solar/antisolar direction. Existence of aerosol particles and/or an underlying water surface further alters the polarization pattern and causes the neutral points to move out of the principal plane.

In the marine light field, the origin of neutral points is more complicated from the combination of in-air multiple scattering, the Fresnel refraction through the air–sea interface, and in-water multiple scattering. The effects of atmospheric scattering can be studied by separating the total in-water polarized light field into components from two sources: a direct part (no scattering in the atmosphere before the sunlight hits the air–sea interface) and a diffuse part (diffuse skylight that enters the ocean). The positions of neutral points can be better understood by showing the neutral lines of the Stokes parameters Q and U, which are zero contour lines in Q and U patterns (Adams and Gray, 2011; Voss et al., 2011). A neutral point appears when the Q and U neutral lines intersect each other. As shown in Voss et al. (2011), the direct part always has two neutral points on the principal plane, while the diffuse part always has two off-axis neutral points. The difference is controlled by the shape of the Q neutral lines in the two parts. Therefore, as illustrated by Fig. 1.6, the balance between the two parts determines the position of the neutral points in the total polarized light field. When the atmospheric optical thickness is small and the direct part dominates, two neutral points are on the principal plane in the in-water upwelling light field. As the atmospheric optical thickness increases, the shape of the Q neutral line changes and the two neutral points move toward each other and, at some point (τ_{Ray} in this case), merge into one point on the principal plane. The dependence on the atmospheric optical thickness for in-water neutral points is opposite to that for the in-air neutral points. As τ_{Ray} increases, the diffuse part dominates and the shape of the Q neutral line in the total light field further changes toward that in the diffuse part. As a result, the two neutral points split and move out of the principal plane. Note, here, that the shapes of the neutral lines

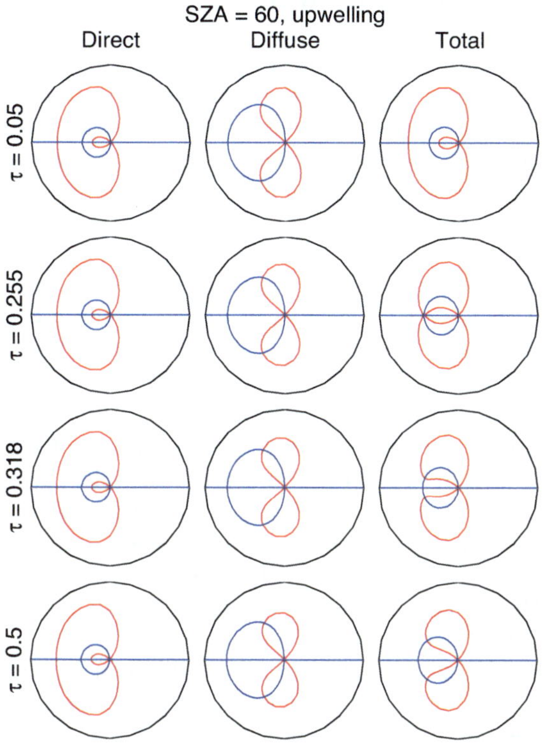

Fig. 1.6. Neutral lines of the Stokes parameters Q (red) and U (blue) in the in-water upwelling polarized light field with various atmospheric optical thicknesses, $\tau = 0.05$, 0.255, 0.318, and 0.5. Redrawn, with modifications, from Voss et al. (2011).

and the positions of neutral points in each part change little. The balance between the two parts determines the positions of neutral points in the total light field.

As pointed out by Voss et al. (2011), off-axis neutral points can appear even when the air–sea interface disappears – that is, when the refractive index of the ocean is set to 1, as long as a highly anisotropic-scattering medium is below the Rayleigh-scattering medium.

To summarize, off-axis in-water neutral points are predicted when: (1) the atmospheric optical thickness is large enough to provide sufficient multiple scattering in the atmosphere; and (2a) an air–sea interface is below the atmosphere; or (2b) an anisotropic-scattering medium is below the atmosphere (medium can be an either an ocean or aerosol layer).

An alternative approach to explain the existence of neutral points was proposed by Berry et al. (2004) using singularity theory. The positions of neutral points in a simple Rayleigh atmosphere predicted by the theory are consistent with those reported by Chandrasekhar (1951). However, to extend the approach to cases where aerosol scattering or a coupled atmosphere–ocean system is considered is not trivial.

1.7 Polarotaxis for marine animals

One important motivation for studies on the polarized nature of underwater light is to understand polarotaxis, or the ability of animals to perceive polarized light and to use the polarization information for navigation and orientation. The honeybee, *Apis mellifera*, was the first species discovered to be polarization-sensitive and to use polarization for navigation (von Frisch, 1948). Other species known to use the polarization compass include beetles, ants, and spiders. Less publicly known is that a great number of marine creatures, including fish, crustaceans, and cephalopods, may use polarotaxis as well. Studies have shown that animals may be sensitive to both the degree and plane of linear polarization, and mantis shrimp are known to be sensitive to circular polarization (Chiou et al., 2008). All types of polarization may be used for orientation.

Polarotaxis may be explained in the following way. When the Sun is away from the zenith, the polarized light field pattern as a function of the viewing azimuth angle will always be symmetric with respect to the solar bearing. Polarization-sensitive animals can find the direction of the Sun by positioning themselves such that the polarization signals seen by the two eyes are symmetric. However, the feasibility of polarotaxis has limitations: (i) it does not work at midday when the Sun is close to zenith and the polarization pattern is almost azimuth-independent; and (ii) it requires a fairly strong polarization signal.

The light field in the natural atmosphere and ocean comes from unpolarized sunlight. Rayleigh scattering of atmospheric molecules makes the light field highly polarized, but may be depolarized when clouds or aerosols are in the atmosphere. The Fresnel transmission through the air–sea interface introduces additional polarization, but scattering in the water further depolarizes the light. The skylight polarization is primarily determined by the atmospheric turbidity. The skylight is highly polarized for a clear atmosphere in which scattering by atmospheric molecules introduces strong polarization signals, but the presence of clouds or aerosols significantly reduces the polarization signals. Therefore, honeybees can orient themselves very well under clear-sky conditions but may have trouble in cloudy situations.

On the other hand, the polarized light field below the ocean surface is more complicated and is determined by multiple factors including the viewing wavelength, solar zenith angle, water turbidity, surface conditions, and viewing depth. Some of polarizing factors may limit the feasibility of using polarized light for navigation: the polarization signal may be too low to be sensed by animals or the polarization pattern may not be azimuth-dependent and no information about the solar position can be extracted. Therefore, studies on underwater polarization and its dependence on various factors are critical in estimating under what conditions the in-water polarization pattern can provide sufficient information for polarization orientation. As a matter of fact, the motivation for some of the earliest measurements of underwater polarization (Waterman, 1954; Waterman and Westell, 1956) was to understand how the in-water polarized light can be used for navigation. In the first decade of the 21st century, multispectral and hyperspectral measurements of underwater polarization were reported (Cronin and Shashar, 2001; Tonizzo et al., 2009).

Of interest is the question of whether the underwater polarization is strong enough to be perceived by animals. Studies have shown that fish can utilize polarization higher than 65% to 75% for orientation, while crustaceans and cephalopods have higher polarization sensitivity. Early polarization measurements by Ivanoff and Waterman (1958) suggested that, in very clear water near Bermuda, underwater linear polarization can be 60% or higher near the surface. Under these conditions, most polarization-sensitive marine animals can utilize the polarization signal for orientation. Cronin and Shashar (2001) reported the underwater degree of polarization measured near the Florida Keys under cloudy skies never exceeded 50%, which could be a challenging task for some fish to perceive. On the other hand, crustaceans and cephalopods likely could make use of the polarization signal. More measurements of the near-surface DoLP include those made in coastal waters in the New York harbor (Tonizzo et al., 2009) and in clear oceanic waters off of Port Aransas, Texas (You et al., 2011). The maximum DoLP was found to be less than 40% and in some cases around 20% in coastal waters and between 50% and 60% in clear waters. In a word, the degree of polarization just beneath the ocean surface is highly sensitive to both the atmospheric conditions and the water turbidity. The presence of clouds and/or aerosols in the atmosphere and increased water turbidity both contribute to a decrease in the in-water DoLP to below the threshold sensitivity of animal photoreceptors, in which case the polarization compass does not work.

The next important aspect of underwater polarization is the depth the polarization signals persist in the water. Bhandari et al. (2011) experimentally showed that the underwater polarization is the result of the refracted downwelling sky polarization and the polarization caused by scattering bodies in the water. Ivanoff and Waterman (1958) showed that, even in clear waters, the degree of polarization may decrease to 30% at a depth of more than 10 m. Therefore, navigation based on polarization patterns must be limited to shallow water (see discussions in Shashar et al. (2004)). Measurements of linear polarization made in clear and turbid waters in the Red Sea by Lerner et al. (2011) suggested that using the DoP pattern as a compass is possible in clear waters but less likely in turbid conditions (e.g., in an algae bloom) at depths as small as 2 m.

As aforementioned, polarotaxis is based on the azimuth dependence of the polarization pattern when the Sun is away from the zenith. Besides decreasing the degree of polarization, another effect from increasing depth is that multiple-scattering events weaken the azimuth dependence of the polarization pattern. At some critical depth, the underwater polarization pattern will become azimuthally symmetric and no information about the solar direction can be extracted. The region below this depth is sometimes called the asymptotic region (Kattawar and Plass, 1976), where the light field approaches an azimuth-independent asymptotic form, regardless of the location of the Sun. In this region, the polarization pattern contains no information about the solar position, which makes polarotaxis impractical.

Another interesting topic is the underwater polarization for large solar zenith angles, namely close to sunrise and sunset. Under those conditions, the contribution from direct sunlight is reduced, while the contribution from diffuse skylight dominates. Additionally, the contribution from the total internal reflection from the lower surface of the sea surface begins to have an effect. As a result, the underwater polarization patterns are distinctive as they are far from the simple Rayleigh-scattering predictions (Sabbah and Shashar, 2007). The unique polarization patterns may provide special information for polarization-sensitive animals. Furthermore, the underwater degree of polarization is fairly insensitive to wavelength (Cronin and Shashar, 2001). As a result, there is not a preferred wavelength for polarization sensing by photoreceptors.

In addition to the degree of polarization, the *e*-vector orientation is also a possible source for polarization orientation. Simulation results (You et al., 2011) suggest that *e*-vector orientation is less affected by the water turbidity and the depth. At 7 m below the surface in turbid waters, the *e*-vector orientation shows a higher azimuth dependence and should be easier to perceive by animals with both sensitivities. The observations by Bhandari et al. (2011) support that the *e*-vector orientation is maintained better than the degree of polarization with increasing depth.

1.8 Application of polarization to atmospheric studies

The polarization state of light is important to atmospheric studies. For example, polarized or unpolarized light transported from the TOA to a cloud or aerosol layer can be reflected back into space to be observed by satellite instruments aimed at measuring global dynamics including changes in clouds, the radiation budget, and other atmospheric processes (Hansen, 1971; van de Hulst, 1981). In cloud and aerosol observations, polarized signals are inevitable and are caused by aerosols, cirrus clouds, and tropical deep convective clouds. The optical and microphysical properties of clouds and aerosols have been extensively studied using radiometric measurements of reflected light (Baum et al., 2000; King et al., 2003; Minnis et al., 2011). The abundant information from three Stokes parameters (Q, U, V), based on airborne or lidar observations, has been gradually phased into atmospheric research (Chepfer et al., 2001; Hu et al., 2003, 2007; Schotland, 1971; Sun et al., 2006) and, at the same time, satellite instruments with polarization capabilities have been launched or proposed. A passive space-borne polarimeter, the POLDER-3 instrument aboard the PARASOL (Polarization and Anisotropy of Reflectances for Atmospheric Sciences coupled with Observations from a lidar) satellite, provided successive multidirectional measurements of up to 16 viewing geometries with nine spectral channels between 0.443 μm and 1.020 μm (Fougnie et al., 2007). Three of the channels (0.490, 0.670, and 0.865 μm) are sensitive to polarized light and measure three Stokes parameters: I, Q, and U. Another satellite, CALIPSO (Cloud-Aerosol Lidar and Infrared Pathfinder Satellite Observations), is the first operational satellite equipped with a polarization lidar that provides high-resolution vertical profiles of clouds and aerosols (Winker et al., 2003). The polarimetric measurements from CALIPSO provide a unique opportunity to study the optical and microphysical properties of clouds and aerosols. For example, Hu et al. (2007) established the relationship between the layer-integrated depolarization ratio and layer-integrated attenuated backscatter and used the information to infer cloud phase and ice-crystal orientation within clouds.

Simulations of polarized radiances by clouds and aerosols have been reported in the literature by a number of authors. Evans and Stephens (2010) reviewed the most popular polarized radiative-transfer models used for satellite applications including the adding–doubling algorithm (Evans and Stephens, 1991), the VDISORT algorithm (Schulz and Stamnes, 2000; Weng, 1992), and the progress of Monte Carlo methods for 3D vector radiative-transfer models (Battaglia and Mantovani, 2005; Cornet et al., 2010; Davis et al., 2005). A polarized radiative-transfer model, not reviewed by Evans and Stephens (2010), was developed by de Haan et al. (1987) and based on the adding and doubling principle for plane-parallel clouds or aerosol layers. The details of the algorithm and computational code can be found in de Haan et al. (1987) and will not be repeated here. The adding–doubling code provides a computationally accurate and efficient tool to simulate all four Stokes parameters. The radiance associated with an optically infinite snow layer under clear sky is evaluated by Xie et al. (2006). Recently, Rozanov et al. (2014) reviewed the radiative-transfer model that is suitable for remote sensing purposes.

In the theoretical simulation of polarized light in the atmosphere, the scattering properties (i.e., single-scattering albedo, extinction coefficient, phase matrix, etc.) of clouds and aerosols are necessary. Substantial effort has been expended toward simulating the light scattering by nonspherical ice cloud particles to which the Lorenz–Mie theory (Wiscombe, 1980) cannot be applied. Interested readers may refer to Baran (2009) for a complete review of the algorithms to compute the scattering properties of ice crystals. Figure 1.7 demonstrates the scattering-phase matrices of randomly oriented ice crystals at $\lambda = 0.532$ μm, which is a wavelength associated with a spectral channel of the CALIPSO Lidar. We consider two typical ice-crystal habits – hexagonal columns and an aggregate of plates, based on *in situ* measurements of cirrus clouds (Heymsfield and McFarquhar, 2002; Heymsfield and Miloshevich, 2003). The scattering properties are computed using a combination of the Amsterdam discrete dipole approximation (ADDA) (Yurkin and Hoekstra, 2009; Yurkin et al., 2007) and an improved geometric optics method (IGOM) (Yang and Liou, 1996b). More details on the ice-crystal numerical model and the computational techniques for scattering properties can be found in Auer and Veal (1970), Xie et al. (2011), and Yang et al. (2005). The definition of effective particle size by Foot (1988) and Francis et al. (1994) and the size distribution parameterized from an *in situ* measurement of clouds (Baum et al., 2005) were used to represent an ensemble of ice crystals within cirrus clouds. In Fig. 1.7, the phase functions of hexagonal columns and aggregates of plates can be seen to have the typical scattering structure of hexagonal particles, namely strong halos associated with internal reflection within the ice crystals. For the aggregate of plates, the maximums of the halos are lower than those of hexagonal columns due to the multiple scattering between the plates. Also evident from Fig. 1.7 are that some non-zero elements of the phase matrices, such as P_{44}, are more sensitive to ice-crystal habit than the phase function, P_{11}. Thus, the use of polarization techniques in satellite instruments can lead to better performance in the retrieval of the microphysical properties of cirrus clouds.

Figure 1.8 illustrates the contour of the Mueller matrices of randomly oriented hexagonal columns and aggregates of plates at $\lambda = 0.532$ μm. In each panel of Fig. 1.8, the solar zenith angle is 30° and the relative azimuth angle varies from 0° to 360° along the angular direction. In the radial directions shown in the panels, the satellite viewing zenith angles varied from 0° to 90°. The Mueller matrices of cirrus

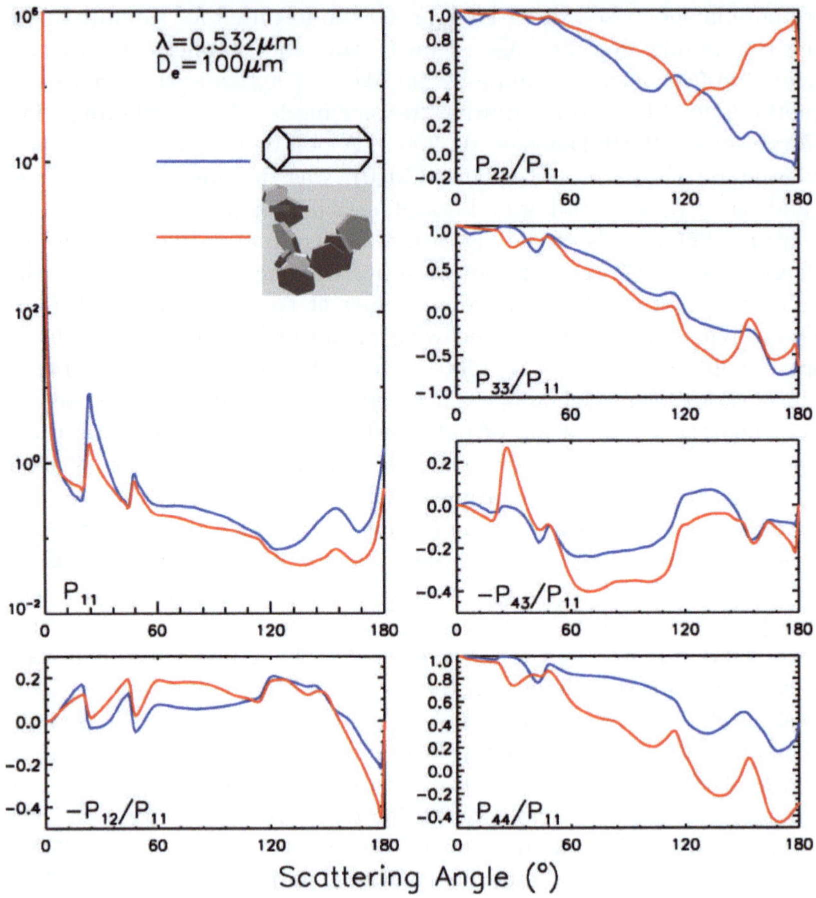

Fig. 1.7. Non-zero elements of the scattering-phase matrices of randomly oriented hexagonal columns and aggregates of plates at $\lambda = 0.532$ μm. The effective particle size De is defined as $De = 3\,\mathrm{V}/(2\,\mathrm{A})$, where V is the total volume and A the total projected area of an ensemble of the particles of interest.

clouds were simulated using the adding–doubling method (de Haan et al., 1987) and a scheme specified by Lawless et al. (2006). Based on the definition of the Mueller matrix, M_{11} represents the bi-directional reflectance of light. As the incident beam of the CALIPSO Lidar has a Stokes state of $(1, 1, 0, 0)$, the first two columns of the Mueller matrices dominate the model simulations. Apparently, some elements of the Mueller matrices, such as M_{21}, are more sensitive to ice-crystal habits as compared with M_{11}, which is capable of providing additional information toward the retrieval of cloud properties.

In addition to revealing particle shape, polarimetric observation is also useful for studying the naturally occurring imperfections in ice particle symmetry. Although an ice cloud particle may consist of, in theory, a single element or multiple elements with highly symmetric hexagonal structures, *in situ* observations show that the symmetry is usually not maintained (Lawson et al., 2006). Inclusion of impurities

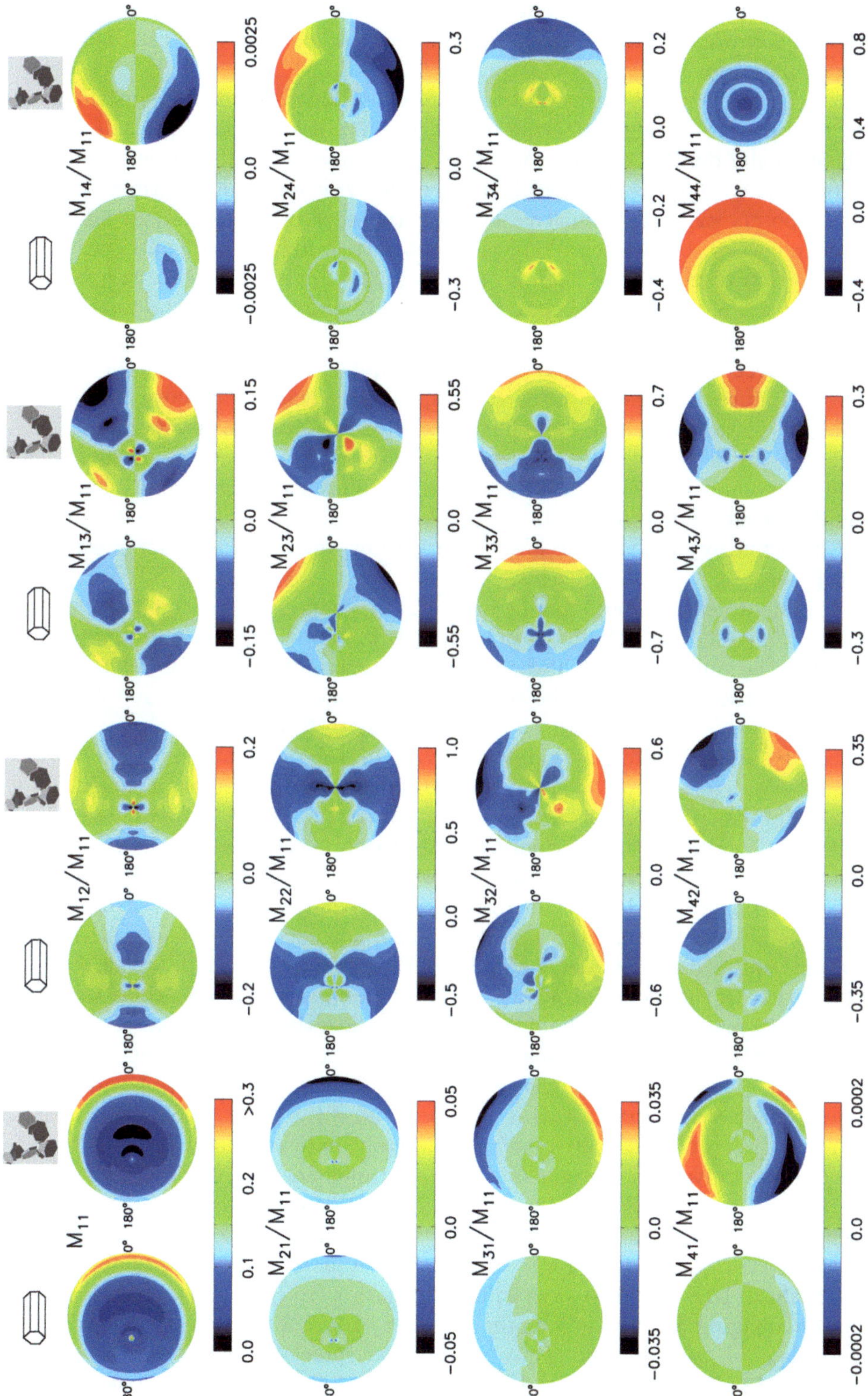

Fig. 1.8. Mueller matrices of randomly oriented hexagonal columns and aggregates of plates at $\lambda = 0.532\,\mu$m.

such as air bubbles, minute roughness on the crystal surface, collision and riming due to cloud processes, and aggregation of multiple elements all can break the monocrystalline symmetry.

Among several ways to incorporate the aforementioned symmetry imperfections into a cloud particle single-scattering model, the use of aggregated particle shapes and theoretically defined surface roughness are computationally efficient. In spite of the fact the two approaches break the symmetry at different length scales, polarimetric observations are sensitive to both types of perturbation. In addition, polarized reflection by a thick cloud is mainly attributed to the contribution by the top portion of the cloud (Goloub et al., 1994). For this reason, polarimetric observations are useful to investigate ice particle randomness near the top of thick ice clouds.

As the effect of aggregation is shown in Fig. 1.7, let us focus on the surface roughness. Figure 1.9 demonstrates the impact of surface roughness on the elements of the scattering matrix; the phase function (P_{11}) is featureless for severely roughened column aggregate particles while preserving some single-element features such as halos at 22° and 46° for smooth column aggregate particles. Note the significant difference seen in the $-P_{12}/P_{11}$ element. The $-P_{12}/P_{11}$ element can be well detected by passive spacecraft observations, including the aforementioned POLDER sensors.

C.-Labonnote et al. (2001) were among the first who inferred the imperfection of ice-crystal symmetry with global polarimetric observations. Their study demonstrated that the single-element hexagonal column model with air bubble inclusions could explain the observations better than the model without air bubble inclusion. A following study by Baran and C.-Labonnote (2006, 2007) also suggested the necessity of using broken symmetry to accurately simulate the globally observed ice cloud reflectivity. Cole et al. (2014) reported an attempt to estimate the frequency distribution of the degree of ice particle surface roughness. Each of the studies employed the residual error approach by calculating the phase matrix with numerical techniques, building look-up tables based on the phase matrix by radiative-transfer modeling, and minimizing fitting errors. The residual error method was chosen because the first two steps, specifically the single-scattering calculations and radiative-transfer simulations, are computationally expensive.

In addition to the global observations, let us look at the applications more related to cloud processes. Ice cloud particles significantly vary in shape, size, and complexity both temporally and spatially. For cirrus clouds, a comprehensive review by Baran (2009) is suggested for the reader. A question which needs to be addressed is: 'Does broken symmetry, which is indispensable to explain global observation, really reflect microphysical asymmetry, or merely indicate cloud heterogeneity?' Figure 1.10 may give some insight into the question's answer. To be more specific, Fig. 1.10 shows a POLDER/PARASOL observation-density map of modified polarized reflectivity (or normalized modified polarized radiance in some literature) over the tropical Western Pacific (15°S–15°N, 120°E–180°E) during September 2005. The PARASOL L1B radiance ($\lambda = 0.865\ \mu m$) data were collocated with Moderate Resolution Imaging Spectroradiometer (MODIS) Level 2 cloud products, and the cloud pixels close enough to the convective center (pixels with a 11 μm-brightness temperature less than 208 K and the pixels in the vicinity of such 'cold' pixels)

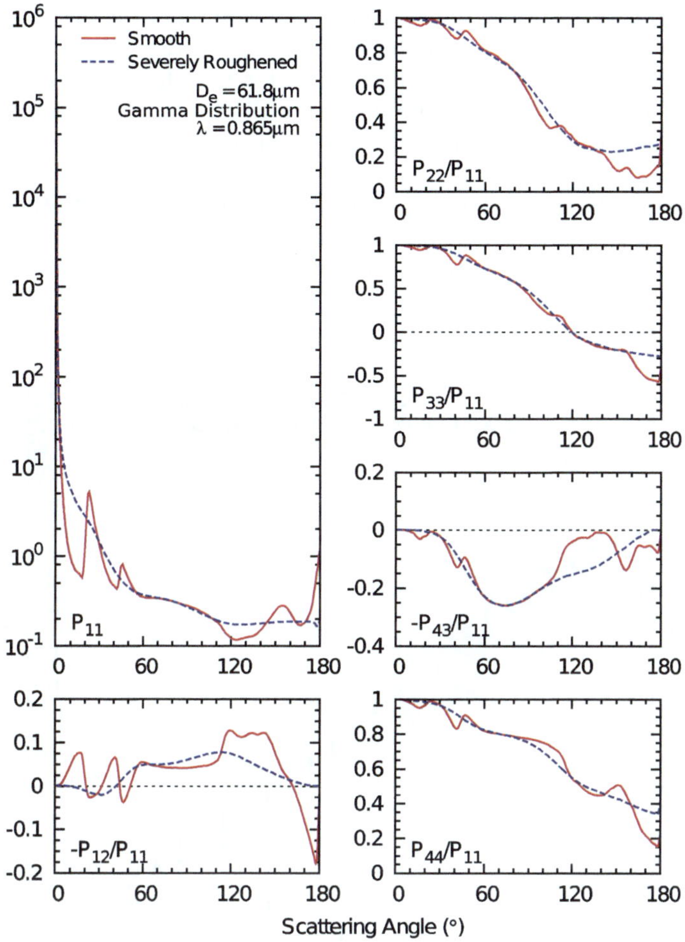

Fig. 1.9. Scattering matrix elements of a smooth and severely roughened column ice particle aggregate with eight elements. The particle size distribution is the Gamma distribution with an effective particle size of 61.8 μm, and an effective variance of 0.1. The wavelength is 0.865 μm.

were selected. The modified polarized reflectivity is defined as

$$L_{nmp} = \frac{\pi L_p}{E_s} \frac{\cos\theta_0 + \cos\theta}{\cos\theta_0}, \tag{1.52}$$

where L_p is the (linearly) polarized radiance defined by $L_p = \sqrt{Q^2 + U^2}$ with Q and U being the second and the third components of the Stokes vector, E_s is the extraterrestrial solar flux, θ_0 is the solar zenith angle, and θ is the viewing zenith angle. The definition is the same as the definition of common reflectivity except that polarized radiance replaces total radiance and is multiplied by the cosine factor $\cos\theta_0 + \cos\theta$. The cosine factor is the inverse of the cosine factor that appears in single-scattering equations. Since the polarized reflection from optically thick clouds has similar zenith-angle dependence to single scattering, the factor makes the modified polarized reflectivity depend strongly on the scattering angle and minimizes the dependence on the zenith angles. Also shown in Fig. 1.10 are the simulated modified polarized reflectivities for two different surface roughness parameters. The reflectivities are simulated for all observation geometries, and the mean and standard deviation in each 3° scattering angle bin are presented with error bars.

As seen from Fig. 1.10, the observed modified polarized reflectivities are better simulated by severely roughened particles than smooth particles. The result suggests that the actual imperfection of a particle's symmetry or small-scale cloud heterogeneity exist in convective clouds over the tropical Western Pacific.

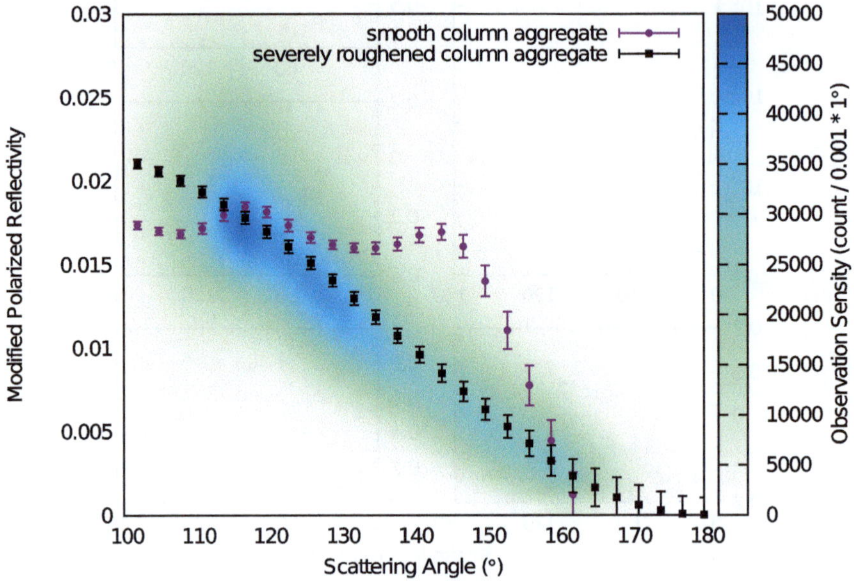

Fig. 1.10. The observation-density map of modified polarized reflectivity obtained by a collocated POLDER–MODIS data set over the tropical Western Pacific in September 2005. Cloud pixels close to the convective center were selected based on MODIS brightness temperatures. Black and purple markers show the simulated modified polarized reflectivity assuming a severely roughened column aggregate particle and a smooth column aggregate particle, respectively.

With a methodology similar to the one demonstrated here, modified polarized reflectivity (normalized modified polarized radiance) is the most widely used measure to infer cloud particle shape and imperfection of symmetry. However, the applicability of the method is limited to optically thick clouds, which return linearly polarized light with a special direction of polarization. The direction of polarization imposed by an optically thick cloud is nearly parallel or perpendicular to the scattering plane. That is, the U component of the Stokes vector is much smaller than the Q component when the scattering plane is taken as a plane of polarimetric reference. This is very similar to the scattering of natural light by an achiral single particle in random orientation from which no U component can appear; the widespread use of modified polarized reflectivity implies that a 'convoluted' single-scattering phase matrix is useful as a surrogate of the bi-directional polarized reflectivity distribution function of optically thick clouds.

With the new generation of instruments with high accuracy, such as the Research Scanning Polarimeter (RSP) (Cairns et al., 1999) and Multiangle SpectroPolarimetric Imager (MSPI) (Diner et al., 2010), the investigation of a weak signal on the U component may be possible in the future. These instruments will bring a dramatic change in the methodology and approach for cloud studies, as the U component does code valuable information about the second or third order of scattering.

To demonstrate how the separate treatment of the U component could benefit cloud property studies, we show the contribution of the U component to the polarized radiance (L_p) at different cloud optical thicknesses. Figure 1.11 shows the simulated polarized radiance and magnitude of the U component at a fixed solar zenith angle ($\theta_0 = 30°$) and a satellite viewing zenith angle ($\theta = 45°$) at the wavelength of $\lambda = 865\ \mu$m. Three sets of simulations over a modeled ocean surface (Cox and Munk, 1954) are presented with each having different ice cloud optical thicknesses; optically thick ice cloud ($\tau = 10$; dashed line), optically thin cloud ($\tau = 2$; solid line), and no ice cloud (dotted line). The point symbols show the polarized radiance and the boxes below the points represent the magnitude of the U component. Because of the sunglint, polarized radiance sharply increases near

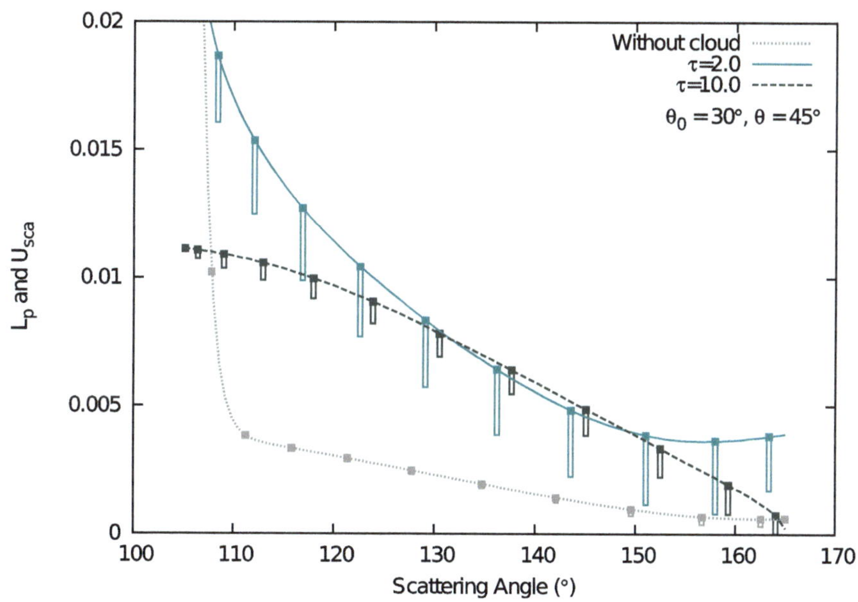

Fig. 1.11. Simulated polarized radiance over ocean with different cloud optical thicknesses. Boxes below the points show the magnitudes of the U component of the Stokes vector. Note that $L_p = \sqrt{Q^2 + U^2}$, so the magnitude of the Q component is not equal to the value at the lower edge of a box. Rather, the boxes show how the relative contribution of the U component changes with different optical thicknesses; the contribution reaches a maximum when the cloud is moderately thick.

105° for the no-cloud case. The contribution of the U component to the polarized radiance is small for the light from optically thick clouds or from the ocean surface, as previously discussed. However, the contribution is significant when the cloud is moderately thick, in which the strongly polarizing ocean surface and cloud cause multiple reflections.

Among the aerosols, the complexity of the mineral dust habit has created the most uncertainty in the retrieval of its scattering properties. Meng et al. (2010) and the literature cited therein discussed the numerical models of mineral dust based on spherical and tri-axial ellipsoidal particles. A database of the scattering properties for mineral dust with a variety of aspect ratios was developed by Meng et al. (2010) using the Lorenz–Mie theory (Bohren and Huffman, 1983), the T-matrix method (Mishchenko et al., 1997), the ADDA (Yurkin and Hoekstra, 2009; Yurkin et al., 2007), and the IGOM (Bi et al., 2009; Yang and Liou, 1996b). Figure 1.12 shows the scattering-phase matrices of randomly oriented mineral dust from a POLDER spectral band centered at $\lambda = 0.670\ \mu$m. The mineral dust is represented by prolate spheroids where a and b are the respective semi-length and semi-width. Thus, the aspect ratio, a/b, of a spheroid denotes the degree of elongation of the particle, whereas a spherical particle has a ratio of $a/b = 1$. Following Deuze et al. (2000), a mono-mode log-normal size distribution was employed in the simulations where the effective particle size and standard deviation are 1.87 μm and 0.724. From Fig. 1.12, the backscattering can be seen to decrease with the increasing aspect ratio of the particle. The phase function shows less variability with aerosol habits when the scattering angle is smaller than 120°, which implies that potential difficulties exist in the study of aerosol properties using radiances from passive instruments without polarimetric sensitivity, such as the MODIS. The other elements of the phase matrices shown in Fig. 1.12 suggest that the polarimetric observations can circumvent the difficulty.

Applying an approach similar to that presented by C.-Labonnote et al. (2001), we investigated the performance of the aerosol model using polarized data from the POLDER instrument. Figure 1.13 shows a comparison between normalized polarized radiances from POLDER measurements made on October 1, 2010, and simulations using aerosol models. According to C.-Labonnote et al. (2001), the normalized modified polarized radiance from POLDER observations is given by Eq. (1.52). Now we replace L_p in Eq. (1.52) with the polarized radiance with sign to better explain the behavior close to backscattering. The polarized radiance is defined as

$$L_p = \pm\sqrt{Q^2 + U^2}\,, \tag{1.53}$$

where the sign is determined by the polarization direction $\vec{P}$ and the normal direction of the scattering plane $\vec{n}$ (see Fig. 1.14). Equation (1.54) is positive when the angle (ξ in Fig. 1.14) between $\vec{P}$ and $\vec{n}$ is within $[0, \pi/4]$ or $[3\pi/4, \pi]$, but is otherwise negative. In the simulation of the polarized radiance, the adding–doubling method was employed when an aerosol layer of $\tau = 1$ was assumed over a black surface. The simulations are compared with observations from October 1, 2010, and over 15,000 pixels were involved. From Fig. 1.13, in comparison to spheres, spheroidal models generally agree better with the observations. The data confirm that a spherical model may introduce significant errors into radiative-transfer simulations and remote sensing applications. From Fig. 1.13, spheroids with $a/b = 2.50$ are found

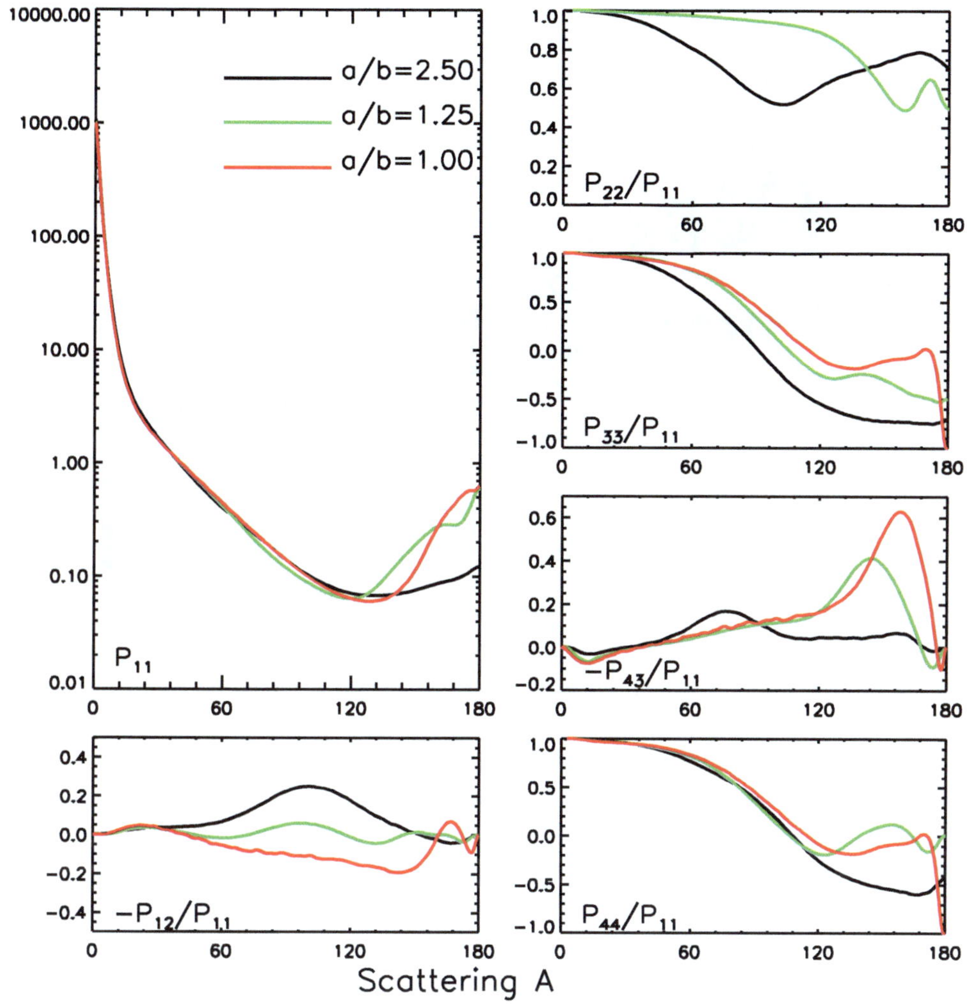

Fig. 1.12. Non-zero elements of the scattering-phase matrices of randomly oriented mineral dust at $\lambda = 0.670\,\mu$m.

to have good agreement with the POLDER data; thus, using this aerosol model in radiative-transfer and climate change studies can lead to relatively smaller biases.

1.9 Summary

We briefly reviewed the history of the study of polarized light, particularly the early-stage development of the concepts of polarization. The concepts and terminology introduced by pioneers are still used in the modern formalism of polarized light. The mathematical formulation is revisited with emphasis on observations of polarized light and simulations of particle single-scattering properties. Furthermore, the applications of the polarization characteristics are briefly elaborated in

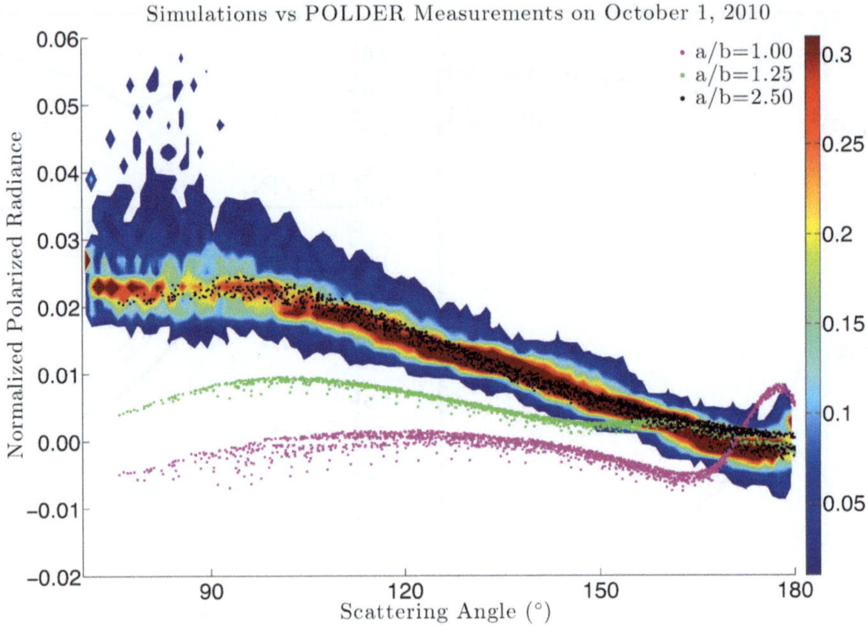

Fig. 1.13. Comparison between normalized polarized radiances from POLDER measurements of October 1, 2010, and simulations using aerosol models. The colors of the observations represent the number densities.

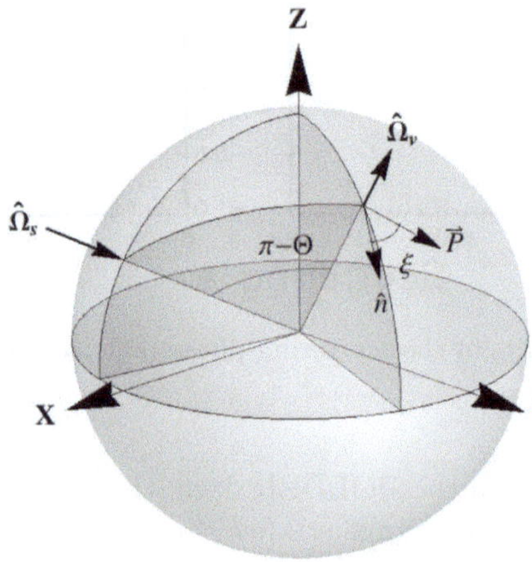

Fig. 1.14. Geometry of the incident and reflected light beams. From C.-Labonnote et al. with modifications.

conjunction with three subjects: neutral points in the atmosphere and ocean; the polarotaxis for marine animals; and the remote sensing of clouds and aerosol particles. It is worth noting that the physics behind the observed polarized light has been gradually revealed by the numerical simulations of light scattering and radiative-transfer processes with important implications to oceanic optics and atmospheric remote sensing.

Acknowledgments

Kattawar's research is supported by the ONR MURI program N00014-09-1-1054 and ONR N00014-11-1-0154, National Science Foundation (NSF) Grant OCE 1130906. Yang acknowledges support by the US NSF under Grant AGS-1338440 and the endowment funds related to the David Bullock Harris Chair in Geosciences at the College of Geosciences, Texas A&M University.

References

Adams, J.T. and Gray, D.J. (2011) 'Neutral points in an atmosphere–ocean system. 2: Downwelling light field,' *Appl. Opt.*, **50**(3), 335–46.

Adams, J.T. and Kattawar, G.W. (1997) 'Neutral points in an atmosphere–ocean system. 1: Upwelling light field,' *Appl. Opt.*, **36**(9), 1976–86.

Adams, J.T., Aas, E., Hojerslev, N.K. and Lundgren, B. (2002) 'Comparison of radiance and polarization values observed in the Mediterranean Sea and simulated in a Monte Carlo model,' *Appl. Opt.*, **41**(15), 2724–33.

Asano, S. (1979) 'Light scattering properties of spheroidal particles,' *Appl. Opt.*, **18**(5), 712–23.

Auer, A.H. and Veal, D.L. (1970) 'Dimension of ice crystals in natural clouds,' *J. Atmos. Sci.*, **27**(6), 919–26.

Babinet, J. (1840) *Sur un nouveau point neutre dans l'atmosphere*, CR, East Lansing, MI, 11, pp. 618–20.

Baran, A.J. (2009) 'A review of the light scattering properties of cirrus,' *J. Quant. Spectrosc. Radiat. Transf.*, **110**(14–16), 1239–60.

Baran, A.J. and C.-Labonnote, L. (2006) 'On the reflection and polarisation properties of ice cloud,' *J. Quant. Spectrosc. Radiat. Transf.*, **100**(1–3), 41–54.

Baran, A.J. and C.-Labonnote, L. (2007) 'A self-consistent scattering model for cirrus. I: The solar region,' *Q. J. R. Meteorol. Soc.*, **133**(629), 1899–902.

Barral, M.J.A. (1858) *Oeuvres de Francois Arago I–V*, Gide, Paris.

Bartholinus, E. (1670) 'An accompt of sundry experiments made and communicated by that learn'd mathematician, Dr. Erasmus Bartholin, upon a chrystal-like body, sent to him out of island,' *Phil. Trans.*, **5**, 2039–48.

Battaglia, A. and Mantovani, S. (2005) 'Forward Monte Carlo computations of fully polarized microwave radiation in non-isotropic media,' *J. Quant. Specrtrosc. Radiat. Transf.*, **5**(3), 285–308.

Baum, B.A., Heymsfield, A.J., Yang, P. and Bedka, S.T. (2005) 'Bulk scattering properties for the remote sensing of ice clouds clouds. Part I: Microphysical data and models,' *J. Appl. Meteorol.*, **44**(12), 1885–95.

Baum, B.A., Kratz, D.P., Yang, P., Ou, S.C., Hu, Y.X., Soulen, P.F. and Tsay, S.C. (2000) 'Remote sensing of cloud properties using MODIS airborne simulator imagery during SUCCESS. Part I: Data and models,' *J. Geophys. Res.*, **105**(D9), 11767–80.

Beardsley, G.F. (1968) 'Mueller Scattering Matrix of Sea Water,' *J. Opt. Soc. Am.*, **58**, 52–7.

Bellver, C. (1987) 'Influence of particulate pollution on the positions of neutral points in the sky at Seville (Spain),' *Atmos. Environ.*, **21**(3), 699–702.

Berry, M.V., Dennis, M.R. and Lee, R.L. (2004) 'Polarization singularities in the clear sky,' *New J. Phys.*, **6**, 162.

Bhandari, P., Voss, K.J., Logan, L. and Twardowski, M. (2011) 'The variation of the polarized downwelling radiance distribution with depth in the coastal and clear ocean,' *J. Geophys. Res.*, **116**, C00H10.

Bi, L. and Yang, P. (2013) 'Physical-geometric optics hybrid methods for computing the scattering and absorption properties of ice crystals and dust aerosols,' in Kokhanovsky, A. (ed.), *Light Scattering Reviews 8*, Springer-Praxis Publishing, Chichester, UK, pp. 69–11.

Bi, L., Yang, P., Kattawar, G.W. and Kahn, R. (2009) 'Single-scattering properties of triaxial ellipsoidal particles for a size parameter range from the Rayleigh to geometric-optics regimes,' *Appl. Opt.*, **48**(1), 114–26.

Bi, L., Yang, P., Kattawar, G.W. and Mishchenko, M.I. (2013) 'Efficient implementation of the invariant imbedding T-matrix method and the separation of variables method applied to large nonspherical inhomogeneous particles,' *J. Quant. Spectrosc. Radiat. Transf.*, **116**, 169–83.

Bohren, C.F. and Huffman, D.R. (1983) *Absorption and Scattering of Light by Small Particles*, Wiley, New York.

Born, M. and Wolf, E. (1964, 2001) *Principles of Optics*, Pergamon Press, New York.

Brewster, D. (1842) 'On the existence of a new neutral point and two secondary neutral points,' *Rep. Brit. Assoc. Advmt. Sci.*, **2**, 13–25.

Brosseau, C. (1998) *Fundamentals of Polarized Light: A Statistical Optics Approach*, Wiley, New York.

C.-Labonnote, L., Brogniez, G., Buriez, J.C., Doutriaux-Boucher, M., Gayet, J.F. and Macke, A. (2001) 'Polarized light scattering by inhomogeneous hexagonal monocrystals: Validation with ADEOS-POLDER measurements,' *J. Geophys. Res.*, **106**(D11), 12139–53.

Cairns, B., Russell, E.E. and Travis, L.D. (1999) 'Research scanning polarimeter: Calibration and ground-based measurements,' *Proc. SPIE*, **3754**, Polarization: Measurement, Analysis, and Remote Sensing II, 186–96.

Chami, M. (2007) 'Importance of the polarization in the retrieval of oceanic constituents from the remote sensing reflectance,' *J. Geophys. Res.*, **112**, C05026.

Chandrasekhar, S. (1950) *Radiative Transfer*, Oxford University Press, Oxford, UK.

Chandrasekhar, S. (1951) 'Polarization of the sunlit sky,' *Nature*, **167**, 51.

Chepfer, H., Goloub, P., Riedi, J., De Haan, J.F., Hovenier, J.W. and Flamant, P.H. (2001) 'Ice crystal shapes in cirrus clouds derived from POLDER/ADEOS-1,' *J. Geophys. Res.*, **106**(D8), 7955–66.

Chiou, T.H., Kleinlogel, S., Cronin, T., Caldwell, R, Loeffler, B., Siddiqi, A., Goldizen, A. and Marshall, J. (2008) 'Circular polarization vision in a stomatopod crustacean,' *Cur. Bio.*, **18**(6), 429–34.

Chowdhary, J., Cairns, B. and Travis, L.D. (2006) 'Contribution of water-leaving radiances to multiangle, multispectral polarimetric observations over the open ocean: Bio-optical model results for case 1 waters,' *Appl. Opt.*, **45**, 5542–67.

Cole, B.H., Yang, P., Baum, B.A., Riedi, J. and C.-Labonnote, L. (2014) 'Ice particle habit and surface roughness derived from PARASOL polarization measurements,' *Atmos. Chem. Phys.*, **14**, 3739–50.

Collett, E. (1971) 'Mathematical formulation of the interference laws of Fresnel and Arago,' *Amer. J. Phys.*, **39**, 1483–95.

Cornet, C., C.-Labonnote, L. and Szczap, F. (2010) 'Three-dimensional polarized Monte Carlo atmospheric radiative transfer model (3DMCPOL): 3D effects on polarized visible reflectances of a cirrus cloud,' *J. Quant. Specrtrosc. Radiat. Transf.*, **111**(1), 174–86.

Cornu, M.A. (1884) 'Observations relatives a la couronne visible actuellement autour de Soleil,' *Compt. Rend.*, **99**, 488−93.

Coulson, K.L. (1983) 'Effects of the El Chichon volcanic cloud in the stratosphere on the polarization of light of light from the sky,' *Appl. Opt.*, **22**, 1036−50.

Coulson, K.L. (1988) *Polarization and Intensity of Light in the Atmosphere*, A. Deepak Publishing, Hampton, VA, USA.

Cox, C. and Munk, W. (1954) 'Measurement of the roughness of the sea surface from photographs of the sun's glitter,' *J. Opt. Soc. Am.*, **44**(11), 838–50.

Cronin, T.W. and Shashar, N. (2001) 'The linearly polarized light field in clear, tropical marine waters: Spatial and temporal variation of light intensity, degree of polarization and e-vector angle,' *J. Exp. Bio.*, **204**(14), 2461–7.

Davis, C., Emde, C. and Harwood, R. (2005) 'A 3D polarized reversed Monte Carlo radiative transfer model for millimeter and submillimeter passive remote sensing in cloudy atmospheres,' *IEEE Trans. Geosci. Remote Sensing*, **43**(5), 1096–101.

de Haan, J.F., Bosma, P.B. and Hovenier, J.W. (1987) 'The adding method for multiple-scattering calculations of polarized-light,' *Astron. Astrophys.*, **183**(2), 371–91.

Deuze, J.L., Goloub, P., Herman, M., Marchand, A., Perry, G., Susana, S. and Tanre, D. (2000) 'Estimate of the aerosol properties over the ocean with POLDER,' *J. Geophys. Res.*, **105**(D12), 15329–46.

Dickey, T.D., Kattawar, G.W. and Voss, K.J. (2011) 'Shedding new light on light in the ocean,' *Physics Today*, **4**, 44–9.

Diner, D.J., Davis, A., Hancock, B., Geier, S., Rheingans, B., Jovanovic, V., Bull, M., Rider, D.M., Chipman, R.A., Mahler, A.-B. and McClain, S.C. (2010) 'First results from a dual photoelastic-modulator-based polarimetric camera,' *Appl. Opt.*, **49**(15), 2929–46.

Doicu, A., Wriedt, T. and Eremin, Y. (2006) *Light Scattering by Systems of Particles*, Springer, Berlin, Heidelberg, New York.

Dougherty, M.J. (2013) *A Dark History: Vikings*, Metro Books, New York.

Draine, B.T. and Flatau, P.J. (1994) 'Discrete-dipole approximation for scattering calculations,' *J. Opt. Soc. Am. A*, **11**, 1491–9.

Evans, K.F. and Stephens, G.L. (1991) 'A new polarized atmospheric radiative-transfer model,' *J. Quant. Specrtrosc. Radiat. Transf.*, **46**(5), 413–23.

Evans, K.F. and Stephens, G.L. (2010) 'Many polarized radiative transfer models,' *J. Quant. Specrtrosc. Radiat. Transf.*, **111**(11), 1686–8.

Farafonov, V.G. (1983) 'Diffraction of a plane electromagnetic wave by a dielectric spheroid,' *(Sov.) Diff. Equat.*, **19**, 1765–77.

Foot, J.S. (1988) 'Some observations of the optical properties of clouds. II: Cirrus,' *Q. J. R. Meteorol. Soc.*, **114**(479), 145–64.

Fougnie, B., Bracco, G., Lafrance, B., Ruffel, C., Hagolle, O. and Tinell, C. (2007) 'PARASOL in-flight calibration and performance,' *Appl. Opt.*, **46**(22), 5435–51.

Francis, P.N., Jones, A., Saunders, R.W., Shine, K.P., Slingo, A. and Sun, Z. (1994) 'An observational and theoretical study of the radiative properties of cirrus: Some results from ICE'89,' *Q. J. R. Meteorol. Soc.*, **120**(518), 809–48.

Fraser, R.S. (1968) 'Atmosphere neutral points over water,' *J. Opt. Soc. Am.*, **58**, 1029.

Gal, J., Horvath, G., Meyer-Rochow, V.B. and Wehner, R. (2001) 'Polarization patterns of the summer sky and its neutral points measured by full-sky imaging polarimetry in Finnish Lapland north of the Arctic Circle,' *Proc. R. Soc. Lond. A.*, **457**(2010), 1385–99.

Gehrels, T. (1974) *Planet's Stars and Nebulae*, University of Arizona Press, Tucson, AZ.

Goloub, P., Deuze, J.L., Herman, M. and Fouquart, Y. (1994) 'Analysis of the POLDER polarization measurements performed over cloud cover,' *IEEE Trans. Geosci. Remote Sensing*, **32**(1), 78–88.

Hansen, J.E. (1971) 'Multiple scattering of polarized light in planetary atmospheres. 1: Doubling Method,' *J. Atmos. Sci.*, **28**(1), 120–5.

Heymsfield, A.J. and McFarquhar, G.M. (2002) 'Mid latitude and tropical cirrus microphysical properties,' in Lynch, D.K., Sassen, K., Starr, D.O. and Stephens, G.L. (eds), *Cirrus*, Oxford University Press, pp. 78–101.

Heymsfield, A.J. and Miloshevich, L.M. (2003) 'Parameterizations for the cross-sectional area and extinction of cirrus and stratiform ice cloud particles,' *J. Atmos. Sci.*, **60**(7), 936–56.

Horvath, G., Bernath, B., Suhai, B., Barta, A. and Wehner, R. (2002) 'First observation of the fourth neutral polarization point in the atmosphere,' *J. Opt. Soc. Am. A*, **19**(10), 2085–99.

Houssier, C. and Sauer, K. (1970) 'Circular dichroism and magnetic circular dichroism of the chlorophyll and protochlorophyll pigments,' *J. Am. Chem. Soc.*, **92**(4), 779–91.

Hu, Y.X., Vaughan, M., Liu, Z.Y., Lin, B., Yang, P., Flittner, D., Hunt, B., Kuehn, R., Huang, J.P., Wu, D., Rodier, S., Powell, K., Trepte, C. and Winker, D. (2007) 'The depolarization-attenuated backscatter relation: CALIPSO Lidar measurements vs. theory,' *Opt. Express*, **15**(9), 5327–32.

Hu, Y.X., Yang, P., Lin, B., Gibson, G. and Hostetler, C. (2003) 'Discriminating between spherical and non-spherical scatterers with lidar using circular polarization: A theoretical study,' *J. Quant. Spectrosc. Radiat. Transf.*, **79**, 757–64.

Ivanoff, A. and Waterman, T.H. (1958) 'Factors, mainly depth and wavelength, affecting the degree of underwater light polarization,' *J. Mar. Res.*, **16**, 283–307.

Ivanoff, A., Jerlov, N. and Waterman, T.H. (1961) 'A comparative study of irradiance, beam transmittance and scattering in the sea near Bermuda,' *Limnol. Oceanogr.*, **6**, 129–48.

Kadyshevich, Y.A. (1977) 'Light-scattering matrices of inshore waters of the Baltic Sea,' *Izv. Acad. Sci. USSR Atmos. Oceanic Phys.*, **13**, 77–8.

Kadyshevich, Y.A., Lyubovtseva, Y.S. and Plakhina, I.N. (1971) 'Measurement of matrices for light scattered by sea water,' *Izv. Acad. Sci. USSR Atmos. Oceanic Phys.*, **7**, 557–61.

Kadyshevich, Y.A., Lyubovtseva, Y.S. and Rozenberg, G.V. (1976) 'Light-scattering matrices of Pacific and Atlantic Ocean waters,' *Izv. Acad. Sci. USSR Atmos. Oceanic Phys.*, **12**, 106–11.

Kahnert, M. (2013) 'The T-matrix code Tsym for homogeneous dielectric particles with finite symmetries,' *J. Quant. Spectrosc. Radiat. Transf.* **123**, 67–82.

Kattawar, G.W. and Adams, C.N. (1989) 'Stokes vector calculations of the submarine light field in an atmosphere–ocean with scattering according to a Rayleigh phase matrix: Effect of interface refractive index on radiance and polarization,' *Limnol. Oceanogr.*, **34**(8), 1453–72.

Kattawar, G.W. and Plass, G.N. (1976) 'Asymptotic radiance and polarization in optically thick media: Ocean and clouds,' *Appl. Opt.*, **15**(12), 3166–78.

Kattawar, G.W., Adams, C.N. and Tanis, F.J. (1988) 'Monte Carlo calculations of the complete stokes vector for an inhomogeneous atmosphere–ocean system with a wind ruffled sea,' *Proc. SPIE*, **925**, Ocean Optics IX, 398–406.

Kattawar, G.W., Plass, G.N. and Guinn, G.A.J. (1973) 'Monte Carlo calculations of the polarization of radiation in the Earth's atmosphere–ocean system,' *J. Phys. Oceanogr.*, **3**, 353–72.

Kawata, Y. and Yamazaki, A. (1998) 'Multiple scattering analysis of airborne POLDER image data over the sea,' *IEEE Trans. Geosci. Remote Sensing*, **36**(1), 51–60.

King, M.D., Menzel, W.P., Kaufman, Y.J., Tanre, D., Gao, B.C., Platnick, S., Ackerman, S.A., Remer, L.A., Pincus, R. and Hubanks, P.A. (2003) 'Cloud and aerosol properties, precipitable water, and profiles of temperature and water vapor from MODIS,' *IEEE Trans. Geosci. Remote Sensing*, **41**(2), 442–58.

Kokhanovsky, A.A. (2003) 'Parameterization of the Mueller matrix of oceanic waters,' *J. Geophys. Res.*, **108**(C6), 3175.

Lawless, R., Xie, Y. and Yang, P. (2006) 'Polarization and effective Mueller matrix for multiple scattering of light by nonspherical ice crystals,' *Opt. Express*, **14**(14), 6381–93.

Lawson, R.P., Baker, B., Pilson, B. and Mo, Q. (2006) 'In situ observations of the microphysical properties of wave, cirrus, and anvil clouds. Part II: Cirrus clouds,' *J. Atmos. Sci.*, **63**(12), 3186–203.

Lerner, A., Sabbah, S., Erlick, C. and Shashar, N. (2011) 'Navigation by light polarization in clear and turbid waters,' *Phil. Trans. R. Soc. B*, **366**(1565), 671–9.

Liou, K.N. (2002) *An Introduction to Atmospheric Radiation*, Academic Press, San Diego.

Logan, N. (1965) 'Survey of some early studies of the scattering of plane waves by a sphere,' *Proc. Inst. Electr. Electron, Eng. (IEEE)*, **53**, 773–85.

Lorenz, L. (1890) 'Lysbevœgelser i og uden for en af plane Lysbolger belyst Kugle,' *Videnskabernes Selskabs Skrifter*, **6**, 2–62.

Mackowski, D.W. and Mishchenko, M.I. (1996) 'Calculation of the T matrix and the scattering matrix for ensembles of spheres,' *J. Opt. Soc. Am. A*, **13**, 2266–78.

Malus, E. (1808) 'Sur une propriété de la lumière réfléchie,' *Mem. Soc. Arcueil*, **2**, 143–58.

Meng, Z.K., Yang, P., Kattawar, G.W., Bi, L., Liou, K.N. and Laszlo, I. (2010) 'Single-scattering properties of tri-axial ellipsoidal mineral dust aerosols: A database for application to radiative transfer calculations,' *J. Aerosol Sci.*, **41**(5), 501–12.

Mie, G. (1908) 'Beiträge zur Optik trüber Medien, speziell kolloidaler Metallösungen, Leipzig,' *Ann. Phys.*, **330**, 377–445.

Minnis, P., Sun-Mack, S., Young, D., Heck, P.W., Garber, D.P., Chen, Y., Spangenberg, D.A., Arduini, R.F., Trepte, Q.Z., Smith, W.L., Ayers, J.K., Gibson, S.C., Miller, W.F., Hong, G., Chakrapani, V., Takano, Y., Liou, K.N., Xie, Y. and Yang, P. (2011) 'CERES edition-2 cloud property retrievals using TRMM VIRS and TERRA and AQUA MODIS data. Part I: algorithms,' IEEE *Trans. Geosci. Remote Sensing*, **49**(11), 4374–400.

Mishchenko, M.I. and Travis, L.D. (1998) 'Capabilities and limitations of a current FORTRAN implementation of the T-matrix method for randomly oriented, rotationally symmetric scatterers,' *J. Quant. Spectrosc. Radiat. Transf.*, **60**, 309–24.

Mishchenko, M.I., Hovenier, J.W. and Travis, L.D. (2000) *Light Scattering by Nonspherical Particles*, Academic Press, San Diego.

Mishchenko, M.I., Rosenbush, V.K., Kiselev, N.N., Lupishko, D.F., Tishkovets, V.P., Kaydash, V.G., Belskaya, I.N., Efimov, Y.S. and Shakhovskoy, N.M. (2010) *Polarimetric Remote Sensing of Solar System Objects*, Akademperiodyka, Kyiv.

Mishchenko, M.I., Travis, L.D., Kahn, R.A. and West, R.A. (1997) 'Modeling phase functions for dustlike tropospheric aerosols using a shape mixture of randomly oriented polydisperse spheroids,' *J. Geophys. Res.*, **102**(D14), 16831–47.

Mueller, Hl. (1948) 'The foundation of optics,' *J. Opt. Soc. Am.*, **38**, 661.

Perrin, F. (1942) 'Polarization of light scattered by isotropic opalescent media,' *J. Chem. Phys.*, **16**, 415–27.

Pospergelis, M.M. (1969) 'Spectroscopic measurements of the four Stokes parameters for light scattered by natural objects,' *Soviet Physics-Astronomy*, **12**, 973–7.

Purcell, E.M. and Pennypacker, C.R. (1973) 'Scattering and absorption of light by nonspherical dielectric grains,' *Astrophys. J.*, **186**, 705–14.

Rayleigh, L. (1871) 'On the scattering of light by small particles,' *Phil. Mag.*, **41**, 447–54.

Roslund, C. and Beckman, C. (1994) 'Disputing Viking navigation by polarized skylight,' *Appl. Opt.*, **33**(21), 4754–5.

Rozanov, V.V., Rozanov, A.V., Kokhanovsky, A.A. and Burrows, J.P. (2014) 'Radiative transfer through terrestrial atmosphere and ocean: Software package SCIATRAN,' *J. Quant. Spectrosc. Radiat. Transf.*, **133**, 13–71.

Sabbah, S. and Shashar, N. (2007) 'Light polarization under water near sunrise,' *J. Opt. Soc. Am. A*, **24**, 2049−56.

Schotland, R. (1971) 'Observations by lidar of linear depolarization ratios for hydrometeors,' *J. Appl. Meteor.*, **10**, 1011–17.

Schulz, F.M. and Stamnes, K. (2000) 'Angular distribution of the Stokes vector in a plane-parallel, vertically inhomogeneous medium in the vector discrete ordinate radiative transfer (VDISORT) model,' *J. Quant. Spectrosc. Radiat. Transf.*, **65**(4), 609–20.

Shashar, N., Sabbah, S. and Cronin, T.W. (2004) 'Transmission of linearly polarized light in seawater: Implications for polarization signaling,' *J. Exp. Bio.*, **207**(20), 3619–28.

Shifrin, K.S. (1951) *Scattering of Light in a Turbid Medium*, Moscow (English translation: NASA TTF-477, Washington, DC, 1968).

Shurcliff, W.A. (1962) *Polarized Light-Production and Use*, Harvard University Press, Cambridge, USA.

Soleillet, P. (1929) 'Sur les paramétres caractérisant la polarization partielle de la lumiére dans les phénoménes de fluorescence,' *Ann. Physique*, **12**, 23–97.

Soret, J.L. (1888) 'Influence des surfaces d'eau sur la polarisation atmospherique et observation de deux points neutres a droite et a guache de Soleil,' *Compt. Rend.*, **107**, 867−70.

Sun, W.B., Loeb, N.G. and Yang, P. (2006) 'On the retrieval of ice cloud particle shapes from POLDER measurements,' *J. Quant. Spectrosc. Radiat. Transf.*, **101**(3), 435–47.

Takashima, T. and Masuda, K. (1985) 'Degree of radiance and polarization of the upwelling radiance from an atmosphere–ocean system,' *Appl. Opt.*, **24**(15), 2423–9.

Timofeeva, V.A. (1974) 'Optics of turbid waters (results of laboratory studies),' in Jerlov, N. and Nielsen, E. (eds), *Optical Aspects of Oceanography*, Academic Press, London and New York, pp. 177–219.

Tonizzo, A., Zhou, J., Gilerson, A., Twardowski, M.S., Gray, D.J., Arnone, R.A., Gross, B.M., Moshary, F. and Ahmed, S.A. (2009) 'Polarized light in coastal waters: Hyperspectral and multiangular analysis,' *Opt. Express*, **17**(7), 5666–82.

Tynes, H.H., Kattawar, G.W., Zege, E.P., Katsev, I.L., Prikhach, A.S. and Chaikovskaya, L.I. (2001) 'Monte Carlo and multicomponent approximation methods for vector radiative transfer by use of effective Mueller matrix calculations,' *Appl. Opt.*, **40**(3), 400–12.

van de Hulst, H.C. (1981) *Light Scattering by Small Particles*, Dover Publications, New York.

von Frisch, K. (1948) 'Gelöste und ungelöste Rätsel der Bienensprache,' *Naturwissenschaften*, **35**, 38–43.

Voss, K.J. and Fry, E.S. (1984) 'Measurement of the Mueller matrix for ocean water,' *Appl. Opt.*, **23**, 4427–39.

Voss, K.J., Gleason, A.C.R., Gordon, H.R., Kattawar, G.W. and You, Y. (2011) 'Observation of non-principal plane neutral points in the in-water upwelling polarized light field,' *Opt. Express*, **19**(7), 5942–52.

Waterman, P.C. (1971) 'Symmetry, unitarity, and geometry in electromagnetic scattering,' *Phys. Rev. D.*, **3**, 825–39.

Waterman, T.H. (1954) 'Polarization patterns in submarine illumination,' *Science*, **120**(3127), 927–32.

Waterman, T.H. and Westell, W.E. (1956) 'Quantitative effect of the Sun's position on submarine light polarization,' *J. Mar. Res.*, **15**, 149−69.

Weng, F.Z. (1992) 'A multilayer discrete-ordinate method for vector radiative-transfer in a vertically-inhomogeneous, emitting and scattering atmosphere: 1. Theory,' *J. Quant. Spectrosc. Radiat. Transf.*, **47**(1), 19–33.

Winker, D.M., Pelon, J.R. and McCormick, M.P. (2003) 'The CALIPSO mission: Spaceborne lidar for observation of aerosols and clouds: Lidar Remote Sensing for Industry and Environment Monitoring III,' in Singh, U.N. et al. (eds), *International Society for Optical Engineering (SPIE Proceedings)*, **4893**, 1–11.

Wiscombe, W. (1980) 'Improved Mie scattering algorithms,' *Appl. Opt.*, **19**, 1505–9.

Xie, Y., Yang, P., Gao, B.C., Kattawar, G.W. and Mishchenko, M.I. (2006) 'Effect of ice crystal shape and effective size on snow bidirectional reflectance,' *J. Quant. Specrtrosc. Radiat. Transf.*, **100**(1–3), 457–69.

Xie, Y., Yang, P., Kattawar, G.W., Baum, B. and Hu, Y.X. (2011) 'Simulation of the optical properties of ice particle aggregates for application to remote sensing of cirrus clouds,' *Appl. Opt.*, **50**, 1065–81.

Yang, P. and Liou, K.N. (1996a) 'Finite-difference time domain method for light scattering by small ice crystals in three-dimensional space,' *J. Opt. Soc. Am. A*, **13**, 2072–85.

Yang, P. and Liou, K.N. (1996b) 'Geometric-optics-integral-equation method for light scattering by nonspherical ice crystals,' *Appl. Opt.*, **35**(33), 6568–84.

Yang, P., Wei, H.L., Huang, H.L., Baum, B.A., Hu, Y.X., Kattawar, G.W., Mishchenko, M.I. and Fu, Q. (2005) 'Scattering and absorption property database for nonspherical ice particles in the near- through far-infrared spectral region,' *Appl. Opt.*, **44**(26), 5512–23.

Yee, S.K. (1966) 'Numerical solution of initial boundary value problems involving Maxwell's equations in isotropic media,' *IEEE Trans. Antennas Propag.*, **14**, 302–7.

You, Y., Tonizzo, A., Gilerson, A.A., Cummings, M.E., Brady, P., Sullivan, J.M., Twardowski, M.S., Dierssen, H.M., Ahmed, S.A. and Kattawar, G.W. (2011) 'Measurements and simulations of polarization states of underwater light in clear oceanic waters,' *Appl. Opt.*, **50**(24), 4873–93.

Yurkin, M.A. and Hoekstra, A.G. (2009) *User Manual for the Discrete Dipole Approximation Code ADDA v.0.79, http://a-dda.googlecode.com/svn/tags/rel_0_79/doc/manual.pdf.*

Yurkin, M.A., Maltsev, V.P. and Hoekstra, A.G. (2007) 'The discrete dipole approximation for simulation of light scattering by particles much larger than the wavelength,' *J. Quant. Spectrosc. Radiat. Transf.*, **106**(1–3), 546–57.

2 Recent developments in the use of light polarization for marine environment monitoring from space

Tristan Harmel

2.1 Introduction

The measurement by satellite sensors of the light field backscattered from the atmosphere–ocean system is a powerful tool to monitor marine ecosystems, carbon cycle, or water quality on the global scale. Indeed, the color of the ocean, defined by the spectral behavior of the radiance exiting through the sea surface, is a good indicator of the phytoplankton pigments (e.g., chlorophyll a) and particulate content (e.g., particulate organic carbon) of the surface layer of the ocean. Estimations from space of the chlorophyll-a concentration and the subsequent estimate of the primary production and carbon stock are currently being based on multispectral measurements of the water-leaving radiance. However, this achievement was marked by the exploitation of the intensity of light, regardless of its state of polarization. This may be explained by the primary difficulty to accurately measure this intensity at sea and to relate it to water content. In a similar fashion, applications to space-borne sensors were complicated by the arduous extraction of the low signal of the water-leaving radiation from that of molecules and aerosols of the atmosphere. Nevertheless, recent investigations have been focused on the exploitation of polarization of light in the water column and exiting the sea surface to improve our capacities of observing and monitoring coastal and oceanic environments. In this direction, new instrumentation and data inversion algorithms are being developed to take advantage of the supplementary piece of information carried by the polarization state of light, sometimes inspired by pioneering works of the mid-20th century.

This chapter attempts to give a brief overview of recent developments on the use of polarization for marine environment monitoring including assessment of aerosol and atmospheric correction, sea state and associated winds, oceanic and coastal water content, and, potentially, estimation of the ocean carbon stock. First, a short historical review of the successive discoveries that punctuated our understanding of light polarization in the marine environment is given. After a description of the transfer of light in the atmosphere–ocean system, impacts of the water constituents (i.e., suspended particles, absorbing material) on polarization are summarized. Recent illustrations of the use and exploitation of light polarization for studying marine environment from laboratory to satellite applications are given. Through the chapter, benefits of polarimetric measurements for monitoring ocean, coastal, or lake environments are discussed in view of the future launch of polarimetric Earth-observing satellite missions.

2.2 Brief history of the discovery of polarization patterns in the marine environment

Far from the modern views on light and its propagation through the atmosphere–ocean system, the Vikings might have been able to determine the azimuth direction of the Sun with the help of skylight polarization, just like some insects. They would have used as polarizers natural crystals available to them, the famous Sunstones described in the sagas, for finding their way across the Atlantic under partially cloudy or foggy atmosphere (Horváth et al., 2011). However, the lack of historical or archaeological evidence prevents conclusions on this poetically appealing discovery and use of polarization. More than six centuries later, in 1669, Erasmus Bartholinus discovered the phenomenon of double refraction of calc-spar (a variety of calcite originating from a former Viking settlement, Iceland). Christian Huygens, after thoroughly analyzing Bartholinus's results, discovered in 1690 the polarization of doubly refracted rays through calcite crystals. Afterwards, Etienne Louis Malus formulated his law in 1808 which identifies the relationship between the orientation of a polarizing filter and the quantity of transmitted light for the restricted case of totally linearly polarized light. In the following years, a fruitful series of discoveries on the nature of light and polarization was made: François Arago coined the term 'neutral point' for a specific pattern of a null polarization state of skylight after discovering that skylight is partially polarized; Jean-Baptiste Biot discovered the high degree of polarization (DOP) of the rainbow; David Brewster documented the relationship between the index of refraction and the angle of incidence at which light is totally converted, after reflection, into linearly polarized light; two supplementary neutral points of skylight were respectively discovered by Jacques Babinet and Brewster (a fourth neutral point was recently found by Horváth et al. (2002), but below the horizon), Augustin-Jean Fresnel, after Arago and Thomas Young's investigations, gave in 1821 a new mathematical description of light (understood as transversal waves). This new theory of light was completely successful in describing totally polarized light, but was still unable to give a satisfactory description of unpolarized and partially polarized light.

In the midpoint of the 19th century,[1] Sir George Gabriel Stokes came up with a completely unique point of view: reformulating light description into a four-component vector of observables – that is, measurable quantities such as intensities (Stokes, 1852b). Stokes showed that his intensity formulation of polarized light was applicable to any state of polarization of light. Those outcomes were rapidly followed by another major paper (Stokes, 1852a) in which Stokes was able to enunciate, after an enormous amount of experimental effort, his now famous law of fluorescence, namely that the wavelength of the emitted fluorescent radiation was greater than the excitation wavelength; he also found that the fluorescence

[1] This part of the historical review was inspired by Edward Collett's Historical Note of the book *Polarized Light* by Dennis Goldstein (CRC Press, 2003).

radiation appeared to be unpolarized (see section 2.6.1 for a recent exploitation of this property). Stokes's findings on polarization, including his four-parameter formulation, were practically forgotten because, by the mid-19th century, classical optics was believed to be complete and physicists had turned their attention to the investigation of the electromagnetic field and the start of the quantum theory. Stokes's formulation of light was later slightly modified, first by Lord Rayleigh in 1902, and definitively adopted in the 1940s by the Nobel laureate Subrahmanya Chandrasekhar to include the effects of polarized light in the equations of radiative transfer (RT). Ironically, it was not until the onset of the 21st century that the RT equation was directly derived from Maxwell's equations governing the electromagnetic field (Mishchenko, 2002, 2003) making the RT theory a legitimate branch of physical optics.

After this short historical survey of the discovery of the polarized nature of light, one can easily understand the view that Alexandre Ivanoff, in 1974, on measuring polarization in the sea, said:

> 'Whereas polarization of light from a clear sky was discovered as early as in 1811 by the French physicist Arago, the underwater daylight polarization, anticipated by scientists such as Le Grand (1939), was observed for the first time in 1954 by the American biologist Waterman [8]. This last discovery illustrates how much more difficult it is to make observations or measurements underwater.'
>
> *(Ivanoff, 1974)*

However, since Waterman's observations, carried out visually during skin diving in the sea around Bermuda, the polarized nature of the underwater light field has been the subject of extensive research. Most of the early efforts were related to the study of the mechanisms for polarization sensitivity and additional ways in which marine animals may utilize underwater polarization. Some of the first measurements of the polarized underwater light field were reported by Waterman (1954), Waterman and Westell (1956), and Ivanoff and Waterman (1958), who pointed out that underwater polarization patterns are important for marine animals and discussed the dependence of the polarization states on the solar position and viewing directions. Some other early measurements of the underwater polarization include Ivanoff et al.'s (1961) measurement of the highly polarized light field (75% to 80%) in clear waters near Bermuda, and laboratory measurements by Valentina Timofeeva (1974), who showed that the degree of linear polarization (DOLP) decreases in turbid waters – that is, highly scattering media (see section 2.5.3).

In an effort to understand the impact of polarization states on marine animals, measurement of multispectral underwater polarization has again become popular in the experimental biology community in recent years (Cronin and Shashar, 2001; Sabbah et al., 2006; Sabbah and Shashar, 2007; Shashar et al., 2004, 2011; Johnsen et al., 2011; Lerner et al., 2011). In these studies, the polarization state of light was measured and analyzed in terms of its dependence on solar positions, viewing angles, wavelengths, and depth in the water column to map the underwater light distribution. In satellite remote sensing, another point of view obviously has to be used even if the underwater light distribution must be fully understood. First, the light field at the top of the atmosphere had to be exactly computed after consideration of the light/particles multiple interactions through the coupled atmosphere–ocean system. Second, mathematical methods to 'inverse' the top-of-atmosphere (TOA) measurements had to be sought to provide unambiguous solutions. This second point is not specific to the use of polarization and is consequently not addressed here; historical notes on 'inverse problems,' a term appearing in the late 1960s, can be found in Argoul (2012). In order to compute the Stokes parameters of the radiation exiting the atmosphere–ocean system, several mathematical and numerical methods have been developed to solve the RT equation.

The first code, fully accounting for polarization and light propagation through the air–sea interface and the water body, was developed based on the Monte Carlo method in the beginning of the 1970s (Kattawar et al., 1973). Analysis of those simulations showed that most of the water-leaving radiation originates from multiple scattering, elliptical polarization is created after transmission through the air–sea interface around the direction of the critical angle (i.e., $\sim 49°$) (Kattawar and Adams, 1989), and polarization must be accurately accounted for in the intensity calculation to avoid significant errors (Kattawar and Adams, 1990; Zhai et al., 2010; Hollstein and Fischer 2012). The Monte Carlo technique had proved to be insufficiently fast for remote sensing purposes and, in the following years up to now, several other methods (discrete-ordinate, adding-doubling, successive order of scattering, etc.) have been developed and used to solve the vector RT equation more rapidly. Most of these codes are freely available, thereby fueling their scientific exploitation and application to remote sensing. Meanwhile, polarimetric instrumentation has been refined, permitting new insights about the scattering properties of the in-water particles (from colloids and virus to large phytoplankton and suspended minerals), analysis of the atmosphere–ocean signal from polarimetric airborne or space-borne (thanks to the unique POLarization and Directionality of the Earth's Reflectances (POLDER) satellite missions) sensors (Harmel and Chami, 2008; Chowdhary et al., 2012), or accurate comparisons between simulated and measured underwater polarized light fields (You et al., 2011b). This instrumental development has even made possible completion of the historical observations of the neutral points but, this time, looking underwater rather than at the sky (Voss et al., 2011). Even though fundamental research on light transfer in the marine environment is still active and harvests new scientific issues by itself, we can certainly conclude with George Kattawar (2013) that the combination of achievement in RT simulation and recently developed instrumentation 'will certainly provide a plethora of new and exciting research topics in ocean optics.'

2.3 Potentialities of polarimetric remote sensing of marine biogeochemical parameters

The measurement by satellite sensors of the light field back or upward scattered from the atmosphere–ocean system is a powerful tool to monitor marine ecosystems, carbon cycle, or water quality on the global scale (McClain, 2009). Radiometry of the 'ocean color,' defined by the spectral behavior of the radiance exiting through the sea surface, provides reliable indicators of the pigment (e.g., chlorophyll *a*), the phytoplankton functional types (Alvain et al., 2005), particulate content (e.g., particulate organic carbon (POC)), or the water transparency (Doron et al., 2007) of the surface layer of the ocean. Estimation from space of chlorophyll-*a* concentration and the subsequent estimate of the primary production are currently based on multispectral measurements of the water-leaving radiance (O'Reilly et al., 1998; Hu et al., 2012; Morel, 1991; Behrenfeld and Falkowski, 1997). In addition, fluorescence measurements may also provide a path for monitoring climate–phytoplankton physiology interactions and improving descriptions of phytoplankton light-use efficiencies in ocean productivity models (Behrenfeld et al., 2009). Interestingly, this chlorophyll fluorescence signal could be retrieved from radiance and polarization measurements of the water-leaving light field based on a polarization discrimination technique which enables one to disentangle the elastically scattered component (partially polarized) and the fluorescence signal (totally unpolarized) (Gilerson et al., 2006) (see section 2.6.1).

On the other hand, knowledge of total POC and subsequent inference of the phytoplankton portion of POC is essential to the development of methods for estimating phytoplankton growth rates and carbon-based net primary production from satellite observations (Behrenfeld et al., 2005). As the turnover time of carbon biomass is relatively short (1–2 weeks), satellite capabilities to monitor changes in particulate carbon pools represent an effective tool for studies related to the biological pump. Measurements of the particulate beam attenuation coefficient at 660 nm, $c_p(660)$, constitute the most informative field data presently available for examining relationships between POC and ocean optical properties (Gardner et al., 1993, 2006). It can also be noted that estimation of c_p is required for determining water transparency from remotely sensed data (Doron et al., 2007). However, this attenuation coefficient cannot be directly inferred from water-leaving radiance measurements (IOCCG, 2006). In comparison, knowledge of the polarization state may play an important role for estimating this critical coefficient following the example of the estimation of the fluorescence signal. Some experimental and theoretical results showed that the DOP of the underwater and water-leaving light field is strongly related to the ratio of the attenuation coefficient over the absorption coefficient (Timofeeva, 1970; Ibrahim et al., 2012). This relationship enables one to contemplate accurate estimation of c_p from radiance and polarization measurements of the water-leaving light field. In other words, estimating c_p would be achievable from above-water or even satellite systems provided that the water-leaving light field can be accurately identified and quantified from the total signal. However, this task remains highly challenging, mainly because of the skylight reflection by the sea surface whose contribution is significantly polarized (Harmel et al., 2012b) (see section 2.5.4).

Estimating the water-leaving radiation from TOA satellite measurements is a complex, and still challenging, task. The main goal of atmospheric correction over open ocean is to remove the atmospheric and sea surface effects from satellite observations in order to retrieve the water-leaving radiance. In the visible part of the spectrum, the atmospheric radiance, along with its component reflected on the sea surface, corresponds to more than 90% of the measured TOA radiance. Those two components need to be accurately estimated by the atmospheric correction procedure in order to retrieve the water-leaving radiance. Moreover, this atmospheric radiance is highly variable due to the high spatiotemporal variability of the optical properties of aerosols (King et al., 1999). Several studies (Mishchenko and Travis, 1997a, 1997b; Chowdhary et al., 2002; Zubko et al., 2007) have highlighted that the remote sensing algorithms for aerosol detection which use both scalar radiance and polarization measurements are less dependent on a priori information used to constrain the retrieval algorithm than those making use of the scalar radiance measurements only. Thanks to these studies, polarization measurements can be now satisfactorily used to enhance the retrieval of aerosol optical thickness, aerosol effective radius, as well as aerosol refractive index, and, therefore, the quality of the remotely sensed water-leaving radiation (Harmel and M. Chami, 2011) (see section 2.6.3). It was also demonstrated that this kind of polarization-based algorithms are sensitive to sea state (i.e., sea surface roughness) enabling accurate estimation of surface wind speeds (Harmel and M. Chami, 2012) (see section 2.6.4).

In recent years, there have been increasing numbers of optical studies performed in coastal waters (Odermatt et al., 2012). This is largely because a significant proportion of global marine primary production occurs on continental shelves and the coastal ocean is most utilized and impacted by humans. In coastal environments, the derivation of marine biogeochemical properties from the upwelling light field can be strongly inaccurate due to high contributions from inorganic and/or colored dissolved organic material. Therefore, assessment of particulate fluxes in the coastal ocean and investigations into the transport and fate of sediments and pollutants on continental shelves, which are of great interest for coastal zone management and other policy decisions, still remain difficult to perform. In this direction, new instrumentations and data inversion algorithms are being developed to take into account the supplementary piece of information carried by the polarization state of light (Voss et al., 2011; Tonizzo et al., 2009). However, estimating the Stokes vector components of the polarized water radiance from above-water measurements is a challenging task, mainly because of their small magnitude and the strong contamination by the polarized skylight reflected from the sea surface. Feasibility of such retrievals and their utility for retrieving inherent optical properties linked with biogeochemical parameters can thus be considered as a preliminary step toward satellite applications, as discussed in the following sections.

2.4 Theoretical basis on radiative transfer through the atmosphere–ocean system

The light coming from the Sun and penetrating Earth's atmosphere is initially unpolarized. The solar radiation is then scattered and absorbed by atmospheric molecules and aerosols, refracted and reflected at the atmosphere–ocean interface, and further scattered and absorbed by sea water molecules and hydrosols (e.g., virus, bacteria, phytoplankton, organic or mineral particles). Secondary interactions of light in the ocean include inelastic (trans-spectral) processes such as fluorescence by dissolved organic matter and phytoplankton pigments, and Raman scattering by the water molecules. Due to the complex journey of light in the atmosphere–ocean system, ocean-color remote sensing methods have been primarily focused on methods to solve the RT equation and documenting the scattering matrix, or, at least, the scattering-phase function, and the absorption coefficient of aerosols and hydrosols. In this section, the formalism and the principal terms and quantities of the vector RT equation are summarized in the case of the atmosphere–ocean system.

The Stokes vector, $\mathbf{S} = [I, Q, U, V]^T$, fully describes electromagnetic radiation, including its polarization state, in terms of observable and additive quantities; the superscript T stands for the transpose of the vector. The first Stokes parameter, I, describes the total (sum of polarized and unpolarized) radiance. The terms Q and U are informative of the linearly polarized radiance: Q quantifies the fraction of linear polarization parallel to the reference plane; U gives the proportion of linear polarization at 45° with respect to the reference plane. The last term, V, quantifies the fraction of circularly polarized radiance. The degree of polarization (DOP) and the degree of linear polarization ($DOLP$) are defined as follows:

$$DOP = \frac{\sqrt{Q^2 + U^2 + V^2}}{I}, \tag{2.1}$$

$$DOLP = \frac{\sqrt{Q^2 + U^2}}{I}. \tag{2.2}$$

The scattering matrix describes the interaction between the incident beam with molecules and the suspended matter within a sample. The Mueller scattering matrix[2] $\mathbf{F}$ is defined based on the Stokes vector formalism, as the matrix which transforms an incident Stokes vector $\mathbf{I}$ into the scattered Stokes vector $\mathbf{I}'$ as:

$$\mathbf{I}' = \begin{pmatrix} I' \\ Q' \\ U' \\ V' \end{pmatrix} = \mathbf{FI} = \begin{pmatrix} F_{11} & F_{12} & F_{13} & F_{14} \\ F_{21} & F_{22} & F_{23} & F_{24} \\ F_{31} & F_{32} & F_{33} & F_{34} \\ F_{41} & F_{42} & F_{43} & F_{44} \end{pmatrix} \begin{pmatrix} I \\ Q \\ U \\ V \end{pmatrix}. \tag{2.3}$$

The above scattering matrix is defined with respect to the scattering plane (i.e., the plane through the incident and scattered beams). In general, the scattering matrix of aerosols or hydrosols can be simplified and expressed for a given scattering angle Θ as follows (Hovenier, 1987; van de Hulst, 1957):

$$\mathbf{F}(\Theta) = \begin{pmatrix} F_{11}(\Theta) & F_{12}(\Theta) & 0 & 0 \\ F_{12}(\Theta) & F_{22}(\Theta) & 0 & 0 \\ 0 & 0 & F_{33}(\Theta) & F_{34}(\Theta) \\ 0 & 0 & -F_{34}(\Theta) & F_{44}(\Theta) \end{pmatrix}. \tag{2.4}$$

Here, $F_{11}(\Theta)$ is the so-called phase function and $-F_{12}(\Theta)/F_{11}(\Theta)$ is the DOLP for incident unpolarized light. The *volume-scattering function* β, widely used in optical oceanography, is given by:

$$\beta(\Theta) = bF_{11}(\Theta), \tag{2.5}$$

where b is the scattering coefficient expressed per meter.

In recent years, a tremendous amount of progress has been made in analytical and computational methods for deriving the scattering matrix of complex-shaped

[2] This formalism was named in honor of H. Mueller, professor at the Massachusetts Institute of Technology, who disseminated among researchers and students the 4×4 matrix first introduced by P. Soleillet in 1929.

particles. Numerous codes are now available utilizing different kinds of numerical methods exhibiting different strengths and weaknesses: T-matrix, discrete-dipole approximation, finite difference time domain, effective medium theory, etc. At the same time, novel instrumentation has been developed to measure the scattering matrix of aerosols (Hovenier et al., 2003) or hydrosols (Volten et al., 1998; Gogoi et al., 2009; Zugger et al., 2008; Voss and Fry, 1984; Lofflus et al., 1992; Witkowski et al., 1993; Slade et al., 2013; Chami et al., 2014). Simulations and measurements are both providing essential information for RT computations and related remote sensing activities.

The RT equation in a medium governed by absorption and scattering mechanisms can be written at a given wavelength as follows:

$$\mu \frac{d\mathbf{S}(\tau,\mu,\phi)}{d\tau} = \mathbf{S}(\tau,\mu,\phi) - \frac{\omega(\tau)}{4\pi} \iint_{\Omega'} \mathbf{P}(\tau,\mu,\phi,\mu',\phi')\mathbf{S}(\tau,\mu,\phi)\, d\mu'\, d\phi' \,, \tag{2.6}$$

where τ is the optical thickness of the medium; μ and ϕ are the cosines of the viewing angle and the azimuth angle, respectively. The superscript prime denotes the incident direction. The second term of the right-hand side of Eq. (2.6), called the source term, depends explicitly on the phase matrix $\mathbf{P}$ and the single-scattering albedo ω. The phase matrix $\mathbf{P}$ is simply obtained by rotating the scattering matrix from the scattering plane to the viewing plane. In the particular case of the atmosphere–ocean system, the matrix $\mathbf{P}$ is replaced at the air–sea interface by the combination of the Fresnel reflection and transmission matrices; the reader is referred to Zhai et al. (2010) and Zhai et al. (2009) for details on the interface problem. As for the scattering matrix computation, numerous numerical codes have been developed or recently upgraded to take the full Stokes vector (i.e., polarization) into account. In order to accurately solve Eq. (2.6) for the coupled atmosphere–ocean system, several points need to be accounted for: the respective vertical distributions of molecules, aerosols, and water constituents; gaseous absorption; sea surface roughness. Another important point when dealing with remote sensing application is to spectrally integrate the Stokes vector onto the spectral band of the polarimetric sensor used. It is also worth mentioning that most codes currently available are based on a series of approximations such as that of a plane-parallel atmosphere (and ocean) which might limit and degrade applications in polar regions, for example.

In several applications of ocean-color remote sensing, it can be useful to split the Stokes vector into its major components. For instance, at the sea surface level, the Stokes vector can be written as follows:

$$\mathbf{S_t} = \mathbf{S_g} + \mathbf{S_{surf}} + \mathbf{S_{wc}} + \mathbf{S_w} \,, \tag{2.7}$$

where $\mathbf{S_t}$, $\mathbf{S_g}$, $\mathbf{S_{surf}}$, $\mathbf{S_{wc}}$, and $\mathbf{S_w}$ stand for the Stokes vectors of the total, sun glint, sea surface reflection of skylight, foam, and water-leaving components, respectively. More precisely, $\mathbf{S_g}$ is the contribution arising from specular reflection of direct sunlight on the sea surface, $\mathbf{S_{surf}}$ is the contribution generated by specular reflection of atmospherically scattered light on the sea surface, $\mathbf{S_{wc}}$ the contribution arising from sunlight and skylight reflected by whitecap-covered areas of the sea surface, and $\mathbf{S_w}$ the contribution of underwater light upwardly scattered after propagation through the sea surface. The same decomposition can be done at the TOA level.

Nevertheless, the different Stokes vector components will obviously include the impact of the RT through the atmosphere and no easy relation can be formulated between the Stokes vectors of the two levels. This is an important difference with the scalar point of view which makes use of the diffuse transmittance to relate the water-leaving component of the radiance at the top of the atmosphere to that at the bottom (Yang and Gordon, 1997).

2.5 Impact of the marine components on polarization state of light

The optical properties of the particulate and dissolved materials are known as the inherent optical properties (IOP) (Preisendorfer, 1965). IOP correspond to the single-scattering properties (e.g., absorption coefficient, scattering matrix) as opposed to the apparent optical properties which are impacted by multiple-scattering interactions and depend on the incoming light field. In biological oceanography, the IOP are primarily used to bridge light measurements to the particulate and dissolved matter in presence (IOCCG, 2006; Zaneveld, 2013). The particulate matter includes a variety of living biological cells (viruses, bacteria, phytoplankton, and zooplankton species) and non-living particulate matter that is also very heterogeneous in its nature and particle size. The non-living particulate matter includes organic detritus (many kinds of biogenous products and debris), inorganic particles (clay minerals, feldspars, quartz, calcite, and many other mineral species), as well as the mixed organic–inorganic types. The dissolved matter is defined based on practical considerations as the matter contained in seawater, which passes through a small pore size filter (usually 0.2–0.4 μm). This definition includes colloids and the so-called dissolved organic matter (DOM). The fraction of DOM that interacts with solar radiation is referred to as chromophoric (or colored) dissolved organic matter (CDOM). CDOM compounds absorb light, elastic scattering is assumed to be null, and a fraction of them are also fluorescent. In optical oceanography, IOP are traditionally partitioned in terms of the additive contributions of the different components of particulate and dissolved matter interacting with light:

$$a(\lambda) = a_w(\lambda) + a_\varphi(\lambda) + a_{CDOM}(\lambda) + a_{sed}(\lambda)\,, \tag{2.8}$$

$$\begin{aligned} b(\lambda)\mathbf{F}(\lambda,\Theta) &= b_w(\lambda)\mathbf{F_w}(\lambda,\Theta) + b_\varphi(\lambda)\mathbf{F}_\varphi(\lambda,\Theta) \\ &\quad + b_{CDOM}(\lambda)\mathbf{F_{CDOM}}(\lambda,\Theta) + b_{sed}(\lambda)\mathbf{F_{sed}}(\lambda,\Theta)\,, \end{aligned} \tag{2.9}$$

where a, b, and $\mathbf{F}$ are the absorption coefficient, the scattering coefficients, and the scattering matrix as defined in Eqs (2.3) and (2.4). The subscripts w, φ, *CDOM*, and *sed* stand for the contributions of sea water molecules, phytoplankton and associated detritus, colored DOM, and suspended sediments, respectively. Understanding the relationships of the dissolved and particulate matter with IOP and, therefore, light radiation is thus of great interest to sound and monitor marine, brackish, or lakes environments. In this section, the impacts on light polarization of three of the main optically active contributors present in ocean (phytoplankton, CDOM, and suspended sediments) are described in light of the recent findings of the scientific community.

2.5.1 Phytoplankton

The first measurements of the scattering matrix of phytoplankton (Voss and Fry, 1984; Fry and Voss, 1985; Quinby-Hunt, 1989) showed values very close to the scattering matrix of sea water itself. In particular, the measured angular distributions of the DOLP M_{12}/M_{11} are mostly bell-shaped, showing a maximum near 90° ranging from 50% to 90% (Volten et al., 1998). Later on, RT studies showed that the polarized contribution of the underwater light field is negligibly small relative to that of the atmosphere and air–water interface in the visible spectral range for most open ocean waters where phytoplankton is the dominant scatterer (Chowdhary et al., 2002, 2006; Chami et al., 2001; Chami, 2007). This theoretical result has then been tested based on analysis of actual airborne and space-borne sensors (Harmel and Chami, 2008; Chowdhary et al., 2012) but excluding very high chlorophyll concentration encountered in bloom conditions (further studies being necessary for such conditions).

The main originalities of the PARASOL mission (by the French space agency, *Centre National d'Etudes Spatiales* (CNES)) are its ability to observe a given ground target at different viewing angles (up to 16 viewing angles) along the satellite track and to perform polarimetric acquisitions of the I, Q, and U terms of the Stokes vector (Deschamps et al., 1994). The Q and U terms are measured at 490, 670, and 865 nm. Harmel and Chami (2008) carried out theoretical RT calculations to help analyze PARASOL data. The TOA Stokes vectors were calculated for various atmospheric, oceanic, and geometrical conditions using the *ordre sucessifs ocean atmosphere* (OSOA) RT model (Chami et al., 2001). The atmospheric layer was modeled using empirical aerosol models, namely the standard maritime M98 and tropospheric T70, as defined by Shettle and Fenn (1979). The computations were carried out for two values of the aerosol optical depth τ_a at 550 nm, namely 0.1 and 0.5, which are representative of the range of variation typically found over open ocean waters. Since this study focuses on the analysis of signal variations over the phytoplankton-dominated water type, only the pure seawater and phytoplankton components were considered here. The IOP of these components were modeled as follows. The scattering and absorption coefficients of pure seawater were taken from Morel (1974) and Pope and Fry (1997), respectively. The absorption coefficient of phytoplankton and co-varying particles a_{ph} was derived by use of the bio-optical model of Bricaud et al. (1998):

$$a_{ph}(\lambda) = A_{ph}(\lambda)[Chl]^{Ep(\lambda)} , \tag{2.10}$$

where Chl is the chlorophyll-a concentration (in mg/m^3), and $A_{ph}(\lambda)$ and $E_p(\lambda)$ are tabulated coefficients. The phytoplankton scattering coefficient was modeled as suggested by Loisel and Morel (1998):

$$b_{ph}(\lambda) = 0.416[Chl]^{0.766}\left(\frac{550}{\lambda}\right) . \tag{2.11}$$

The refractive index of phytoplankton relative to water was set to 1.05. The size distribution of phytoplankton cells was assumed to follow the Junge hyperbolic function, which is often used for natural waters (Bader, 1970), with a Junge exponent value of −4. The minimum and maximum radii of the size distribution were

0.1 μm and 50 μm, respectively. The Mueller scattering matrix of phytoplankton was computed by means of Mie theory (Stramski et al., 2004).

The polarized reflectance ρ_{pol} at the TOA level is then calculated as follows:

$$\rho_{pol} = \frac{\pi\sqrt{Q^2 + U^2}}{E_d} , \tag{2.12}$$

where E_d is the downwelling irradiance (in W/m^2) and $\sqrt{Q^2 + U^2}$ is the polarized radiance. The polarized reflectance was analyzed at 490 nm, which is consistent with the PARASOL polarized visible band. The concentrations of chlorophyll a used in the calculations were 0.03, 0.1, 0.3, 0.5, 1, 3, and 10 mg/m^3.

The influence of phytoplankton particles on the TOA polarized reflectance was theoretically analyzed for a large set of viewing geometries. The contribution from atmospheric molecules to the polarized TOA reflectance, which was calculated using Rayleigh theory, has been subtracted for each Stokes parameter to emphasize the polarized signature of aerosols and/or hydrosols. The polarized reflectance obtained after subtracting the atmospheric molecules effects is called the Rayleigh-corrected polarized reflectance in the rest of this section. To evaluate the sensitivity of the Rayleigh-corrected polarized reflectance, hereafter referred to as ρ_{pol_rc}, with the turbidity of the water mass, the absolute difference $\Delta\rho_{pol\_rc}$ between the Rayleigh-corrected polarized reflectance calculated for a given chlorophyll-a concentration and for a reference chlorophyll-a concentration:

$$\Delta\rho_{pol\_re}(Chl) = |\rho_{pol\_rc}(Chl) - \rho_{pol\_rc}(Chl_{ref})| , \tag{2.13}$$

where Chl_{ref} is the reference chlorophyll-a concentration which represents typical residual content of open ocean, namely 0.03 mg/m^3 (Morel et al., 2007).

The simulations were carried out for a clear (i.e., τ_a(550 nm) $= 0.1$, Fig. 2.1) and a moderately turbid atmosphere (i.e., τ_a(550 nm) $= 0.5$, not shown) using the M98 aerosol model. The azimuth angle shown in Fig. 2.1 corresponds to the azimuth difference between the solar and observation planes. As an example, an azimuth difference of 0° means that the half planes containing the Sun and the sensor are similar, while an azimuth difference of 180° means that the Sun and the sensor are located in two opposite planes. For a clear atmosphere (Fig. 2.1), the absolute difference $\Delta\rho_{pol\_rc}(Chl)$ is lower than 8×10^{-4} whatever the observation geometry and solar zenith angle when chlorophyll-a concentrations are smaller than 1 mg/m^3. The influence of marine particles on ρ_{pol_rc} increases with their concentration and is dependent on the observation geometry. As an example, the principal impact of the marine particles, when the values of chlorophyll-a concentration and θ_s are 3.0 mg/m^3 and 30°, respectively, is limited to the range of azimuth angles 120° to 180°. Note that these observation geometries are often sun glint contaminated (Deuze et al., 2000), and thus they are not necessarily exploited for remote sensing studies over the ocean. The influence of water content on TOA signal is reduced for higher solar zenith angle (e.g., $\theta_s = 50°$). This is because the atmospheric path length of the radiation is longer in this latter case and the impact of the atmosphere increases, thus reducing significantly the effect of the hydrosol signature. For similar reasons, the variations in ρ_{pol_rc} with the chlorophyll-a concentration observed in the azimuth range 120° to 180° decrease as the atmosphere gets more turbid. The variations remain smaller than 8×10^{-4} whatever the viewing geometries when the aerosol optical depth is 0.5. This latter result confirms the prevailing effect of aerosols on the polarization in comparison to the phytoplankton content of the oceanic layer. The computations made using the T70 aerosol model showed similar features to those obtained for the M98 aerosol model.

Fougnie et al. (2007) showed, based on an in-flight calibration study, that the noise-equivalent polarized reflectance of the PARASOL sensor at 490 nm is 8.5×10^{-4} for ocean targets, which is consistent with preflight estimation and specification of the instrument. Therefore, from an instrumental viewpoint, the polarized reflectance at the TOA is practically not sensitive to optically active marine particles in the case of a turbid atmosphere. In the case of a clear atmosphere, similar conclusions can be drawn only for waters showing *Chl* concentrations lower than 1 mg/m^3 (Fig. 2.1), which represent more than 90% of global ocean waters (Antoine and Morel, 1996). As reported by Chami (2007), the contribution to TOA polarized reflectance is predominantly from the atmospheric particles and, to a lesser extent, the skylight Fresnel reflection at the sea surface, which both significantly reduce the already weak polarization effects induced by biogenic marine particles (Chami et al., 2001). Those results showed that the sole aerosol polarized signature is sufficiently appreciable, including in the case of a clear atmosphere, to mask the influence of the hydrosols on the TOA polarized signal, despite the strong influence of the atmospheric molecular scattering on the polarization features.

A rigorous approach was then conducted to validate the theoretical calculations. A statistical analysis of geometries of observation exclusive to the PARASOL sensor was performed to determine the degree of applicability of RT modeling computations for real world conditions. The analysis showed that 87% of the targets viewed by PARASOL are observed under geometries for which predictions indicate that the TOA polarized signal is insensitive to phytoplankton biomass. Second, PARASOL-derived geophysical products such as atmospheric parameters and chlorophyll-*a* concentration were used to test the property of invariance of TOA polarized signals with oceanic constituents. The PARASOL images were analyzed at both regional and global scales. The regional-scale study showed that the variations in the TOA polarized reflectance at 490 nm ρ_{pol_rc} are within the noise-equivalent polarized reflectance of the PARASOL instrument when the targets exhibited similar aerosol optical properties and strongly variable chlorophyll-*a* concentrations. The variations in ρ_{pol_rc}, which were greater than the instrumental noise, were attributed to variations in aerosol optical properties. Similar results were obtained when the analysis was performed on a global area such as the Atlantic Ocean, thus providing a rigorous validation of theoretical predictions.

The innovative concept of the PARASOL instrument was exploited by various aerosol studies over the ocean (Deuze et al., 2000; Herman et al., 2005; Waquet et al., 2005; Tanré et al., 2011). However, these studies did not account for the directional and polarized measurements in the visible band because of the uncertainty in the polarized water-leaving radiance. The presented results on the significantly low variability of the TOA polarized signal with phytoplankton highlighted that the radiometric performances of the polarized visible band (490 nm) of the PARASOL satellite sensor can be used either for aerosol detection or atmospheric correction algorithms over most open ocean water, regardless of the biomass concentration.

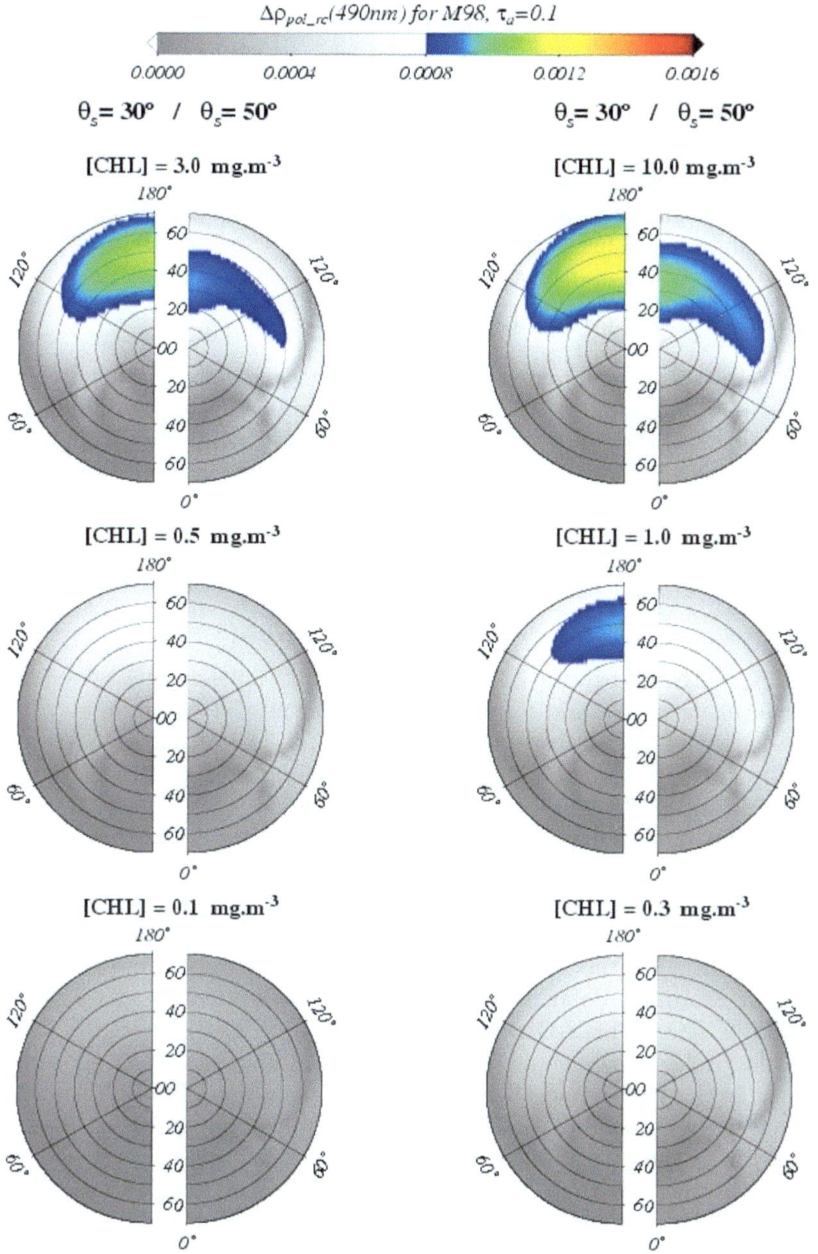

Fig. 2.1. Polar diagrams of absolute difference $\Delta\rho_{pol\_rc}$ between the Rayleigh-corrected polarized reflectance calculated for any given chlorophyll-a concentration and the one calculated for a chlorophyll-a concentration of 0.03 mg/m^3. The circular dot lines represent the viewing angles by step of 10° (numbered from 0° to 60° in the figure). The solar zenith angles θ_s are 30° and 50° and the Shettle and Fenn (1979) aerosol model M98 is used. The calculations are shown for a clear atmosphere (τ_a(550 nm) = 0.1). Note that a gray color scale is used when $\Delta\rho_{pol\_rc}$ is lower than PARASOL noise-equivalent polarized reflectance (i.e., 8.5×10^{-4}).

Note that the contribution of ocean body scattering to polarized reflectance observed at large azimuth angles does need to be properly accounted for at this wavelength, even though the dependence on chlorophyll concentration is negligible. On the other hand, it has been shown that this invariance property of the TOA polarized light is no longer valid when wavelengths are lower than 470 nm (Chowdhary et a., 2012). In this sense, it can be hypothesized that meaningful information on phytoplankton could be retrieved from polarimetric measurements in the violet and deep-blue parts of the spectrum.

2.5.2 Colored dissolved organic matter (CDOM)

CDOM is a ubiquitous component of the dissolved matter pool of the open ocean and coastal or inland waters. CDOM is important owing to its influence on the optical properties of the water column, its role in photochemistry and photobiology, and its utility as a tracer of deep ocean biogeochemical processes and circulation (Nelson and Siegel, 2013). CDOM is most often characterized by features of its absorption spectra considering a negligible scattering. The absorption spectra are modeled based on exponential parameters over a discrete wavelength interval such that:

$$a_{CDOM}(\lambda) = a_{CDOM}(\lambda_{ref}) \exp\left(-S_{CDOM}(\lambda - \lambda_{ref})\right) , \qquad (2.14)$$

where a_{CDOM} is the absorption coefficient (per m) at wavelength λ (given in nm) or reference wavelength λ_{ref}, and S_{CDOM} is the exponential slope parameter (per nm). S_{CDOM} values for CDOM absorption spectra in the visible and UV-A wavebands are typically in the range of 0.015–0.03/nm for open ocean spectra (Bricaud et al., 1981; Swan et al., 2013).

A few recent studies explicitly used this exponential relationship of Eq. (2.14) to assess the impacts of CDOM on the polarization state of the water-leaving light field (Chowdhary et al., 2012; Ibrahim et al., 2012). Nevertheless, only Chowdhary et al. (2012) isolated from RT simulations the impact of CDOM in their analysis of actual airborne measurements. Those actual data were obtained using the Research Scanning Polarimeter (RSP) instrument (Cairns et al., 1999) during the MILAGRO/INTEX-B campaign in March of 2006. The RSP instrument measures the Stokes parameters I, Q, and U simultaneously at nine narrow-band wavelengths in the visible and short-wave near-infrared and for 152 viewing angles covering an angular range of $+60°$ to $-70°$. Flights were conducted over a patch of open ocean off the coast of Veracruz (Mexico) at low (65 m) and mid (4.1 km) altitudes, with relatively low CDOM concentrations. Note that further studies are still needed to investigate the impact of CDOM on polarization in bloom and post-bloom conditions or in coastal waters where absorption and scattering properties can be significantly different (Organelli et al., 2014; Babin et al., 2003b; Stedmon et al., 2000). First, the authors showed that the RSP measurements of the multidirectional Stokes parameters are accurately reproduced by the RT simulations when accounting for contributions of phytoplankton and CDOM. Afterward, they examined the sensitivity of I, Q, and U to variations in CDOM concentrations. It was shown that variations with CDOM of polarized water-leaving radiance for wavelengths greater than 550 nm can be ignored altogether for average waters of the open ocean. At shorter wavelengths down to 410 nm, the changes in polarized reflectance caused by the natural variability in CDOM absorption are comparable to or less than the uncertainty in their observations, whereas changes on the radiance I are known to be substantial.

This previous study was extended to the configuration of the future ocean-color-observing missions such the NASA/Pre-ACE (Aerosols-Clouds-Ecosystem) mission focusing on UV-A spectral acquisitions (Chowdhary et al., 2013). Analysis of the TOA signal showed that the unpolarized part of radiance, which is obtained by subtracting the polarized TOA radiance from the total TOA radiance, is more sensitive to changes in CDOM absorption than the total radiance (i.e., unpolarized + polarized radiance). This property can be understood as follows: stronger the CDOM absorption, the lower the total radiance (I) for a virtually constant polarized radiance (i.e., $\sqrt{Q^2 + U^2}$). This can be reformulated based on the definition of DOLP (Eq. (2.2)): the stronger the CDOM absorption, the higher the DOLP. This impact on DOLP has been recently investigated by Ibrahim et al. (2012) for coastal waters where scattering by suspended sediment must be accounted for.

2.5.3 Suspended sediments

In coastal waters, hydrosols are mainly composed of two types of particles: living and non-living, sometimes called non-algal particles. Algal particles with high water content have a low refractive index (approximately 1.06) relative to that of water and are therefore poor distinctive scatterers (Harmel and Chami, 2008; Stramski et al., 2001) and might impact polarization fields mostly through their absorption features. Non-algal particles (NAP), such as minerals and detritus, are more effective scatterers due to their high relative refractive index, typically around 1.18 (Babin et al., 2003a; Chami and McKee, 2007). These particles can significantly decrease DOLP of the water-leaving radiance and should be more easily detectable than organic particles in the open ocean (Chami and McKee, 2007). Although the DOLP is highly sensitive to scattering (Ivanoff et al., 1961; Timofeeva, 1970), absorbing properties of the water column may also impact the polarized light field: increase in absorption corresponds usually to a decrease in the number of scattering events and leads to an increase in the DOLP. Thus, as we have seen section 2.5.2, CDOM might increase the DOLP of the water-leaving light.

Several approaches were proposed for the retrieval of water parameters from polarized observations (Chami et al., 2001; Chami, 2007; Chami and McKee, 2007; Loisel et al., 2008; Lotsberg and Stamnes 2010) but have not been fully implemented yet, for various technical reasons. For example, Chami and McKee (2007) suggest that it is possible to retrieve the suspended particulate matter (SPM) from measurements of DOLP at the Brewster angle using a empirical relationship. The signal measured from the POLarization and Directionality of the Earth's Reflectances sensor (POLDER-2) over turbid areas has been found to be in excellent agreement with theory with regard to its variability with the bulk particulate matter (Loisel et al., 2008). Ibrahim et al. (2012) proposed a relevant theoretical relationship between the DOP of the light and the ratio between the particulate attenuation (hereafter referred to as c_p) and the absorption coefficient.

Remembering the 1970s' work of Timofeeva (1970), Ibrahim et al. (2012) explored in detail the relation between DOLP and c/a for various water compositions typical of coastal waters. Indeed, Timofeeva (1970) analyzed polarized radiance scattered in milky solutions and found a relationship between the DOLP and 'the parameter T which is equal to the ratio of the attenuation coefficient of the scattered light flux to the direct light flux.' This parameter is in turn well connected to the attenuation/absorption ratio c/a, where c is equal to $a + b$, with b the scattering coefficient. Ibrahim and collaborators carried out RT computations with the RayXP code (Tynes et al., 2001) assuming a Rayleigh (molecular) atmosphere and a homogenous optically deep ocean. The optical properties of the oceanic layer were generated using the bio-optical model detailed in Gilerson et al. (2007) and references therein. The scattering properties were calculated through Mie theory computations. The radii of both phytoplankton particles and NAP were assumed to range from 0.1 to 50 μm following a Junge-type particle size distribution (PSD).

A comprehensive set of simulations was generated (about 10,000 different cases) by permuting the different values of the input parameters: chlorophyll concentration, NAP concentration, CDOM absorption, as well as slope of the PSD. Figure 2.2 shows the dependence of the DOLP on the c/a ratio for the complete range of water composition. It is readily visible that higher c/a ratios are associated with lower DOLP. This can be understood as follows. Higher c/a ratios mean that the medium is less absorbing and more scattering. The probability of multiple scattering is then increased. The highly polarized incoming light will be altered by this multiple scattering through its propagation, resulting in a depolarizing effect – that is, a decrease in the DOLP. Conversely, a lower c/a ratio results in higher DOLP of the upwelling radiation. It can also be noticed that the variation in DOLP with a/c is not linear (color gradation and gray scale of Fig. 2.2). The DOLP versus a/c relationship was parameterized using a power–law function. This parameterization demonstrated that it is possible to fit the relationship between the DOLP and the c/a ratio with satisfactory correlation ($R^2 > 0.9$, regardless of the viewing geometry) and with a relatively weak dependency on the PSD.

This type of study, along with analysis of actual data, is still necessary to make polarization-based algorithms operational in open and coastal ocean contexts. Nonetheless, polarization has been proved to be highly relevant to improve the performance of inverse algorithms dedicated to retrieval of hydrosols' optical and biogeochemical properties. Relationships such as that identified by Ibrahim et al. (2012) or Chami and McKee (2007) will surely be a great aid in the retrospective or future analysis of the polarimetric data of satellite sensors such as the POLDER instrument on PARASOL (Fougnie et al., 2007) and the planned 3MI instrument (Marbach et al., 2013) having multidirectional and polarized measurement capability.

2.5.4 Air–sea interface: impacts on above-water radiometric measurements

The impact of the sea surface on polarization is discussed here in the specific framework of ocean-color radiometry, one of whose issues is related to accuracy and reliability of the measured water-leaving radiation. The estimation of the water-leaving radiance from satellites remains complex due to the presence of the atmosphere between the water body and the sensor (Gordon, 1997). As a result, this estimation needs sophisticated data processing which is particularly challenging in coastal

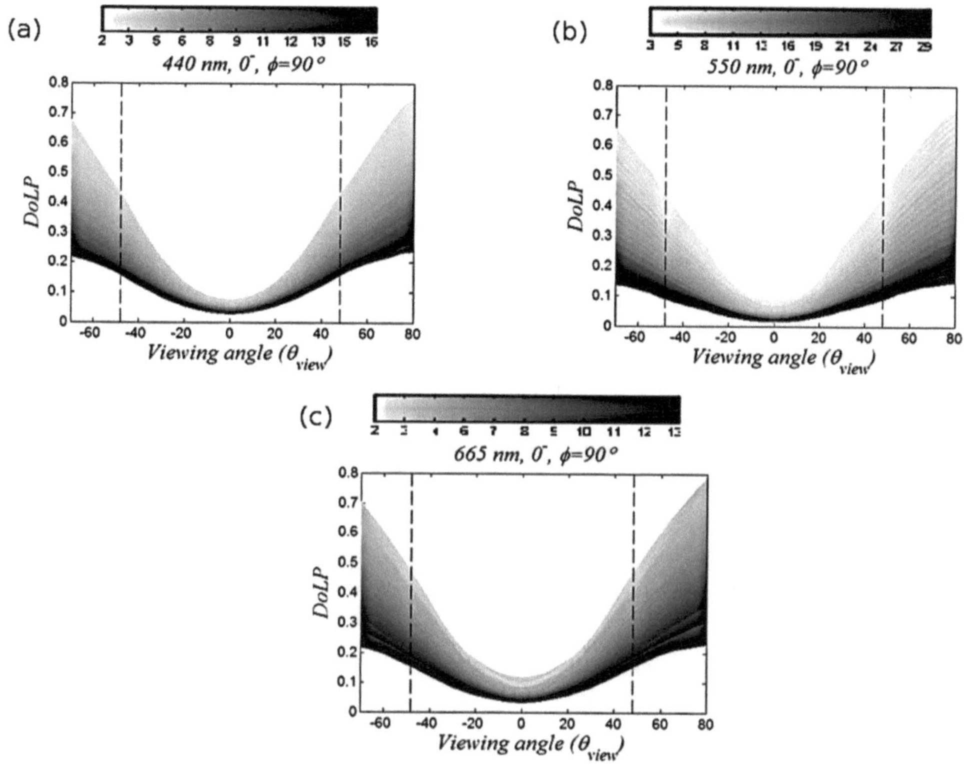

Fig. 2.2. Degree of linear polarization (DOLP), just below surface, for different viewing angles (0° for nadir direction) and c/a ratios at (a) 440, (b) 550, and (c) 665 nm. Simulations were performed for a Sun zenith angle of 30° and relative azimuth of 90° for positive viewing angle and −90° otherwise. The dashed vertical lines correspond to the critical angle; beyond this angle, transmission of light through the water–air interface is null. Modified from Ibrahim et al. (2012).

water areas. A mandatory effort is then dedicated to validation of the satellite retrievals on the basis of comparison with field-truth measurements (Antoine et al., 2008; Zibordi et al., 2009; Clark et al., 2003). The water-leaving radiance can be derived from underwater or above-water instrumentation. The latter method seeks to derive the water-leaving radiance from radiometric measurement of the sea surface from the deck of a boat or from instrumentation installed on a platform. The Sun and skylight, which have undergone specular reflection on the sea surface in the direction toward the radiometer, must be subtracted prior to water-leaving radiance estimation. This procedure corresponds to the most critical step in above-water radiometry achievement (Harmel et al., 2011a; Zibordi, 2012; Harmel et al., 2012a).

Generally, the data processing of above-water measurement permits filtering out any data contaminated by prohibitive amounts of sun glint or foam contribution (Zibordi et al., 2009; Koepke, 1984). Consequently, Eq. (2.7) of the Stokes vector can be reformulated, at the sea surface level, with respect to the surface reflection of diffuse skylight, S_{surf}, and the water-leaving contribution, S_w:

$$\mathbf{S_t} = \mathbf{S_{surf}} + \mathbf{S_w} + \varepsilon\,, \tag{2.15}$$

where the term ε accounts for residual sun glint or foam which might remain even after data filtering. In neglecting ε, the main issue to solve in above-water radiometry is to accurately estimate $\mathbf{S_{surf}}$ in order to retrieve $\mathbf{S_w}$ from measurements.

The polar diagrams of Fig. 2.3 give a synoptic view of the I, Q, and U parameters of $\mathbf{S_{surf}}$ and its associated DOLP. In those diagrams, azimuth is increasing counterclockwise and concentric circles are plotted for each viewing angle of 10° increment from the center. It can be readily seen in this figure that the impact of the sea state is variable with respect to the viewing configuration. This is particularly the case for the U parameter when the azimuth is close to 90°. It is worth mentioning that the DOLP is equal to unity for any azimuth when the viewing angle is close to 53°, corresponding to the Brewster angle when the sea surface is assumed to be flat (i.e., wind speed = 0 m/s). However, this characteristic is

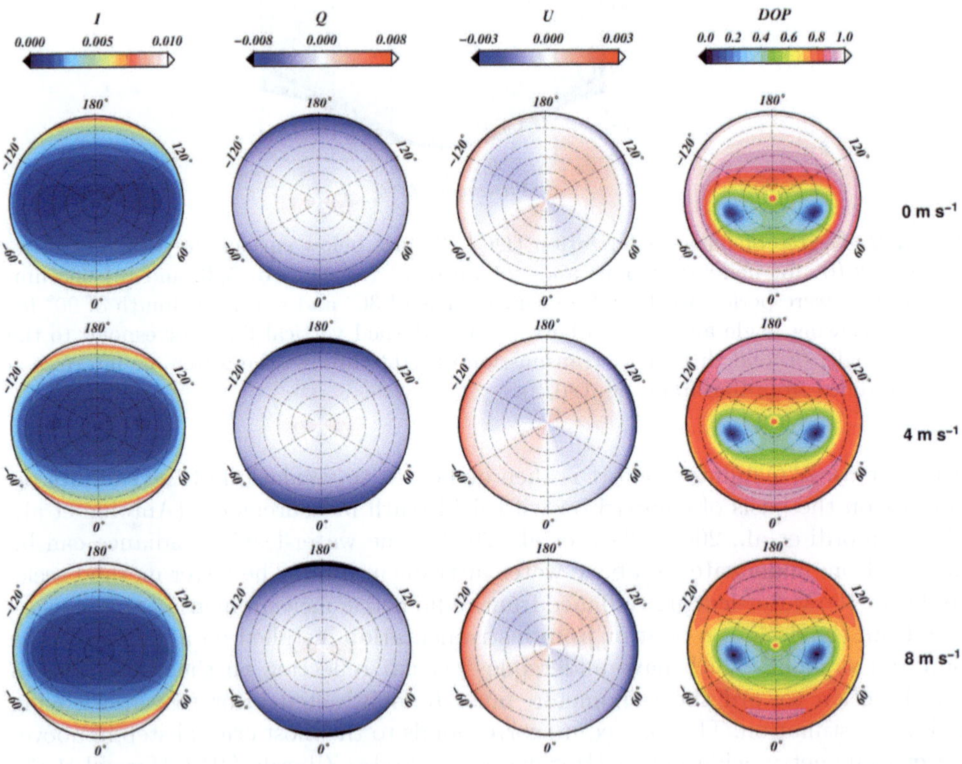

Fig. 2.3. Polar diagrams of the Stokes parameters I, Q, and U and the resulting degree of polarization (DOP) of the sky radiation reflected by the sea surface as observed just above the surface. Results are shown for three different wind speeds (0, 4, and 8 m/s) on which the state of the ruffled sea surface is taken into account. Viewing angles correspond to the concentric circles plotted with a 10° increment from the center. The Sun angle is equal to 60°.

washed out when wind speed increases and no characteristics of the Brewster angle are discernable any longer. On the other hand, two neutral points, corresponding to null DOLP values, are noticeable in the DOLP diagrams of Fig. 2.3. The location of these neutral points is in line with results of other studies (Adams and Kattawar, 1997). The characteristics of those neutral points are not significantly modified by changes in the sea state.

On the other hand, it has been noticed that neglecting polarization might significantly bias the I component of $\mathbf{S_{surf}}$ and thereby the estimation of the water-leaving radiance (Harmel et al., 2012b). A quantitative view of the surface polarization effects can be obtained by computing the relative difference between I_{surf}(scalar), computed through scalar RT, and I_{surf}(vector) accounting for polarization as follows:

$$\Delta I_{surf} = 100 \frac{I_{surf}(scalar) - I_{surf}(vector)}{I_{surf}(vector)} \% . \tag{2.16}$$

A synoptic view of the impact of atmospheric polarization on the sea surface reflectance can be obtained by plotting the relative difference ΔI_{surf} for all the azimuth and solar angles and, for the viewing angle of 40° commonly used in above-water radiometry, see Fig. 2.4. In Fig. 2.4a, calculations were done at 550 nm for an aerosol optical thickness of 0.1 and for flat sea surface. For instance, it can be seen in this figure that the relative difference at 90° azimuth is first negative for solar angles smaller than 35° and then becomes positive for greater solar angles. It can also be noted that the large negative differences occur in the vicinity of the 40° solar angle and 0° azimuth. However, this region corresponds to the zone of influence of the sun glint and cannot be used for above-water radiometry purposes.

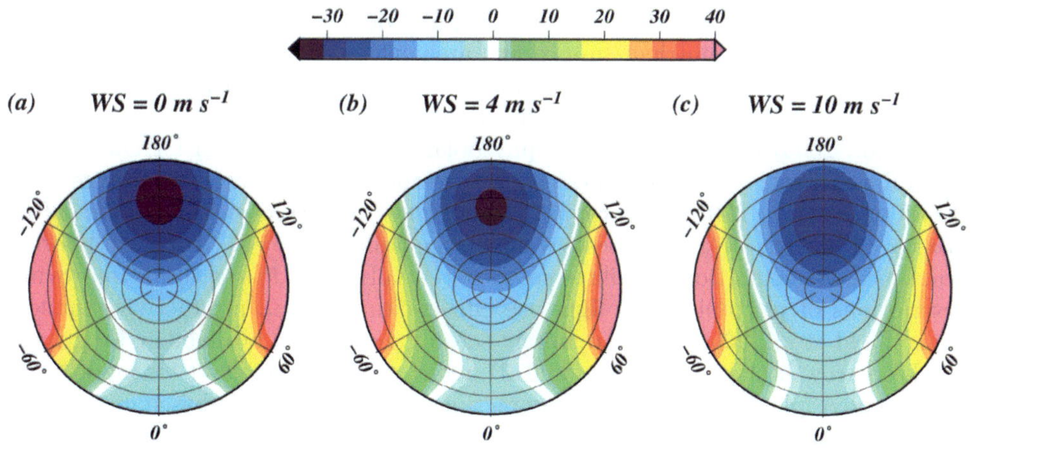

Fig. 2.4. Relative difference ΔI_{surf} at 550 nm and an aerosol optical thickness of 0.1 for different wind-speed conditions: (a) 0, (b) 4, and (c) 10 m/s. The polar diagrams account for all the azimuth and solar angles for a given viewing angle of 40°. The concentric circles represent the solar zenith angles by step of 10° (from 0° to 70°) and angles in the polar diagrams represent the relative azimuth with the Sun. Modified from Harmel et al. (2012b).

For wind speeds of 4 and 10 m/s (Fig. 2.4b and 2.4c), the sea surface ruffles and becomes a rough surface on which myriads of wave facets can reflect light coming from various directions of the sky region, thereby creating depolarization of the signal by beam superposition. As a result, it could be expected that the impact of skylight polarization on surface reflection is smoothed away when wind speed increases. Interestingly, this behavior does occur, but the relative difference ΔI_{surf} remains significant over the viewing configurations. Based on data (Harmel et al., 2012b), it can therefore be safely concluded that the polarization state of the skylight reflected on the sea surface must be accurately taken into account for estimating the sea surface reflectance and in turn the water-leaving radiance.

2.6 Benefits of polarization measurements for field and satellite remote sensing

2.6.1 Estimation of chlorophyll fluorescence through the polarization discrimination technique

A small fraction (~2%) of the solar energy absorbed by phytoplankton for photosynthesis is re-radiated by chlorophyll pigments through fluorescence. Under natural sunlight conditions, chlorophyll fluorescence can be detected in the subsurface and above-surface upwelling radiance spectrum (Morel and Prieur, 1977; Neville and Gower, 1977). This Sun-induced natural fluorescence feature has been proposed as a tool for deriving phytoplankton photosynthetic rates (Kiefer, 1973; Falkowski and Kiefer, 1985) and physiological variability (Letelier et al., 1997; Shallenberg et al., 2008). In addition, independent measurements of chlorophyll fluorescence and chlorophyll-*a* concentration can provide additional insight into photosynthetic activity and variability of fluorescence quantum yield, which can be linked to the differences in species, nutrient supply, and ambient light levels (Babin et al. 1996). In this manner, measurements of the fluorescence properties of phytoplankton in the red part of the spectrum can help in the detection of harmful algae blooms (Hu et al., 2005) or disentangle the phytoplankton component of CDOM-rich or mineral-rich waters, which is especially important in coastal water remote sensing (Dall'Olmo et al., 2005). Chlorophyll fluorescence quantum yields provide information on specific nutrient stressors on phytoplankton growth. Accurate fluorescence measurements from satellites could potentially provide a path for improving descriptions of phytoplankton light-use efficiencies in ocean productivity models (Behrenfeld et al., 2009) or for remotely sensing global iron availability for phytoplankton growth (Westberry et al., 2013), which would be an asset for monitoring climate–phytoplankton physiology interactions. However, studies based on light intensity measurements have encountered several issues mainly due to the large variability of water compositions and the intrinsic difficulty to decouple chlorophyll fluorescence from absorption. This has left room for improvement where polarization-based techniques could play their own part in furthering remote sensing of the marine environment.

The fluorescence phenomenon results from inelastic scattering and therefore presents a frequency shift between the incident light and scattered light. In addition to this trans-spectral process, the fluorescence signal is totally unpolarized (i.e., $Q = U = V = 0$). The polarization discrimination technique (Gilerson et al., 2006), described in this section, is based on this property to separate the unpolarized chlorophyll fluorescence signal from the elastic scattering signal, partially polarized, of the water-leaving radiance. Such a technique has been used and validated based on laboratory and field measurements. Nevertheless, it remains to be applied to polarimetric satellite sensors which might be contemplated as a step forward in improving our capacity to monitor the complex aquatic ecological systems from space.

The light elastically scattered by algae will show various degrees of linear polarization, depending on the polarization of the incident light and on the scattering angle (Mishchenko et al., 2006) whereas fluorescence emitted by these same algae is totally unpolarized (Mobley, 1994). Based on this premise, Gilerson et al. (2006) proposed a specific experimental scheme to decouple the unpolarized fluorescence signal from three independent polarimetric measurements: the total reflectance R_{tot}^{mes}, and the two components R_{P}^{mes} and $R_{\perp}^{mes}$ of this reflectance that are totally polarized alongside and perpendicularly to the scattering plane, respectively. Practically, those measurements can be carried out with a standard spectrometer, mounted with a rotating polarizer, looking at the water surface. The total reflectance R_{tot}^{mes} is simply given by the sum of the elastically scattered component plus the fluorescence component:

$$R_{tot}^{mes}(\lambda) = R^{elas}(\lambda) + Fluo(\lambda)\,. \tag{2.17}$$

The elastic component can be further decomposed into the perpendicular and parallel polarization components as follows:

$$R^{elas}(\lambda) = R_{\perp}^{elas}(\lambda) + R_{\mathrm{P}}^{elas}(\lambda)\,. \tag{2.18}$$

Likewise, the two polarized measurements can be written as:

$$\begin{cases} R_{\perp}^{mes}(\lambda) = R_{\perp}^{elas}(\lambda) + 0.5Fluo(\lambda) \\ R_{\mathrm{P}}^{mes}(\lambda) = R_{\mathrm{P}}^{elas}(\lambda) + 0.5Fluo(\lambda) \end{cases} \tag{2.19}$$

On the other hand, Ahmed et al. (2004) showed that the difference between $R_{\perp}^{mes}$ and R_{P}^{mes} can be advantageously expressed as a linear function of the elastic signal when calculated outside of the spectral range of the fluorescence emission (~670–750 nm), that is:

$$\begin{aligned} &\Delta R^{mes}(\lambda) = R_{\perp}^{mes}(\lambda) - R_{\mathrm{P}}^{mes}(\lambda) \\ &\Delta R^{mes}(\lambda)\colon AR^{elas}(\lambda) + B; \quad \forall\lambda \notin [670\text{ nm};\ 750\text{ nm}] \end{aligned} \tag{2.20}$$

where the two constants A and B are estimated by linear regression performed in the [450nm;670nm] range. Based on Eqs (2.17) and (20), the fluorescence component is simply retrieved as follows:

$$Fluo(\lambda) = R_{tot}^{mes}(\lambda) - \frac{\Delta R^{mes}(\lambda) - B}{A}\,. \tag{2.21}$$

The polarization discrimination technique has been tested on laboratory cultures of phytoplankton (Fig. 2.5a) and mixtures of phytoplankton and clay. After successful laboratory experiments, this technique was applied to field measurements acquired in Chesapeake Bay. The measured and processed spectra are displayed in Fig. 2.5b for subsurface spectrometer acquisitions. It can first be noticed that the fluorescence signal has been properly isolated, demonstrating the efficacy of the technique. It was found that the retrieval algorithm gives good results for both open ocean and coastal waters, permitting accurate extraction of fluorescence with less than 10% error, depending on the chlorophyll concentration and the water composition. Nevertheless, some residuals can be seen around 570 nm which might be due to some unconstrained sources of polarization. Such limitations of the method were particularly encountered when bottom reflectance and highly scattering suspended mineral are major contributors. Improvements of retrieval accuracy under such circumstances require a change in viewing angle such that the underwater light single-scattering angles becomes closer to 90° where polarization of scattered light is the largest. Other ways of investigation could take place in a reformulation of the above equations into the Stokes vector formalism or in taking advantage of multidirectional measurements of the Stokes vector, of which several underwater (Tonizzo et al., 2009; Voss and Souaidia, 2010) or space-borne instruments (Deschamps et al., 1994) are now capable.

Gilerson et al. (2006) also noticed that applications to above-water measurements, as a prerequisite to satellite applications, necessitates an accurate correction for the skylight reflection
indexskylight reflectionindexreflection on the sea surface that introduces other sources of polarization (see section 2.5.4). Provided that accurate correction for atmospheric signal and skylight reflection is achieved, use of polarization would be a significant aid to discriminate the fluorescence component of the signal from satellite measurements over oceanic and coastal waters. Moreover, methodologies such as the polarization discrimination technique could be generalized to study and remotely sense other source of fluorescence such as that of the DOM, whose

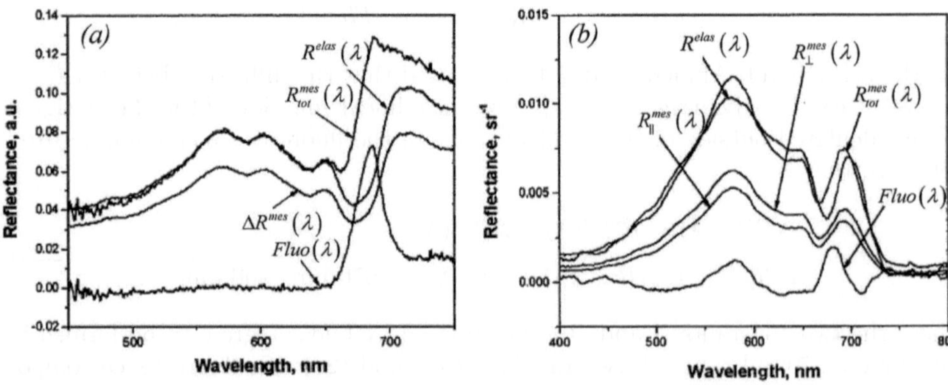

Fig. 2.5. Measured and processed spectra for (a) laboratory cultures of algae Isochrysis sp. and (b) field measurements in Chesapeake Bay, Sun zenith angle of 20°, and an underwater viewing angle of 45°. Modified from Gilerson et al. (2006).

fluorescence emission mostly takes place in the UV and blue part of the spectrum, depending on the matter present (e.g., humic or fulvic acid, amino acids) (Hudson et al., 2007). In a similar fashion, polarization techniques could be seen as a potential asset to discriminate the elastically scattered signal from the Raman scattered light, which produces significant effects on the polarization of the in-water light field (Kattawar and Xu, 1994) and can contribute as much as 25% of the solar backscattered radiance at wavelengths greater than 500 nm (Gordon, 1999). Application of those methods to polarimetric satellite missions remains today an important way in the aim of fully monitoring the ocean radiation and refining the estimation of the biogeochemical parameters from water-leaving radiance measurements (Morel et al., 2002; Gordon, 2014).

2.6.2 Measuring the polarization state of water-leaving radiation

As mentioned in section 2.5, the in-water dissolved and particulate matter may significantly impact the polarization state of the water-leaving radiation mainly in coastal waters or blooming ocean waters. The possibility of extracting useful information from polarimetric measurements performed at above-water levels (e.g., ships, satellites) relies on the faculty to isolate and accurately quantify the water-leaving component of the measured signal. The feasibility of measuring the water-leaving polarization field in the coastal environment from the above-water system is therefore a preliminary step toward satellite applications.

In this direction, the polarization state of skylight must be accounted for in the retrieval of the water-leaving radiance but also for estimation of the full Stokes vector of the water-leaving radiation when measured from above water. Harmel et al. (2012b) proposed an original algorithm to correct for the aerosol-dependent signal of the reflected skylight. This algorithm was applied to the Long Island Sound Coastal Observatory (LISCO) data set (Harmel et al., 2011a). The LISCO platform is located at around 3 km from the shore of Long Island near Northport, NY, USA. The platform combines a multispectral SeaPRISM system (CIMEL ELECTRONIQUE, France) which is now part of AERONET Ocean Color Network (Zibordi et al., 2009; Holben et al., 1998; Zibordi et al., 2004), with a collocated hyperspectral HyperSAS system (Satlantic, Canada). The instruments are positioned on a retractable tower on the LISCO platform with an elevation of 12 m. HyperSAS systems (HyperSAS-POL) were added to the LISCO platform enabling polarization measurements. In addition, regular field campaigns provide in-water measurements of the IOP (e.g., absorption and attenuation coefficients).

The DOLP of the water-leaving light retrieved from the LISCO measurements was recently analyzed to assess its consistency and its representativeness of the in-water composition (Harmel et al., 2011b). It was recently shown that DOLP patterns are quite stable even when the sea surface is ruffled by strong winds and can be well predicted by vector RT simulations based on known IOPs (Tonizzo et al., 2009; Adams et al., 2002; You et al., 2011a). On the other hand, Timofeeva (1970) demonstrated the strong relationship of the underwater DOLP with respect to the ratio of the total absorption coefficient over total attenuation coefficient, denoted as a/c. This relationship was expressed as a simple power law as follows:

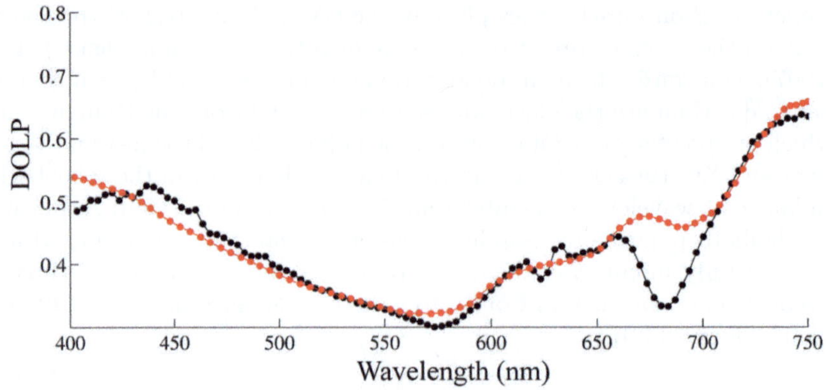

Fig. 2.6. Spectral degree of linear polarization (DOLP) retrieved from the above-water HyperSAS-POL measurements (black line) and from Timofeeva's parameterization (red line) based on the actual values of the a over c ratio measured at the LISCO site. The scattering angles are 106° and 126° above and below the surface, respectively. Adapted from Harmel et al. (2011b).

$$DOLP_{sim}(\lambda) = p\left(\frac{a(\lambda)}{c(\lambda)}\right)^q, \tag{2.22}$$

where p and q depend on the viewing configuration (i.e., Sun position and viewing direction). These parameters were calculated to fit $DOLP_{sim}$ from the actual measured value of a/c and then compared to the $DOLP_{LISCO}$ retrieved from the LISCO measurements (Fig. 2.6).

$DOLP_{sim}$ and $DOLP_{LISCO}$ exhibit satisfactory agreement, with discrepancies limited to a few percent over the major part of the spectrum (Fig. 2.6). However, a noticeable exception occurs in the red part of the spectrum (650 nm $< \lambda <$ 700 nm). In this spectral region, a certain amount of light is absorbed and re-emitted by phytoplankton fluorescence phenomenon. Moreover, the fluorescence emission is totally unpolarized by definition. Subsequently, the actual DOLP is greatly disturbed by fluorescence radiation and exhibits lower values. The corresponding trough in DOLP around the fluorescence peak at 683 nm is not handled in the Timofeeva parameterization, which does not consider light reemission phenomena as fluorescence, but only effects of absorption and elastic scattering. This explains the observed discrepancies between simulated and retrieved DOLP.

Although the relationship of Eq. (2.22) is a rough parameterization, it has been successfully used to model the spectral shape of the DOLP retrieved from above-water measurement. Thus, the achievability of accurate DOLP measurements from above-water instrumentation is demonstrated. RT computations are required to confirm Timofeyeva's observations; in particular, if the coefficients p and q are, indeed, only dependent on the viewing geometry, tabulated values of those coefficients could be used to retrieve the attenuation coefficient of the water body using data obtained with above-water polarization sensors (the absorption coefficient is routinely estimated from the remote sensing reflectance using well-established algorithms (Lee et al., 2002)). Extensive RT calculations are required to assess the dependence of the fitting coefficients on the specific geometrical configuration and water-constituent optical and biophysical properties (size distribution, refractive index, shape, etc.). Nonetheless, the feasibility of in-water polarization retrieval from above-water measurements has been demonstrated based on an approach very similar to that of the atmospheric correction scheme of satellite images. In other words, the results of this study gave preliminary evidence of the potentiality of the observation of water-leaving polarization signals from space.

2.6.3 Aerosol determination and atmospheric correction

The main goal of atmospheric correction over open ocean is to remove the atmospheric and sea surface effects from satellite observations in order to retrieve the water-leaving radiance. The atmospheric correction step is of paramount importance in ocean-color remote sensing because of the weak contribution of the water-leaving radiance (at most 10%) to TOA signals relative to the atmospheric component (IOCCG, 2010). In recent decades, numerous satellite missions dedicated to ocean-color radiometry purposes have been successfully launched such as the Sea-Viewing Wide Field-of-View Sensor (SeaWiFS/NASA), the Moderate Resolution Imaging Spectro-radiometer (MODIS/NASA), and the Medium Resolution Imaging Spectrometer (MERIS/ESA), POLDER/CNES), to mention only a few. Note that the three generations of the POLDER sensor (POLDER 3 on the PARASOL satellite) were the sole sensors being able to provide multidirectional and polarimetric measurements over the oceans from space.

The development of even more accurate atmospheric correction algorithms remains a challenging task. The application of atmospheric correction algorithms to satellite data provides aerosol optical property retrievals whose accuracy can also be significantly improved (Ahmad et al., 2010). Current atmospheric correction procedures rely on the fact that the water-leaving radiance can be assumed to be negligible in the red and infrared spectral bands because of the high absorption coefficient of seawater. Thus, the signal measured by a satellite sensor at these wavelengths carries information on the atmosphere layer only and can be used to estimate the aerosol optical properties (Gordon, 1997; Antoine and Morel, 1999; Fukushima et al., 1998; Gordon and Wang, 1994). Based on these derived optical properties, the atmospheric signal is extrapolated from the red to the shorter wavelengths (i.e., visible bands). However, such an extrapolation might lead to significant uncertainties in the retrieved water-leaving radiances because of inaccurate assessment of atmospheric optical properties at the visible wavelengths. Recent studies, based on comparison of satellite and *in situ* data, demonstrated the need to improve performances of the atmospheric correction algorithms over open ocean and coastal waters (Antoine et al., 2008; Mélin et al., 2010; Mélin and Zibordi, 2010).

Previous work showed that satellite polarized radiance is nearly insensitive to variations in phytoplankton concentration in open ocean waters both in the near-infrared and the blue/green parts of the spectrum (Harmel and Chami, 2008; Chowdhary et al., 2002; Chami, 2007) (see sections 2.5.1 and 2.5.2). As a result, the polarization signal measured over the open ocean at the TOA can be used to characterize the optical properties of aerosols regardless of the optical characteristics of the observed water mass. Several studies (Mishchenko and Travis, 1997a,

1997b; Chowdhary et al., 2002; Zubko et al., 2007) have highlighted that the remote sensing algorithms for aerosol detection which use both scalar radiance and polarization measurements are less dependent on a priori information used to constrain the retrieval algorithm than those making use of the scalar radiance measurements only. Consequently, the exploitation of polarimetric data has recently been reported as one of the main perspectives of aerosol detection from space (Mishchenko et al., 2007).

Until recently, TOA polarized information has been scarcely exploited to derive water-leaving radiances for ocean-color purposes (Harmel and Chami, 2011; He et al., 2014; Chowdhary et al., 2001). Here, the recent algorithm POLarization-based Atmospheric Correction (POLAC) (Harmel and Chami, 2011) is succinctly described with a focus on its ability to retrieve the water-leaving radiances and the aerosol optical characteristics simultaneously. Such an algorithm is currently dedicated to the sole satellite sensor measuring the multidirectional polarized radiance (POLDER 3 on PARASOL, hereafter referred to as 'PARASOL'). Note that the algorithm could be potentially adjusted to the future satellite missions capable of multidirectional and polarimetric measurements such as NASA's ACE and its preparatory mission (PACE) or the 'Multi-directional, Multi-polarization and Multispectral (3MI)' mission (ESA/EUMETSAT), which are both scheduled for launch around 2020.

The POLAC algorithm is composed of two principal phases, hereafter noted as phases (P1) and (P2), dealing with the retrieval of the aerosol optical properties and the water-leaving radiance, respectively. The two phases make use of an optimization scheme which is performed to retrieve the geophysical parameters of interest. The optimization process is achieved between PARASOL measurements and simulations carried out using a vector RT model which was designed for simulating the light field including polarization in the atmosphere–ocean system. The optical properties of the aerosols are determined based on the radiance (i.e., Stokes parameter I) in the near-infrared (NIR), namely at 865 nm, as well as the Stokes parameters Q and U in the visible spectrum and NIR as measured by PARASOL, namely at Q and U at 490, 670, and 865 nm. It is worth remembering that Q and U are supposed to be insensitive to variations in the in-water-constituent concentrations with some limitation concerning very high phytoplankton conditions (cf. section 2.5.1).

The inverse method of POLAC is based on direct RT calculations for the coupled atmosphere–ocean system whose results are stored in look-up tables for reducing central processing unit (CPU) time. In the RT model, a three-component oceanic layer is considered: pure seawater, phytoplankton pigments, and CDOM. The IOP of these components are modeled as follows. The scattering and absorption coefficients of pure seawater are taken from Morel (1974) and Pope and Fry (1997), respectively. The absorption coefficient of phytoplankton and CDOM is derived from the bio-optical model of Bricaud et al. (1998). The phytoplankton scattering coefficient is modeled as suggested by Loisel and Morel (1998). The bulk refractive index of phytoplankton relative to water is 1.05. The size distribution of phytoplankton cells is assumed to follow the *Junge* hyperbolic distribution, which is commonly used for the oceanic environment (Bader, 1970; Morel, 2009), with *Junge* exponent values of -4.5, -4, and -3.5.

The atmospheric layer is considered a mixture of molecules and aerosols. Aerosols are assumed to follow a bimodal log-normal distribution representing a mixture of a fine mode with a coarse mode (Deuze et al., 2000; Kaufman et al., 2001; Francis et al., 1999; Dubovik et al., 2000; Veselovskii et al., 2004). The microphysical properties of the aerosol modes were provided by the *Laboratoire d'Optique Atmospherique* (Université de Lille, France) based on the results derived by Dubovik et al. (2002) from AERONET network observations. In addition, the nonspherical aerosol model measured by Volten et al. (2001) is used. Note that this model is assumed to be spectrally flat. For each of the fine and coarse modes, simulations have been generated and stored in look-up tables (LUT) for aerosol optical thicknesses $\tau_a(\lambda = 550$ nm) ranging from 0 (purely molecular atmosphere) to 1 (very turbid atmosphere). On the one hand, the Stokes parameters I, Q, and U at the TOA, calculated for a chlorophyll-a concentration (*Chl*) value of 0.03 mg/m, have been stored in the so-called LUT-atmosphere for the wavelengths which are insensitive to *Chl* variations, namely in the red-infrared part of the spectrum (for I, Q, and U) and in the visible (for Q and U only).

The POLAC inversion scheme is based on the minimization of a multidirectional cost function using the Levenberg–Marquardt method. At each step of the algorithm, the variables which need to be retrieved (e.g., aerosol optical thickness, fine mode, water-leaving radiance, etc.) are represented as a vector, denoted $\mathbf{x}$. A mono-directional and multidirectional cost functions are defined with respect to $\mathbf{x}$ to take into account the different dynamics exhibited by the values of the three Stokes parameters acquired at different wavelengths and viewing geometries. The cost function J is defined for a given viewing geometry configuration, denoted $\boldsymbol{\Psi}$, as follows:

$$J(\boldsymbol{\Psi}, \mathbf{x}) = \sum_{i}^{N_{cri}} p_i(\boldsymbol{\Psi}) \frac{\left(S_i^{mes}(\boldsymbol{\Psi}) - S_i^{sim}(\boldsymbol{\Psi}, \mathbf{x})\right)^2}{\sigma_i^2(\boldsymbol{\Psi})}, \tag{2.23}$$

where N_{cri} is the number of criteria used in the retrieval procedure; here a criterion is defined as one of the Stokes parameters I, Q, or U for a given wavelength. S_i^{mes} and S_i^{sim} stand for the measured and simulated values of the ith criterion, respectively. The parameter σ_i is the absolute uncertainty affecting the measurement of S_i^{mes}; here, σ_i is taken as the noise-equivalent normalized radiance as calculated by Fougnie et al. (2007). Finally, p_i is the variance of S_i^{sim} calculated over the entire set of $\mathbf{x}$ used in the simulations; more explicitly, p_i is calculated for a given viewing geometry as follows:

$$p_i = \frac{1}{N_{LUT} - 1} \sum_{j=1}^{N_{LUT}} \left(S_i^{sim}(\mathbf{x}_j) - \langle S_i^{sim} \rangle\right)^2, \tag{2.24}$$

where N_{LUT} is the total number of values taken by $\mathbf{x}$ in the LUT generation and $\langle S_i^{sim} \rangle$ is the average of the simulated S_i over the set of $\mathbf{x_j}$. The p factor is an estimator of the amount of information contained in a given Stokes parameter that is effectively used to retrieve the desired variable $\mathbf{x}$. In other words, the greater p is, the more informative S is.

The multidirectional cost function J^* is then given by:

$$J^*(\mathbf{x}) = \sum_{k}^{N_{dir}} J(\mathbf{\Psi_k}, \mathbf{x}), \tag{2.25}$$

where N_{dir} is the number of viewing configurations available. In the algorithm, the values of the vector $\mathbf{x}$ are obtained after minimization of the appropriate cost function, either J or J^*.

The entire inversion scheme of the PARASOL data leads to the retrieval of bimodal aerosol optical parameters (aerosol optical thickness, Angström exponent) and the marine reflectance. Each step of the algorithm takes specifically into account the directional information of the PARASOL data, enabling estimation of the associated uncertainties. Figure 2.7 summarizes the main characteristics of these steps. The phase (P1) is subdivided into two iterative steps; Q and U are primarily used in a first step for retrieving the best bimodal aerosol model. A bimodal aerosol model is defined as a couple of a fine mode and a coarse mode as well as their respective proportion. In a second step, the scalar radiance I is used to derive the aerosol optical thickness. The convergence of the iterative procedure is typically obtained after two or three iterations. Based on the retrieved values of aerosol optical properties, the phase (P2) is activated to derive the water-leaving radiance in the visible spectrum (i.e., PARASOL bands centered on 490, 565, and 670 nm). The water-leaving radiance is retrieved by matching the measurements of TOA radiance with RT simulations computed for various hydrosol compositions of the oceanic layer. An example of the chlorophyll-a concentration estimated based on the retrieved radiances is given in Fig. 2.8.

Harmel and Chami (2011) showed that POLAC theoretical accuracies, including realistic instrumental noise, for deriving the Angström exponent are around 3% and 1% in clear (τ_a(550 nm) = 0.1) and turbid atmospheric (τ_a(550 nm) = 0.5) conditions, respectively. The use of the sole polarized information at 490 nm is efficient to estimate the Angström exponent with an uncertainty lower than 4%. The spectral water-leaving radiances are retrieved within 3% accuracy. The algorithm

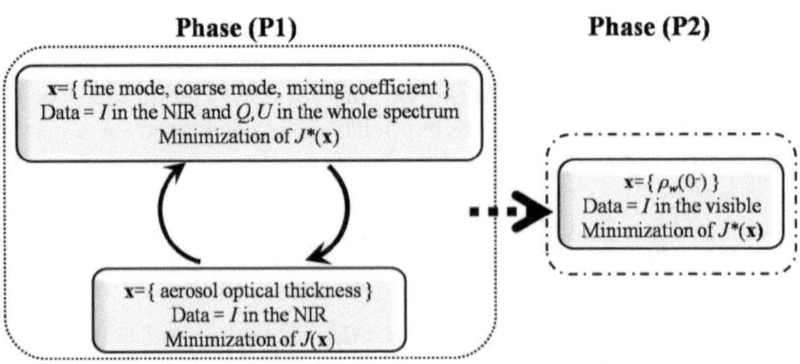

Fig. 2.7. Schematic diagram of the different steps of the POLAC algorithm. For each step, the vector $\mathbf{x}$ of the parameters which need to be retrieved, the PARASOL data, and the directional cost function used are mentioned in the figure. NIR means the near-infrared part of the spectrum, ρ_w is the nadir marine reflectance just beneath the sea surface, and the functions J and J^* are the mono- and multidirectional cost functions, respectively.

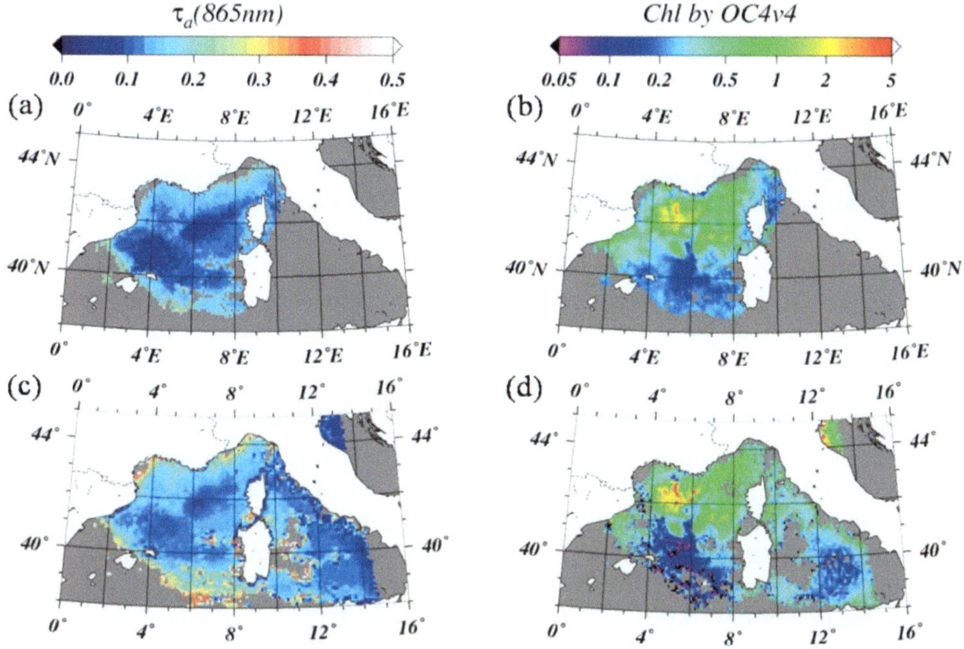

Fig. 2.8. Level 2 satellite images acquired on May 5, 2006, over the Mediterranean Sea. (a) SeaWiFS Aerosol optical thickness at 865 nm and (b) SeaWiFS chlorophyll concentration in mg/m^3, (c) aerosol optical thickness at 865 nm as derived by POLAC applied to the level 1 PARASOL image, (d) chlorophyll concentration in mg/m^3 as derived by POLAC applied to the level 1 PARASOL image. The chlorophyll concentration is calculated using the SeaWiFS OC4v4 (OC2v2 for PARASOL data) algorithm based on the retrieved water-leaving radiances.

was applied to numerous actual level 1 PARASOL images. The influence of the polarized information on the retrieval of geophysical products was studied based on these images. It was shown that the use of the polarimetric data increases significantly the performance of detection of the aerosols, thus confirming the theoretical sensitivity study. Consequently, the atmospheric signal is reappraised, improving de facto the water-leaving radiance estimation.

2.6.4 Near-surface wind-speed estimation

The sparkling patch of sunlight reflected on the ocean can be readily seen in any optical images acquired from the shore or the deck of a ship as well as from a plane or a satellite platform. The peak of intensity due to the sunlight reflection on the rough sea surface, commonly referred to as sun glint, occurs in the vicinity of the theoretical location of the specular reflection spot that would be obtained for a flat sea. The whole area where sun glint is present is called sun glitter and its extent is spread by a myriad of glints generated by reflection of sunlight on the facets of the water surface which are tilted toward the observer (Torrance et al., 1966). Hulbert (1934) worked out the geometry of sun glitter as it relates to sun elevation and wave slope, but did not consider the detailed optics of single glints. Later investigations conducted by Cox and Munk (1954a, 1954b) led to quantitative and statistical results about the link between sun glint and sea state.

Knowledge of surface winds is critically important for budgeting energy transport, oceanic primary productivity, and studies of ocean acidification (Ohlmann and Siegel, 2000; Dickey and Falkowski, 2002). Active and passive satellite remote sensing sensors that measure in the microwave spectral domain are commonly used to derive the wind-speed values from space over the oceans with a spatial resolution varying between 25 km and 50 km and a typical accuracy of 1 m/s (Bourassa et al., 2010). Wald and Monget (1983) demonstrated the feasibility of deriving the synoptic field of wind magnitude from measurements of the glitter pattern from space. Bréon and Henriot (2006) showed using the previous generation of PARASOL sensor, namely the POLDER sensor on the ADEOS platform, that the radiance measured at observation geometries for which the radiance is the most highly sensitive to sun glint may be exploited to accurately characterize the sea surface wind. Harmel and Chami (2012, 2013) expanded those results by including the exploitation of polarization in addition to the radiance signal to estimate sea surface wind speed at the spatial resolution of the PARASOL pixels, namely 6 km to 7 km, for more than 75% of the PARASOL swath.

The algorithm developed by Harmel and Chami (2012) first relies on the POLAC algorithm (Harmel and Chami, 2011) to estimate the atmospheric contribution ($\mathbf{S_{atm}}$) and the underwater contribution ($\mathbf{S_w}$) of the TOA Stokes vector. These values are then used to determine the Stokes vector of the sun glint plus the foam contribution ($\mathbf{S_{g+wc}}$) at the sea level for each viewing direction of PARASOL (Harmel and Chami, 2013). Note that $\mathbf{S_{g+wc}}$ is derived from the PARASOL data without any a priori assumptions on the sea surface conditions. The method used for estimating the surface wind-speed values relies on the comparison between the estimated sun glint/whitecaps contribution $\mathbf{S_{g+wc}}$ and the corresponding simulated values modeled based on a series of parameters that include wind-speed values.

The forward model that is used to simulate the sun glint/whitecap contributions and the inverse method can be summarized as follows. Based on measurements that covered a wind-speed range varying from 0 m/s to 14 m/s, Cox and Munk (1954) showed that the ocean surface can be modeled using a distribution of small facets which are oriented following a near-Gaussian distribution, namely a *Gram–Charlier* series. This distribution can be expressed as a function of the crosswind and upwind components of the wave slope z_c and z_u, respectively, which are related to wind direction:

$$\begin{aligned} p(z_c, z_u) = {} & \frac{1}{2\pi\sigma_c\sigma_u} \exp\left(-\frac{\xi^2+\eta^2}{2}\right) \left[1 - \frac{C_{21}}{2}(\xi^2-1)\eta - \frac{C_{03}}{6}(\eta^2-3)\eta \right. \\ & \left. + \frac{C_{40}}{24}(\xi^4 - 6\xi^2 + 3) + \frac{C_{22}}{4}(\xi^2-1)(\eta^2-1) + \frac{C_{04}}{24}(\eta^4 - 6n^2 + 3)\right] . \end{aligned} \tag{2.26}$$

Here, $\xi = z_c/\sigma_c$ and $\eta = z_u/\sigma_u$, where σ_c and σ_u are the crosswind and upwind root-mean-square components to the total variance of the slope distribution, respectively. The C_{ij} coefficients quantify the non-Gaussian nature of the distribution. The C_{ij} coefficients recalculated by Bréon and Henriot (2006) are used. The sun glint Stokes vector $\mathbf{S_g}$ is obtained using the distribution function expressed in Eq. (2.26).

The fraction of the ocean surface that is covered by sea foam is defined as whitecap coverage (hereafter noted as f_f). The modeling of f_f as a function of the surface wind-speed variable (ws, in m/s) is often used (Anguelova and Webster, 2006). The optimal power-law formula obtained by (Monahan and O'Muircheartaigh, 1980) is used:

$$f_f = 2.95 \times 10^{-6}\ ws^{3.52}\,. \tag{2.27}$$

The Stokes vector of the sun glint/whitecaps contributions at sea level is modeled as:

$$\mathbf{S_{g+wc}} = (1 - f_f)\mathbf{S_g} + f_f\mathbf{S_{wc}}\,. \tag{2.28}$$

The radiance term I_{wc} of the $\mathbf{S_{wc}}$ vector is calculated by considering the average foam reflectance assumed to be equal to 0.13 at 865 nm (Frouin et al., 1996; Kokhanovsky, 2004). The foam reflectance is assumed to be totally unpolarized (i.e., $Q_{wc} = 0$ and $U_{wc} = 0$).

Based on the estimation of the glint/whitecaps contributions for each viewing direction of PARASOL data, hereafter noted as $\mathrm{S_{g+wc}}$, and using the forward model of $\mathbf{S_{g+wc}}$, we can construct the following cost function $\phi(\mathbf{x})$:

$$\Phi(\mathbf{x}) = \sum_{\substack{viewing\\directions}} \left\|\mathbf{S^*_{g+wc}} - \mathbf{S_{g+wc}}(\mathbf{x})\right\|^2\,. \tag{2.29}$$

The minimization of such a cost function permits to retrieve the parameter $\mathbf{x}$ which here corresponds to the wind speed. The Levenberg–Marquardt damped least-squares method is used to minimize the cost function. The derived uncertainty of the parameter $\mathbf{x}$, denoted as σ, expresses the sensitivity of the cost function to a variation of $\mathbf{x}$ lower than σ around the solution $\mathbf{x}$.

The inverse method was applied to PARASOL level 1 images and the performances were evaluated through comparisons with concurrent remote sensing and field data. An example of retrievals is given Fig. 2.8 for the wind-speed values and their associated uncertainties retrieved from a PARASOL image acquired on May 5, 2006, over the north-west Mediterranean basin. Several wind patterns are readily visible in the retrieved products showing small-scale variability. The zones where high-wind-speed values are retrieved generally exhibit higher uncertainties (i.e., ~0.7 m/s). These higher uncertainties are likely due to the presence of whitecaps. A zone of very weak wind-speed values (ws < 2 m/s) is noticeable in the northern part of the image (~43°N) between 4°E and 5°E. The uncertainties associated to the weak wind-speed area are generally smaller than 0.3 m/s (Fig. 2.8b). Thus, the sun glint multidirectional and polarized radiation is sufficiently sensitive to the wind-speed parameter to allow the method to identify low-wind-speed area with satisfactory performances. Therefore, knowledge of wind speeds at a spatial resolution as fine as PARASOL full resolution may have some important implications for modeling purposes. Typically, the assimilation of high spatial resolution wind-speed data into meteorological models should improve the predictions related to the weather forecast. A higher spatial resolution of wind-speed data could also improve the representation of dynamic structure like eddies or specific upwelling cells in coastal regions where the dynamic processes are complex (Schaeffer et al., 2010).

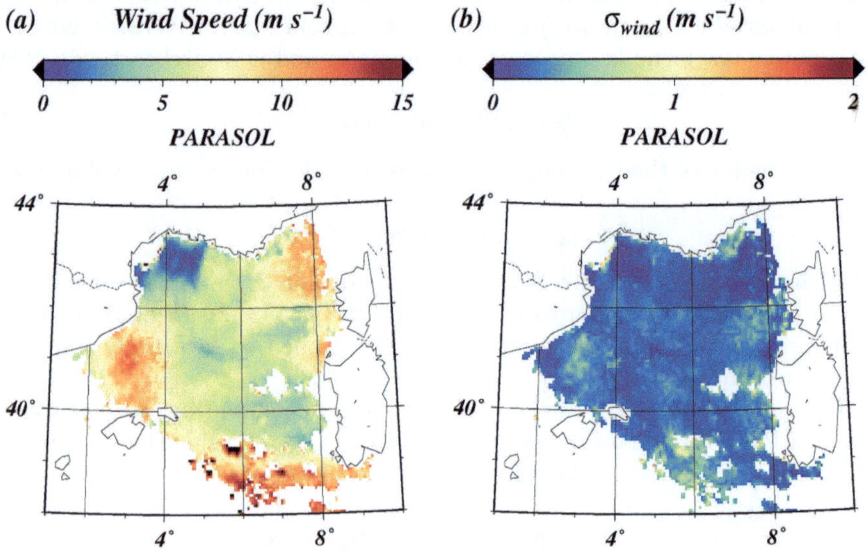

Fig. 2.9. (a) Wind-speed values and (b) their associated uncertainties, σ_{wind} (in m/s), retrieved from the PARASOL image acquired on May 5, 2006, over the north-west part of the Mediterranean Sea.

The PARASOL wind products were quantitatively compared over the same area of Fig. 2.2 with the wind speed measured by two buoys (Météo-France organization; (43.4°N, 7.8°E) and (42.1°N, 4.7°E)) for one entire year, namely 2006 (146 match-ups). The comparisons (Fig. 2.3a) show a coefficient of correlation greater than 0.96. The value of the slope is 0.96 and the root-mean-square error is satisfactory (1.1 m/s). Therefore, the comparisons between the PARASOL wind-speed product and *in situ* measurements from buoys confirm that the PARASOL space-based measurements do not exhibit any systematic errors. The wind-speed values retrieved with PARASOL were compared with concurrent wind-speed data derived by the passive microwave satellite sensor AMSR-E (NASA) (Wentz and Meissner, 2000). The AMSR-E sensor is part of the A-Train, thus permitting virtually coincident observations with the PARASOL sensor. To perform relevant comparisons with AMSR-E products, which were rigorously validated using *in situ* measurements, the PARASOL wind-speed data were re-projected into the same 0.25° × 0.25° grid (i.e., spatial resolution of 25 km) as the AMSR-E data. The comparison was performed at global scale for PARASOL and AMSR-E data acquired over three successive days, which correspond to the total revisit time of the PARASOL mission. Figure 2.10a shows that significant correlation ($r = 0.84$) is obtained between both sensors for a slope of the regression line in agreement within 1%.

The number of pixels used for wind-speed retrieval represents almost 80% of the cloud-free measurements within the PARASOL swath. Such a wide spatial coverage indicates that the wind-speed product could be combined with coincident

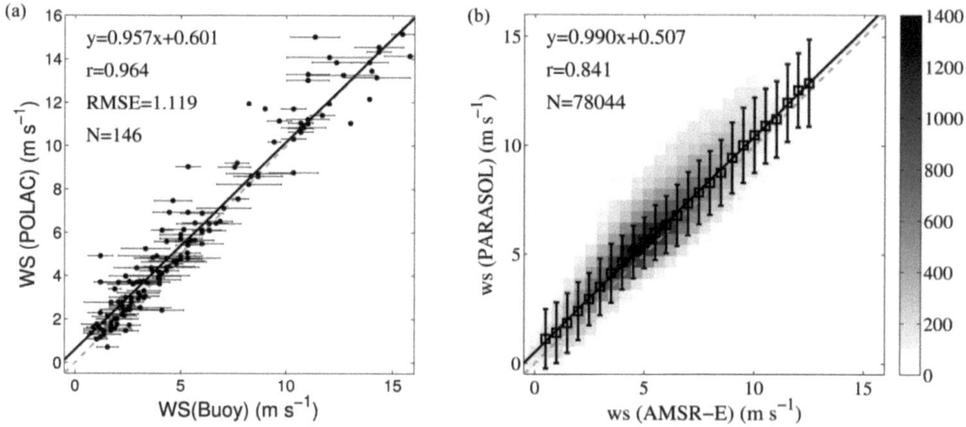

Fig. 2.10. (a) Comparison of the wind-speed values *ws* as retrieved from the PARASOL data with meteorological buoys data acquired over the year 2006 in the Mediterranean Sea. (b) Comparison of the wind-speed values over the entire globe derived from PARASOL and AMSR-E measurements after re-projection of PARASOL data into the 0.25° × 0.25° AMSR-E grid. Equation of the regression line (black line) is given in the left upper corner; r is the correlation coefficient; N is the number of comparisons.

aerosols and ocean-color products. Thus, the wind-speed product derived from satellite sensors measuring at short wavelengths should allow further investigations in various multidisciplinary topics such as studies dedicated to aerosol transport and marine primary productivity. The implications for determining the wind speed at a high spatial resolution are important as well for modeling and data assimilation purposes. The applications could concern the weather forecast predictions or the oceanic dynamic.

2.7 Conclusion

Polarization tends to provide information that is largely uncorrelated with spectral and intensity signal. Thus, polarimetric remote sensing has the potential to enhance many fields of optical oceanography. Historically, the measurement of the polarization state of light at sea has been proved to be an even more burdensome challenge than that of intensity. Since the midpoint of the 20th century, efforts have been dedicated to measuring and understanding the full nature of light propagating through the atmosphere–ocean system. These efforts have recently led to new insights on light propagation within the natural environment and efficient tools for both theoretical and instrumental applications. Thanks to those scientific advances, measurements and interpretation of the polarization state of light may be considered as an important step toward even more accurate remote sensing of the marine and lake environments. In particular, polarimetric remote sensing could provide useful information on fluorescence and Raman scattered light, the respective optical signature of organic and inorganic matter, the attenuation coefficient (in link with the multiple-scattering regime), the sea surface roughness, and the correlated overlying winds. In addition, polarization characteristics must be accounted for to fully commensurate the water-leaving radiance when measured from above-water or space-borne sensors. Reversely, polarization measurements are of paramount importance for retrieving the optical and microphysical properties of aerosols which, in turn, have to be accurately considered to estimate the water-leaving radiance through the atmospheric correction process of satellite images.

In this chapter, impacts on light polarization of the various components of the in-water matter have been described with the objective to give the contemporary view on the respective contributions of dissolved matter, phytoplankton, and suspended sediments. Polarized light exiting the atmosphere, and potentially captured by a satellite sensor, has been established to be slightly impacted by open ocean dissolved matter and phytoplankton in non-blooming conditions. This almost total invariance of the ocean-polarized signal was later exploited to characterize aerosols from polarimetric and multidirectional acquisitions at wavelengths greater than 470 nm. On the other hand, suspended mineral matter and high bloom conditions produce significant changes in the polarization state of light which could be potentially measured and exploited for monitoring such water conditions from space through, for instance, estimation of the attenuation coefficient. Notwithstanding these recent scientific achievements, the full scattering matrix of hydrosols are currently poorly documented and further efforts are needed to provide deeper understanding and complete agreement with theory. Such advances would potentially lead to new developments in inverse methods to diagnose phytoplankton concentration, physiology, or types from polarimetric measurements of light. In addition, measuring polarization around the fluorescence peaks of chlorophyll pigments, or of other fluorescent matter of interest, would be an important asset to estimate the fluorescence signal and thereby physiology and light-use efficiency of phytoplankton as a link to ocean productivity. Another interest of multidirectional and polarimetric satellite sensors, discussed in this chapter, relies on the capacity of isolating and quantifying the sun glint contribution with potentialities to estimate surface wind speeds or absorption of aerosols. Exploitation of polarimetric data for remotely sensing marine or lake environments is therefore highly promising and will depend on the future developments of polarization-based algorithms to make the most of the planned polarimetric satellite missions, whose launches are scheduled for the start of the next decade.

Acknowledgments

Studies of the PARASOL data for ocean-color applications were supported by the French *Centre National d'Etudes Spatiales* (CNES) and by ACRI-ST Company. Implementation of the wind-speed algorithm has been achieved through the ICARE organization (*Laboratoire Optique Atmospherique* and CNES). Analysis of the polarization impacts on above-water radiometry was partially supported by grants from the Office of Naval Research (ONR) and the National Oceanographic and Atmospheric Administration (NOAA).

References

Adams, J.T. and Kattawar, G.W. (1997) 'Neutral points in an atmosphere–ocean system. 1: Upwelling light field,' *Appl. Opt.*, **36**, 1976–86.

Adams, J.T., Aas, E., Hjoerslev, N.K. and Lundgren, B. (2002) 'Comparison of radiance and polarization values observed in the Mediterranean Sea and simulated in a Monte Carlo model,' *Appl. Opt.*, **41**, 2724–33.

Ahmed, S., Gilerson, A., Gill, A., Gross, B.M., Moshary, F. and Zhou, J. (2004) 'Separation of fluorescence and elastic scattering from algae in seawater using polarization discrimination,' *Optics Communs.*, **235**, 23–30.

Ahmad, Z., Franz, B.A., McClain, C.R., Kwiatkowska, E.J., Werdell, J., Shettle, E.P. and Holben, B.N. (2010) 'New aerosol models for the retrieval of aerosol optical thickness and normalized water-leaving radiances from the SeaWiFS and MODIS sensors over coastal regions and open oceans,' *Appl. Opt.*, **49**, 5545–60.

Alvain, S., Moulin, C., Dandonneau, Y. and Bréon, F.M. (2005) 'Remote sensing of phytoplankton groups in case 1 waters from global SeaWiFS imagery,' *Deep-Sea Res. Pt I*, **52**, 1989–2004.

Anguelova, M.D. and Webster, F. (2006) 'Whitecap coverage from satellite measurements: A first step toward modeling the variability of oceanic whitecaps,' J. *Geophys. Res.*, **111**, C03-17.

Antoine, D. and Morel, A. (1996) 'Oceanic primary production. 1. Adaptation of a spectral light-photosynthesis model in view of application to satellite chlorophyll observations,' *Global Biogeochem. Cy.*, **10**, 43–55.

Antoine, D. and Morel, A. (1999) 'A multiple scattering algorithm for atmospheric correction of remotely sensed ocean colour (MERIS instrument): Principle and implementation for atmospheres carrying various aerosols including absorbing ones,' *Int. J. Remote Sens.*, **20**, 1875–916.

Antoine, D., d'Ortenzio, F., Hooker, S.B., Bécu, G., Gentili, B., Tailliez, D. and Scott, A.J. (2008) 'Assessment of uncertainty in the ocean reflectance determined by three satellite ocean color sensors (MERIS, SeaWiFS and MODIS-A) at an offshore site in the Mediterranean Sea (BOUSSOLE project),' *J. Geophys. Res.*, **113**, C07013.

Argoul, P. (2012) 'Overview of inverse problems,' *cel.archives-ouvertes.fr/cel-00781172.*

Babin, M., Morel, A. and Gentili, B. (1996) 'Remote sensing of sea surface Sun-induced chlorophyll fluorescence: Consequences of natural variations in the optical characteristics of phytoplankton and the quantum yield of chlorophyll a fluorescence,' *Int. J. remote Sens.*, **17**, 2417–48.

Babin, M., Morel, A., Fournier-Sicre, V., Fell, F. and Stramski, D. (2003a) 'Light scattering properties of marine particles in coastal and open ocean waters as related to the particle mass concentration,' *Limnol. Oceanogr.*, **48**, 843–59.

Babin, M., Stramski, D., Ferrari, G.M., Claustre, H., Bricaud, A., Obolensky, G. and Hoepffner, N. (2003b) 'Variations in the light absorption coefficients of phytoplankton, nonalgal particles, and dissolved organic matter in coastal waters around Europe,' *Journal of Geophysical Research: Oceans*, **108**, C7, 3211.

Bader, H. (1970) 'Hyperbolic distribution of particle sizes,' *J. Geophys. Res.*, **75**, 2822–30.

Behrenfeld, M.J. and Falkowski, P.G. (1997) 'Photosynthetic rates derived from satellite-based chlorophyll concentration,' *Limnol. Oceanogr.*, **42**, 1–20.

Behrenfeld, M.J., Boss, E., Siegel, D.A. and Shea, D.M. (2005) 'Carbon-based ocean productivity and phytoplankton physiology from space,' *Global Biogeochem. Cy.*, **19**, GB1006.

Behrenfeld, M.J., Westberry, T.K., Boss, E.S., O'Malley, R.T., Siegel, D.A., Wiggert, J.D., Franz, B.A., McClain, C.R., Feldman, G.C., Doney, S.C., Moore, J.K., Dall'Olmo,

G., Milligan, A.J., Lima, I. and Mahowald, N. (2009) 'Satellite-detected fluorescence reveals global physiology of ocean phytoplankton,' *Biogeosciences*, **6**, 779–94.

Bourassa, M., Bonekamp, H., Chang, P., Chelton, D., Courtney, J., Edson, R., Figa, J., He, Y., Hersbach, H., Hilburn, K., Lee, T., Liu, W., Long, D., Kelly, K., Knabb, R., Lindstorm, E., Perrie, W., Portabella, M., Powell, M., Rodriguez, E., Smith, D., Stoffelen, A., Swail, V. and Wentz, F. (2010) 'Remotely sensed winds and wind stresses for marine forecasting and ocean modeling,' *Proceedings of the 'OceanObs'09: Sustained Ocean Observations and Information for Society' Conference*, Vol. 2, Venice, Italy, September 21–25, 2009, ESA.

Bréon, F.M. and Henriot, N. (2006) 'Spaceborne observations of ocean glint reflectance and modeling of wave slope distributions,' *J. Geophys. Res.-Oceans*, **111**, C06005.

Bricaud, A., Morel, A. and Prieur, L. (1981) 'Absorption by dissolved organic matter of the sea (yellow substance) in the UV and visible domains,' *Limnol. Oceanogr.*, **26**, 43.

Bricaud, A., Morel, A., Babin, M., Allali, K. and Claustre, H. (1998) 'Variations of light absorption by suspended particles with chlorophyll a concentration in oceanic (case 1) waters: Analysis and implications for bio-optical models,' *J. Geophys. Res.*, **103**, 31033–44.

Cairns, B., Russell, E.E. and Travis, L.D. (1999) 'Research scanning polarimeter: Calibration and ground-based measurements,' presented at SPIE's International Symposium on Optical Science, Engineering, and Instrumentation, 186–96.

Chami, M. (2007) 'Importance of the polarization in the retrieval of oceanic constituents from the remote sensing reflectance,' *J. Geophys. Res.-Oceans*, **112**, C05026.

Chami, M. and McKee, D. (2007) 'Determination of biogeochemical properties of marine particles using above water measurements of the degree of polarization at the Brewster angle,' *Opt. Express*, **15**, 9494–509.

Chami, M., Santer, R. and Dilligeard, E. (2001) 'Radiative transfer model for the computation of radiance and polarization in an ocean–atmosphere system: Polarization properties of suspended matter for remote sensing,' *Appl. Opt.*, **40**, 2398–416.

Chami, M., Thirouard, A. and Harmel, J. (2014) 'POLVSM (Polarized Volume Scattering Meter) instrument: An innovative device to measure the directional and polarized scattering properties of hydrosols,' *Opt. Express*, **22**, 26403–28.

Chowdhary, J., Cairns, B. and Travis, L.D. (2002) 'Case studies of aerosol retrievals over the ocean from multiangle, multispectral photopolarimetric remote sensing data,' *J. Atmos. Sci.*, **59**, 383–97.

Chowdhary, J., Cairns, B. and Travis, L.D. (2006) 'Contribution of water-leaving radiances to multiangle, multispectral polarimetric observations over the open ocean: Bio-optical model results for case 1 waters,' *Appl. Opt.*, **45**, 5542–67.

Chowdhary, J., Cairns, B., Mishchenko, M. and Travis, L. (2001) 'Retrieval of aerosol properties over the ocean using multispectral and multiangle photopolarimetric measurements from the Research Scanning Polarimeter,' *Geophys. Res. Lett.*, **28**, 243–6.

Chowdhary, J., Cairns, B., Waquet, F., Knobelspiesse, K., Ottaviani, M., Redemann, J., Travis, L. and Mishchenko, M. (2012) 'Sensitivity of multiangle, multispectral polarimetric remote sensing over open oceans to water-leaving radiance: Analyses of RSP data acquired during the MILAGRO campaign,' *Remote Sens. Environ.*, **118**, 284–308.

Chowdhary, J., Frouin, R. and Cairns, B. (2013) 'Multi-angle, multi-spectral, photo-polarimetric radiance observed from space in the UV: Sensitivities to natural variations in scattering and absorption properties of open oceans,' presented at AGU Fall Meeting Abstracts, 06.

Clark, D.K., Yarbrough, M.A., Feinholz, M.E., Flora, S., Broenkow, W., Kim, Y.S.., Johnson, B.C., Brown, S.W., Yuen, M. and Mueller, J.L. (2003) 'MOBY, a radiometric buoy for performance monitoring and vicarious calibration of satellite ocean color sensors: Measurement and data analysis protocols,' *Ocean Optics Protocols for Satellite Ocean Color Sensor Validation, Revision*, **4**, 3–34.

Cox, C. and Munk, W. (1954a) 'Measurement of the roughness of the sea surface from photographs of the suns glitter,' *J. Opt. Soc. Am.*, **44**, 838–50.

Cox, C. and Munk, W. (1954b) 'Statistics of the sea surface derived from sun glitter,' *Journal of Marine Research*, **13**, 198–227.

Cronin, T.W. and Shashar, N. (2001) 'The linearly polarized light field in clear, tropical marine waters: Spatial and temporal variation of light intensity, degree of polarization and e-vector angle,' *J. Expl. Biol.*, **204**, 2461–7.

Dall'Olmo, G., Gitelson, A.A., Rundquist, D.C., Leavitt, B., Barrow, T. and Holz, J.C. (2005) 'Assessing the potential of SeaWiFS and MODIS for estimating chlorophyll concentration in turbid productive waters using red and near-infrared bands,' *Remote Sens. Environ.*, **96**, 176–87.

Deschamps, P.Y., Bréon, F.M., Leroy, M., Podaire, A., Bricaud, A., Buriez, J.C. and Seze, G. (1994) 'The Polder Mission: Instrument characteristics and scientific objectives,' *IEEE T. Geosci. Remote*, **32**, 598–615.

Deuze, J.L., Goloub, P., Herman, M., Marchand, A., Perry, G., Susana, S. and Tanre, D. (2000) 'Estimate of the aerosol properties over the ocean with POLDER,' *J. Geophys. Res.-Atmos.*, **105**, 15329–46.

Dickey, T. and Falkowski, P. (2002) 'Solar energy and its biological–physical interactions in the sea,' in Robinson, A.R., McCarthy, J.J. and Rothschild, B.J. (eds), *The Sea*, Vol. 12, John Wiley, New York, pp. 401–40.

Doron, M., Babin, M., Mangin, A. and Hembise, O. (2007) 'Estimation of light penetration, and horizontal and vertical visibility in oceanic and coastal waters from surface reflectance,' *J. Geophys. Res.-Oceans*, **112**, C06003.

Dubovik, O., Holben, B., Eck, T.F., Smirnov, A., Kaufman, Y.J., King, M.D., Didier, T. and Slutsker, I. (2002) 'Variability of absorption and optical properties of key aerosol types observed in worldwide locations,' *J. Atmos. Sci.*, **59**, 590–608.

Dubovik, O., Smirnov, A., Holben, B.N., King, M.D., Kaufman, Y.J., Eck, T.F. and Slutsker, I. (2000) 'Accuracy assessments of aerosol optical properties retrieved from Aerosol Robotic Network (AERONET) Sun and sky radiance measurements,' *J. Geophys. Res.*, **105**, 9791–806.

Falkowski, P. and Kiefer, D.A. (1985) 'Chlorophyll-*a* fluorescence in phytoplankton: Relationship to photosynthesis and biomass,' *Journal of Plankton Research*, **7**, 715–31.

Fougnie, B., Bracco, G., Lafrance, B., Ruffel, C., Hagolle, O. and Tinell, C. (2007) 'PARASOL in-flight calibration and performance,' *Appl. Optics*, **46**, 5435–51.

Francis, P.N., Hignett, P. and Taylor, J.P. (1999) 'Aircraft observations and modeling of sky radiance distributions from aerosol during TARFOX,' *J. Geophys. Res.*, **104**, 2309–19.

Frouin, R., Schwindling, M. and Deschamps, P.-Y. (1996) 'Spectral reflectance of sea foam in the visible and near-infrared: *In situ* measurements and remote sensing implications,' *J. Geophys. Res.*, **101**, 14361–71.

Fry, E.S. and Voss, K.J. (1985) 'Measurement of the Mueller matrix for phytoplankton,' *Limnol. Oceanogr.*, **30**, 1322–6.

Fukushima, H., Higurashi, A., Mitomi, Y., Nakajima, T., Noguchi, T., Tanaka, T. and Toratani, M. (1998) 'Correction of atmospheric effects on ADEOS/OCTS ocean color data: Algorithm description and evaluation of its performance,' *J. Oceanogr.*, **54**, 417–30.

Gardner, W.D., Mishonov, A.V. and Richardson, M.J. (2006) 'Global POC concentrations from in-situ and satellite data,' *Deep Sea Research Part II: Topical Studies in Oceanography*, **53**, 718–40.

Gardner, W.D., Walsh, I.D. and Richardson, M.J. (1993) 'Biophysical forcing of particle production and distribution during a spring bloom in the North Atlantic,' *Deep Sea Research Part II: Topical Studies in Oceanography*, **40**, 171–95.

Gilerson, A., Zhou, J., Hlaing, S., Ioannou, I., Schalles, J., Gross, B., Moshary, F. and Ahmed, S. (2007) 'fluorescence component in the reflectance spectra from coastal waters: Dependence on water composition,' *Opt. Express*, **15**, 15702–21.

Gilerson, A., Zhou, J., Oo, M., Chowdhary, J., Gross, B.M., Moshary, F. and Ahmed, S. (2006) 'Retrieval of chlorophyll fluorescence from reflectance spectra through polarization discrimination: Modeling and experiments,' *Appl. Opt.*, **45**, 5568–81.

Gogoi, A., Buragohain, A.K., Choudhury, A. and Ahmed, G.A. (2009) 'Laboratory measurements of light scattering by tropical fresh water diatoms,' *J. Quant. Spectrosc. Radiat. Transfer*, **110**, 1566–78.

Gordon, H.R. (1997) 'Atmospheric correction of ocean color imagery in the Earth Observing System era,' *J. Geophys. Res.*, **102**, 17081–106.

Gordon, H.R. (1999) 'Contribution of Raman scattering to water-leaving radiance: A reexamination,' *Appl. Opt.*, **38**, 3166–74.

Gordon, H.R. (2014) 'Influence of Raman scattering on the light field in natural waters: A simple assessment,' *Opt. Express*, **22**, 3675–83.

Gordon, H.R. and Wang, M. (1994) 'Retrieval of water leaving radiance and aerosol optical thickness over the oceans with SeaWiFS: A preliminary algorithm,' *Appl. Opt.*, **33**, 443–58.

Harmel, T. and Chami, M. (2008) 'Invariance of polarized reflectance measured at the top of atmosphere by PARASOL satellite instrument in the visible range with marine constituents in open ocean waters,' *Opt. Express*, **16**, 6064–80.

Harmel, T. and Chami, M. (2011) 'Influence of polarimetric satellite data measured in the visible region on aerosol detection and on the performance of atmospheric correction procedure over open ocean waters,' *Opt. Express*, **19**, 20960–83.

Harmel, T. and Chami, M. (2012) 'Determination of sea surface wind speed using the polarimetric and multidirectional properties of satellite measurements in visible bands,' *Geophys. Res. Lett.*, **39**, L19611.

Harmel, T. and Chami, M. (2013) 'Estimation of the sun glint radiance field from optical satellite imagery over open ocean: Multidirectional approach and polarization aspects,' *J. Geophys. Res.*, **118**, 1–15.

Harmel, T., Gilerson, A., Hlaing, S., Tonizzo, A., Legbandt, T., Weidemann, A., Arnone, R. and Ahmed, S. (2011a) 'Long Island Sound Coastal Observatory: Assessment of above-water radiometric measurement uncertainties using collocated multi and hyperspectral systems,' *Appl. Opt.*, **50**, 5842–60.

Harmel, T., Gilerson, A., Hlaing, S., Weidemann, A., Arnone, R. and Ahmed, S. (2012a) 'Long Island Sound Coastal Observatory: Assessment of above-water radiometric measurement uncertainties using collocated multi and hyper-spectral systems: Reply to comment,' *Appl. Opt.*, **51**, 3893–9.

Harmel, T., Gilerson, A., Tonizzo, A., Chowdhary, J., Weidemann, A., Arnone, R. and Ahmed, S. (2012b) 'Polarization impacts on the water-leaving radiance retrieval from above-water radiometric measurements,' *Appl. Opt.*, **51**, 8324–40.

Harmel, T., Tonizzo, A., Ibrahim, A., Gilerson, A., Chowdhary, J. and Ahmed, S. (2011b) 'Measuring underwater polarization field from above-water hyperspectral instrumentation for water composition retrieval,' presented at *SPIE*, Prague.

He, X., Pan, D., Bai, Y., Wang, D. and Hao, Z. (2014) 'A new simple concept for ocean colour remote sensing using parallel polarisation radiance,' *Scientific Reports*, **4**, doi:10.1038/srep03748.

Herman, M., Deuze, J.L., Marchand, A., Roger, B. and Lallart, P. (2005) 'Aerosol remote sensing from POLDER/ADEOS over the ocean: Improved retrieval using a nonspherical particle model,' *J. Geophys. Res.-Atmos.*, **110**, 1–11.

Holben, B.N., Eck, T.F., Slutsker, I., Tanre, D., Buis, J.P., Setzer, A., Vermote, E., Reagan, J.A., Kaufman, Y.J. and Nakajima, T. (1998) 'AERONET: A federated instrument network and data archive for aerosol characterization,' *Remote Sens. Environ.*, **66**, 1–16.

Hollstein, A. and Fischer, J. (2012) 'Effects of salinity, temperature, and polarization on top of atmosphere and water leaving radiances for case 1 waters,' *Appl. Opt.*, **51**, 8022–33.

Horváth, G., Barta, A., Pomozi, I., Suhai, B., Hegedüs, R., Åkesson, S., Meyer-Rochow, B. and Wehner, R. (2011) 'On the trail of Vikings with polarized skylight: Experimental study of the atmospheric optical prerequisites allowing polarimetric navigation by Viking seafarers,' *Philosophical Transactions of the Royal Society B: Biological Sciences*, **366**, 772–82.

Horváth, G., Bernáth, B., Suhai, B., Barta, A. and Wehner, R. (2002) 'First observation of the fourth neutral polarization point in the atmosphere,' *JOSA A*, **19**, 2085–99.

Hovenier, J.W. (1987) 'A unified treatment of polarized light emerging from a homogeneous plane-parallel atmosphere,' *Astron. Astrophys.*, **183**, 363–70.

Hovenier, J.W., Volten, H., Munoz, O., Van der Zande, W.J. and Waters, L. (2003) 'Laboratory studies of scattering matrices for randomly oriented particles: Potentials, problems, and perspectives,' *J. Quant. Spectrosc. Radiat. Transfer*, **79**, 741–55.

Hu, C., Lee, Z. and Franz, B. (2012) 'Chlorophyll *a* algorithms for oligotrophic oceans: A novel approach based on three-band reflectance difference,' *J. Geophys. Res.*, **117**, C01011.

Hu, C., Muller-Karger, F.E., Taylor, C.J., Carder, K.L., Kelble, C., Johns, E. and Heil, C.A. (2005) 'Red tide detection and tracing using MODIS fluorescence data: A regional example in SW Florida coastal waters,' *Remote Sens. Environ.*, **97**, 311–21.

Hudson, N., Baker, A. and Reynolds, D. (2007) 'fluorescence analysis of dissolved organic matter in natural, waste and polluted waters: A review,' *River Research and Applications*, **23**, 631–49.

Hulburt, E.O. (1934) 'Polarization of light at sea,' *J. Opt. Soc. Am.*, **24**, 35–42.

Ibrahim, A., Gilerson, A., Harmel, T., Tonizzo, A., Chowdhary, J. and Ahmed, S. (2012) 'The relationship between upwelling underwater polarization and attenuation/absorption ratio,' *Opt. Express*, **20**, 25662–80.

IOCCG (2006) 'Remote sensing of inherent optical properties: Fundamentals, tests of algorithms, and applications,' in Lee, Z.P. (ed.), *Reports of the International Ocean Colour Coordinating Group No. 5*, International Ocean-Colour Coordinating Group, Dartmouth, Nova Scotia.

IOCCG (2010) 'Atmospheric correction for remotely-sensed ocean-colour products,' in Wang, M., Antoine, D., Deschamps, P.Y., Frouin, R., Fukushima, H., Gordon, H.R., Morel, A., Nicolas, J.M. and Stuart, V. (eds), *Reports of the International Ocean-Colour Coordinating Group, No. 10*, IOCCG.

Ivanoff, A. (1974) 'Polarization measurements in the sea,' in Jerlov, N.G. and Neilsen, E.S. (eds), *Optical Aspects of Oceanography*, Academic Press, London, pp. 151–75.

Ivanoff, A. and Waterman, T.H. (1958) 'elliptical polarization of submarine illumination,' *J. Mar. Res.*, **16**, 255–82.

Ivanoff, A., Jerlov, N. and Waterman, T.H. (1961) 'A comparative study of irradiance, beam transmittance and scattering in the sea near Bermuda,' *Limnol. Oceanogr.*, **6**, 129–48.

Johnsen, S., Marshall, N.J. and Widder, E.A. (2011) 'Polarization sensitivity as a contrast enhancer in pelagic predators: Lessons from *in situ* polarization imaging of transparent zooplankton,' *Philosophical Transactions of the Royal Society B: Biological Sciences*, **1556**, 655–70.

Kattawar, G.W. (2013) 'Genesis and evolution of polarization of light in the ocean,' *Appl. Opt.*, **52**, 940–8.

Kattawar, G.W. and Xu, X. (1994) 'Detecting Raman scattering in the ocean by use of polarimetry,' presented at *Ocean Optics XII*, 222–33.

Kaufman, Y.J., Smirnov, A., Holben, B.N. and Dubovik, O. (2001) 'Baseline maritime aerosol: Methodology to derive the optical thickness and scattering properties,' *Geophys. Res. Lett.*, **28**, 3251–6.

Kiefer, D.A. (1973) 'Chlorophyll a fluorescence in marine centric diatoms: Responses of chloroplasts to light and nutrient stress,' *Marine Biology*, **23**, 39–46.

King, M.D., Kaufman, Y.J., Tanre, D. and Nakajima, T. (1999) 'Remote sensing of tropospheric aerosols from space: Past, present, and future,' *Bull. Am. Met. Soc.*, **80**, 2229–59.

Koepke, P. (1984) 'Effective reflectance of oceanic whitecaps,' *Appl. Opt.*, **23**, 1816–24.

Kokhanovsky, A.A. (2004) 'Spectral reflectance of whitecaps,' *J. Geophys. Res.*, **109**, C05021.

Lee, Z.P., Carder, K.L. and Arnone, R.A. (2002) 'Deriving inherent optical properties from water color: A multiband quasi-analytical algorithm for optically deep waters,' *Appl. Opt.*, **41**, 5755–72.

Lerner, A., Sabbah, S., Erlick, C. and Shashar, N. (2011) 'Navigation by light polarization in clear and turbid waters,' *Philosophical Transactions of the Royal Society B: Biological Sciences*, **366**, 671–9.

Letelier, R.M., Abbott, M.R. and Karl, D.M. (1997) 'Chlorophyll natural fluorescence response to upwelling events in the Southern Ocean,' *Geophys. Res. Lett.*, **24**, 409–12.

Lide, D.R. (2004) *CRC Handbook of Chemistry and Physics*, CRC Press.

Lofflus, K.D., Quinby-Hunt, M.S., Hunt, A.J., Livolant, F. and Maestre, M. (1992) 'Light scattering by *Prorocentrum micans*: A new method and results,' *Appl. Opt.*, **31**, 2924–31.

Loisel, H. and Morel, A. (1998) 'Light scattering and chlorophyll concentration in case 1 waters: A reexamination,' *Limnol. Oceanogr.*, **43**, 847–58.

Loisel, H., Duforet, L., Dessailly, D., Chami, M. and Dubuisson, P. (2008) 'Investigation of the variations in the water leaving polarized reflectance from the POLDER satellite data over two biogeochemical contrasted oceanic areas,' *Opt. Express*, **16**, 12905–18.

Lotsberg, J.K. and Stamnes, J.J. (2010 'Impact of particulate oceanic composition on the radiance and polarization of underwater and backscattered light,' *Opt. Express*, **18**, 10432–45.

Marbach, T., Phillips, P., Lacan, A. and Schlüssel, P. (2013) 'The Multi-Viewing, -Channel, -Polarisation Imager (3MI) of the EUMETSAT Polar System – Second Generation (EPS-SG) dedicated to aerosol characterisation,' *Proc. SPIE*, **8889**, Sensors, Systems, and Next-Generation Satellites XVII.

McClain, C.R. (2009) 'A decade of satellite ocean color observations,' *Annual Review of Marine Science*, **1**, 19–42.

Mélin, F. and Zibordi, G. (2010) 'Vicarious calibration of satellite ocean color sensors at two coastal sites,' *Appl. Opt.*, **49**, 798–810.

Mélin, F., Clerici, M., Zibordi, G., Holben, B.N. and Smirnov, A. 'Validation of SeaWiFS and MODIS aerosol products with globally distributed AERONET data,' Remote Sens. Environ. **114**, 230-250 (2010).

Mishchenko, M.I. (2002) 'Vector radiative transfer equation for arbitrarily shaped and arbitrarily oriented particles: A microphysical derivation from statistical electromagnetics,' *Appl. Opt.*, **41**, 7114–34.

Mishchenko, M.I. (2003) 'Microphysical approach to polarized radiative transfer: Extension to the case of an external observation point,' *Appl. Opt.*, **42**, 4963–7.

Mishchenko, M.I. and Travis, L.D. (1997a) 'Satellite retrieval of aerosol properties over the ocean using measurements of reflected sunlight: Effect of instrumental errors and aerosol absorption,' *J. Geophys. Res.-Atmos.*, **102**, 13543–53.

Mishchenko, M.I. and Travis, L.D. (1997b) 'Satellite retrieval of aerosol properties over the ocean using polarization as well as intensity of reflected sunlight,' *J. Geophys. Res.-Atmos.*, **102**, 16989–7013.

Mishchenko, M.I., Geogdzhayev, I.V., Cairns, B., Carlson, B.E., Chowdhary, J., Lacis, A.A., Liu, L., Rossow, W.B. and Travis, L.D. (2007) 'Past, present, and future of global aerosol climatologies derived from satellite observations: A perspective,' *J. Quant. Spectrosc. Radiat. Transfer*, **106**, 325–47.

Mishchenko, M.I., Travis, L.D. and Lacis, A.A. (2006) *Multiple Scattering of Light by Particles: Radiative Transfer and Coherent Backscattering*, Cambridge University Press, New York.

Mobley, C.D. (1994) *Light and Water: Radiative Transfer in Natural Waters*, Academic Press, New York.

Monahan, E.C. and O'Muircheartaigh, I. (1980) 'Optimal power-law description of oceanic whitecap coverage dependence on wind speed,' *J. Phys. Oceanogr.*, **10**, 2094–9.

Morel, A. (1974) 'Optical properties of pure water and pure seawater,' in Jerlov, N.G. and Nielsen, E.S. (eds), *Optical Aspects of Oceanography*, Academic Press, New York, pp. 1–24.

Morel, A. (1991) 'Light and marine photosynthesis: A spectral model with geochemical and climatological implications,' *Prog. Oceanogr.*, **26**, 263–306.

Morel, A. (2009) 'Are the empirical relationships describing the bio-optical properties of case 1 waters consistent and internally compatible?' *J. Geophys. Res.*, **114**, C01016.

Morel, A. and Prieur, L. (1977) 'Analysis of variations in ocean color,' *Limnol. Oceanogr.*, **22**, 709–22.

Morel, A., Antoine, D. and Gentili, B. (2002) 'Bidirectional reflectance of oceanic waters: Accounting for Raman emission and varying particle scattering phase function,' *Appl. Opt.*, **41**, 6289–306.

Morel, A., Gentili, B., Claustre, H., Babin, M., Bricaud, A., Ras, J. and Tieche, F. (2007) 'Optical properties of the "clearest" natural waters,' *Limnol. Oceanogr.*, **52**, 217–29.

Nelson, N.B. and Siegel, D.A. (2013) 'The global distribution and dynamics of chromophoric dissolved organic matter,' *Annual Review of Marine Science*, **5**, 447–76.

Neville, R.A. and Gower, J.F.R. (1977) 'Passive remote-sensing of phytoplankton via chlorophyll a fluorescence,' *J. Geophys. Res.*, **82**, 3487–93.

O'Reilly, J.E., Maritorena, S., Mitchell, G., Siegel, D.A., Carder, K.L., Garver, D.L., Kahru, M. and McClain, C.R. (1998) 'Ocean color chlorophyll algorithms for SeaWiFS,' *J. Geophys. Res.*, **103**, 24937–50.

Odermatt, D., Gitelson, A., Brando, V.E. and Schaepman, M. (2012) 'Review of constituent retrieval in optically deep and complex waters from satellite imagery,' *Remote Sens. Environ.*, **118**, 116–26.

Ohlmann, J.C. and Siegel, D.A. (2000) 'Ocean radiant heating. Part II: Parameterizing solar radiation transmission through the upper ocean,' *J. Phys. Oceanogr.*, **30**, 1849–65.

Organelli, E., Bricaud, A., Antoine, D. and Matsuoka, A. (2014) 'Seasonal dynamics of light absorption by Chromophoric Dissolved Organic Matter (CDOM) in the NW Mediterranean Sea (BOUSSOLE site),' *Deep-Sea Res. Pt I*, **91**, 72–85.

Pope, R.M. and Fry, E.S. (1997) 'Absorption spectrum (380–700 nm) of pure water. 2. Integrating cavity measurements,' *Appl. Opt.*, **36**, 8710–23.

Preisendorfer, R.W. (1965) *Radiative Transfer on Discrete Spaces*, Pergamon, Oxford.

Quinby-Hunt, M.S., Hunt, A.J., Lofftus, K. and Shapiro, D. (1989) 'Polarized-light scattering studies of marine Chlorella,' *Limnol. Oceanogr.*, **34**(8), 1587–600.

Sabbah, S. and Shashar, N. (2007) 'Light polarization under water near sunrise,' *Journal of the Optical Society of America A*, **24**, 2049–56.

Sabbah, S., Barta, A., Gál, J., Horváth, G. and Shashar, N. (2006) 'Experimental and theoretical study of skylight polarization transmitted through Snell's window of a flat water surface,' *Journal of the Optical Society of America A*, **23**(8), 1978–88.

Schaeffer, A., Garreau, P., Molcard, A., Fraunié, P. and Seity, Y. (2011) 'Influence of high-resolution wind forcing on hydrodynamic modeling of the Gulf of Lions,' *Ocean Dynamics*, **61**, 1823–44.

Shallenberg, C., Lewis, M.R. Kelley, D.E. and Cullen, J.J. (2008) 'Inferred influence of nutrient availability on the relationship between sun-induced chlorophyll fluorescence and incident irradiance in the Bering Sea,' *J. Geophys. Res.*, **113**, C07046.

Shashar, N., Johnsen, S., Lerner, A., Sabbah, S., Chiao, C.C., Mathger, L.M. and Hanlon, R.T. (2011) 'Underwater linear polarization: Physical limitations to biological functions,' *Philosophical Transactions of the Royal Society B: Biological Sciences*, **366**, 649–54.

Shashar, N., Sabbah, S. and Cronin, T.W. (2004) 'Transmission of linearly polarized light in seawater: Implications for polarization signaling,' *J. Expl. Biol.*, **207**, 3619–28.

Shettle, E.P. and Fenn, R.W. (1979) 'Models for the aerosols of the lower atmosphere and the effect of humidity variations on their optical properties,' in Tsipouras, P. and Garrett, H.B. (eds), *Environmental Research Paper Air Force Geophysics Lab.*, Hanscom AFB, MA, Optical Physics Div.

Slade, W.H., Agrawal, Y.C. and Mikkelsen, O.A. (2013) 'Comparison of measured and theoretical scattering and polarization properties of narrow size range irregular sediment particles,' presented at *Oceans*, San Diego, 23–7.

Stedmon, C.A., Markager, S. and Kaas, H. (2000) 'Optical properties and signatures of chromophoric dissolved organic matter (CDOM) in Danish coastal waters,' *Estuarine, Coastal and Shelf Science*, **51**, 267–78.

Stokes, G.G. (1852a) 'On the change of refrangibility of light,' *Phil. Trans. R. Soc.*, **142**, 463–562.

Stokes, G.G. (1852b) 'On the composition and resolution of streams of polarized light from different sources,' *Trans. Cambridge Philos. Soc.*, **3**, 233–59.

Stramski, D., Boss, E., Bogucki, D. and Voss, K.J. (2004) 'The role of seawater constituents in light backscattering in the ocean,' *Prog. Oceanogr.*, **61**, 27–56.

Stramski, D., Bricaud, A. and Morel, A. (2001) 'Modeling the inherent optical properties of the ocean based on the detailed composition of the planktonic community,' *Appl. Opt.*, **40**, 2929–45.

Swan, C.M., Nelson, N.B., Siegel, D.A. and Fields, E.A. (2013) 'A model for remote estimation of ultraviolet absorption by chromophoric dissolved organic matter based on the global distribution of spectral slope,' *Remote Sens. Environ.*, **136**, 277–85.

Tanré, D., Bréon, F.M., Deuzé, J.L., Dubovik, O., Ducos, F., François, P., Goloub, P., Herman, M., Lifermann, A. and Waquet, F. (2011) 'Remote sensing of aerosols by using polarized, directional and spectral measurements within the A-Train: The PARASOL mission,' *Atmospheric Measurement Techniques Discussions*, **4**, 2037–69.

Timofeeva, V.A. (1970) 'The degree of light polarization in turbid media,' *Izv. Atmos. Ocean Phys.*, **6**, 513–22.

Timofeeva, V.A. (1974) 'Optics of turbid waters (results of laboratory studies),' in Jerlov, N. and Nielsen, E. (eds), *Optical Aspects of Oceanography*, Academic, London, pp. 177–219.

Tonizzo, A., Zhou, J., Gilerson, A., Twardowski, M.S., Gray, D.J., Arnone, R.A., Gross, B.M., Moshary, F. and Ahmed, S.A. (2009) 'Polarized light in coastal waters: Hyperspectral and multiangular analysis,' *Opt. Express*, **17**, 5666–83.

Torrance, K.E., Sparrow, E.M. and Birkebak, R.C. (1966) 'Polarization, directional distribution, and off-specular peak phenomena in light reflected from roughened surfaces,' *JOSA*, **56**, 916–24.

Tynes, H.H., Kattawar, G.W., Zege, E.P., Katsev, I.L., Prikhach, A.S. and Chaikovskaya, L.I. (2001) 'Monte Carlo and multicomponent approximation methods for vector radiative transfer by use of effective Mueller matrix calculations,' *Appl. Opt.*, **40**, 400–12.

van de Hulst, H.C. (1957) *Light Scattering by Small Particles*, John Wiley & Sons, New York, reprinted in 1981, Dover Publications, New York.

Veselovskii, I., Kolgotin, A., Griaznov, V., Müller, D., Franke, K. and Whiteman, D.N. (2004) 'Inversion of multiwavelength Raman lidar data for retrieval of bimodal aerosol size distribution,' *Appl. Opt.*, **43**, 1180–95.

Volten, H., De Haan, J.F., Hovenier, J.W., Schreurs, R., Vassen, W., Dekker, A.G., Hoogenboom, H.J., Charlton, F. and Wouts, R. (1998) 'Laboratory measurements of angular distributions of light scattered by phytoplankton and silt,' *Limnol. Oceanogr.*, **43**(6), 1180–97.

Volten, H., Munoz, O., Rol, E., de Haan, J.F., Vassen, W., Hovenier, J.W., Muinonen, K. and Nousiainen, T. (2001) 'Scattering matrices of mineral aerosol particles at 441.6 nm and 632.8 nm,' *J. Geophys. Res.-Atmos.*, **106**, 17375–401.

Voss, K.J. and Fry, E.S. (1984) 'Measurement of the Mueller matrix for ocean water,' *Appl. Opt.*, **23**, 4427–39.

Voss, K.J. and Souaidia, N. (2010) 'POLRADS: Polarization radiance distribution measurement system,' *Opt. Express*, **18**, 19672–80.

Voss, K.J., Gleason, A.C.R., Gordon, H.R., Kattawar, G.W. and You, Y. (2011) 'Observation of non-principal plane neutral points in the in-water upwelling polarized light field,' *Opt. Express*, **19**, 5942–52.

Wald, L. and Monget, J.-M. (1983) 'Sea surface winds from sun glitter observations,' *Journal of Geophysical Research: Oceans*, **88**, 2547–55.

Waquet, F., Leon, J.F., Goloub, P., Pelon, J., Tanre, D. and Deuze, J.L. (2005) 'Maritime and dust aerosol retrieval from polarized and multispectral active and passive sensors,' *J. Geophys. Res.-Atmos.*, **110**, D10S10.

Waterman, T.H. (1954) 'Polarization patterns in submarine illumination,' *Science*, **120**(3127), 927–32.

Waterman, T.H. and Westell, W.E. (1956) 'Quantitative effects of the sun's position on submarine light polarization,' *Journal of Marine Research*, **16**, 283–307.

Wentz, F.J. and Meissner, T. (2000) *AMSR Ocean Algorithm*, Algorithm Theor. Basis Doc. 121599A-1, Remote Sens. Syst., Santa Rosa, CA.

Westberry, T.K., Behrenfeld, M.J., Milligan, A.J. and Doney, S.C. (2013) 'Retrospective satellite ocean color analysis of purposeful and natural ocean iron fertilization,' *Deep Sea Research I*, **73**, 1–16.

Witkowski, K., Wolinski, L., Turzynski, Z., Gedziorowska, D. and Zielinski, A. (1993) 'The investigation of kinetic growth of *Chlorella vulgaris* cells by the method of integral and dynamic light scattering,' *Limnol. Oceanogr.*, **38**, 1365–72.

Yang, H. and Gordon, H.R. (1997) 'Remote sensing of ocean color: Assessment of the water-leaving radiance bidirectional effects on the atmospheric diffuse transmittance,' *Appl. Opt.*, **36**, 7887–97.

You, Y., Kattawar, G.W., Voss, K.J., Bhandari, P., Wei, J., Lewis, M., Zappa, C.J. and Schultz, H. (2011a) 'Polarized light field under dynamic ocean surfaces: Numerical modeling compared with measurements,' *J. Geophys. Res.*, **116**, C00H05.

You, Y., Tonizzo, A., Gilerson, A.A., Cummings, M.E., Brady, P., Sullivan, J.M., Twardowski, M.S., Dierssen, H.M., Ahmed, S.A. and Kattawar, G.W. (2011b) 'Measurements and simulations of polarization states of underwater light in clear oceanic waters,' *Appl. Opt.*, **50**, 4873–93.

Zaneveld, J.R.V. (2013) 'Fifty years of inherent optical properties,' *Methods in Oceanography*, **7**, 3–20.

Zhai, P.-W., Hu, Y., Chowdhary, J., Trepte, C.R., Lucker, P.L. and Josset, D.B. (2010) 'A vector radiative transfer model for coupled atmosphere and ocean systems with a rough interface,' *J. Quant. Spectrosc. Radiat. Transfer*, **111**, 1025–40.

Zhai, P.-W., Hu, Y., Trepte, C.R. and Lucker, P.L. (2009) 'A vector radiative transfer model for coupled atmosphere and ocean systems based on successive order of scattering method,' *Opt. Express*, **17**, 2057.

Zibordi, G. (2012) 'Comment on "Long Island Sound Coastal Observatory: Assessment of above-water radiometric measurement uncertainties using collocated multi and hyperspectral systems",' *Appl. Opt.*, **51**, 3888–92.

Zibordi, G., Holben, B.N., Slutsker, I., Giles, D., D'Alimonte, D., Mélin, F., Berthon, J.F., Vandemark, D., Feng, H., Schuster, G., Fabbri, B., Kaitala, S. and Seppala, J. (2009) 'AERONET-OC: A network for the validation of ocean color primary radiometric products,' *J. Atmos. Ocean. Technol.*, **26**, 1634–51.

Zibordi, G., Mélin, F., Hooker, S.B., D'Alimonte, D. and Holben, B. (2004) 'An autonomous above-water system for the validation of ocean color radiance data,' *IEEE T. Geosci. Remote*, **42**, 401–15.

Zubko, V., Kaufman, Y.J., Burg, R.I. and Martins, J.V. (2007) 'Principal component analysis of remote sensing of aerosols over oceans,' *IEEE T. Geosci. Remote*, **45**, 730–45.

Zugger, M.E., Messmer, A., Kane, T.J., Prentice, J., Concannon, B., Laux, A. and Mullen, L. (2008) 'Optical scattering properties of phytoplankton: Measurements and comparison of various species at scattering angles between 1 (degree) and 170 (degree),' *Limnol. Oceanogr.*, **53**, 381.

3 Polarimetry in terrestrial applications

Sergey N. Savenkov

3.1 Introduction

To develop polarimetric methods for object identification and classification, one needs to understand the relation between polarimetric and physical properties of the objects. As electromagnetic radiation interacts with an object under study, its polarization state and intensity are changed. Polarization properties of the scattered radiation contain extensive information on morphological and functional properties of the object. For example, since depolarization of scattered radiation depends on the morphological and physical parameters of scatterers (i.e., density, size, distribution, shape, refractive index, etc.) present in the studied object (Elachi, 1987; Boerner, 1992; Bohren and Huffman, 1983; Lee and Pottier, 2009; Cloude, 2010), this information can be utilized for making the object identification techniques. Many constituents of an object also exhibit polarization properties such as birefringence, dichroism, depolarization, etc., which might serve to discriminate between surface and volume scattering as well.

The enormous importance of the matrix polarimetry, optical and radar, is that it contains all the information that one can obtain from a scattering scene (Bohren and Huffman, 1983; Brosseau, 1998; Azzam and Bashara, 1977; Collett, 1993; Shurcliff, 1962). The matrix polarimetry has many useful applications in such diverse fields as interaction with various optical systems (Shurcliff, 1962; Azzam and Bashara, 1977; Collett, 1993; Brosseau, 1998), cloud diagnostics (van de Hulst, 1957; Bohren and Huffman, 1983; Mishchenko et al., 2000, 2002; Kokhanovsky, 2003c), remote sensing of the ocean, atmosphere, and planetary surfaces (Boerner, 1992; Kokhanovsky, 2001, 2003b, 2003c; Muttiah, 2002; Mishchenko et al., 2010), and biological tissue optics (Priezzhev et al., 1989; Tuchin, 2002; Tuchin et al., 2006) and others.

The methods of interpretation of the Jones and Mueller matrices in optical polarimetry and target decompositions in radar polarimetry have been developed by many authors (Hurwitz and Jones, 1941; Huynen, 1970; Whitney, 1971; Cloude, 1986; Gil and Bernabeu, 1987; Krogager, 1990; Lu and Chipman, 1994, 1996; Mar'enko and Savenkov, 1994; Freeman and Durden, 1998; Yamaguchi et al., 2005; Savenkov et al., 2005, 2006, 2007b).

In this chapter, we intend to illustrate the fact that polarization contributes reliably in a wide scope of terrestrial problems, which are rather unexpected on occasion. An ample discussion of important polarimetric applications which do not involve the matrix or full polarization measurement is beyond the scope of our discussion. In any case, our reference list should by no means be considered exhaustive and is merely intended to provide initial reference points for the interested reader.

3.2 Mueller matrices of deterministic and depolarizing objects

In the Mueller matrix calculus, the polarization state of light can be completely characterized by a Stokes vector, while the polarization transforming properties of a medium can be completely characterized by a Mueller matrix:

$$\mathbf{S}^{out} = \mathbf{M}\mathbf{S}^{imp}, \tag{3.1}$$

where the four-component Stokes column vector (with '*out*' and '*inp*' denoting the Stokes vectors of the output and input light, respectively) consists of the following parameters:

$$\mathbf{S} = \begin{pmatrix} I \\ Q \\ U \\ V \end{pmatrix} = \begin{pmatrix} s_1 \\ s_2 \\ s_3 \\ s_4 \end{pmatrix} = \begin{pmatrix} \langle |E_x|^2 + |E_y|^2 \rangle \\ \langle |E_x|^2 - |E_y|^2 \rangle \\ \langle E_x^* E_y + E_x E_y^* \rangle \\ i\langle E_x^* E_y + E_x E_y^* \rangle \end{pmatrix}, \tag{3.2}$$

with $\mathrm{i} = (- - 1)^{1/2}$. Among the pioneering contributions to this field of research, we note those by Solleillet (1929), Perrin (1942), Mueller (1948), and Parke (1948, 1949).

The Stokes parameter I is proportional to the total energy flux of the light beam. The Stokes parameters Q and U represent the differences between two components of the flux in which the electric vectors oscillate in mutually orthogonal directions. The Stokes parameter V is the difference between two oppositely circularly polarized components of the flux. As indicated by the angular brackets, the Stokes parameters s_i are ensemble averages (or time averages in the case of ergodic, stationary processes). This implies that no coherence effects are considered.

The Stokes vectors and Mueller matrices represent operations on intensities and their differences, namely incoherent superpositions of light beams; they are not adequate to describe either interference or diffraction effects. However, they are well suited to describe partially polarized and unpolarized light. Extensive lists of various Mueller matrices have been presented by several authors (e.g., Shurcliff, 1962; Kliger et al., 1990; Gerrard and Burch, 1975).

The Stokes parameters obey the inequality

$$s_1^2 \geq s_2^2 + s_3^2 + s_4^2 . \tag{3.3}$$

This inequality is called the Stokes—Verdet criterion and is a consequence of the Schwartz (or Couchy–Buniakovski) theorem (Barakat, 1963). The degree of polarization p is defined by

$$p = \sqrt{s_2^2 + s_3^2 + s_4^2}/s_1\,. \tag{3.4}$$

In Eq. (3.3), the equality holds for a completely polarized (pure) beam of light. In this case, $p = 1$ Another limiting case, $p = 0$ occurs when $s_2^2 + s_3^2 + s_4^2 = 0$, namely when the electric vector vibrates in all directions randomly and with no preferential orientation. An intermediate case, $0 < p < 1$ implies that light contains both polarized and depolarized components and is, therefore, called partially polarized.

The inequality in Eq. (3.3) plays an important role in polarimetry because it allows one to classify the character of the light–medium interaction. Assume first that the input light is completely polarized. In this case, the equality in Eq. (3.3) implies that the medium is non-depolarizing. Note that the terms 'non-depolarizing' and 'deterministic' or 'pure' are not, in general, identical. The term 'deterministic' means that the Mueller matrix describing such a medium can be derived from the corresponding Jones matrix (Simon, 1982; Gil and Bernabeu, 1985; Anderson and Barakat, 1994; Gopala Rao et al., 1998b). Hereinafter, we call this class of matrices pure Mueller matrices (Hovenier, 1994). If the output light results in an inequality in Eq. (3.3), then the scattering medium is not deterministic. If, in addition, the transformation matrix in Eq. (3.1) can be represented as a convex sum of deterministic Mueller matrices (Cloude, 1986; Gil, 2000, 2007) then the result is a depolarizing Mueller matrix (hereinafter Mueller matrix); otherwise, the result is a Stokes transformation matrix – that is, the transformation matrix ensures the fulfillment of the Stokes–Verdet criterion only. The properties of matrices transforming Stokes vectors into Stokes vectors, namely those satisfying the Stokes–Verdet criterion, have been studied by many authors (Xing, 1992; van der Mee, 1993; van der Mee and Hovenier, 1992; Sridhar and Simon, 1994; Nagirner, 1993; Givens and Kostinski, 1993; Gopala Rao et al., 1998a).

Any pure Mueller matrix $\mathbf{M}$ can be transformed to the corresponding Jones matrix $\mathbf{T}$ using the following relation (Parke, 1949; Azzam and Bashara, 1977; Dubois and Norikane, 1987):

$$\mathbf{M} = \mathbf{A}(\mathbf{T} \otimes \mathbf{T}^*)\mathbf{A}^{-1}\,, \tag{3.5}$$

where the asterisk denotes the complex-conjugate value,

$$\mathbf{T} = \begin{pmatrix} t_1 & t_4 \\ t_3 & t_2 \end{pmatrix}, \tag{3.6}$$

$$\mathbf{A} = \begin{pmatrix} 1 & 0 & 0 & 1 \\ 1 & 0 & 0 & -1 \\ 0 & 1 & 1 & 0 \\ 0 & \mathrm{i} & -\mathrm{i} & 0 \end{pmatrix}, \tag{3.7}$$

where the t_i are, in general, complex and $\otimes$ is the tensorial (Kronecker) product.

Since the element m_{11} is a gain for unpolarized incident light, it must satisfy the following inequality:

$$m_{11} > 0\,. \tag{3.8}$$

Furthermore, the elements of the Mueller matrix must obey the following conditions:

$$m_{11} \geq |m_{ij}|\,, \tag{3.9}$$

$$\mathrm{Tr}(\mathbf{M}) \geq 0\,, \tag{3.10}$$

$$\mu\mathbf{T} \sim |\mu|^2\mathbf{M}\,, \tag{3.11}$$

where Tr denotes the trace operation and μ is an arbitrary real or complex constant.

Note that the last relation defines the ability of the Jones and Mueller matrices to represent a 'physically realizable' medium (Lu and Chipman, 1994; Anderson and Barakat, 1994; Gil, 2007) and implies the physical restriction according to which the ratio g of the intensities of the emerging and incident light beams (the gain or intensity transmittance) must always be in the interval $0 \leq g \leq 1$ This condition is called the gain or transmittance condition and can be written in terms of the elements of the Mueller matrix as follows (Barakat, 1987):

$$\begin{aligned} m_{11} + (m_{12}^2 + m_{13}^2 + m_{14}^2)^{1/2} &\leq 1\,, \\ m_{11} + (m_{21}^2 + m_{31}^2 + m_{41}^2)^{1/2} &\leq 1\,. \end{aligned} \tag{3.12}$$

While a Jones matrix has generally eight independent parameters, the absolute phase is lost in Eq. (3.5), yielding only seven independent elements for a pure Mueller matrix. Evidently, this results in the existence of interrelations for the elements of a general pure Mueller matrix. This fact was pointed out for the first time, although without a derivation of their explicit form, by van de Hulst (1957). Since then, this subject has been studied by many authors (e.g., Abhyankar and Fymat, 1969; Fry and Kattawar, 1981; Hovenier et al., 1986). In the most complete and refined form, these interrelations are presented in Hovenier (1994).

In particular, one can derive the following important equation for the elements of a pure Mueller matrix:

$$\sum_{i=1}^{4}\sum_{j=1}^{4} m_{ij}^2 = 4m_{11}^2\,. \tag{3.13}$$

This equality was obtained for the first time by Fry and Kattawar (1981). However, the question of whether this is a sufficient condition for $\mathbf{M}$ to be a pure Mueller matrix has been the subject of extensive discussions (see, e.g., Simon, 1982, 1987; Hovenier, 1994; Kim et al., 1987; Kostinski, 1992; Kostinski et al., 1993; Gil and Bernabeu, 1985; Anderson and Barakat, 1994; Brosseau, 1990; Brosseau et al., 1993). Under the premise that the Mueller matrix in question can be represented as a convex sum of pure Mueller matrices, Eq. (3.13) is both a necessary and a sufficient condition for $\mathbf{M}$ to be a pure Mueller matrix (Gil, 2007).

In addition to the equalities presented above, a set of inequalities can be derived to characterize the structure of the pure Mueller matrix, as follows (Hovenier et al., 1986):

$$\begin{aligned} &m_{11} + m_{22} + m_{12} + m_{21} \geq 0\,, \\ &m_{11} - m_{22} - m_{12} + m_{21} \geq 0\,, \\ &m_{11} + m_{22} - m_{12} - m_{21} \geq 0\,, \\ &m_{11} - m_{22} + m_{12} - m_{21} \geq 0\,, \\ &m_{11} + m_{22} + m_{33} + m_{44} \geq 0\,, \\ &m_{11} + m_{22} - m_{33} - m_{44} \geq 0\,, \\ &m_{11} - m_{22} + m_{33} - m_{44} \geq 0\,, \\ &m_{11} - m_{22} - m_{33} + m_{44} \geq 0\,. \end{aligned} \tag{3.14}$$

Equation (3.5) can be used to derive interrelations between the structures of a Jones matrix and the corresponding pure Mueller matrix. For example, the successive application of transposition and sign reversal for the off-diagonal elements of the Jones matrix in Eq. (3.6) yields

$$\begin{pmatrix} t_1 & -t_3 \\ -t_4 & t_2 \end{pmatrix} \leftrightarrow \begin{pmatrix} m_{11} & m_{21} & -m_{31} & m_{41} \\ m_{12} & m_{22} & -m_{32} & m_{42} \\ -m_{13} & -m_{23} & m_{33} & -m_{43} \\ m_{14} & m_{24} & -m_{34} & m_{44} \end{pmatrix}. \tag{3.15}$$

Physical reasons for the above relations are quite clear. Indeed, Eq. (3.15) originates from the operation of interchanging the incident and emerging light beams, the principle of reciprocity (Saxon, 1955; Sekera, 1966; Vansteenkiste et al., 1993; Potton, 2004), and mirror symmetry (Hovenier, 1969, 1970).

The effect of the symmetry of the individual scatterers and collections of scatterers on the structure (number of independent parameters) of the Mueller matrix has been considered by van de Hulst (1957). In particular, he demonstrated that the collection of scatterers containing equal numbers of particles and their mirror particles possesses the following Mueller matrix:

$$\begin{pmatrix} m_{11} & m_{12} & 0 & 0 \\ m_{21} & m_{22} & 0 & 0 \\ 0 & 0 & m_{33} & m_{34} \\ 0 & 0 & m_{43} & m_{44} \end{pmatrix}. \tag{3.16}$$

If in Eq. (3.6), $t_3 = t_4$, then $m_{11}^2 - m_{12}^2 - m_{33}^2 - m_{34}^2 = 0$, $m_{22} = m_{11}$, $m_{33} = m_{44}$, and $m_{34} = -m_{43}$ (see Eq. (3.5)).

If, in addition, the collection of scatterers contains equal numbers of particles in positions described by Eqs (3.6) and (3.15) and those corresponding to the transposition and sign reversal of the off-diagonal elements of the Jones matrix in Eq. (3.6), then $m_{12} = m_{21}$.

The Mueller matrix of Eq. (3.16) plays a key role in many light-scattering applications. Some of them will be discussed later in this section. The structure of Eq. (3.16) can be caused by a symmetry of individual particles and a collection of particles in single and multiple scattering (van de Hulst, 1957; Mishchenko and Travis, 2000) as well as by illumination–observation geometries for backward (Zubko et al., 2004) and forward (Savenkov et al., 2007a) scattering.

The model of a medium described by the Mueller matrix of Eq. (3.16) has been used in studies of optical characteristics of oceanic water (Voss and Fry, 1984; Kokhanovsky, 2003b); ensembles of identical, but randomly oriented fractal particles (Kokhanovsky, 2003a); dense spherical particle suspensions in the multiple-scattering regime (Kaplan et al., 2001); ice clouds consisting of nonspherical ice crystals in the multiple-scattering regime (Lawless et al., 2006); polydisperse, randomly oriented ice crystals modeled by finite circular cylinders with different size distributions (Xu et al., 2002); cylindrically shaped radially inhomogeneous particles (Manickavasagam and Menguc, 1998). Other applications included measurements of the complex refractive index of isotropic materials as matrices of isotropic and ideal metal mirror reflections (Deibler and Smith, 2001); the development of a symmetric three-term product decomposition of a Mueller–Jones matrix (Ossikovski, 2008); and the description of very general and practically important cases of (i) randomly oriented particles with a plane of symmetry (Hovenier and van der Mee, 2000) and/or (ii) equal numbers of particles and their mirror particles (Mishchenko et al., 2002). This list of applications can be extended significantly.

An example of the situation in which the Mueller matrix has the structure of Eq. (3.16) and contains information on the strong dependence of depolarization and depolarization on the polarization state of the input light is the exact forward scattering of polarized light by a slab of inhomogeneous linear birefringent medium (Savenkov et al., 2007a).

The scattering angles 0° (exact forward direction) and 180° (exact backward direction) deserve special attention owing to their importance in numerous practical applications, including the scattering by biological tissues. For the first time, the general form of Mueller matrices for these scattering angles was derived by van de Hulst (1957). Hu et al. (1987) presented a comprehensive study of forward and backward scattering by an individual particle in a fixed orientation. For forward scattering, they distinguished 16 different symmetry shapes which were classified into five symmetry classes; for backward scattering, four different symmetry shapes were identified and classified into two symmetry classes. A large number of relations were derived in this way. The structures of Mueller matrices for various collections of particles in the cases of forward and backward scattering can be found elsewhere (van de Hulst, 1957; Hovenier and Mackowski, 1998).

It is important to note that, although analyses of the internal structure of a general pure Mueller matrix, the symmetry relations between matrix elements caused by interchanging the incident and emerging light beams, and the principle of reciprocity have historically been carried out in the framework of light scattering by discrete particles, these results are also relevant to pure Mueller matrices in the continuous-medium approximation.

In this section, we consider the problem of Mueller matrix interpretation in the framework of the approach wherein the medium studied is modeled as a medium with a continuous (and possibly random) distribution of optical parameters. The polarization of light changes if the amplitudes and phases of the components of the electric vector **E** change separately or simultaneously (Shurcliff, 1962; Azzam and Bashara, 1977; Brosseau, 1998). It is, therefore, customary to distinguish between the corresponding classes of anisotropic media: dichroic (or possessing amplitude anisotropy), influencing only the amplitudes; birefringent (or possessing phase anisotropy), influencing only the phases; and 'all other' (possessing both amplitude and phase anisotropy) affecting both the amplitudes and the phases of the components of the electric field vector. Among these classes, four types of anisotropic mechanisms are recognized as basic or, after Jones, elementary (Jones, 1941, 1942,

1947, 1956; Hurwitz and Jones, 1941): linear and circular phase and linear and circular amplitude anisotropies.

Linear birefringence is described by the following pure Mueller matrix:

$$\mathbf{M}^{LP}=\begin{pmatrix}1 & 0 & 0 & 0\\ 0 & \cos^2 2\alpha+\sin^2 2\alpha\cos\Delta & \cos 2\alpha\sin 2\alpha(1-\cos\Delta) & -\sin 2\alpha\sin\Delta\\ 0 & \cos 2\alpha\sin 2\alpha(1-\cos\Delta) & \sin^2 2\alpha+\cos^2 2\alpha\cos\Delta & \cos 2\alpha\sin\Delta\\ 0 & \sin 2\alpha\sin\Delta & -\cos 2\alpha\sin\Delta & \cos\Delta\end{pmatrix}, \tag{3.17}$$

where Δ is the phase shift between two orthogonal linear components of the electric field vector and α is the azimuth of the anisotropy.

The Mueller matrix describing linear dichroism is

$$\mathbf{M}^{LA}=$$

$$\begin{pmatrix}1+P & (1-P)\cos 2\gamma & (1-P)\sin 2\gamma & 0\\ (1-P)\cos 2\gamma & \cos^2 2\gamma(1+P)+2\sin^2 2\gamma\sqrt{P} & \cos 2\gamma\sin 2\gamma(1-\sqrt{P})^2 & 0\\ (1-P)\sin 2\gamma & \cos 2\gamma\sin 2\gamma(1-\sqrt{P})^2 & \sin^2 2\gamma(1+P)+2\cos^2 2\gamma\sqrt{P} & 0\\ 0 & 0 & 0 & 2\sqrt{P}\end{pmatrix}, \tag{3.18}$$

where P is the relative absorption of two linear orthogonal components of the electric vector and γ is the azimuth of the anisotropy.

The Mueller matrix describing circular birefringence is

$$\mathbf{M}^{CP}=\begin{pmatrix}1 & 0 & 0 & 0\\ 0 & \cos 2\varphi & \sin 2\varphi & 0\\ 0 & -\sin 2\varphi & \cos 2\varphi & 0\\ 0 & 0 & 0 & 1\end{pmatrix}, \tag{3.19}$$

where φ is the induced phase shift between two orthogonal circular components of the electric vector.

Finally, in terms of the Mueller matrix calculus, circular amplitude anisotropy is described by the following matrix:

$$\mathbf{M}^{CA}=\begin{pmatrix}1+R^2 & 0 & 0 & 2R\\ 0 & 1-R^2 & 0 & 0\\ 0 & 0 & 1-R^2 & 0\\ 2R & 0 & 0 & 1+R^2\end{pmatrix}, \tag{3.20}$$

where R is the magnitude of anisotropy, namely the relative absorption of two orthogonal circular components of the electric vector. The six quantities α, Δ, P, γ, φ, and R are called anisotropy parameters.

It can be seen that the matrices describing linear and circular birefringence belong to the class of unitary matrices (in the case of matrices with real-valued elements – orthogonal matrices). The matrices of linear Eq. (3.18) and circular Eq. (3.20) dichroism belong to the class of Hermitian matrices (in the case of matrices with real-valued elements – symmetric matrices).

The Mueller matrices of Eqs (3.17)–(3.20) represent media exhibiting individual types of anisotropy. Experimental measurements of these matrices or of the corresponding informative matrix elements allow one to interpret and characterize anisotropy properties of media. However, more often, two or more types of anisotropy are exhibited by a medium simultaneously. Evidently, such cases require the development of more sophisticated polarimetric matrix models (Hurwitz and Jones, 1941; Cloude, 1986; Gil and Bernabeu, 1987; Lu and Chipman, 1996; Savenkov et al., 2006; Ossikovski, 2008, 2009).

The matrix model that is used most extensively in optical polarimetry for decoupling constituent polarization properties of optical medium is the polar decomposition proposed by Lu and Chipman (1996). This model is based on the so-called polar decomposition theorem (Lancaster and Tismenetsky, 1985), according to which an arbitrary matrix $\mathbf{M}$ can be represented by a product

$$\mathbf{M} = \mathbf{M}_P \mathbf{M}_R \quad \text{or} \quad \mathbf{M} = \mathbf{M}_R \mathbf{M}'_P \,, \tag{3.21}$$

where $\mathbf{M}_P$ and $\mathbf{M}'_P$ are Hermitian matrices and $\mathbf{M}_R$ is a unitary one. The Hermitian matrix is associated with amplitude anisotropy, while the unitary matrix describes phase anisotropy (Whitney, 1971). The matrices $\mathbf{M}_P$ and $\mathbf{M}_R$ are called the dichroic and the phase polar forms (Whitney, 1971; Gil and Bernabeu, 1987; Lu and Chipman, 1996).

The polar decomposition was first employed by Whitney (1971) without finding explicit expressions for $\mathbf{M}_P$ and $\mathbf{M}_R$ They were proposed later, independently by Gil and Bernabeu (1987) and Lu and Chipman (1996). Alternatively, the dichroic and phase polar forms can be derived using spectral methods of linear algebra (Azzam and Bashara, 1977).

The phase polar form $\mathbf{M}_R$ (using notation from Lu and Chipman, 1996) is given by

$$\begin{aligned} \mathbf{M}_R &= \begin{pmatrix} 1 & \vec{0}^T \\ \vec{0} & \mathbf{m}_R \end{pmatrix}, \\ (m_R)_{ij} &= \delta_{ij} \cos R + a_i a_j (1 - \cos R) + \sum_{k=1}^{3} \varepsilon_{ijk} a_k \sin R \,, \end{aligned} \tag{3.22}$$

where $\vec{0}$ is the 3×1 zero vector; $[1\ a_1\ a_2\ a_3]^T = [1\ \hat{R}^T]^T$ is the normalized Stokes vector for the fast axis of $\mathbf{M}_R$; δ_{ij} is the Kronecker delta; ε_{ijk} is the Levi–Civita permutation symbol, $\mathbf{m}_R$ is the 3×3 submatrix of $\mathbf{M}_R$ obtained by striking out the first row and the first column of $\mathbf{M}_R$ and R is the birefringence given by

$$R = \arccos\left(\frac{1}{2} \operatorname{Tr} \mathbf{M}_R - 1\right). \tag{3.23}$$

The dichroic polar form $\mathbf{M}_P$ is as follows:

$$\begin{aligned} \mathbf{M}_P &= T_u \begin{pmatrix} 1 & \vec{\mathbf{D}}^T \\ \vec{\mathbf{D}} & \mathbf{m}_P \end{pmatrix}, \\ \mathbf{m}_P &= \sqrt{1 - D^2}\, \mathbf{I} + \left(1 - \sqrt{1 - D^2}\right) \hat{\mathbf{D}} \hat{\mathbf{D}} \,, \end{aligned} \tag{3.24}$$

where $\mathbf{I}$ is the 3×3 identity matrix; $\hat{\mathbf{D}} = \vec{\mathbf{D}}/|\vec{\mathbf{D}}|$ is the unit vector in the direction of the diattenuation vector $\vec{\mathbf{D}}$; T_u is the transmittance for unpolarized light; and the value of diattenuation can be obtained as

$$D = \left(1 - 4|\det(\mathbf{T})|^2/[\mathrm{Tr}(\mathbf{T}^*\mathbf{T})]^2\right)^{1/2}. \tag{3.25}$$

The models of anisotropic media based on the polar decomposition contain six independent parameters: three for the phase polar form $\mathbf{M}_R$ and three for the dichroic polar form $\mathbf{M}_P$ It can be seen that the phase polar form is a unitary (orthogonal) matrix and the dichroic polar form is a Hermitian (symmetric) matrix. Note that unitarity (orthogonality) of the phase polar form, Eq. (3.22), is in complete agreement with the first Jones equivalence theorem (Hurwitz and Jones, 1941) and is a general model of elliptically birefringent media. The situation with the dichroic polar form is more complex (Savenkov et al., 2005, 2007b). Mathematically, the complexity originates from the fact that, in contrast to unitary matrices, the product of Hermitian matrices is generally not a Hermitian matrix (Lancaster and Tismenetsky, 1985).

If the incident light is fully polarized and the output light is characterized by an inequality in Eq. (3.3), then the equalities for matrix elements obtained in Hovenier (1994) and Eq. (3.13), which determine the structure of the Mueller matrix as a deterministic matrix, are lost. In this case, the output light is composed of several incoherent contributions, and the medium as a whole cannot be represented by a Jones matrix. However, the medium can be considered as a parallel set of deterministic media, each one being described by a well-defined Jones matrix, in such a way that the light beam is shared among these different media. It is important to point out that the same result could be obtained by considering the medium as an ensemble (Kim et al., 1987) so that each realization i, characterized by a well-defined Jones matrix $\mathbf{T}^i$ occurs with a probability p^i.

If a Mueller matrix can be represented by a convex sum of pure Mueller matrices (Cloude, 1986; Simon, 1987; Cloude and Pottier, 1995; Gil, 2007), then it is called a depolarizing Mueller matrix. It is important to note that this class of matrices does not coincide with the class of matrices, called Stokes matrices, satisfying the Stokes–Verdet criterion – that is, matrices transforming Stokes vectors into Stokes vectors (see Eq. (3.3)). Any physical Mueller matrix is a Stokes matrix, but the converse is not, in general, true (Gil, 2007). On the other hand, no method has been quoted to physically realize a Stokes matrix that cannot be represented as a convex sum of deterministic Mueller matrices.

Linear inequalities for the elements of a pure Mueller matrix are also valid for a depolarizing Mueller matrix $\mathbf{M}$ (Hovenier and van der Mee, 2000), in particular:

$$\begin{aligned}
&m_{11} \geq 0; \quad m_{11} \geq |m_{ij}|, \\
&m_{11} + m_{22} + m_{12} + m_{21} \geq 0, \\
&m_{11} + m_{22} - m_{12} - m_{21} \geq 0, \\
&m_{11} - m_{22} + m_{12} - m_{21} \geq 0, \\
&m_{11} - m_{22} - m_{12} + m_{21} \geq 0.
\end{aligned} \tag{3.26}$$

In this case, the following quadratic inequalities are also valid (Fry and Kattawar, 1981):

$$
\begin{aligned}
(m_{11}+m_{12})^2-(m_{21}+m_{22})^2 &\geq (m_{31}+m_{32})^2+(m_{41}+m_{42})^2\,,\\
(m_{11}-m_{12})^2-(m_{21}-m_{22})^2 &\geq (m_{31}-m_{32})^2+(m_{41}-m_{42})^2\,,\\
(m_{11}+m_{21})^2-(m_{12}+m_{22})^2 &\geq (m_{13}+m_{23})^2+(m_{14}+m_{24})^2\,,\\
(m_{11}-m_{21})^2-(m_{12}-m_{22})^2 &\geq (m_{13}-m_{23})^2+(m_{14}-m_{24})^2\,,\\
(m_{11}+m_{22})^2-(m_{12}+m_{21})^2 &\geq (m_{33}+m_{44})^2+(m_{34}-m_{43})^2\,,\\
(m_{11}-m_{22})^2-(m_{12}-m_{21})^2 &\geq (m_{33}-m_{44})^2+(m_{34}+m_{43})^2\,,
\end{aligned}
\tag{3.27}
$$

while Eq. (3.13) becomes an inequality as well:

$$
\sum_{i=1}^{4}\sum_{j=1}^{4} m_{ij}^2 \leq 4m_{11}^2\,. \tag{3.28}
$$

The study and characterization of depolarization is of considerable importance owing to the fact that depolarization phenomena are encountered in many theoretical and experimental applications of polarimetry to discrete random media and media with bulk and surface inhomogeneities. Note that the light–medium interaction with depolarization is heretofore studied in considerably less detail than the problem described by Mueller–Jones matrices discussed above.

Depolarization is the result of decorrelation of the phases and the amplitudes of the electric field vectors and/or selective absorption of polarization states (Brosseau, 1998). Depolarization can be observed in both single and multiple light scattering and depends on geometrical and physical characteristics of the scattering particles: shape, morphology, refractive index, size parameter (ratio of the particle circumference to the wavelength of the incident light), and orientation with respect to the reference frame (Mishchenko and Travis, 2000). Furthermore, multiple scattering results in depolarization of the output light even in the case of a collection of spherically symmetric particles and often reinforces depolarization caused by particle non-sphericity (Mishchenko and Travis, 2000; Mishchenko et al., 2006). Our purpose here is to consider the depolarization phenomenon using the Mueller matrix formalism; in particular, we intend to discuss single-number depolarization metrics and Mueller matrices of depolarization. The notion of depolarization Mueller matrix factorization has long existed in the literature and will be considered in this section below.

Depolarization metrics provide a single scalar number that varies from zero, thereby corresponding to a totally depolarized output light, to a certain positive number corresponding to a totally polarized output light. All intermediate values are associated with partial polarization.

The depolarization index was introduced by Gil and Bernabeu (1985, 1986):

$$
DI(\mathbf{M}) = \sqrt{\sum_{i,j=1}^{4} m_{ij}^2 - m_{11}^2} \Big/ (\sqrt{3}m_{11})\,. \tag{3.29}
$$

The depolarization index is bounded according to $0 \leq DI(\mathbf{M}) \leq 1$. The extreme values of $DI(\mathbf{M})$ correspond to the case of unpolarized and totally polarized output light, respectively.

An 'analog' to the degree of polarization, Eq. (3.4), for linearly polarized input light in terms of Mueller matrix elements, the so-called index of linear polarization, was introduced by Bueno (2001):

$$G_L = \frac{\sqrt{3}}{2m_{11}} \left(m_{21}^2 + m_{31}^2 + \frac{1}{3} \sum_{i=1}^{4} (m_{2i}^2 + m_{3i}^2) \right)^{1/2} . \tag{3.30}$$

It can be seen that G_L is the ratio of the mean of the sum of the squares of matrix elements corresponding to linear polarization of the output light and the value of the corresponding averaged intensity normalized by the maximum value of this ratio which occurs for a linear polarizer: $(G_L)_{\max} = 2/3^{1/2}$ The former implies the following range of variation: $0 \leq G_L \leq 1$.

The average degree of polarization was defined by Chipman (2005) as follows:

$$Average\ DoP(\mathbf{M}) = \frac{1}{4\pi} \int_0^{\pi} \int_{-\pi/2}^{\pi/2} p[\mathbf{MS}(\varepsilon, \zeta)] \cos \varepsilon \, d\varepsilon \, d\zeta \, . \tag{3.31}$$

The term $\cos \varepsilon \, d\varepsilon \, d\zeta$ scans the incident polarization state over the Poincaré sphere, with the latitude ε and longitude ζ. The Stokes vector $\mathbf{S}(\varepsilon, \zeta)$ is a function of ellipticity and orientation azimuth of the polarization ellipse of light:

$$\mathbf{S}(\varepsilon, \zeta) = \begin{bmatrix} 1 & \cos 2\varepsilon \cos 2\zeta & \cos 2\varepsilon \sin 2\zeta & \sin 2\varepsilon \end{bmatrix}^T , \tag{3.32}$$

where T stands for 'transposed.'

The so-called $Q(\mathbf{M})$ metric is defined as follows (Espinosa-Luna and Bernabeu, 2007):

$$Q(\mathbf{M}) = \left(3[DI(\mathbf{M})]^2 - [D(\mathbf{M})]^2\right) \Big/ \left(1 + [D(\mathbf{M})]^2\right) , \tag{3.33}$$

where $D(\mathbf{M}) = (m_{12}^2 + m_{13}^2 + m_{14}^2)^{1/2}$ is the diattenuation parameter and $0 \leq D(\mathbf{M}) \leq 1$. The metric $Q(\mathbf{M})$ is bounded according to $0 \leq Q(M) \leq 3$. Specifically, $Q(\mathbf{M}) = 0$ corresponds to a totally depolarizing medium; $0 < Q(M) < 1$ describes a partially depolarizing medium; $1 \leq Q(M) < 3$ represents a partially depolarizing medium if, in addition, $0 < DI(M) < 1$; otherwise, it represents a non-depolarizing diattenuating medium; finally, $Q(M) = 3$ for a non-depolarizing non-diattenuating medium.

Thus, the depolarization metrics provide a summary of the depolarizing property of a medium via a single number. The depolarization index $DI(\mathbf{M})$ and the $Q(\mathbf{M})$ metrics are directly related to the Mueller matrix elements only and, in contrast to the average degree of polarization *Average DoP*, require no scan of the whole Poincaré sphere of the input polarizations. Furthermore, $Q(\mathbf{M})$ provides more detailed information about depolarization properties of a medium.

Quantities referring to the intrinsic depolarization properties of light have had wide applications in polarimetry as well. These are the linear, δ_L and circular, δ_C, depolarization ratios defined according to Mishchenko and Hovenier (1995) and Mishchenko and Travis (2000) as

$$\delta_L = (s_1 - s_2)/(s_1 + s_2) \, , \tag{3.34}$$

$$\delta_C = (s_1 + s_4)/(s_1 - s_4) \, . \tag{3.35}$$

The interest in these parameters is explained by the fact that they are susceptible to particle non-sphericity. Indeed, for spherical particles, both ratios are equal to zero identically, whereas, for nonspherical scatterers, both δ_L and δ_C can substantially deviate from zero (Mishchenko and Hovenier, 1995). The former means that, if the incident light is linearly polarized, then the backscattered light is completely linearly polarized in the same plane, whereas, if the incident light is circularly polarized, then the backscattered light is completely circularly polarized in the opposite sense. For nonspherical particles, this is generally not the case.

Chipman (1995) introduced somewhat different versions of the degrees of linear and circular polarization:

$$DoLP = \sqrt{s_2^2 + s_3^2}/s_1 \,, \tag{3.36}$$

$$DoCP = s_4/s_1 \,. \tag{3.37}$$

These parameters turn out to be very useful for applications in meteorology, astronomy, ophthalmology, optical fibers, etc. (e.g., Bueno, 2001, and references therein).

Some media depolarize all polarization states equally. Other depolarizing media partially depolarize most polarization states but may not depolarize one or some incident states. Depolarization depends significantly on the polarization state of the input light in the multiple-scattering regime (Bicout et al., 1994; Rojas-Ochoa et al., 2004; Kim et al., 2006, and references therein). In particular, Bicout et al. (1994) studied numerically and experimentally how depolarization evolves for linear and circular input polarizations as the size of the particles increases from very small (Rayleigh regime) to large (Mie regime) in the case of a forward-scattering geometry.

A single-number metric providing a summary of depolarization by a medium cannot give detailed information about all features of depolarization. Such information can only be obtained from Mueller matrix models of depolarization. The case when for all polarizations of the input light the degree of polarization p of the output light is the same is called isotropic depolarization. When the degree of polarization of the output light is a function of parameters of the input polarization, one speaks of anisotropic depolarization.

There seems to be a consensus regarding the form of the Mueller matrix model describing isotropic depolarization (Brosseau, 1998; Chipman, 1999):

$$\mathrm{diag}\begin{bmatrix}1 & p & p & p\end{bmatrix} . \tag{3.38}$$

It can be seen that the properties of this type of depolarization are the following:

1. The transmittance is the same for all polarizations of the incident light.
2. p of the output light is the same for all input polarizations.

On the other hand, there is no consensus in the literature concerning the Mueller matrix for the case of anisotropic depolarization. Apparently, one of the most accepted forms of the Mueller matrix describing the dependence of p of the output light on the incident polarization is the following (Shindo, 1995; Brosseau, 1998; Chipman, 1999; Ossikovski, 2009):

$$\mathrm{diag}\begin{bmatrix}1 & a & b & c\end{bmatrix} . \tag{3.39}$$

The elements a, b, and c are interpreted physically in the following manner: a and b are the degrees of linear depolarization, while c is the degree of circular depolarization. If $a = b = c = 0$, then the Mueller matrix represents an ideal depolarizer. Bicout et al. (1994) discussed the depolarization arising in multiple scattering of light by spherical scatterers in the Rayleigh regime and given by Eq. (3.39) with $a = b$.

The most general expression for the Mueller matrix describing depolarization was suggested by Lu and Chipman (1996):

$$\begin{pmatrix} 1 & \vec{\mathbf{O}}^T \\ \vec{\mathbf{P}}_\Delta & \mathbf{m}_\Delta \end{pmatrix}, \quad \mathbf{m}_\Delta^T = \mathbf{m}_\Delta\,, \tag{3.40}$$

where $\vec{\mathbf{P}}_\Delta$ denotes the so-called polarizance vector. The polarizance vector describes the state of polarization generated by this Mueller matrix from unpolarized incident light. The Mueller matrix of Eq. (3.40) has nine degrees of freedom, and this is of interest because this matrix, along with a generalized deterministic Mueller matrix, is jointly characterized by 16° of freedom. This means that, in this way, one obtains the generalized Mueller matrix of an arbitrary medium that has 16° of freedom and linearly interacts with polarized light.

The product of Mueller matrices of the polar forms (Eqs (3.22) and (3.24)) and the depolarizing Mueller matrix (Eq. 3.40))

$$\mathbf{M} = \mathbf{M}_\Delta \mathbf{M}_R \mathbf{M}_P\,, \tag{3.41}$$

is the generalized polar decomposition and a multiplicative matrix model of an arbitrary Mueller matrix (Lu and Chipman, 1996; Gil, 2000, 2007).

The product of the phase polar form and the depolarizing matrices can then be obtained as

$$\mathbf{M}_\Delta \mathbf{M}_R = \mathbf{M}' = \mathbf{M}\mathbf{M}_P^{-1}\,. \tag{3.42}$$

Then

$$\vec{\mathbf{P}}_\Delta = (\vec{\mathbf{P}} - \mathbf{m}\vec{\mathbf{D}})/(1 - D^2)\,, \tag{3.43}$$

where $\vec{\mathbf{P}} = (1/m_{11}) \begin{bmatrix} m_{21} & m_{31} & m_{41} \end{bmatrix}^T$ and $\mathbf{m}$ is the submatrix of the initial matrix $\mathbf{M}$. The $\mathbf{m}'$ is the submatrix of $\mathbf{M}'$ and can be written as

$$\mathbf{m}' = \mathbf{m}_\Delta \mathbf{m}_R\,. \tag{3.44}$$

The submatrix $\mathbf{m}_\Delta$ can be calculated as follows:

$$\begin{aligned} \mathbf{m}_\Delta &= \pm \left[\mathbf{m}'(\mathbf{m}')^T + \left(\sqrt{\lambda_1\lambda_2} + \sqrt{\lambda_2\lambda_3} + \sqrt{\lambda_1\lambda_3}\right)\mathbf{I}\right]^{-1} \\ &\quad \times \left[\left(\sqrt{\lambda_1} + \sqrt{\lambda_2} + \sqrt{\lambda_3}\right)\mathbf{m}'(\mathbf{m}')^T + \sqrt{\lambda_1\lambda_2\lambda_3}\mathbf{I}\right]\,, \end{aligned} \tag{3.45}$$

where λ_1 are the eigenvalues of $\mathbf{m}'(\mathbf{m}')^T$ The sign '+' or '−' is determined by the sign of the determinant of $\mathbf{m}'$. The net depolarization coefficient Δ can be calculated according to

$$\Delta = 1 - \frac{1}{3}|\operatorname{Tr}(\mathbf{M}_\Delta) - 1|\,. \tag{3.46}$$

In the conclusion of this section, we consider the additive Mueller matrix model of depolarizing object suggested by Cloude (1986) and extensively employed in optical and radar polarimetry (see, e.g., Savenkov et al., 2003; Savenkov and Muttiah, 2004; Munoz et al., 2001, 2002, 2004; Volten et al., 2001; Cloude and Pottier, 1996, 1997). The Cloude coherence matrix **J** is derived from the corresponding arbitrary Mueller matrix as follows:

$$\begin{array}{ll}
J_{11} = (1/4)(m_{11}+m_{22}+m_{33}+m_{44}) & J_{12} = (1/4)(m_{12}+m_{21}-im_{34}+im_{43}) \\
J_{13} = (1/4)(m_{13}+m_{31}+im_{24}-im_{42}) & J_{14} = (1/4)(m_{14}-im_{23}+im_{32}+m_{41}) \\
J_{21} = (1/4)(m_{12}+m_{21}+im_{34}-im_{43}) & J_{22} = (1/4)(m_{11}+m_{22}-m_{33}-m_{44}) \\
J_{23} = (1/4)(im_{14}+m_{23}+m_{32}-im_{41}) & J_{24} = (1/4)(-im_{13}+im_{31}+m_{24}+m_{42}) \\
J_{31} = (1/4)(m_{13}+m_{31}-im_{24}+im_{42}) & J_{32} = (1/4)(-im_{14}+m_{23}+m_{32}+im_{41}) \\
J_{33} = (1/4)(m_{11}-m_{22}+m_{33}-m_{44}) & J_{34} = (1/4)(im_{12}-im_{21}+m_{34}+m_{43}) \\
J_{41} = (1/4)(m_{14}+im_{23}-im_{32}+m_{41}) & J_{42} = (1/4)(im_{13}-im_{31}+m_{24}+m_{42}) \\
J_{43} = (1/4)(-im_{12}+im_{21}+m_{34}+m_{43}) & J_{44} = (1/4)(m_{11}-m_{22}-m_{33}+m_{44})
\end{array} \tag{3.47}$$

It can be seen that coherence matrix **J** is positive semi-definite Hermitian and, hence, always has four real eigenvalues. The eigenvalues of the coherence matrix, λ_i, can be combined to form a quantity that is a measure of the depolarization, depolarization metric, of the studied medium. This quantity is called entropy and is defined as:

$$H = -\sum_{i=1}^{N} \left(\lambda_i \Big/ \sum_j \lambda_j \right) \log_N \left(\lambda_i \Big/ \sum_j \lambda_j \right) . \tag{3.48}$$

Given eigenvalues λ_i of coherence matrix **J**, we have for the initial Mueller matrix:

$$\mathbf{M} = \sum_{k=1}^{4} \lambda_k \mathbf{M}_D^k; \quad \mathbf{M}_D^k \Leftrightarrow \mathbf{T}^k , \tag{3.49}$$

where $\mathbf{M}_D^k$ are the pure Mueller matrices obtained from the Jones matrices by Eq. (3.5).

The Jones matrix, **T**, in turn, is obtained in the following manner:

$$\begin{array}{l}
t_{11}^{(k)} = \Psi_1^{(k)} + \Psi_2^{(k)}, \; t_{12}^{(k)} = \Psi_3^{(k)} - i\Psi_4^{(k)} \\
t_{21}^{(k)} = \Psi_3^{(k)} + i\Psi_4^{(k)}, \; t_{22}^{(k)} = \Psi_1^{(k)} - \Psi_2^{(k)} \quad k = \overline{1,4} ,
\end{array} \tag{3.50}$$

where $\mathbf{\Psi}^{(k)} = \begin{pmatrix} \Psi_1 & \Psi_2 & \Psi_3 & \Psi_4 \end{pmatrix}_k^T$ is k-th eigenvector of coherence matrix **J**.

Thus, the substance of Cloude's coherency matrix concept, which, in essence, is an additive matrix model of depolarizing Mueller matrix, Eq. (3.49), is the representation of the initial depolarizing Mueller matrix as a weighted convex sum of four pure Mueller matrices.

If three of the eigenvalues of $\mathbf{J}$ vanish, then the entropy $H = 0$ and initial matrix $\mathbf{M}$ is a deterministic Mueller–Jones matrix. If all four eigenvalues of $\mathbf{J}$ are not equal to zero and $H \leq 0.5$, then the pure Mueller matrix, which corresponds to the maximal eigenvalue, is the dominant type of deterministic polarization transformation of the studied object. So, this model gives the possibility to study the anisotropy properties of depolarizing objects on the one hand and, on the other hand, is a necessary and sufficient criterion for a given 4×4 real matrix to be a Mueller matrix (the case when all four eigenvalues of $\mathbf{J}$ are non-negative) and a pure Mueller matrix (the case when three of the eigenvalues vanish) (Munoz et al., 2001, 2002, 2004; Volten et al., 2001).

3.3 Mueller matrix polarimetry

The aim of this section is to discuss the general concept of the Mueller matrix measurement. Ample practical schemes of the Stokes and Mueller polarimeters can be found elsewhere (Hauge, 1980; Azzam, 1997; Chipman, 1995).

The Mueller matrix polarimeter at visible and infrared is composed of a polarization state generator (PSG) and polarization state analyzer (PSA), as shown in Fig. 3.1.

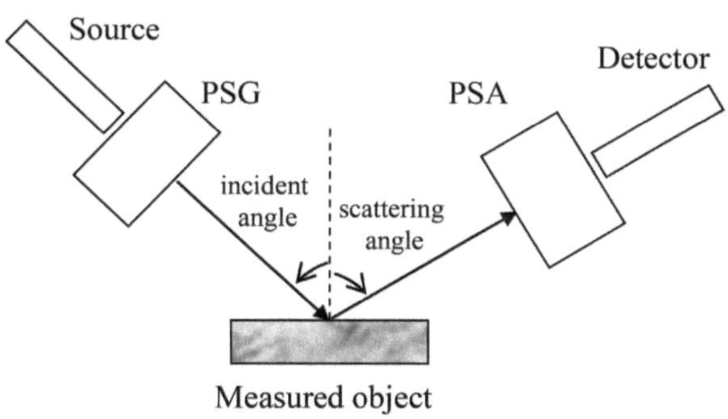

Fig. 3.1. Schematic overview to measure the Mueller matrices.

The PSG forms the particular polarization state of incident light on the studied object. The PSA is operated to measure either the full Stokes vector or some of the Stokes parameters of the scattered light. Both PSG and PSA consist of retarders and diattenuators that are capable of analyzing the polarization state of the scattered beam.

Nearly all existing Mueller matrix polarimeters are configured so that the entire Mueller matrix has to be measured (Chipman, 1995). Such an approach is required to make up the conditioned set of 16 equations for matrix elements in order to reconstruct the full 4×4 Mueller matrix. The approach has been determined by the structure of the so-called characteristic or data-reduction matrix of the generalized measurement equation. The data-reduction matrix describes the conversion of a set of polarized-intensity measurements into the Mueller matrix which is represented as a 16 × 1 vector. For any PSG and PSA, the total flux measured by the detector is

$$g = \mathbf{QML} = \sum_{i=1}^{4}\sum_{j=1}^{4} q_i m_{ij} l_j \,, \tag{3.51}$$

where **L** is the Stokes vector produced by PSG, **M** is the object Mueller matrix, and **Q** is the Stokes vector corresponding to the first row of the Mueller matrix representing the PSA.

To measure the full Mueller matrix, $N \geq 16$ flux measurements (Eq. (3.51)) are required. Flattening the Mueller matrix **M** into 16 × 1 Mueller vector of the form $\vec{\mathbf{M}} = \begin{bmatrix} m_{11} & m_{12} & m_{13} & m_{14} & \cdots & m_{43} & m_{44} \end{bmatrix}^T$, the polarimetric measurement equation can be represented as

$$\mathbf{G} = \mathbf{W}\vec{\mathbf{M}} = \begin{pmatrix} q_1^1 l_1^1 & q_1^1 l_2^1 & q_1^1 l_3^1 & \cdot & q_4^1 l_4^1 \\ q_1^2 l_1^2 & q_1^2 l_2^2 & q_1^2 l_3^2 & \cdot & q_4^2 l_4^2 \\ q_1^3 l_1^3 & q_1^3 l_1^3 & q_1^3 l_3^3 & \cdot & q_4^3 l_4^3 \\ \cdot & \cdot & \cdot & \cdot & \cdot \\ q_1^N l_1^N & q_1^N l_2^N & q_1^N l_3^N & \cdot & q_4^N l_4^N \end{pmatrix} \begin{pmatrix} m_{11} \\ m_{12} \\ m_{13} \\ \cdot \\ m_{44} \end{pmatrix}, \tag{3.52}$$

where **G** is the $N \times 1$ vector, whose components are the fluxes measured by detector; and **W** is the $N \times 16$ general characteristic or data-reduction matrix with elements $w_{ij}^N = q_i^N l_j^N$.

Equation (3.52) is a system of generally N algebraic equations for Mueller matrix elements m_{ij}. The simplest case of the system in Eq. (3.52) occurs when 16 independent measurements are performed. In this case, $N = 16$, **W** is of rank 16, and inverse matrix $\mathbf{W}^{-1}$ is unique. Then, all 16 Mueller matrix elements are

$$\vec{\mathbf{M}} = \mathbf{W}^{-1}\mathbf{G}\,. \tag{3.53}$$

Most Mueller matrix polarimeters are configured so that $N > 16$. This makes $\vec{\mathbf{M}}$ overdetermined, and $\vec{\mathbf{W}}^{-1}$ does not exist. The optimal (least-squares) estimation of $\vec{\mathbf{M}}$ can be obtained using the pseudoinverse matrix $\tilde{\mathbf{W}}$ of **W** (Lancaster and Tismenetsky, 1985):

$$\vec{\mathbf{M}} = \tilde{\mathbf{W}}\mathbf{G} = \left(\mathbf{W}^T\mathbf{W}\right)^{-1}\mathbf{W}^T\mathbf{G}\,. \tag{3.54}$$

In mathematics, there exist a variety of pseudoinverse matrices (e.g., one-sided inverse, Drazin inverse, group inverse, Bott–Duffin inverse, etc.). Here, we use the so-called Moore–Penrose pseudoinverse matrix (Moore, 1920; Bjerhammar, 1951; Penrose, 1955). Note that the characteristic matrix **W** in Eq. (3.7) is or can evidently be reduced to those of the full column rank.

This approach is named the complete Mueller polarimetry. The theory of operation and calibration of Mueller matrix polarimetry was developed in Chipman (1995), and the general formalism has been applied by many authors to the optimization of Mueller matrix polarimeters in the presence of noise and measurement error (Savenkov, 2002; Smith, 2002; De Martino et al., 2003; Twietmeyer and Chipman, 2008). This procedure is repeated at different scattering angles in order to determine the angular profile of the Mueller matrix.

However, in many applications, reconstruction of the full Mueller matrix is not necessary (Savenkov, 2002, 2007; Tyo et al., 2010; Oberemok and Savenkov, 2003). First of all, some subsets of matrix elements might completely describe scattering which is of interest and hence these subsets can be considered as initial information for the solution of corresponding classes of inverse problems. Another reason making the measurement of the complete Mueller matrix unnecessary is matrix symmetry. An illustrative example is the pure Mueller matrix with symmetry determining by the first Jones equivalence theorem (Hurwitz and Jones, 1941; Hovenier, 1994). This matrix is widespread in the literature (Tang and Kwok, 2001; Swami et al., 2006) because it describes linear crystal optics without absorption. This approach is termed incomplete or partial Mueller polarimetry (Savenkov, 2007; Tyo et al., 2010).

The exact sets of matrix elements, namely structures of incomplete Mueller matrices that can be measured in the framework of any of measurement strategies (time-sequential, dynamic, etc.), are also determined by structure of the data-reduction matrix of generalized measurement equation Eq. (3.51). This corresponds to the third case in Eq. (3.51) occurring when $N < 17$ and $\mathbf{W}$ is of rank less than 16. The optimal estimation of $\vec{\mathbf{M}}$ is again obtained using the pseudoinverse matrix $\tilde{\mathbf{W}}$. However, only 15 or fewer Mueller matrix elements can be determined from the system Eq. (3.54) – that is, polarimetry is 'incomplete' or 'partial.'

3.4 Radar polarimetry

In this section, we discuss the polarimetric metrics and target decompositions used for data interpretation in radar polarimetry. In so doing, we restricted ourselves only by those which are used in the sections to follow. The comprehensive review of this subject can be found, for example, in an excellent book by Lee and Pottier (2009).

Radar polarimetry is a special class of equipment for carrying out polarization measurements at micro- and radiowave ranges. Monostatic radars use the same antenna to transmit and receive electromagnetic radiation and are limited to measurements at the exact backscattering direction. Bistatic radars use one or more additional receiving antennas which provide supplementary polarization information.

Radar polarimetry is active sensors operating in the microwave electromagnetic (EM) range, which traditionally is denoted by the letters shown in Table 3.1.

Different frequencies are characterized by different 'penetration capabilities' as schematically shown in Fig. 3.2.

Radar polarimetry has different polarization configurations. The most widely used are the linear polarizations indicated as HH, VV, HV, and VH, where the first term refers to the polarization of input (emitted) and the latter to the received radiations, respectively. In this sense, the radar polarimetry can have different polarization levels:

- single polarization – HH or VV or HV or VH;
- dual polarizations – HH and HV, VV and VH, or HH and VV;
- four (quad) polarizations – HH, VV, HV, and VH.

Table 3.1. Radar polarimetry bands and frequencies (after Lasaponara and Masini, 2013).

Name	Nominal frequency range	Wavelength range	Specific bands used
VHF	30–300 MHz	10–1 m	138–144 MHz, 216–225 MHz
P (UHF)	300–1000 MHz	100–30 cm	420–450 MHz, 890–942 MHz
L	1–2 GHz	30–15 cm	1.215–1.4 GHz
S	2–4 GHz	17–7.5 cm	2.3–2.5 GHz, 2.7–3.7 GHz
C	4–8 GHz	7.5–3.75 cm	5.25–5.925 GHz
X	8–12 GHz	3.75–2.5 cm	8.5–10–68 GHz
Ku	12–18 GHz	2.5–1.67 cm	13.4–14.0 GHz, 15.7–17.7 GHz
K	18–27 GHz	1.67–1.11 cm	24.05–24.25 GHz
Ka	27–40 GHz	1.11–0.75 cm	33.4–36.0 GHz
V	40–75 GHz	0.75–0.40 cm	59–64 GHz
W	75–110 GHz	0.40–0.27 cm	76–81 GHz, 92–100 GHz
Millimetre	110–300 GHz	2.7–1.0mm	

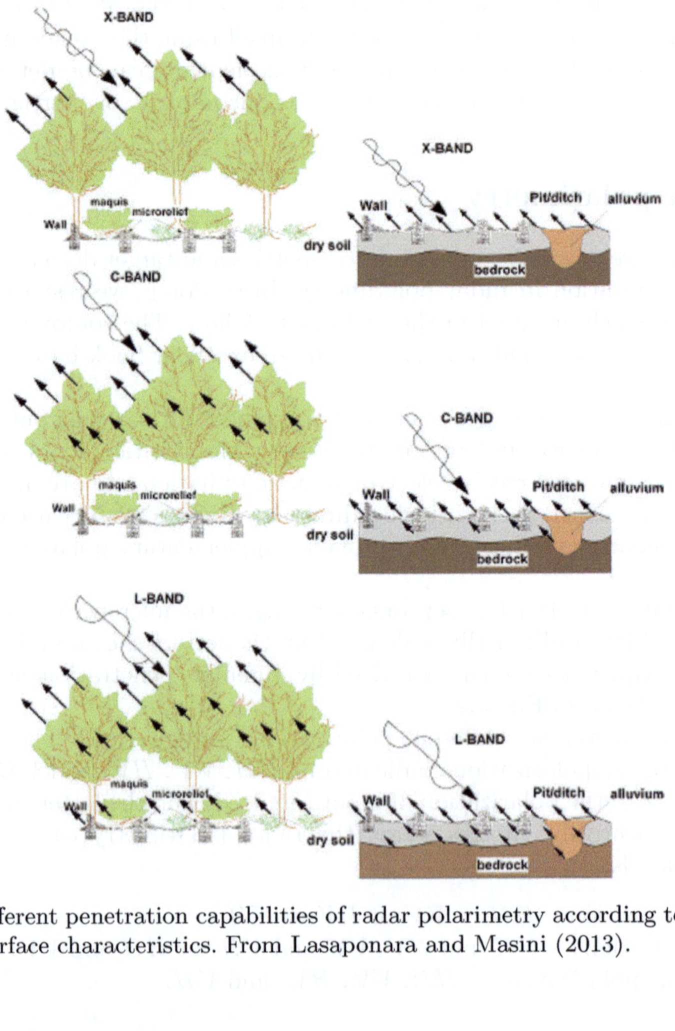

Fig. 3.2. Different penetration capabilities of radar polarimetry according to bands, land cover, and surface characteristics. From Lasaponara and Masini (2013).

The modes HV or VH are termed cross-polarization, whereas HH and VV modes are denoted as standard polarization. Quad-polarization polarimetry provides the four polarizations HH, VV, HV, and VH, and also measures the difference in the magnitudes and phase between channels. Fully polarimetric sensors provide data that can be created using all possible combinations of transmitting and receiving polarizations, not the standard HH and VV only (Evans et al., 1988; Boerner et al., 1998).

For radar polarimetry, the Stokes vector is not the most effective way to characterize the data, since there are effectively two measurements of polarization to quantify – one for each of the orthogonal transmitted waves. The radar transmits a horizontal polarized wave, measures the echo polarization, transmits a vertical polarized wave, and measures the polarization of that echo. At least two Stokes vectors would then be required (Woodhouse, 2006). Since the polarimetric measurements of the echoes are made as orthogonal measurements, it is convenient to define an alternative notion of the Jones matrix (Eq. (3.6)) named in radar polarimetry as scattering matrix (Mott, 2007; Cloude, 2010)

$$\mathbf{S} = \begin{pmatrix} s_{VV} & s_{VH} \\ s_{HV} & s_{HH} \end{pmatrix}, \tag{3.55}$$

where s_{HV} denotes a transmitting antenna of a horizontal polarization and a receiving antenna of a vertical polarization. $\mathbf{S}$ becomes symmetric if the target is reciprocal, $s_{HV} = s_{VH}$.

In Eqs (3.6) and (3.55), there are two major conventions currently used for coordinate systems: the forward-scatter alignment (FSA) convention and the backscattered alignment (BSA) convention (van Zyl and Zebker, 1990; Guissard, 1994; Boerner et al., 1998). If te BSA convention is used, then the scattering matrix defined in Eq. (3.18) relates the scattered wave viewed approaching the receiving antenna to the incident wave viewed receding from the transmitting antenna (van Zyl et al., 1987). In optical or transmission polarimetry, the FSA convention is used.

The $\mathbf{S}$ matrix, which is expressed in the BSA coordinates, is referred to as the Sinclair matrix (Sinclair, 1950; Kennaugh, 1952; Zebker and van Zyl, 1991; Guissard, 1994; Boerner et al., 1998).

When we say in Eq. (3.1) that the input and output Stokes vectors are connected through the Mueller matrix, it assumes utilization of the FSA convention. Using the BSA convention, the Stokes vector of the backscattered wave is related to the incident-wave Stokes vector through the Kennaugh matrix, $\mathbf{K}$ (Kennaugh, 1951; van de Hulst, 1957; van Zyl et al., 1987; Boerner et al., 1998). Thus, the Mueller matrix $\mathbf{M}$ and the Kennaugh matrix $\mathbf{K}$ are formally related by (Guissard, 1994; Lüneburg, 1995; Boerner et al., 1998)

$$\mathbf{M} = \mathrm{diag}\begin{bmatrix} 1 & 1 & 1 & -1 \end{bmatrix} \cdot \mathbf{K}. \tag{3.56}$$

For symmetrical targets, the matrices $\mathbf{S}$, $\mathbf{K}$, and $\mathbf{M}$ are characterized by only five parameters.

In addition to $\mathbf{M}$ and $\mathbf{K}$, two matrices named the target covariance matrix and the target coherence matrix might be used for characterization of partially polarized waves.

The scattering matrix of Eq. (3.55) can be represented in the vector form as

$$\vec{\mathbf{S}} = \begin{bmatrix} s_{HH} & s_{HV} & s_{VH} & s_{VV} \end{bmatrix}^T . \tag{3.57}$$

An ensemble average of the complex product between $\vec{\mathbf{S}}$ and $\vec{\mathbf{S}}^{*T}$ leads to the so-called covariance matrix $\mathbf{C}$ (Borgeaud et al., 1987; van Zyl and Ulaby, 1990):

$$\mathbf{C} = \langle \vec{\mathbf{S}} \cdot \vec{\mathbf{S}}^{*T} \rangle . \tag{3.58}$$

The Hermitian positive semi-definite matrix $\mathbf{C}$ has precisely the same elements as the Kennaugh matrix $\mathbf{K}$ and the Mueller matrix $\mathbf{M}$ but with different arrangements. The full established properties of Hermitian matrices make convenient the use of $\mathbf{C}$ in some applications (Borgeaud et al., 1987; Kong et al., 1987; Novak and Burl, 1990; van Zyl, 1992; Touzi and Lopes, 1994).

The coherency matrix Eq. (3.47) introduced by Cloude (1986) can per analogy with Eq. (3.58) be written as

$$\mathbf{J} = \langle \vec{\mathbf{k}} \cdot \vec{\mathbf{k}}^{*T} \rangle , \tag{3.59}$$

where $\vec{\mathbf{k}}$ is the target scattering vector (Cloude, 1986) and is given by

$$\vec{\mathbf{k}} = \begin{bmatrix} s_{HH} + s_{VV} & s_{HH} - s_{VV} & s_{HV} + s_{HV} & i(s_{HV} - s_{VH}) \end{bmatrix}^T . \tag{3.60}$$

Note that $\mathbf{C}$ and $\mathbf{J}$ are unitarily similar. The two matrices carry the same information; both are Hermitian positive semi-definite and both have the same eigenvalues but different eigenvectors (Cloude and Pottier, 1996).

Once reciprocity, $s_{HV} = s_{VH}$, has been assumed, the elements of $\mathbf{S}$ may be stacked into a three-element vector $\vec{\mathbf{k}} = \begin{bmatrix} s_{VV} & s_{HV} & s_{HH} \end{bmatrix}^T$. The linear basis is not always the most efficient way of dealing with the analysis of polarimetric data, and the Pauli basis of the target vector

$$\vec{\mathbf{k}}_P = \begin{bmatrix} s_{HH} - s_{VV} & 2s_{HV} & s_{HH} + s_{VV} \end{bmatrix}^T \tag{3.61}$$

is for many applications more useful, as it helps to emphasize the phase difference between the HH and VV terms. Double interactions are dominated by the first term, multiple (volume) scattering dominates the second term, and direct scattering is dominated by the second term.

Equation (3.59) gives in these cases the 3×3 coherency matrix and the corresponding three-component decomposition of the target Mueller matrix. Using normalized eigenvalues of coherency matrix $p_i = \lambda_i / \sum_{i=1}^{3} \lambda_i$ (Cloude et al., 2001), one more important polarimetric metric, the scattering anisotropy A, can be introduced:

$$A = (p_2 - p_3)/(p_2 + p_3) . \tag{3.62}$$

Scattering anisotropy A varies between zero and one, defining the relation between the second and the third eigenvalues, namely the difference of the secondary scattering mechanisms. entropy for smooth surfaces becomes zero, implying a non-depolarizing scattering process described by a single-scattering matrix, and increases with surface roughness. Depolarizing surfaces are characterized by non-zero entropy values. However, A can be zero even for rough surfaces. For surfaces characterized by intermediate entropy values, a high scattering anisotropy A indicates the presence of only one strong secondary scattering process, while a low anisotropy indicates the appearance of two equally strong scattering processes. For azimuthally symmetric surfaces, $p_2 = p_3$ and A becomes zero (Cloude and Pottier, 1996). In this sense, the anisotropy provides complementary information to the entropy and facilitates the interpretation of the surface scatterer.

Finally, eigenvalue–eigenvector decomposition of the coherency matrix in Eq. (3.59) provides one more metric: the alpha angle α, representing the dominant scattering mechanism and calculated from the eigenvectors and eigenvalues of $\mathbf{J}$:

$$\alpha = \sum_{i=1}^{3} (\alpha_i p_i)\,, \tag{3.63}$$

where α_i are the scattering mechanisms represented by the three eigenvectors: $\alpha = 0$ corresponds to a surface scattering, $\alpha = 45°$ to a volume scattering, and $\alpha = 90°$.

Another set of polarimetric metrics to characterize the scattering scene can be deduced from Pauly decomposition of the scattering matrix in Eq. (3.55) (Lee and Pottier, 2009). In accordance with this decomposition, the scattering matrix $\mathbf{S}$ is represented as the complex sum of the Pauli matrices. The elementary scattering mechanisms are associated with corresponding basis matrices:

$$\mathbf{S} = \frac{a}{\sqrt{2}}\begin{pmatrix} 0 & 0 \\ 0 & 1 \end{pmatrix} + \frac{b}{\sqrt{2}}\begin{pmatrix} 0 & 0 \\ 0 & -1 \end{pmatrix} + \frac{c}{\sqrt{2}}\begin{pmatrix} 0 & 1 \\ 1 & 0 \end{pmatrix} + \frac{d}{\sqrt{2}}\begin{pmatrix} 0 & -i \\ i & 0 \end{pmatrix}, \tag{3.64}$$

where a, b, c, and d are complex and given by

$$\begin{aligned} a &= (s_{HH} + s_{VV})/\sqrt{2}, \quad b = (s_{HH} - s_{VV})/\sqrt{2} \\ c &= (s_{HV} + s_{VH})/\sqrt{2}, \quad d = i(s_{HV} - s_{VH})/\sqrt{2}. \end{aligned} \tag{3.65}$$

In the monostatic case, where $s_{HV} = s_{VH}$, the Pauli matrix basis can be reduced to the first three matrices, yielding $d = 0$.

The Pauli decomposition determines following three scattering mechanisms characterizing the target under consideration: $|a|^2$ determines the power scattered by the targets characterized by single- or odd-bounce; $|b|^2$ determines the power characterized by double- or even-bounce; $|c|^2$ determines the power characterized by a deplane oriented at 45°. The scattering mechanism represented by $|c|^2$ refers to backscattering with orthogonal polarization. One of the character examples of this is the volume scattering produced by the forest canopy.

The scattering patterns are described by four independent variables: the ellipticity and orientation of the incident wave, and the ellipticity and orientation of the backscattered wave. However, using all four of these variables would result in a response that would be too cumbersome and complicated to interpret. Therefore, for clearness, only two variables are used at a time to interpret the scattering patterns – the ellipticity and orientation angle of the incident wave. This yields two signatures – the co-polarization and cross-polarization signatures introduced in van Zyl et al. (1987), Agrawal and Boerner (1989), and McNairn et al. (2002)

(see Fig. 3.3). In the co-polarization case, the polarization of the scattered wave is the same as the polarization of the incident wave, while, in the cross-polarization case, the polarization of the scattered wave is orthogonal to the polarization of the incident wave.

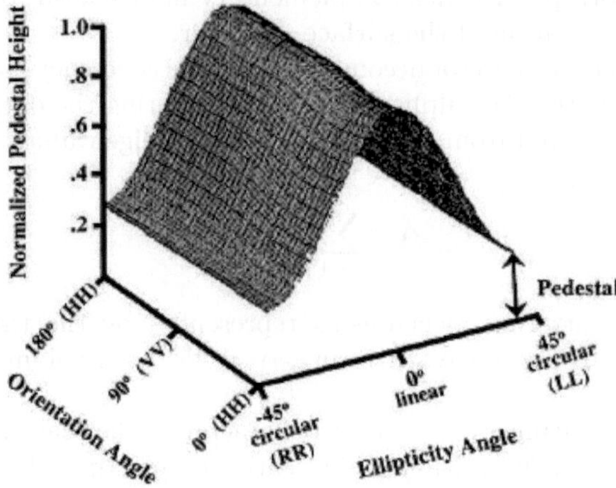

Fig. 3.3. Example of the polarization signature. From McNairn et al. (2002).

The height of the pedestal (see Fig. 3.3) is an indicator of the presence of an unpolarized scattering component, and thus the degree of polarization of a scattered wave (van Zyl et al., 1987). Signatures with significant pedestals are typical of targets that are dominated by volume scattering or multiple surface scattering. Evans et al. (1988) reported that pedestal height was directly proportional to vegetation density. Ray et al. (1992) and van Zyl (1989) noted that pedestal height was related to surface roughness, with increases in roughness resulting in higher pedestals.

Although not unique, a polarization signature capturing many scattering characteristics of the target, at all polarizations, is informative and can indicate the scattering mechanisms dominating the target response. Since different scattering mechanisms give different polarization signatures, they could be extracted from the measured Mueller matrix. The 'building blocks' of such interpretation of the measured Mueller matrix are the following scattering mechanisms: (i) double-bounce scattering; (ii) Bragg scattering; (iii) single (odd) bounce scattering; (iv) cross-scattering (see, e.g., Dong et al., 1996). Since measurements for independent scattering mechanisms can be added incoherently (van de Hulst, 1957; Kim et al., 1987), the total Mueller matrix can be represented as a sum of the above basic scattering mechanisms. A number of additive decompositions of the scattering and Mueller matrix can be found elsewhere (Cloude and Pottier, 1996, 1997; Touzi et al., 2004).

Double-bounce scattering models are typically the scattering from the dihedral-corner-reflector-like structures such as the trunk-ground structure in forested areas and the wall-ground structures in urban areas. It has been shown (Dong and

Richards, 1995b) that the scattering matrix for this mechanism can be written as

$$\mathbf{S}_1 = \begin{pmatrix} 1 & 0 \\ 0 & (1/\sqrt{\alpha})\exp(i\delta) \end{pmatrix}, \tag{3.66}$$

where α and δ are referred to as the polarization index, defined as the ratio of *HH* to *VV* polarization response, and polarization phase difference, defined as the phase difference between *HH* and *VV* backscattered responses. For the trunk-ground structure, $\alpha \approx 4-6$ and $\delta \approx 140-160°$ (Dong and Richards, 1995a). The Mueller matrix for the double-bounce scattering has the form

$$\mathbf{M}_1 = \begin{pmatrix} (\alpha+1)/2\alpha & (\alpha-1)/2\alpha & 0 & 0 \\ (\alpha-1)/2\alpha & (\alpha+1)/2\alpha & 0 & 0 \\ 0 & 0 & (1/\sqrt{\alpha})\cos\delta & -(1/\sqrt{\alpha})\sin\delta \\ 0 & 0 & -(1/\sqrt{\alpha})\sin\delta & -(1/\sqrt{\alpha})\cos\delta \end{pmatrix}. \tag{3.67}$$

Bragg scattering models the scattering from slightly rough surfaces, such as sea surface (Valenzuela, 1967; Elachi, 1987). The Mueller matrix of the Bragg scattering is

$$\mathbf{M}_2 = \begin{pmatrix} (\alpha+1)/2\alpha & (\alpha-1)/2\alpha & 0 & 0 \\ (\alpha-1)/2\alpha & (\alpha+1)/2\alpha & 0 & 0 \\ 0 & 0 & (1/\sqrt{\alpha}) & 0 \\ 0 & 0 & 0 & -(1/\sqrt{\alpha}) \end{pmatrix}, \tag{3.68}$$

where the mean polarization index value $\alpha < 1$, the mean polarization phase difference value δ is zero, and the Bragg scattering undergoes a single bounce.

The single-bounce scattering models are direct specular reflections from the ground surfaces or from building roofs perpendicular to incident waves, etc. The Mueller matrix for this mechanism is

$$\mathbf{M}_3 = \begin{pmatrix} 1 & 0 & 0 & 0 \\ 0 & 1 & 0 & 0 \\ 0 & 0 & 1 & 0 \\ 0 & 0 & 0 & -1 \end{pmatrix}. \tag{3.69}$$

The co-polarized response from forest crown volume backscattering can be included in this mechanism. If the orientations of leaves, twigs, and small branches are assumed to be uniformly distributed, the backscattering response will be independent of polarization, giving the same *HH* and *VV* responses. The backscattering from the trihedral-corner-reflector-like, wall–wall–ground structures can also be classified into this mechanism, since the scattering undergoes odd bounces (Dong et al., 1996).

The polarimetric response of a point or distributed target can generally consist of the co- and cross-polarized responses. If the total cross-polarized component is of interest, it can be assumed theoretically that the total cross-polarized response is caused by a hypothetical cross-scattering mechanism whose scattering matrix is

$$\mathbf{S}_4 = \begin{pmatrix} 0 & 1 \\ 1 & 0 \end{pmatrix}, \tag{3.70}$$

and the corresponding Mueller matrix is

$$\mathbf{M}_4 = \begin{pmatrix} 1 & 0 & 0 & 0 \\ 0 & -1 & 0 & 0 \\ 0 & 0 & 1 & 0 \\ 0 & 0 & 0 & 1 \end{pmatrix} . \tag{3.71}$$

3.5 Applications of optical and radar polarimetry

Optical and radar polarimetry have recently demonstrated their unique abilities in identification, characterization, and classification of anisotropic media and point or distributed targets of various natures. Some of their applications will be discussed in this section. However, a comprehensive survey would here be impossible through, on the one hand, naturally restricted space of the chapter and, on the other, all applications considered below actually deserve to date the individual books to be discussed systematically. Thus, we select a few representative examples taken from different terrestrial problems. Additionally, in all cases below, we concentrate on polarimetric results literally. For more calculation, technical details, etc., one should refer to the corresponding bibliography.

3.5.1 Vegetation

Complete radar polarimetry has evidently provided inherent advantages over optical polarimetry in vegetation monitoring due to its ability to penetrate the vegetation cover and obtain information of the underlying soil surface. Several experimental and theoretical studies have been carried out in the last few decades to investigate the sensitivity of microwave polarization sensors to the vegetation and soil parameters. In this subsection, we concentrate on results of radar polarimetry of different types of vegetation and trees. Some recent results obtained in optical range, particularly Mueller polarimetry of the different types of leaves at visible levels, are discussed as well.

Imhoff et al. (1986) and Hess et al. (1990) demonstrated that, when the incident wave propagates through the entire canopy and reaches an underlying highly reflecting surface, mangrove forests and flooded vegetation exhibit enhancement of the backscattering coefficient for the co-polarizations *HH* and *VV* by a factor of up to 10 dB, depending on the magnitude of canopy attenuation and thus on radar parameters and vegetation type. This originates from double-bounce returns or multiple scattering between the water surface and forest components (Engheta and Elachi, 1982; Richards et al., 1987; Wang and Imhoff, 1993; Wang and al., 1995). At the same time, the cross-polarization *HV* originating from multiple scattering within the canopy layer does not exhibit any particular enhancement.

In Mougin et al. (1999), experimental results obtained over mangrove forests of French Guiana with the NASA/JPL AIRSAR have been presented. The further challenge is to provide a physically based interpretation of observed polarimetric radar signatures of mangrove forests (Proisy et al., 2000). To this end, it seems reasonable to use the polarimetric scattering model of Karam et al. (1995) to simulate the response of mangrove canopies aimed at identifying the dominant scattering mechanisms in the radar–forest interactions. In compliance with this model, the forest is treated as a multilayer medium over a rough surface. The layers represent the canopy volume, and the rough surface delineates the soil interface. Each layer contains the tree constituents, namely trunks, branches, and leaves. The branches and trunks are modeled as randomly oriented finite cylinders and the leaves are modeled as randomly oriented elliptic discs. All the scatterers are assumed to be uniformly oriented in the azimuth direction.

The study site, named Crique Fouillee (52°19′W, 4°52′N), is a 2.5×32.5 km area allocated along the coast of French Guiana. On this test site, three development stages of mangrove forests are present: pioneer, mature, and declining stages. The pioneer stage consists of a very homogeneous canopy dominated by the gray mangrove (*Laguncularia racemosa*). Tree density is high, ranging from about 10,000 to 40,000 stems per hectare. Mean tree height lies between 0.8 and 7.7 m. The mature stage is dominated by the white mangrove (*Avicennia germinans*), with a tree density ranging between 500 and 2,000 stems per hectare. Mean tree height is about 15 m, reaching a maximum of 25 m for the dominant species. The declining stage shows more heterogeneous canopies, including two strata: a high single-species stratum composed of the white mangrove and a lower stratum of the red mangrove (*Rhizophora* ssp.). Tree density is low, from about 300 to 600 stems per hectare. Overall, the considered forest stands consist of closed canopies. The topography of the study site is nearly flat. (More details about the study site can be found in Fromard et al. (1998), Mougin et al. (1999), and Proisy et al. (2000).)

From the delivered data, the average Mueller matrix is formulated, and then used to determine the following quantities for each frequency: the backscattering coefficients σ_{VV}, σ_{HH}, and σ_{HV} (for the *HH*, *VV*, and *HV* polarizations, respectively), the co-polarized σ_{VV}/σ_{HH} ratio, and the two cross-polarized σ_{HH}/σ_{HV} and σ_{VV}/σ_{HV} ratios.

Preliminary experimental observations on the relations between mangrove parameters and backscattering coefficients were reported in Mougin et al. (1999). Results show that, for all frequencies, there is a positive relation between σ_{ii} and total biomass, with the largest sensitivity to biomass found at *P*-band (0.44 GHz) *HV* and *L*-band (1.225 GHz) *HV*. Moreover, strong differences are observed between polarizations at *L*-band and *P*-band below a biomass value of about 100–150 tDM/ha (tons of dry matter per hectare). Above this threshold, co- and cross-polarization ratios reach small and constant values.

The mangrove stands are modeled as multilayer media above a rough semi-infinite interface. The young stands are modeled with one or two layers, whereas three layers are necessary for the oldest stands (Fig. 3.4).

Figure 3.5a shows the comparison between simulated backscattering coefficients versus total biomass and observed backscatter at *C*-, *L*-, and *P*-band. The three main contributions to the total response also are plotted, namely the soil-scattering component, the volume-scattering component, and the double-bounce scattering component (see Fig. 3.5b). The soil term is the surface response attenuated by the canopy layer. The volume term consists of the scattering from the tree components up to the second order. The double-bounce term corresponds to the interaction between the tree components and the ground.

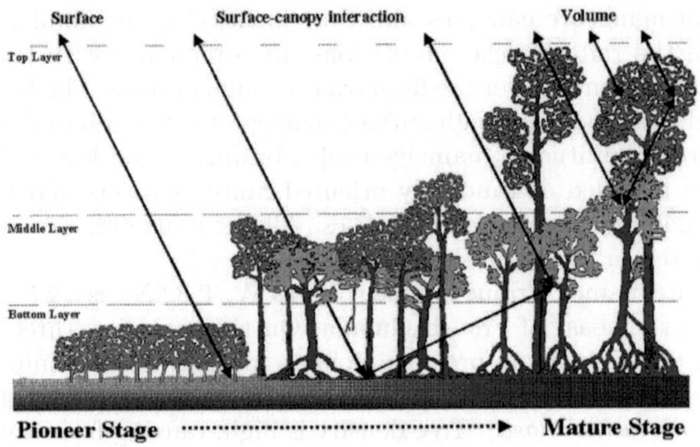

Fig. 3.4. Geometry of the mangrove forest model. From Proisy et al. (2000).

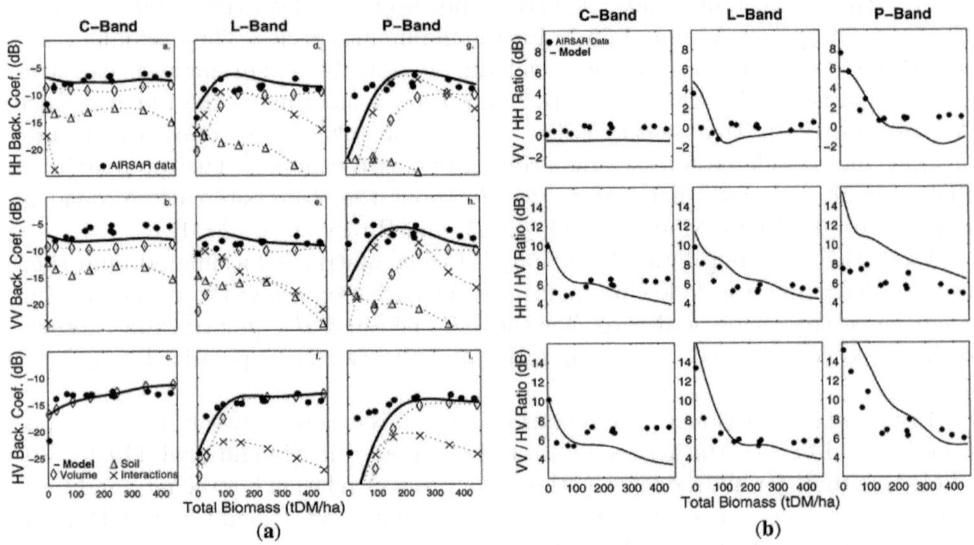

Fig. 3.5. Simulated (–) and experimental data (•) versus total biomass at C-, L-, and P-band: (a) backscattering coefficients and (b) polarization ratios. From Proisy et al. (2000).

On the whole, P-band provides the most pronounced polarimetric patterns. Among the polarimetric parameters, the polarization ratio is found to be most informative and useful for analyzing scattering mechanisms for discriminating between various forest stages.

Validation of the above-mentioned three-layer canopy model is carried out for another type of canopy – coniferous trees (Karam et al., 1995). This is done in two steps. First, computed backscattering coefficients are compared to experimental data. The computed backscattering coefficient is broken down into individual contributions from different scattering mechanisms in order to develop an understanding of the relation between the backscattering coefficients and different forest biophysical parameters. The microwave data were retrieved from multipolarization, multifrequency synthetic aperture radar (SAR) images acquired over a coniferous forest in Les Landes Forest, in south-western France, near Bordeaux, during the MAESTRO-I campaign. The data were acquired with the NASA 1 JPL AIRSAR system, which operates at P-band (0.44 GHz), L-band (1.225 GHz), and C-band (5.3 GHz) (Held et al., 1988).

This test site is the largest pine plantation forest in Europe. The forest is almost totally formed of maritime pine (*Pinus pinaster*). It is managed in a consistent fashion in order to ensure that the canopy remains homogeneous. The test site under consideration consists of an area 7×10 km, comprising quasi-uniform large stands with a mean area of about 25 hectares. The stands are rectangular in shape and delimited by fire protection tracks. Most of the stands are artificially sown, generally in rows of 4 m spacing. The rows follow an east–west direction on the test site. The test area includes many clear cuts and a range of classes varying from seedlings to stands over 46 years old. Consequently, these stands provide a wide range of variability of the relevant forest parameters to the model, such as density, diameter at breast height (dbh), tree height, and standing biomass. In particular, standing biomass ranges from 0 to 150 tons of dry matter per hectare.

As an example, Fig. 3.6a shows the prediction of the three-layer model against experimental data for P-bands. Each figure shows VV, VH, and HH polarizations. Figure 3.6b shows the individual contributions from those scatterers, which, according to the model, most affect the total backscattering coefficient. The contributions shown are limited to those from bare soil; single-bounce scattering from the needles, branches, and trunks; and double-bounce scattering involving the needles, branches, trunks, and the ground.

In Fig. 3.6a, more dynamic range is obtained in the case of cross-polarization than co-polarization. Figure 3.6b shows that, for younger trees, both the single-bounce scattering from branches and the double-bounce scattering of the trunk–ground interaction make significant contributions. As the trees grow older, the trunk–ground contribution becomes less significant, and the contribution from the branch dominates. This dominance is more obvious in the case of VV polarization than HH. On the other hand, the VH backscattering coefficient is due almost solely to the long branches, for all ages. Therefore, for P-band and VH polarization, the dynamic range observed reflects the physiological changes in the long branches as the canopy age increases.

An empirical relation between biomass and age reflecting the fact that the growth of trees tends to slow down as the canopy becomes older has been suggested in the form

$$Biomass = \sqrt{463.85\ age - 2938}\,. \tag{3.72}$$

The relation in Eq. (3.72) was used to obtain the total aboveground biomass for the ages (excluding the clear cuts) for which ground truth was available (see Fig. 3.7 for biomass as a function of tree age).

The Lear Area Index (LAI) is a dimensionless variable that measures the amount of leaf area per area unit ground surface, and may be used to describe the health of the forest after an insect attack, for example (e.g., Salberg et al., 2009; Solberg et al., 2009). The use of polarimetric radar data has shown great potential for obtaining precise estimates of the LAI of boreal forests (Manninen et

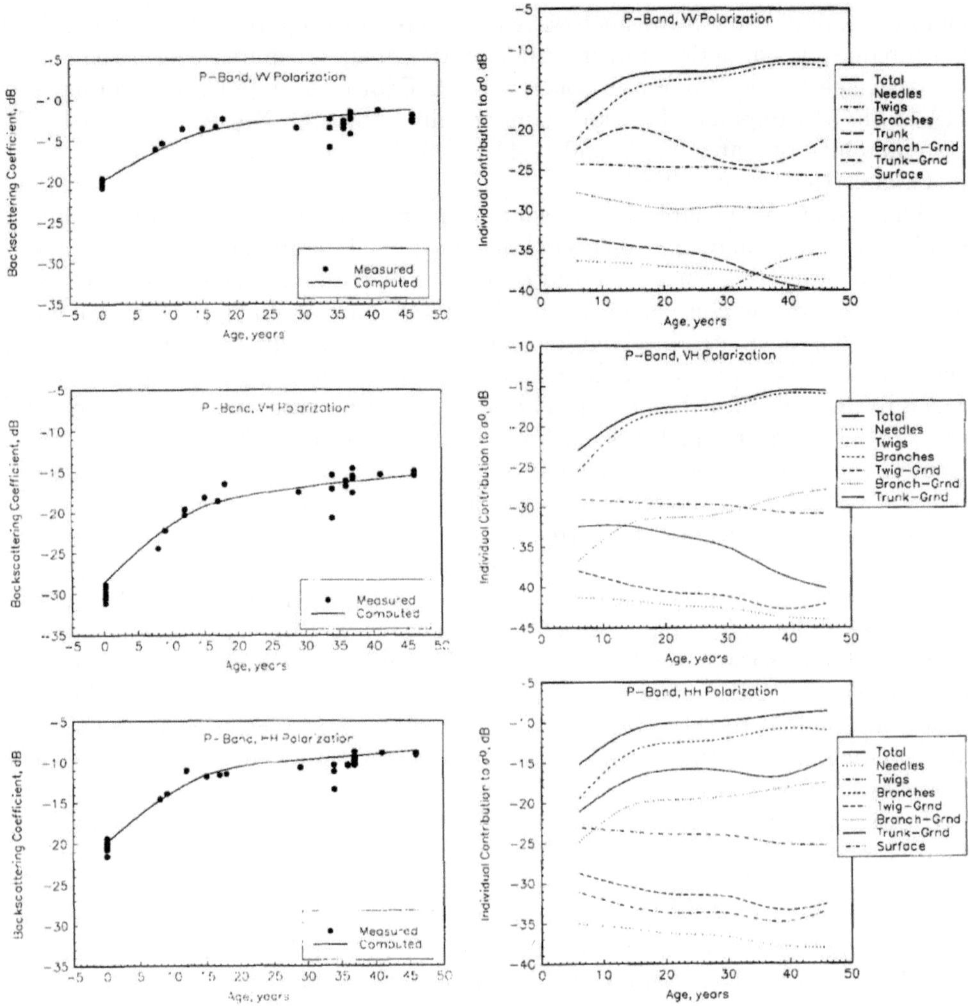

(a) **(b)**

Fig. 3.6. (a) Comparison between computed and measured values of the backscattering coefficients at P-band as a function of the tree age; (b) backscattering coefficients at P-band as a function of the tree age. The contributions are broken down into single-bounce and double-bounce scattering. From Karam et al. (1995).

al., 2005), rice fields (Durden et al., 1995), and corn, sorgum, and wheat (Ulaby et al., 1984).

There exist a few models of the backscatter as a function of LAI for forest vegetation. These models are in general very complex, and an inverse function (i.e., a function describing LAI as a function of radar backscatter) is more or less impossible to obtain (Freeman and Durden, 1998). Manninen et al. (2005) showed that LAI of Norway spruce and Scots pine were well explained by the backscattering ratio VV/HH. In their study, they obtain for single measurement

$$LAI = a_0 + a_1(\sigma_{VV}/\sigma_{HH}) \tag{3.73}$$

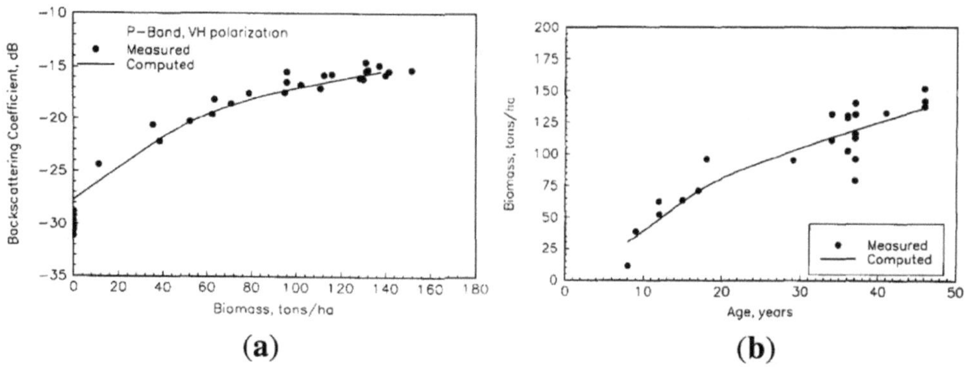

Fig. 3.7. (a) Comparison between computed and measured backscattering coefficient as a function of total tree biomass showing VH polarization and (b) biomass as a function of tree age both at P-band. From Karam et al. (1995).

and for multiple measurements

$$LAI = a + b_1(\sigma_{VV1}/\sigma_{HH1}) + \cdots + b_k(\sigma_{VVK}/\sigma_{HHK}), \tag{3.74}$$

where K is the measurement number.

In Salberg et al. (2009), a simple quadratic model to predict the LAI from the backscatter measurements has been used:

$$LAI = a_0 + a_1\sigma_{HH} + a_2\sigma_{HH}^2 + a_3\sigma_{XX} + a_4\sigma_{XX}^2 + a_5\sigma_{VV} + a_6\sigma_{VV}^2, \tag{3.75}$$

where σ_{HH}, σ_{VV}, and σ_{XX} denote the backscatter with horizontal, vertical, and cross-polarization, respectively.

The area of investigation is situated at Palokangas east in Finland, about 62°52′23′N, 30°54′10′E. The image of the test site acquired on August 28, 2007, is shown in Fig. 3.8.

Figure 3.9 shows the estimates of the LAI from backscatter HH and VV (Fig. 3.8). Here, the cross-polarization terms are not included. The main elements within a forest canopy are individual trees, leaves, branches, and trunks. All of these elements act as scatters and may strongly influence the backscatter patterns. The magnitude of each of these backscatter components depends on wavelength, polarization, angle of incidence, and many more terrain and canopy parameters. Therefore, scatterings by these elements are individual problems, which are important for studying the scattering and polarization properties of these elements by themselves and play a crucial role in understanding canopy scattering as a whole. Further, we consider the scattering of microwaves and optic radiation by different trees and leaves.

In Narayanan et al. (1988), Borel and McIntosh (1990), Mead and McIntosh (1990, 1991), and Chang et al. (1992), measurements of the Mueller matrices at 95 GHz (3.07 mm) and 225 GHz (1.3 mm) of four individual trees – eastern cottonwood and weeping willow which have vertically oriented leaves (erectophil), silver maple which has horizontally oriented leaves (planophil), and white pine, a conifer having needles which are less than 0.5λ in diameter – have been carried out.

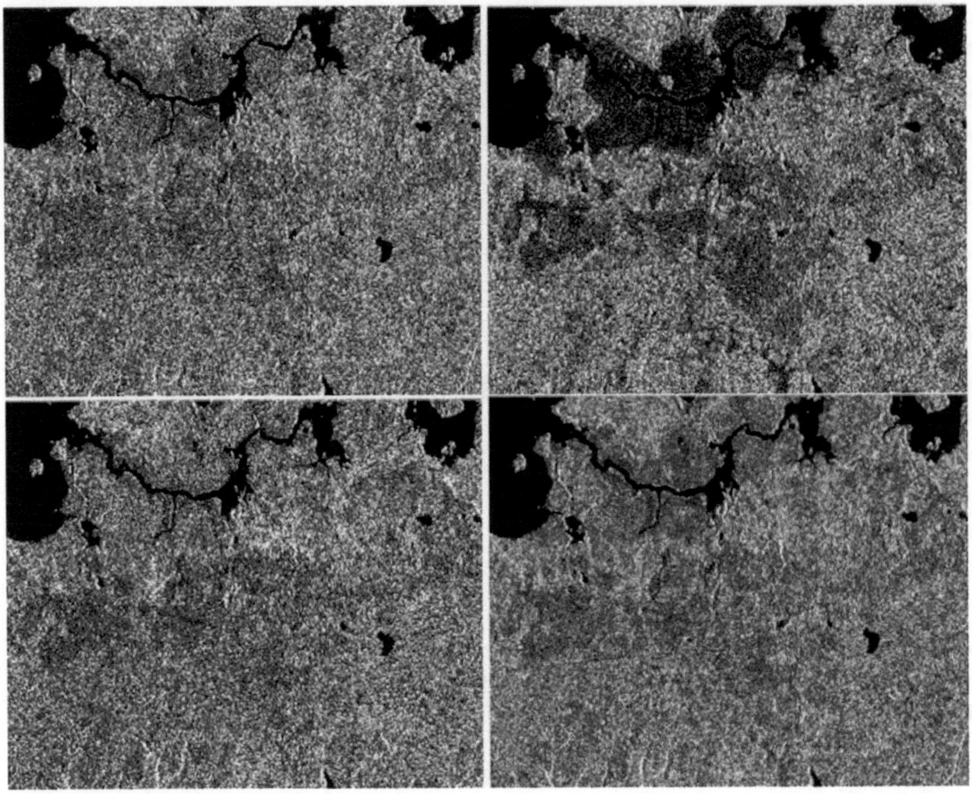

Fig. 3.8. Images of the scene with 20×20 m pixel resolution. Upper/left: *HH*-polarized. Upper/right: *HV*-polarized. Lower/left: *VV*-polarized. Lower-right: Linear decomposition RGB = (*HV*, *HH*, *VV*). From Salberg et al. (2009).

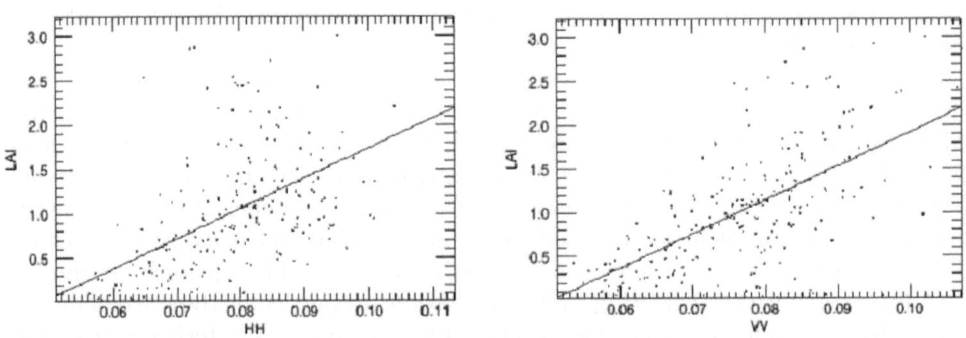

Fig. 3.9. Lear Area Index (LAI) versus backscatter *HH* (right) and *VV* (left) (August 28, 2008). From Salberg et al. (2009).

Measured co-polarized and cross-polarized signatures for all four trees are shown in Fig. 3.10a–h. Note that the signatures at $\tau = \pm 45°$ are non-zero as a result of unpolarized energy in the scattered wave. The greater unpolarized energy in the white pine's response is illustrated in Fig. 3.11d, where, for all four trees, the power-

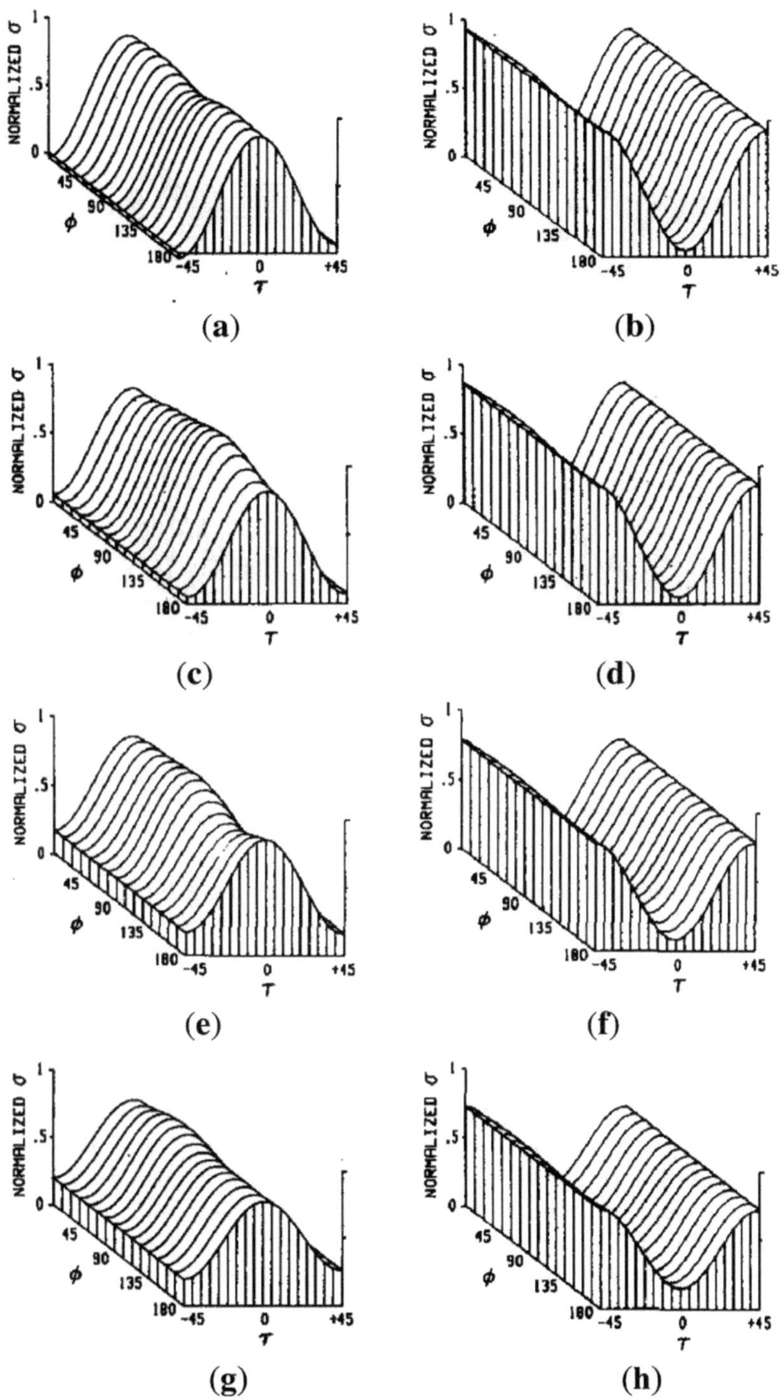

Fig. 3.10. Co-polarized and cross-polarized signatures: (a) and (b) eastern cottonwood; (c) and (d) weeping willow; (e) and (f) silver maple; and (g) and (h) white pine trees, respectively. From Mead and MacIntosh (1991).

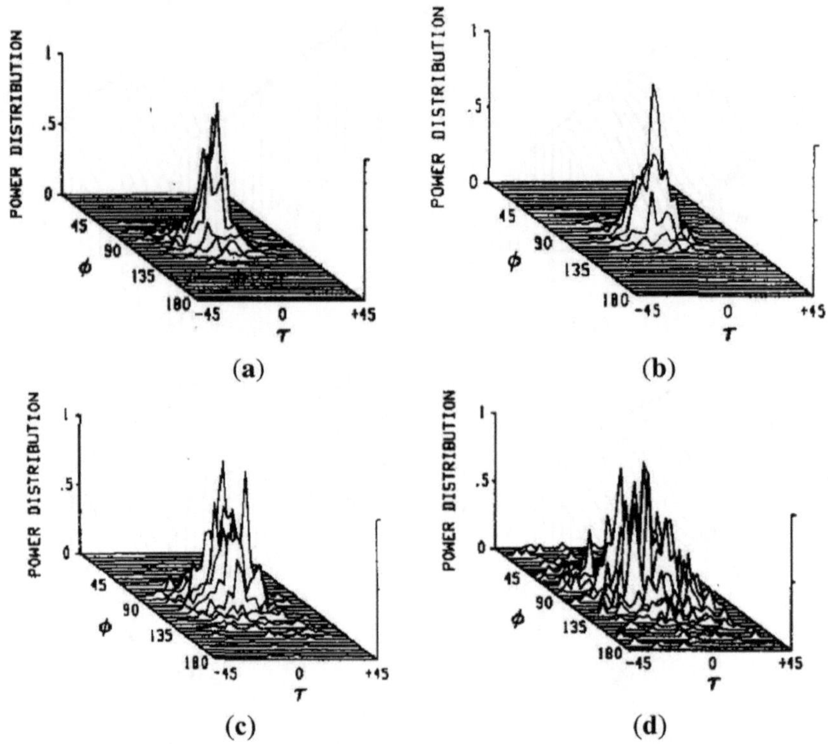

Fig. 3.11. Scatter plots of received polarization states: (a) eastern cottonwood; (b) weeping willow; (c) silver maple; and (d) white pine. From Mead and MacIntosh (1991).

weighted histograms of the receive polarization state as a function of azimuth φ and ellipticity τ for vertical transmit polarization are presented.

As can be seen, at 225 GHz, the two erectophil trees (cottonwood and willow) are the most highly polarized, the white pine (conifer) is the least polarized, and the silver maple (planophil) falls roughly in between.

These results are of interest to compare with complete polarimetric data presented in Dong et al. (1996) for eucalypts at *P*-, *L*-, and *C*-bands measured in Gippsland, Australia (see Fig. 3.12).

It can be seen that pedestals are high for all bands. The authors explained this by the significance of the responses of *RR* (right-handed transmission and right-handed receiving) and *LL* (left-handed transmission and receiving) circular polarizations. The *VH* component contributes also to the pedestals. Eucalypts in the studied site are quite sparse, with less than 25% closure of their crowns. This can manifest in double-bounce scattering between underlying vegetation and the ground surface.

To study the dependence of scattering on leaf orientation, in Mead and MacIntosh (1991), the variability of percent polarization versus footprint location within a single tree canopy was measured. Figure 3.13 shows the approximate locations of the center of the radar footprints (A–F) within the tree crown of an eastern cottonwood tree which was approximately 20 m tall by 10 m wide and results for

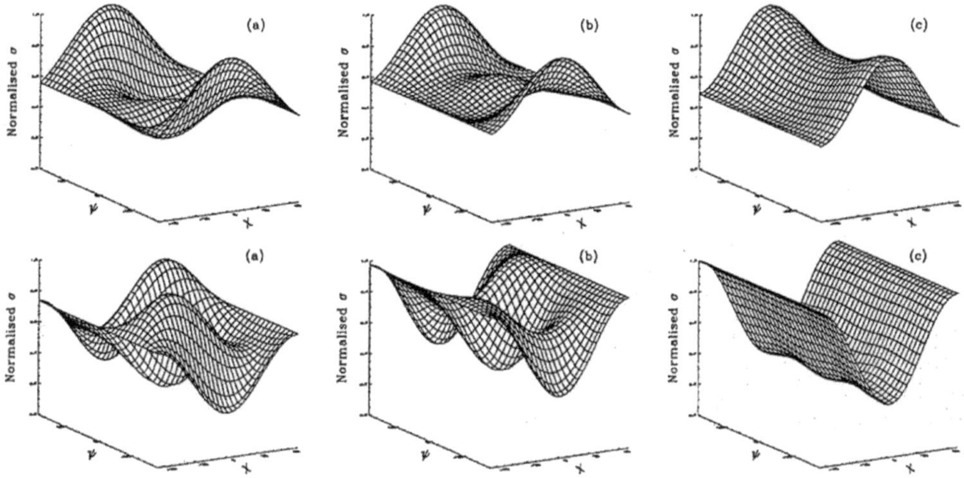

Fig. 3.12. Co- (top) and cross-polarization (bottom) signatures for eucalypts at (a) *P*-band; (b) *L*-band; and (c) *C*-band. From Dong et al. (1996).

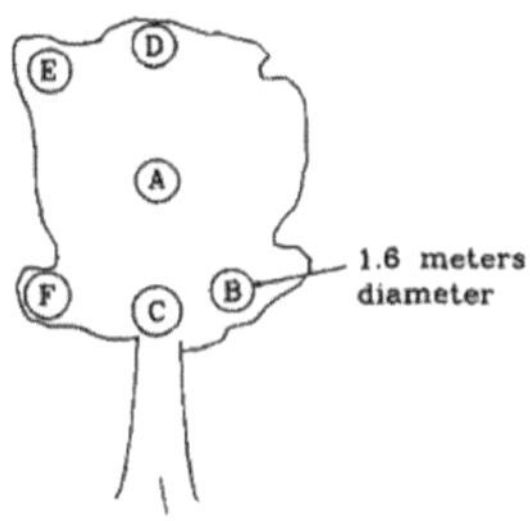

(a)

location	V	H	-45^0	RC
A	93.2	92.4	94.4	88.9
B	89.7	88.7	92.3	84.3
C	89.1	91.0	88.6	83.1
D	87.7	88.4	89.8	82.3
E	87.8	85.1	89.9	81.5
F	84.5	83.7	87.0	78.6

(b)

Fig. 3.13. (a) Locations of the center of the footprints (A–F) within the tree crown of an eastern cottonwood tree; (b) results for the percent polarization for the three linear transmit polarizations V, H, $-45°$, and right-handed circular polarization. From Mead and MacIntosh (1991).

the percent polarization for the three linear transmit polarizations V, H, $-45°$, and right-handed circular polarization.

From Fig. 3.13, there is a considerable difference in percent polarization for the central measurement point A versus the measurement point F at the edge of the crown. This difference relates apparently to the predominant orientation of leaves for these points: at point A, leaves are perpendicular to the direction of the propagation, whereas the average leaf normal near the edge of the crown (points F and E) is more likely to be at a large angle with respect to the direction of propagation. At the same time, for point A, the percent polarization measured for the eastern cottonwood and weeping willow trees, which have significantly different leaf shapes, turned to be nearly equal. This suggests that the angular distribution of the leaves is dominant for scattering compared to the leaf shape and size, provided that the leaves are significantly larger than a wavelength in length and width.

Thus, the polarization characteristics of the trees are varied according to species and location on crown. A further important aspect is whether there exists any variability of the polarimetric behavior for a given tree species. Evidently, it would be reasonable to make such measurements for coniferous trees, because white pine has an angular distribution of needles that is practically uniform and the scattering is expected to depend on the footprint location. The Mueller matrices of 11 independent footprints from seven individual white pine trees were measured (Mead and MacIntosh, 1991). The averaged Mueller matrix and the standard deviation matrix are

$$\begin{pmatrix} 1.044 & -0.006 & 0.007 & -0.011 \\ -0.097 & 0.739 & 0.020 & -0.005 \\ -0.021 & -0.027 & 0.749 & 0.144 \\ 0.003 & -0.014 & 0.088 & -0.532 \end{pmatrix} \text{ and } \begin{pmatrix} 0.0073 & 0.0129 & 0.0129 & 0.0110 \\ 0.0085 & 0.0097 & 0.0093 & 0.0073 \\ 0.0076 & 0.0091 & 0.0091 & 0.0076 \\ 0.0040 & 0.0123 & 0.0082 & 0.0081 \end{pmatrix}. \quad (3.76)$$

This indicates little variability between the polarimetric behavior for a given tree species.

Light scattering from leaves has been the subject of many previous studies (see, e.g., Martin et al. 2010, and references therein) but only a few (e.g., Pospergelis, 1969; Krishnan and Nordine, 1994; Savenkov et al., 2003, 2004; Savenkov and Muttiah, 2004; Martin et al., 2010) address the Mueller matrix measurements and these have mainly been single-wavelength (633 nm), multiple-incidence angle measurements.

Savenkov et al. (2003, 2004) reported Mueller matrices at a fixed wavelength (633 nm) with normal incidence and backscattering angles for English oak and wheat leaves. Krishnan and Nordine (1994) reported Mueller matrices for yew and maple leaves at 633 nm with a fixed incidence angle of 40° measuring the reflected light from the leaf surface. In Martin et al. (2010), measurement geometry assumes the angle of incidence on the sample is 45° and the light scattered is measured at a fixed angle of 135°. The spectral range is from 345 to 1,020 nm, with a gap at 510–550 nm for three leaf samples: *Arabidopsis thaliana* (aka Thale Cress), English oak (*Quercus robur*), and *Ficus benjamina*. Results for English oak presented in Savenkov et al. (2003) are consistent with those measured in Martin et al. (2010), at 630 nm.

Structures of the Mueller matrices for transmitted and backscattered light reported in Savenkov et al. (2003) have the forms

$$\mathbf{M}^{tr} = \begin{pmatrix} m_{11}^{tr} & 0 & 0 & 0 \\ 0 & m_{22}^{tr} & 0 & 0 \\ 0 & 0 m_{33}^{tr} & 0 & \\ 0 & 0 & 0 & m_{44}^{tr} \end{pmatrix} \text{ and } \mathbf{M}^{bsc} = \begin{pmatrix} m_{11}^{bsc} & m_{12}^{bsc} & 0 & 0 \\ m_{21}^{bsc} & m_{22}^{bsc} & 0 & 0 \\ 0 & 0 m_{33}^{bsc} & 0 & \\ 0 & 0 & 0 & m_{44}^{bsc} \end{pmatrix}. \quad (3.77)$$

It is interesting that Krishnan and Nordine (1994) obtained the same structure of the Mueller matrices as Eq. (3.77) for yew and maple leaves, but with considerably greater (negative) values for elements m_{12} and m_{21}. The degree of polarization of light reflected by plant-leaf surfaces in their experiments was largely explained as the result of higher external and smaller internal scattering by the leaves.

It turned out that the most informative parameters for leaf type identification (coniferous or deciduous) (Krishnan and Nordine, 1994), for estimation of changes to water vapor concentration (Savenkov et al., 2003), and for testing of leaf virus infection (Savenkov et al., 2004) were the single depolarization metrics: degree of polarization (linear and/or circular, Eq. (3.4)) and Cloude's entropy (Eq. (3.48)). Figures 3.14 and 3.15 show the degree of polarization as a function of input light for a fresh leaf and a day-old leaf for transmitted (observation angle 0°) and backscattered (observation angle 170°) light, respectively. Figure 3.16 shows Cloude's entropy for wheat infected by WSMV (the wheat streak mosaic virus) with various infection developments for forward- and backscattered light.

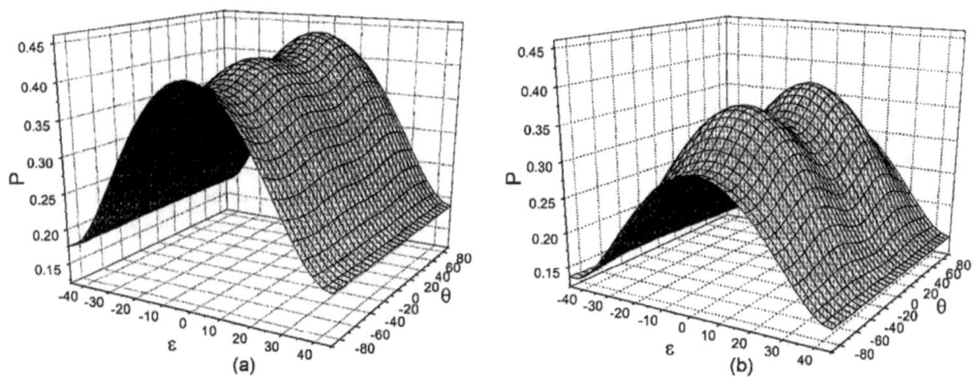

Fig. 3.14. Degree of polarization as a function of input light azimuth and ellipticity for (a) a fresh leaf and (b) a day-old leaf for transmitted light at observation angle 0°. From Savenkov et al. (2003).

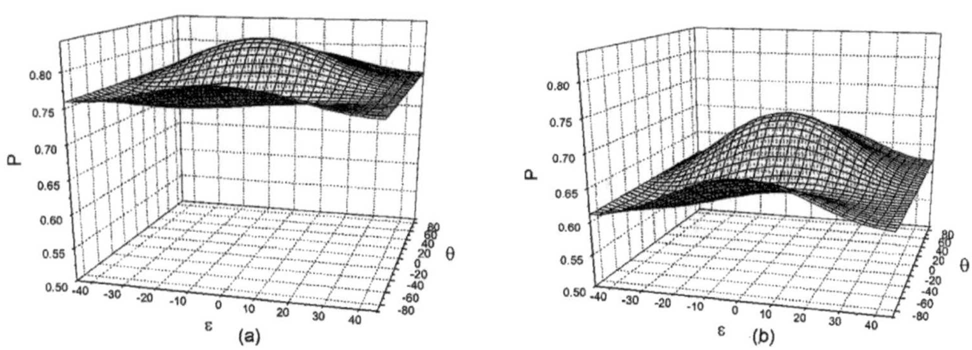

Fig. 3.15. Degree of polarization as a function of input light azimuth and ellipticity for (a) a fresh leaf and (b) a day-old leaf for backscattered light at observation angle 170°. From Savenkov et al. (2003).

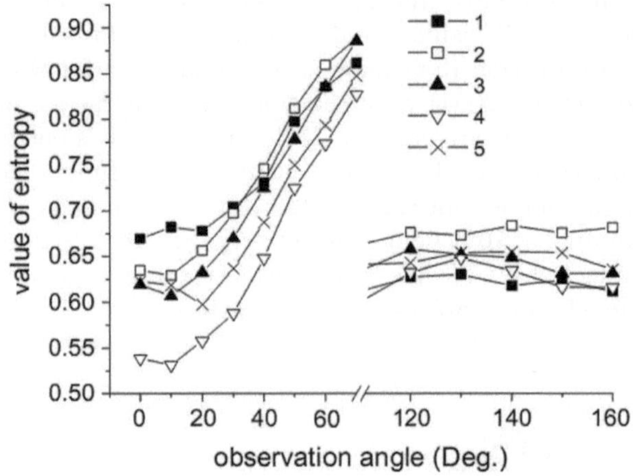

Fig. 3.16. Dependencies of the entropy versus the observation angle for wheat leaves grown under field conditions: (1) wheat infected with the wheat streak mosaic virus (WSMV) (infection development four points); (2) wheat infected with WSMV (infection development three points); (3) wheat infected with WSMV (infection development two points); (4) healthy wheat; (5) wheat infected with WSMV (infection development one point). From Savenkov et al. (2004).

The backscattering coefficients for a tall-grass field, the biomasses, leaf moisture contents, and soil moisture contents are measured in Oh (2008) with a polarimetric L-band scatterometer. For computing the radar backscattering coefficients of vegetated surfaces, the radiative-transfer model was suggested. This model includes five basic scattering mechanisms as shown in Fig. 3.17: (1) directly backscattering from the vegetation canopy, (2) forward-scattering from the vegetation layer and then reflecting from the soil surface, (3) reflecting from the soil surface and then forward-scattering from the vegetation layer, (4) reflecting from the soil surface, then backscattering from the vegetation layer, and reflecting again from the soil surface, and (5) direct backscatter contribution of the underlying soil surface with two-way attenuations through the vegetation layer.

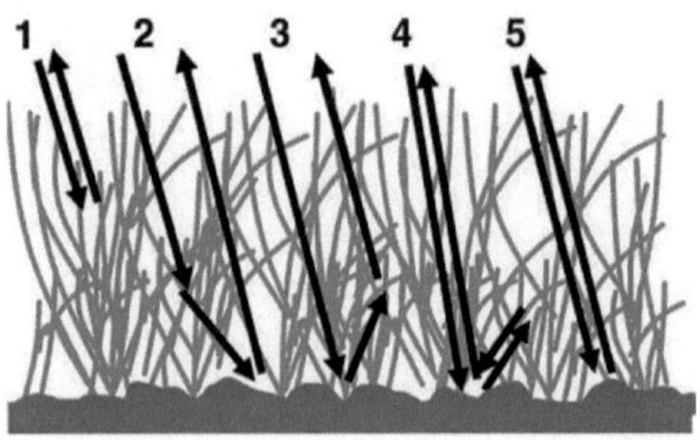

Fig. 3.17. Scattering mechanisms of the radiative-transfer model. From Oh (2008).

A proposed model was used for analysis of a set of polarimetric measurement data acquired using the Hongik Polarimetric Scatterometer system at 1.85 GHz from a tall-grass field at an incidence angle of 40° for a period of 2 months in 2006 with various soil moisture conditions and a fixed-surface roughness. The measured ground-truth data have the following values: the vegetation height $h = 72$ cm, averaged leaf length $L_{leaf} = 60$ cm, averaged leaf width $W_{leaf} = 2$ cm, leaf density $N_{leaf} = 800/\text{m}^2$, surface root-mean-square (r.m.s.) height $s = 2.35$ cm, and surface correlation length $l = 34.5$ cm. The measured volumetric soil moisture contents m_v have a range from 0.17 cm^3/cm^3 to 0.38 cm^3/cm^3 during the experiment period.

The measured co-polarized ratio $p = \sigma_{HH}/\sigma_{VV}$ and cross-polarized ratio $q = \sigma_{HV}/\sigma_{VV}$ of the tall-grass field, as well as the HH-, VV-, and HV-polarized backscattering coefficients (σ_{VV}, σ_{HH}, and σ_{HV}), were used to retrieve the soil moisture and surface roughness. Figure 3.18 demonstrates that the scattering model agrees quite well with the measured parameters.

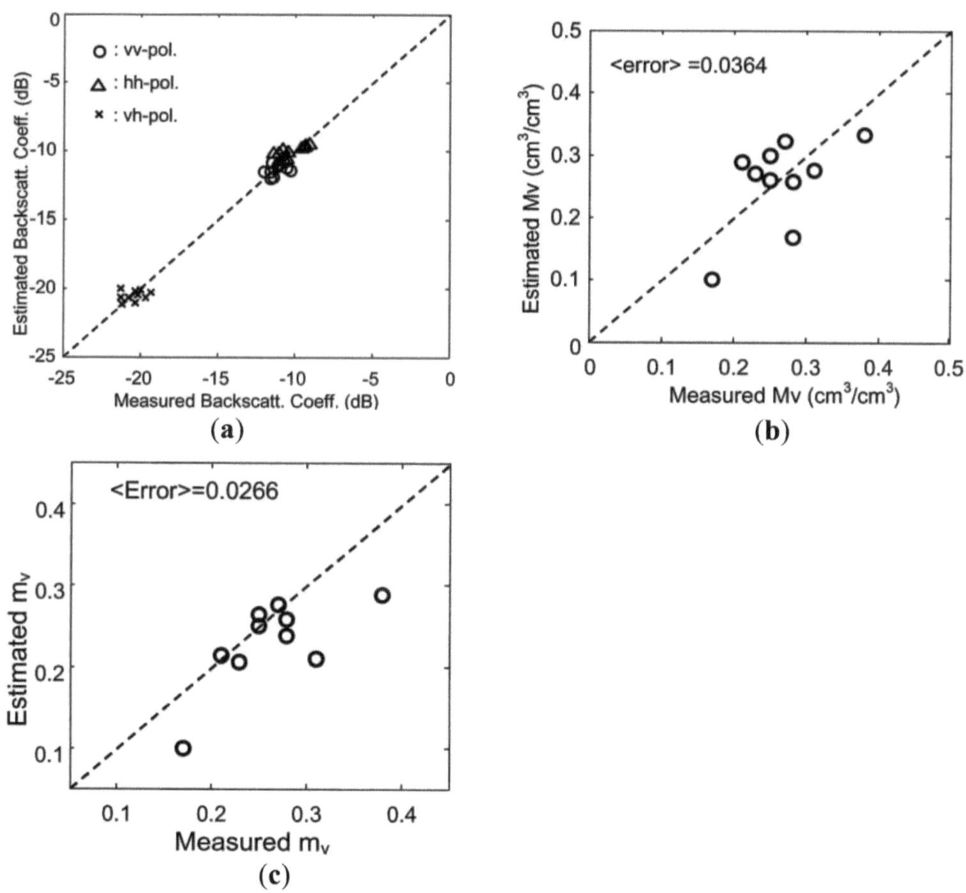

Fig. 3.18. Comparison between the estimated and measured (a) backscattering coefficients; soil moisture contents using (b) co- and cross-polarized measurements; and (c) only co-polarized measurements. From Oh (2008).

The crop residue classification by radar polarimetry was addressed in McNairn et al. (2002). This is of great importance for land use, economy, and agricultural production. In this analysis, polarimetric data (*HH*, *VV*, and *HV*) for the Altona site, southern Manitoba, acquired on April 12 and October 5, 1994, were used. In April, the Altona test site consisted of bare agricultural fields with varying surface roughness and amounts of crop residue cover. The site conditions in October were similar, although some crops had not yet been harvested.

The incidence angle was approximately 44° at the center of the site for the April and 51° for the October measurements. Results for *L*-band co-polarization signatures are presented in Fig. 3.19.

From Fig. 3.19a, it can be seen that these surfaces are not rough enough and do not have enough vegetation to cause significant multiple or volume scattering. These are the typical cases of surface scattering. The cases of no-till residue fields (Fig. 3.19b) and senesced crops (Fig. 3.19c) are dominated by multiple and volume scattering. Evidently, one of the most informative parameters for given test field characterization is the pedestal height. Larger pedestals are associated with standing crops and no-till fields. This demonstrates the sensitivity of pedestal height to multiple and volume scattering.

The potential of polarimetric SAR data in the *L*- and *P*-bands was examined over a Mediterranean vineyard site near Bordeaux in south-western France (Baghdadi et al., 2006). RAMSES images were evaluated to investigate the capacity of the *L*- and *P*-bands to discriminate between the main types of cover present at the study site: vineyards, grasslands, forests, and bare soil. Forests have a bright tone that corresponds to volume and multiple scattering (Fig. 3.20).

For this purpose, authors have used the polarimetric metrics frequently employed in the literature: backscattering coefficients, co-polarization and depolarization ratios, entropy, alpha angle, and anisotropy.

The difference between vines with wooden stakes and vines with metal stakes is most marked for *L*-band and *VV* polarization or the VV/HH ratio (Fig. 3.21a, 3.21f, 3.21c, and 3.21h).

The results also show that, in *L*-band, the backscattering coefficient in *HV* polarization and anisotropy enables the distinction between forests, grasslands, and vineyards, and bare soils (see Fig. 3.21b and 3.21g). Vineyards and bare soils have similar behavior of polarimetric metrics used, except for α (see Fig. 3.21d and 3.21i). The other polarimetric metrics show insignificant potential for the discrimination of land cover types.

The results obtained in the *P*-band show that the potential for discrimination between the various classes is much lower than that of for the *L*-band (Fig. 3.21). Also, the *P*-band demonstrates considerable variation among the metrics considered. For example, the *HV* polarization for vines ranges from about 16 dB in *P*-band to 5 dB in *L*-band (Fig. 3.21b and 3.21g). This may result from the strong sensitivity of the *P*-band (high wavelength) to the physical parameters of the environment.

The potential of *L*- and *P*-band polarimetric SAR for the characterization of vineyards was evaluated: first by row direction (north-east and north-west) and second by age (young or old). Results are presented in Figs. 3.22 and 3.23.

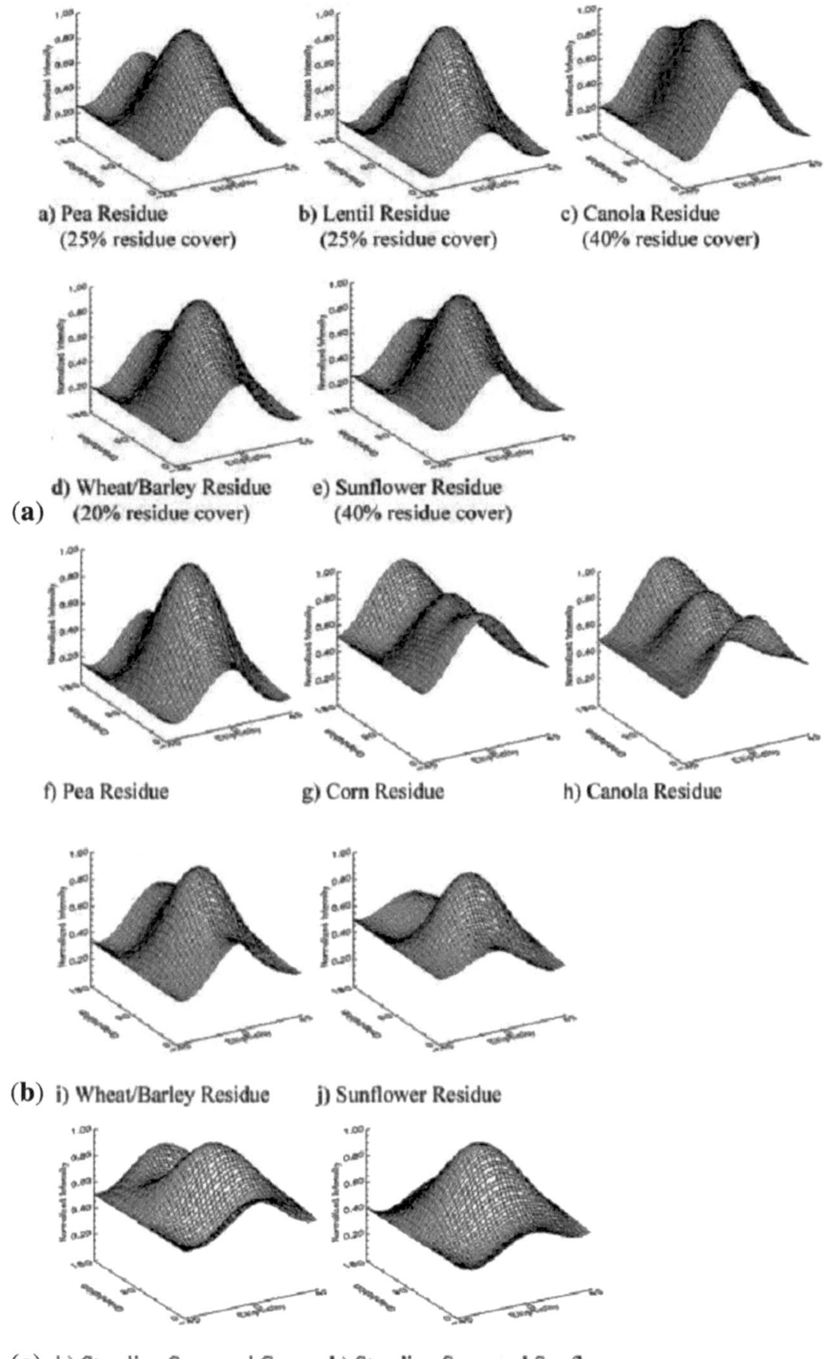

Fig. 3.19. L-band co-polarization signatures: (a) tilled fields with lower amounts of residue; (b) no-till fields; (c) senesced crop prior to harvest. From McNairn et al. (2002).

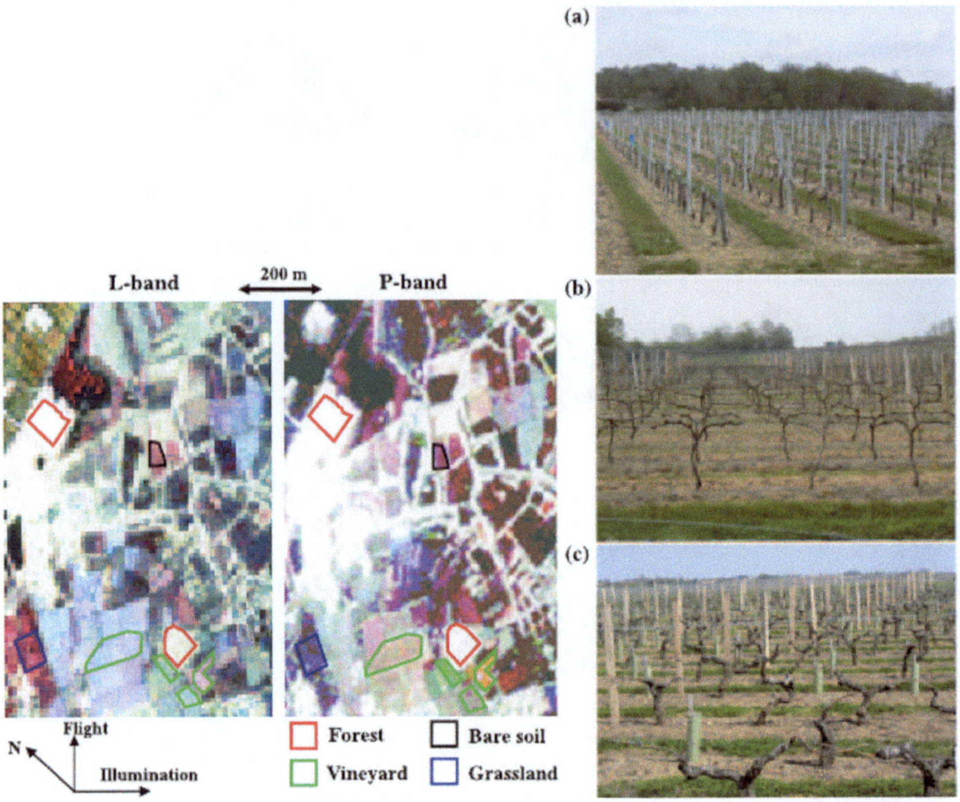

Fig. 3.20. Left: the polarimetric images of the study site in *L*- and *P*-bands (*HH* in red, *HV* in green and *VV* in blue). Right: (a) young vines with metal stakes; (b) young vines with wooden stakes; (c) old vines with wooden stakes. From Baghdadi et al. (2006).

It can be seen from the results presented above that the *L*- and *P*-band SAR data are not suitable for distinguishing either between young vines and old vines or between the two principal row directions of the vines (parallel and perpendicular to flight direction), despite the slight differences observed.

Note, however, that the vines considered in this experiment have no foliage and it is mainly the vine trunks (5 cm in diameter for the young vines and 10–15 cm for the old vines) and the ground that cause backscattering. Because of that, it should be relevant to re-examine this when the vines have their leaves, since the increase in biomass may enable better discrimination between the two age groups.

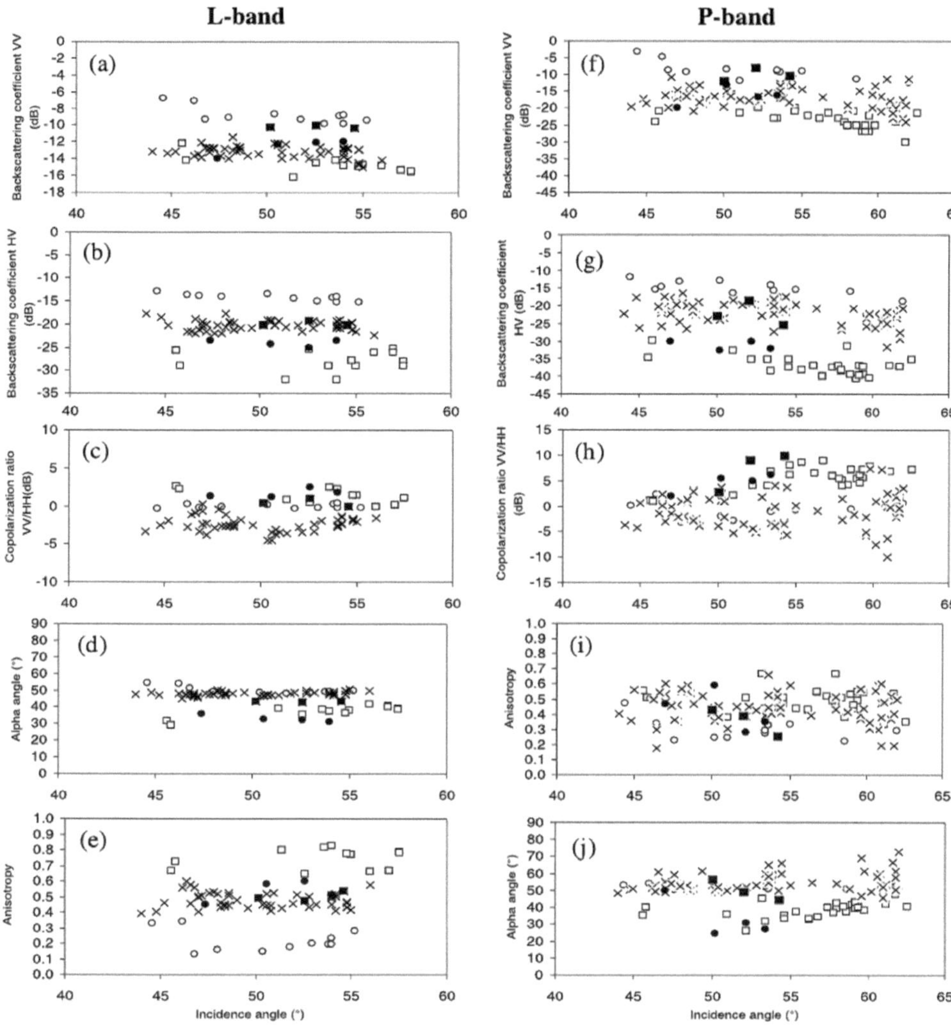

Fig. 3.21. The values of polarimetric metrics *HH*, *VV*/*HH*, *HV*/*HH*, α, entropy, and anisotropy in the *L*- and *P*-bands for the principal types of land cover (open square, grassland; solid square, vines with metal stakes; open circle, forest; solid circle, bare soils; cross, vines with wooden stakes). From Baghdadi et al. (2006).

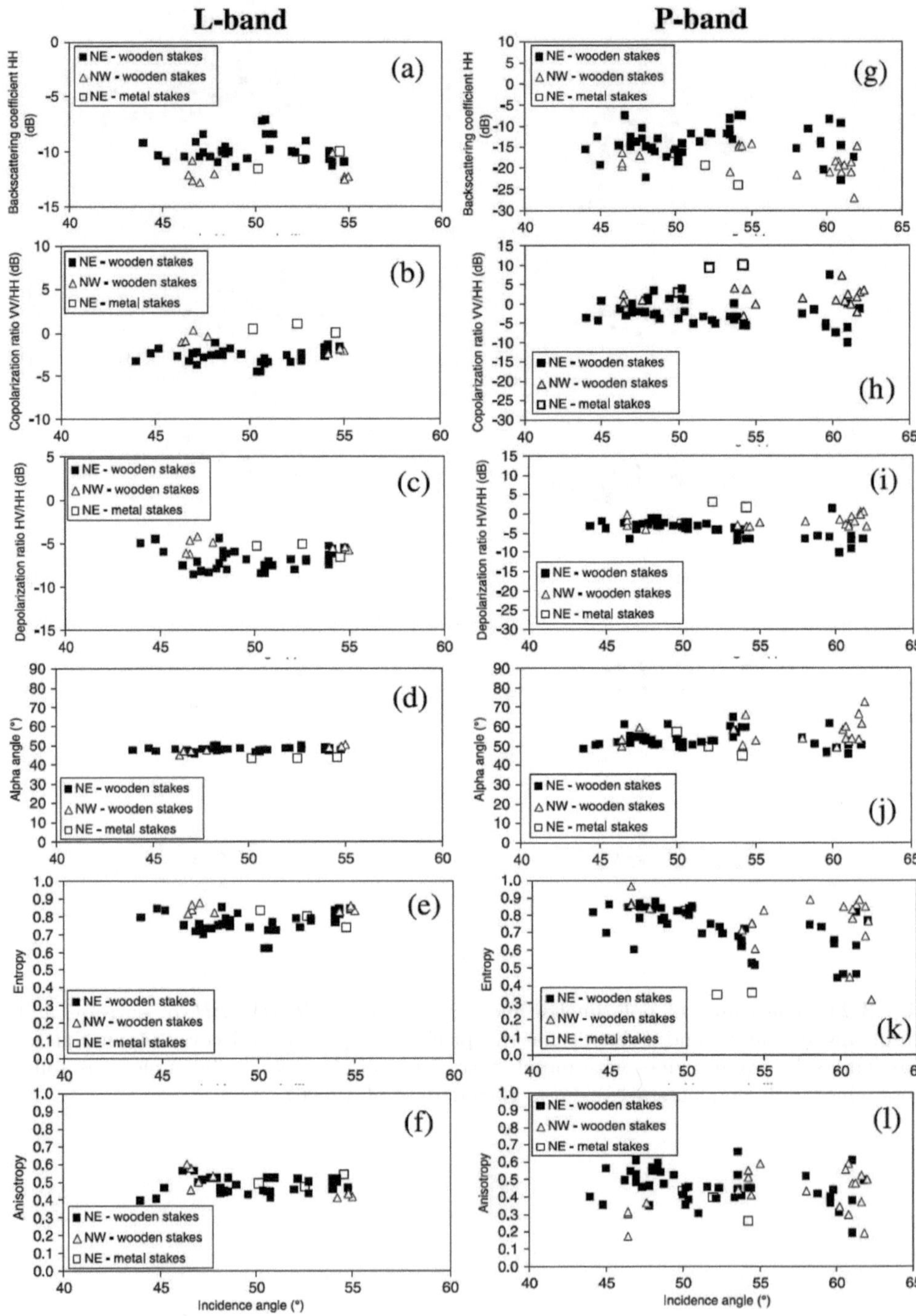

Fig. 3.22. The values of polarimetric metrics HH, VV/HH, HV/HH, α, entropy, and anisotropy in the L- and P-bands as functions of the row direction of the vines (north-east and north-west). From Baghdadi et al. (2006).

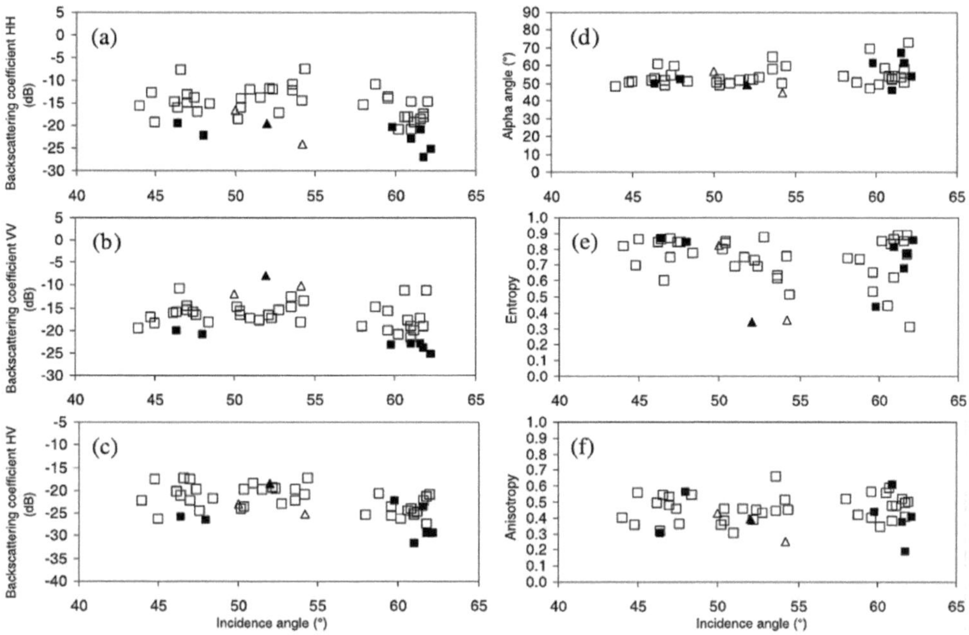

Fig. 3.23. The values of the polarimetric metrics *HH*, *VV*/*HH*, *HV*/*HH*, α, entropy, and anisotropy in the *P*-band as a function of vineyard age (solid square, young vines with wooden stakes; open square, old vines with wooden stakes; solid triangle, young vines with metal stakes; open triangle, old vines with metal stakes). From Baghdadi et al. (2006).

3.5.2 Soil

The sensitivity of microwaves to the dielectric properties and geometrical surface structure of bare soil surfaces makes radar remote sensing applicable to a wide range of environmental issues relating to soil. In particular, the potential to retrieve soil moisture with high spatial and/or temporal resolution represents a significant contribution to predictive modeling in hydrology and meteorological modeling, as well as to economical optimization of agricultural procedures. A major problem in the quantitative estimation of either soil moisture or surface roughness is the separation of their individual contributions to the backscattered radar signal. In this subsection, we consider the use of fully polarimetric measurements to obtain independent estimates of surface roughness and soil moisture.

Soil surfaces are characterized in terms of their material (dielectric) and geometrical properties. One of the key material parameters is the volumetric soil moisture content that affects the radar backscatter primarily through its strong influence to the dielectric constant. Soil moisture is determined by gathering soil samples with a probe of known volume, weighing them with a portable scale and drying them in a microwave oven. By using wet M_1 and dry M_2 soil masses and the volume V, the volumetric soil moisture values is calculated as $((M_1 - M_2)/V) \cdot 100\%$. On the other hand, the geometric properties of surfaces are described by the spatial roughness scales in both the horizontal and vertical directions, through the surface r.m.s. height and the surface correlation length, respectively.

In the absence of any simple relationship between the backscattered signal and the surface parameters, polarimetry plays an important role allowing a direct or indirect separation of roughness- and moisture-induced effects.

In Hajnsek et al. (2003), the quantitative estimation of roughness and dielectric constant over a wide range of natural bare surfaces from polarimetric data based on the model which is an extension of the small perturbation model have been carried out. The used model is of two components including a Bragg scattering term and a roughness-induced rotation symmetric disturbance. In order to decouple the real part of the dielectric constant from surface roughness, the model is addressed in terms of the polarimetric scattering entropy H, scattering anisotropy A, and alpha angle α that are derived from the eigenvalues and eigenvectors of the polarimetric coherency matrix (Cloude and Pottier, 1997).

The performance of the proposed inversion algorithm is tested using fully polarimetric L-band (*HH*, *VV*, *HV*, *VH*) data against ground measurements collected over two different test sites: the flat terrain of the river Elbe, characterized by dry and rough surface conditions, and the hilly terrain of the Weiherbach watershed, characterized by wet and slight to moderate surface roughness. Experimental polarimetric data with a spatial resolution of the single-look complex data in azimuth of about 0.75 m and in range of about 1.5 m were acquired in April and August 1997 along two 15 km long and 3.2 km wide strips over two test sites.

The experimental scattering matrix data are transformed into a coherency matrix form. Then the coherency matrix data were eigendecomposed to compute the entropy H (Eq. (3.48)), anisotropy A (Eq. (3.62)), and alpha angle α (Eq. (3.63)) (Cloude and Pottier, 1997). This allows identifying dominant surface scattering areas. The areas for which $H > 0.5$ and alpha $\alpha > 45°$ are not considered in the following. Results for surface roughness ks and volumetric soil moisture m_v are presented in Figs 3.24 and 3.25.

The application of the model to experimental data shows good agreement between the inverted values and ground measurements for surface roughness ks and soil moisture m_v. The inversion accuracy is high enough to point out the seasonal variation effects: the r.m.s. errors for the Weiherbach test site are much smaller for the upper soil layer (flight campaign in March), while for the Elbe-Auen test site the r.m.s. error is smaller in the deeper soil layer (flight campaign in August).

In Hajnsek (2001), it was shown that the model permits robust roughness estimation largely independently of topographic variations of test areas. However, the challenging problem and the main limitation for surface parameter estimation from polarimetric data is the presence of vegetation. This leads to overestimations of the surface roughness and complex dielectric constant – the former due to increasing of the entropy and decreasing the anisotropy and the last through increasing of the alpha angle.

The most popular scattering model for vegetated surfaces is the radiative-transfer model (Tsang et al., 1985). This scattering model, however, is too complicated to be used for the inversion algorithm. In De Roo et al. (2001), for retrieving soil moisture from radar response, the regression models for the backscattering coefficients of a specific vegetation canopy has been used. An approach assuming an inversion of a simplified scattering model, so-called water-cloud model, which represents the vegetation canopy as uniformly distributed water particles like a cloud,

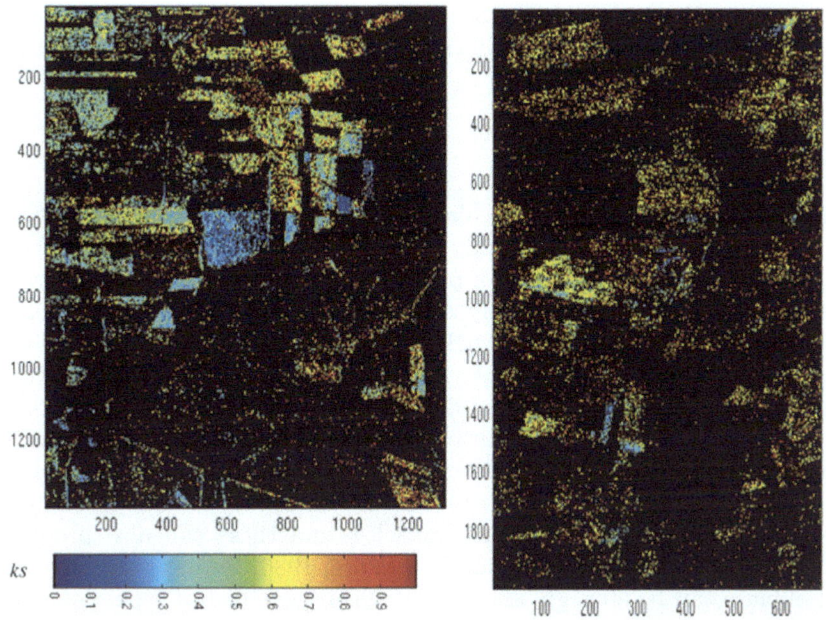

Fig. 3.24. Estimated surface roughness ks, ranging from 0 to 1; not valid areas are indicated with black. (a) Elbe-Auen test site; (b) Weiherbach test site. From Hajnsek et al. (2003).

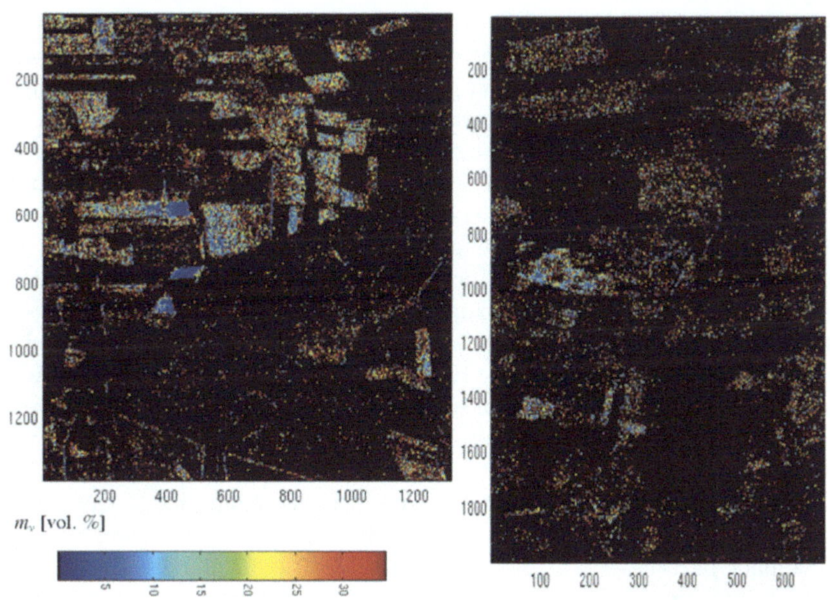

Fig. 3.25. Estimated volumetric soil moisture m_v ranging from 0 to 40 (vol.%); not valid areas are indicated with black. (a) Elbe-Auen test site; (b) Weiherbach test site. From Hajnsek et al. (2003).

was proposed by Bindlish and Barros (2001) and Sikdar et al. (2005). The parameters of the water-cloud model are derived by fitting the model with experimental data.

3.5.3 Atmosphere

Evidently, Solar's radiation budget plays a key role for all living beings on Earth and it can be dramatically affected by volcanic eruptions injecting into the atmosphere active gases and solid aerosol particles. They can remain there for months to several years, changing the composition of the atmosphere (Oberbeck et al., 1983; Snetsinger et al., 1987; Pueschel et al., 1994; Tsitas and Yung, 1996). The composition and size distribution of the volcanic aerosols can significantly affect the climate changes (Hansen et al., 1992).

After eruption of Eyjafjallajökull in Iceland, April 2010, it is well known that volcanic ash clouds that are formed right after the eruption can disrupt the air traffic of the whole continent. Ash clouds are extremely dangerous to jets. Jet engines are affected by volcanic ash particles, and these can also produce acid abrasion of windshields and other surfaces (Rose, 1986; Bernard and Rose, 1990).

The climate can be strongly affected by anthropogenic atmospheric aerosols as well, in particular, by fly ash clouds which is a by-product (consisting mainly of clays) of the combustion of coal in electricity power plants (Charlson et al., 1992).

Therefore, studying the scattering and polarization properties of ash particles is important and useful in order to estimate their effect on Earth's climate as a whole and, in particular, in industrial regions where the concentration of anthropogenic aerosols can be high. Unfortunately, theoretical and numerical techniques are not suited for realistic polydispersions of volcanic ash particles, because of their wide ranges of sizes and shapes. Therefore, experimental studies remain an important source of information on scattering properties of irregular ash particles.

In Volten et al. (2001) and Munoz et al. (2002, 2004), laboratory measurements of complete Mueller matrices as functions of the scattering angle in the range from 5° to 173° of several different samples of volcanic ashes and in Munoz et al. (2001) of fly ash particles at 633 nm wavelength have been reported.

The samples of volcanic ashes include the following: (i) a sample of dacitic pyroclastic flow from the 1989–90 eruption of Mount St. Helens, Washington; (ii) andesitic samples from the 1989–90 eruption of Redoubt Volcano, Alaska; (iii) andesitic ashes from the August 18, 1992, eruption of Crater Peak, Mount Spurr, Alaska; (iv) two andesitic ash samples from the September 17–18, 1992, eruption of Crater Peak, Mount Spurr, Alaska; and some others. Details of the volcanic ashes and fly ash particles' parameters and of the measurement procedure are in Munoz et al. (2001, 2004). Particles presented are the examples of the two classes of small irregular particles (Lumme, 2000): fly ash (aggregates) and the volcanic ashes (compact particles).

Figure 3.26 presents elements of the complete Mueller matrices as functions of the scattering angle for the randomly oriented particles of fly ash. The same for the Mount St. Helens sample of volcanic ash has been presented in Fig. 3.27.

For all samples considered, the elements F_{13}/F_{11}, F_{14}/F_{11}, F_{23}/F_{11}, and F_{24}/F_{11} are zero within the experimental errors in the complete measured scat-

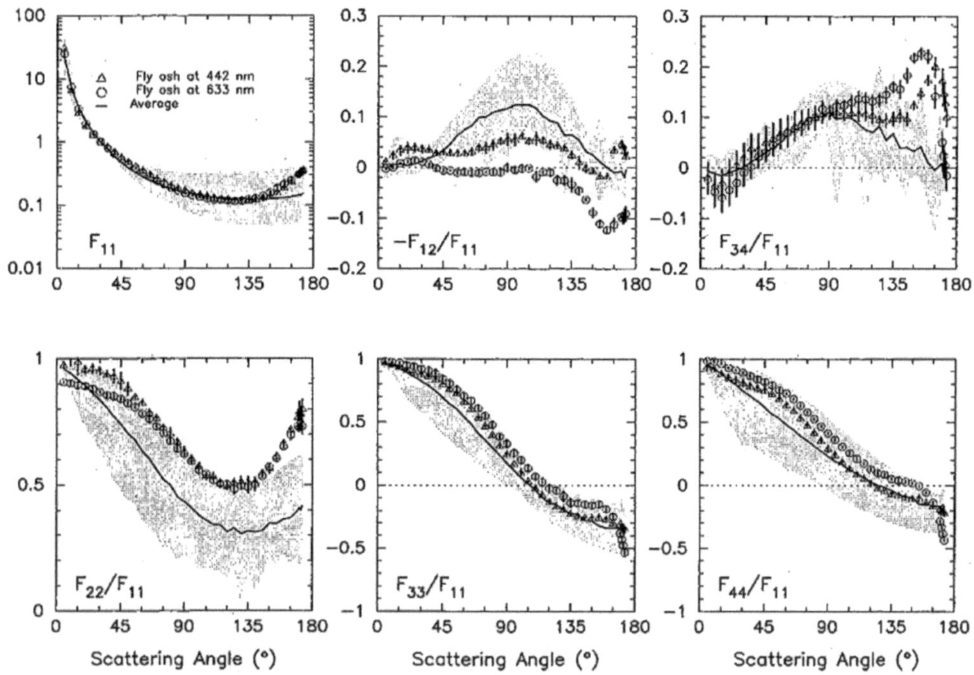

Fig. 3.26. Measured Mueller matrix as functions of the scattering angle for fly ash particles at 442 nm (triangles) and 633 nm (circles). Solid lines correspond to the average mineral scattering matrix for compact particles. The domains occupied by the aerosol measurements presented by Volten et al. (2001) are indicated by shading. From Munoz et al. (2001).

tering angle range. Therefore, their Mueller matrices have a block-diagonal structure (see Eq. (3.16)). This allows assuming that these scattering samples consist of randomly oriented particles with equal amounts of particles and their mirror particles.

The phase functions F_{11} for the fly ash particles show a relatively strong increase at backscattering angles, which seems to be related to their shape, compared to experimental results obtained for all samples of volcanic ashes. For all volcanic ashes, the phase functions F_{11} are smooth functions of the scattering angle showing a strong forward peak; they are featureless and flat at side-scattering angles and have almost no structure at backscattering angles. This scattering behavior can probably be considered a general property of ensembles of mineral irregular particles (Jaggard et al., 1981; West et al., 1997; Mishchenko et al., 2000; Volten et al., 2001). The measured $-F_{12}/F_{11}$ curves for fly ash particles are found to differ appreciably from the ones for volcanic ashes. This means that the values of the degree of linear polarization for the fly ash are generally lower. The shape of F_{22}/F_{11} with an increase around 130° observed for fly ash was also obtained by Mackowski and Mishchenko (1996) for monodisperse aggregates of N spheres with a size parameter of the single sphere of five and N ranging from 2 to 5.

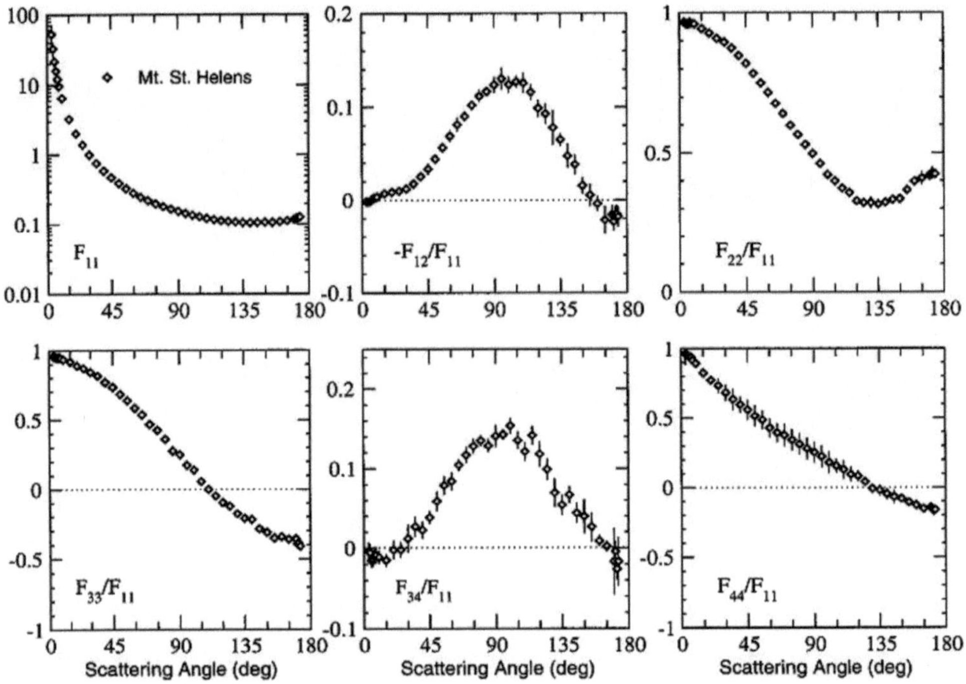

Fig. 3.27. Measured scattering matrix elements as functions of the scattering angle at 632.8 nm for the Mount St. Helens sample. Error is indicated by bars or is within the size of the symbols. From Munoz et al. (2004).

The Mueller matrix elements for the volcanic ash samples show relatively few differences. Nevertheless, disposable differences cannot evidently be attributed to the shapes of the volcanic particles rather to the size and color of these samples. Through the relatively small observed differences, authors (Munoz et al., 2004) suggested constructing an average Mueller matrix for volcanic ashes that can be used for many purposes, particularly in studies of climatic effects of volcanic eruptions when the actual properties of the volcanic ash are unknown.

One more factor which can make a significant radiative impact on climate is dust clouds. In many regions of Earth, the atmospheric aerosol component is dominated by mineral dust. The main source of mineral dust over the globe is probably Saharan dust. Indeed, large quantities of dust are transported from sources in North Africa across the Atlantic, covering very large areas (Li et al., 1996; Prospero and Lamb, 2003; Chiapello et al., 2005).

In Munoz et al. (2007), the complete Mueller matrices as a function of the scattering angle of a Sahara sand sample were measured. The samples of particles were collected from the upper part of a dune in Libya. Thus, these samples mainly consist of large particles, since small particles were blown up by the wind. The Libyan sand particles in this experiment have the effective radius $r_{eff} = 124.75$ μm and effective variance $v_{eff} = 0.15$ and a wide variety of irregular, round, and elongated shapes with occasional sharp edges (Koren et al., 2001; Nousiainen et al., 2003; Kalashnikova and Sokolik, 2004). Such particles are composed of a variety of dif-

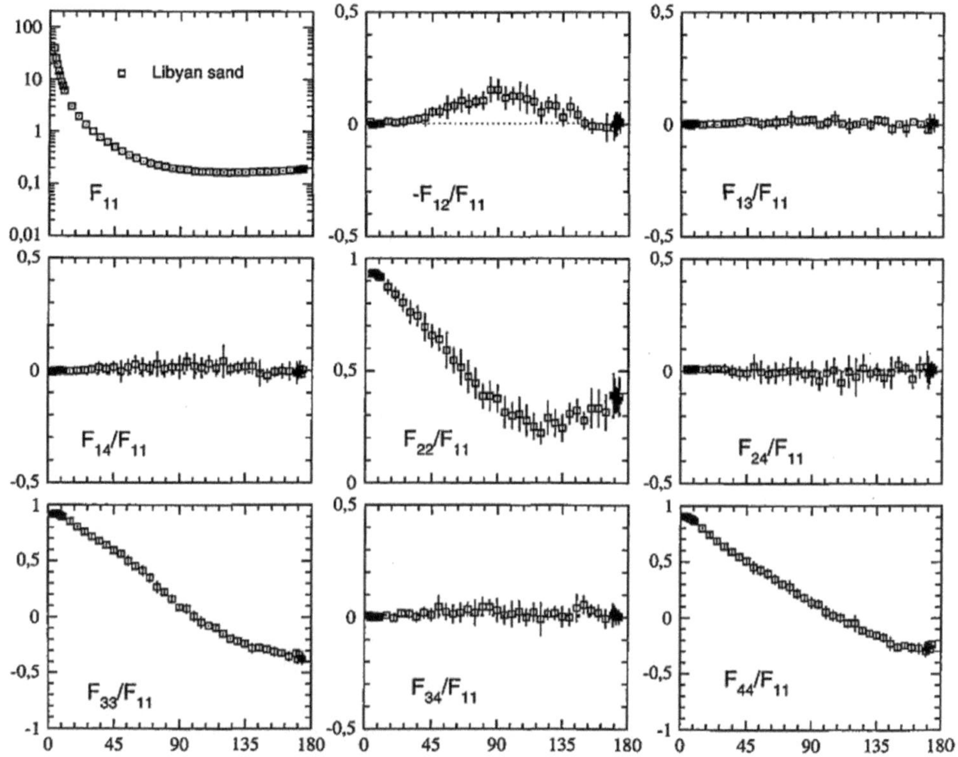

Fig. 3.28. Measured scattering matrix elements as functions of the scattering angle at 632.8 nm for the Libyan sand sample. Errors are indicated by error bars or are within the symbols. From Munoz et al. (2007).

ferent minerals. Taking into account previous data (Patterson et al., 1977; Koepke et al., 1997; Dubovik et al., 2002), it is assumed for the sample's refractive indices an average value $m = 1.5 + 4 \times 10^{-3}\, i$ at 632.8 nm. Results of the Mueller matrix element measurements are presented in Fig. 3.28.

In spite of considerable differences in size, the measured Mueller matrix elements for the Libyan sand sample, similarly to volcanic ash samples, follow the general trends presented by irregular mineral particles (see, e.g., Mishchenko et al., 2000; Volten et al., 2001). Numerical simulations show that the single-scattering properties of the Libyan sand particles can be reasonably modeled, accounting for the effects of surface roughness, namely very spiky shapes of the particles combined with Lambertian surface elements.

Obtained results are expected to improve the accuracy of remote sensing of desert surfaces. Besides, it is substantial for the understanding of the scattering properties of asteroid surfaces apparently composed of particles that are large compared to the visible wavelength (Muinonen et al., 2002).

3.5.4 Contamination

Until now, there has been limited research performed on the capability of radar polarimetry to detect green macroalgae. Results presented in Shen et al. (2014) are based on RADARSAT-2 full polarization images which were captured on July 11, 2012, with spatial resolution of 4.7 m in the radar range direction and 4.8 m in the azimuth. The spatial coverage of the image is the northern part of the Yellow Sea, with the coastal region including an area that has suffered invasion by green macroalgae. The outbreak of these blooms is generally related to coastal eutrophication resulting from excessive nutrients and pollutants that are dumped into the ocean from agriculture, aquaculture, industries, and urbanization (Hu et al., 2010). Thus, early detection of the outbreak of a green macroalgae bloom is vitally important for decision-making and remedial measures.

Figure 3.29 shows the images for the four full polarizations. It can be seen, that co-polarization images (HH and VV) show stronger radar returns over the open water area than cross-polarization (cross-pol) images (HV and VH). At that, all images exhibit clear bright slicks which are related to the green macroalgae patches

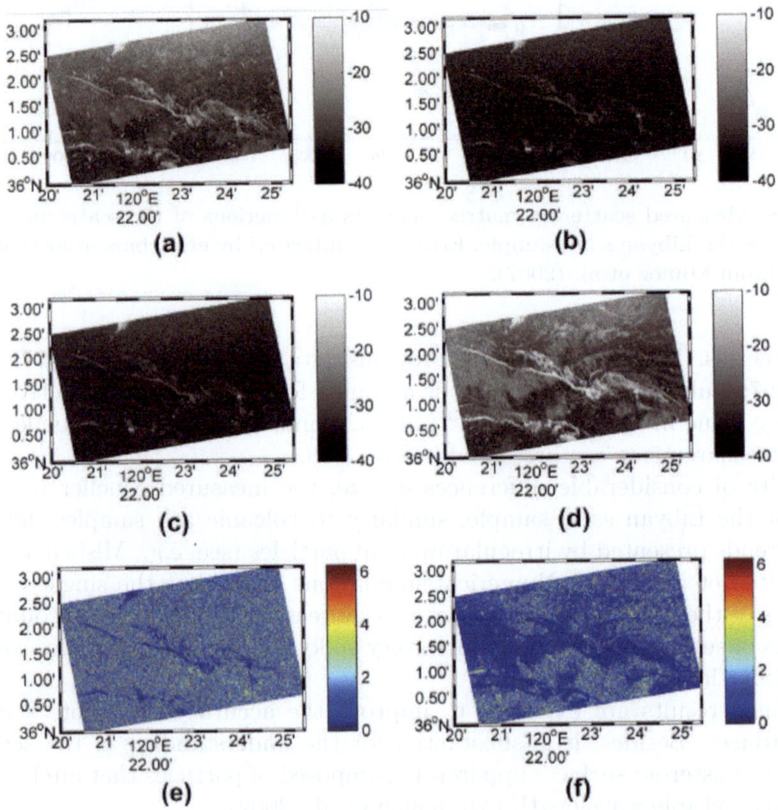

Fig. 3.29. A set of polarization images captured on July 11, 2012: (a) HH, (b) HV, (c) VH, (d) VV polarization; (e) and (f) for phase differences of $\theta_{VH} - \theta_{HV}$ and $\theta_{HH} - \theta_{VV}$, respectively. From Shen et al. (2014).

on the ocean surface. Interestingly, the comparison of images for co- and cross-polarizations yields that, although backscatter from the macroalgae is weaker in the cross-polarizations (volume scattering) than that in the co-polarizations (Bragg scattering), they are more clearly evident due to the strong contrast against the background in the cross-polarizations. The phase difference data, namely $\theta_{VH} - \theta_{HV}$ and $\theta_{HH} - \theta_{VV}$, give similar patterns that result from the amplitudes *HH*, *HV*, *VH*, and *VV*, as shown in Fig. 3.29a–d.

For further analysis of the images in Fig. 3.29, the authors suggested generating the following co- and cross-polarization metrics:

$$IF_{co} = \text{Re}\left(\frac{s_{HH} - s_{VV}}{s_{VV}}\right), \tag{3.78}$$

$$IF_{cross} = \text{Re}\left(\frac{2s_{HV} - s_{VH}}{s_{VH}}\right), \tag{3.79}$$

where s_{ij} represents the elements of corresponding complex scattering matrix for *HH*, *HV*, *VH*, and *VV* combinations of polarizations and 'Re' denotes the real part of the complex number.

The sign differences of IF_{co} and IF_{cross} between green macroalgae areas and open water resulting from data in Fig. 3.29 make these metrics a perspective for unsupervised green macroalgae detection. The results generated for the metrics in Eqs (3.78) and (3.79) are presented in Fig. 3.30.

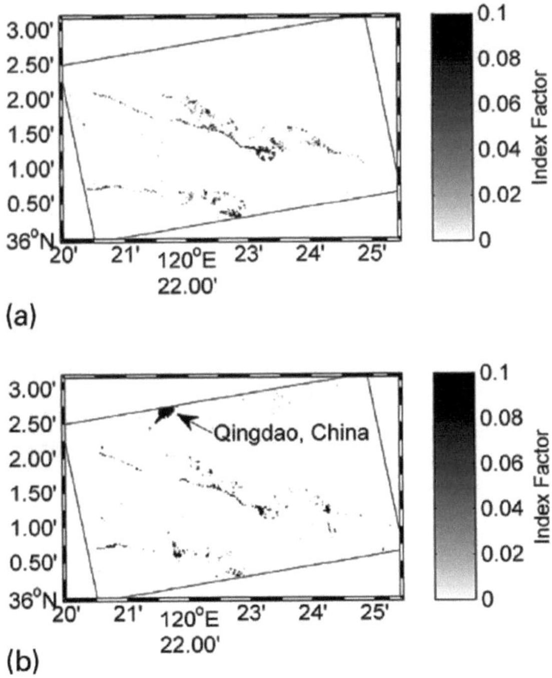

Fig. 3.30. Coverage of green macroalgae bloom patches detected by metrics defined in Eq. (3.3) from: (a) co-polarization image; (b) cross-polarization image. From Shen et al. (2014).

Comparing the two proposed metrics for co- and cross-polarization images, co-polarization has a better signal-to-noise ratio than that for cross-polarization and also has the ability to distinguish land from the macroalgae-covered ocean surface.

Furthermore, Fig. 3.30 gives a possibility to obtain an estimate of the total amount of the green macroalgae on the ocean surface. Indeed, of the 22.95 km^2 total spatial coverage that is observed by the image, 0.36 km^2 is detected to be covered by green macroalgae from the co-polarization data (Fig. 3.30a) and 0.20 km^2 from the cross-polarization data (Fig. 3.30b). Assuming a conservative density of algae of 1.0 kg/m^2 (Hu et al., 2010), this suggests a total of 360 tons and 200 tons of algae, respectively, for the area under consideration.

In Huang et al. (2014), the results on the scattering properties of supraglacial debris in *L*- and *C*-bands has been presented. The site under consideration is the Koxkar (Koxicar) glacier, which lies on the south slope of Mountain Tuomuer, West Tianshan, China (see Fig. 3.31).

It is a large mountain valley glacier, with an altitude ranging from 6,342 to 3,020 m. The glacier is 25 km long and covers an area of 83.6 km^2. The Koxkar glacier and other large valley glaciers in the Mountain Tuomuer area are characterized by the presence of debris covering most of the ablation zones. The debris thickness on the Koxkar glacier increases from zero at approximately 3,900 m to over 2 m near the front (Xie et al., 2007). The debris is very loose and mainly composed of fragments of broken stones.

The data considered are the Phased-Array *L*-band SAR (PALSAR) of the Advanced Land Observing Satellite (ALOS) and the *C*-band RADARSAT-2 quad-polarization SAR. For analysis of the data to determine the average or dominant scattering mechanism, the target decomposition is performed (Cloude and Pottier, 1996).

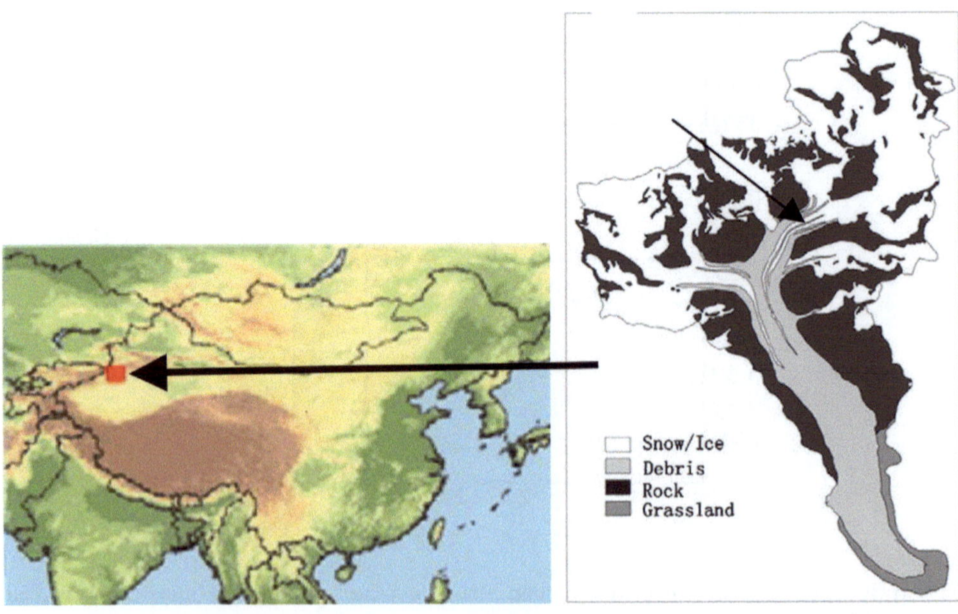

Fig. 3.31. The Koxkar glacier. From Han et al. (2010).

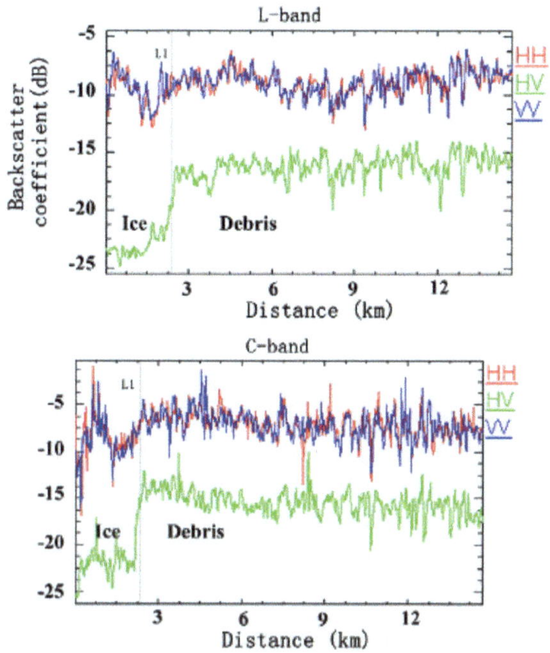

Fig. 3.32. The backscatter coefficients acquired from Advanced Land Observing Satellite (ALOS) and RADARSAT-2 polarimetric synthetic aperture radar (SAR) images along the central flow line of the Koxkar glacier. From Huang et al. (2014).

Backscatter coefficients of the Koxkar glacier along central flow line (Fig. 3.31) in the ablation zone for the L- and C-band are shown in Fig. 3.32.

It can be seen that, in co-polarizations (HH and VV), the backscatter coefficients of the ice and debris are similar in both the L- and C-bands, while cross-polarization (HV) demonstrates a distinct contrast between ice and debris.

The target decomposition enables obtaining additional information on the glacial scattering mechanisms. The results of $H/A/\alpha$ decomposition are presented in Fig. 3.33.

Indeed, according to Fig. 3.33, at the distance between 0 km and $L1$ ($L1$ is the same as for Fig. 3.32), the alpha values of the ice are lower than 30 and the entropy is approximately 0.3 in both the L- and C-bands. This indicates that surface scattering is dominant in this range of distance. In what follows, the values of alpha and entropy behave differently in L- and C-bands. In the L-band, the alpha value increases from approximately 20 to 40 and the entropy also increases between $L1$ and $L2$ – that is, in debris-covered zones. This means that volume scattering increases. The main reason causing this is the change in debris thickness. In the C-band, the change from surface scattering to volume scattering is not observable.

So, the volume scattering rises with the debris thickness when the debris is thinner than its penetration depth determined by radar wavelength and the extinction coefficient of the debris (Ulaby et al., 1986).

Figure 3.34 presents the results of the Pauli decomposition.

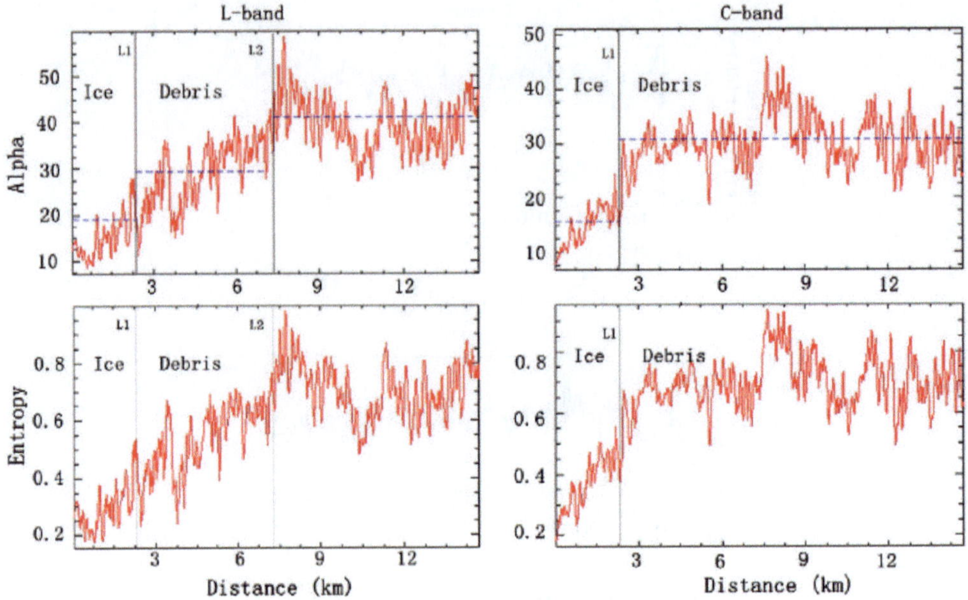

Fig. 3.33. The values of alpha and entropy extracted along the central flow line of the glacier. The horizontal dashed line represents the average value in different regions. From Huang et al. (2014).

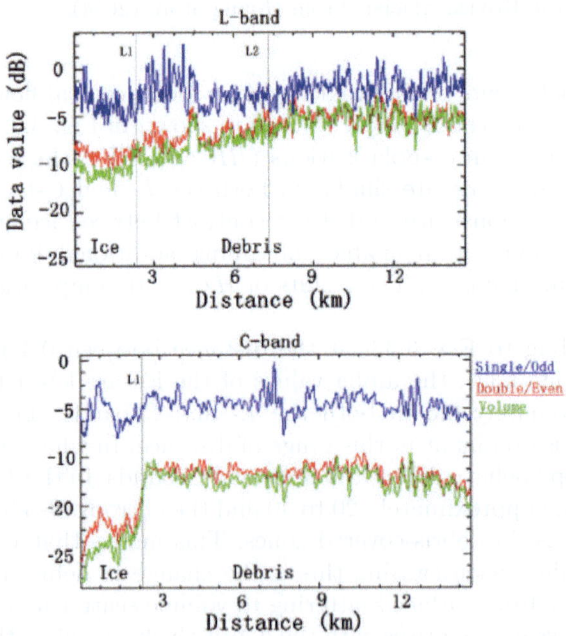

Fig. 3.34. The Pauli decomposition parameters extracted along the central flow line of the glacier. $L1$ and $L2$ correspond to their position in Fig. 3.33. From Huang et al. (2014).

In the L-band, the value of volume scattering in the ice zone is –10 dB and this value increases slightly between $L1$ and $L2$ and remains practically constant further. In the C-band, the sharp changing between the ice and debris zones is observed.

Polarimetric SAR images are promised for debris mapping and, in future work, more detailed parameters (such as the moisture, fragment size, and component) of the debris are needed to evaluate their influence on debris thickness and extent recognition.

3.5.5 Sea ice and water

In Geldsetzer and Yackel (2009), the utility of satellite-based dual-polarization C-band imagery for discriminating sea ice types and open water in winter (i.e., at air temperatures of less than $-5°C$) was investigated. The topicality of this problem results from the necessity to monitor the Arctic ecosystem and marine cryosphere, consisting of sea ice and its snow cover, whose variability considerably impacts the climatic, oceanographic, biological, cultural, and economic systems of the Arctic.

In this research, the following nomenclature for sea ice is adopted: all new, nilas, and young ice types are united under the term 'thin ice', and all old ice types are referred to as 'multi-year sea ice' (MYI) and first-year sea ice (FYI). The study area encompasses the waters surrounding Cornwallis Island, Nunavut, Canada (see Fig. 3.35).

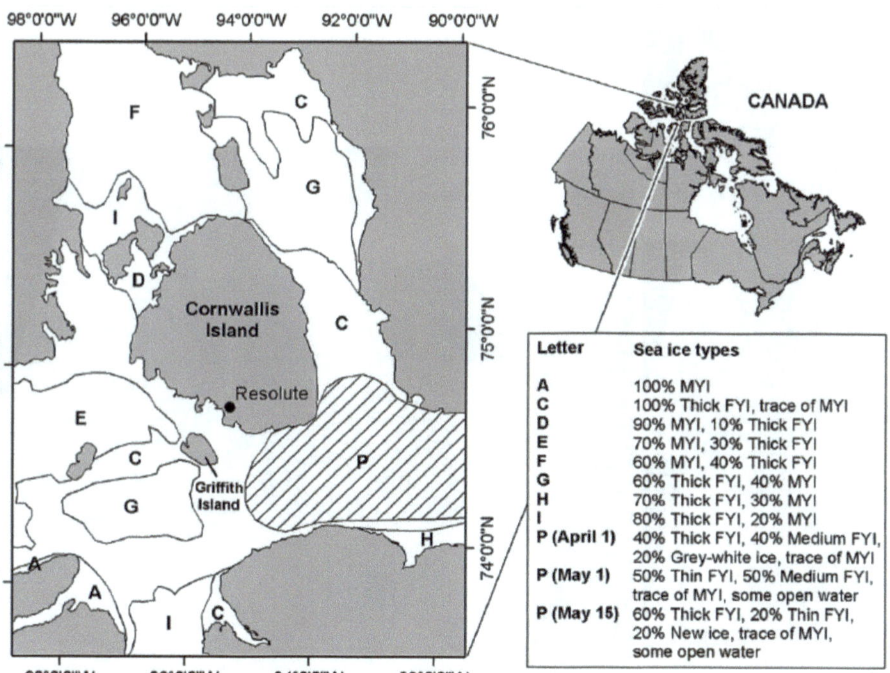

Fig. 3.35. Study area around Cornwallis Island, Nunavut, Canada. Letter codes refer to ice-type classifications delineated by the Canadian Ice Service for April 1, 2004. From Geldsetzer and Yackel (2009); source: Canadian Ice Service, Environment Canada (http://ice-glaces.ec.gc.ca).

Data for Fig. 3.35 are compiled by human operators based on *in situ* visual observations from shore, ship, or reconnaissance aircraft.

Figure 3.36 presents co-polarization signatures σ^0_{HH}, σ^0_{VV}, and the co-polarization ratio

$$\gamma_{co} = \frac{\sigma^0_{VV}}{\sigma^0_{HH}}, \tag{3.80}$$

generated for study area data (Fig. 3.35). The co-polarization ratio is routinely used to quantify the difference between backscatters of vertical and horizontal polarizations. All images are enhanced via Gaussian stretch.

Comparative interpretation of the composite image (right part of Fig. 3.36), together with associated ice chart (Fig. 3.35) shows that the study area represents a broad nomenclature of ice types and open water. At that, composite image provides more visually distinct and varied backscatter signatures compared with single-polarization imaging (Fig. 3.36a and 3.36b) and the addition of co-polarization ratio data enhances ice and water features that have polarization differences.

MYI existing as floes frozen within FYI (gray areas) is characterized by high values of σ^0_{HH} and σ^0_{VV}. The values of σ^0_{HH} and σ^0_{VV} for rough FYI are lower than that of MYI, while values of ratio γ_{co} are almost equal (Fig. 3.36c). Smooth FYI appears as both dark blue (with positive γ_{co} values) and brown areas (with negative γ_{co} values) located vertically through the center of the image, in the north-east corner, and in isolated areas between the MYI and rough FYI.

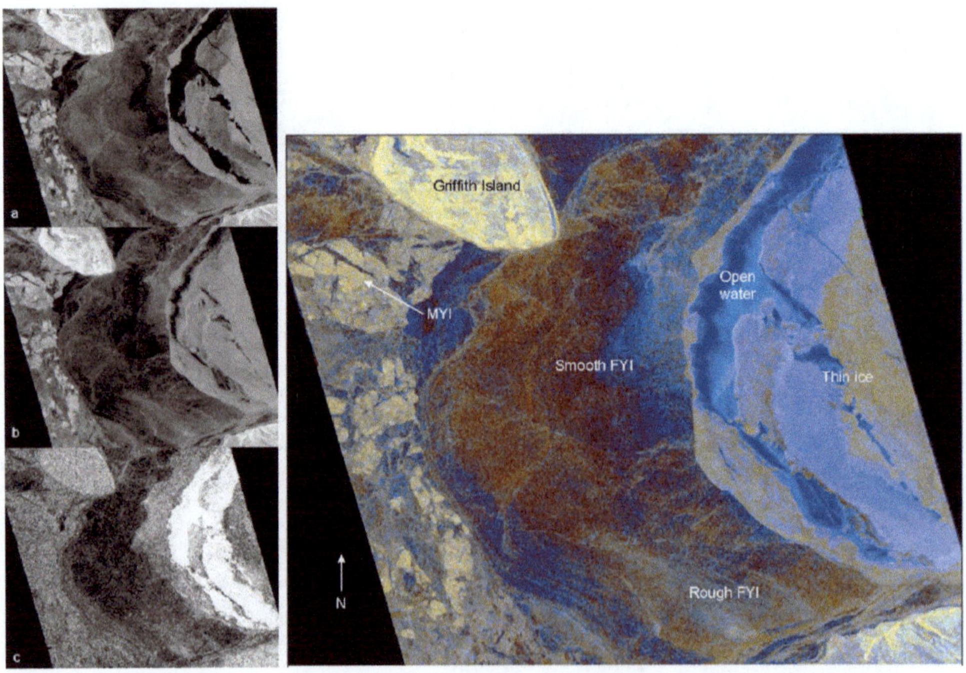

Fig. 3.36. Dual-polarization images for a portion of the study area, April 2004, 0345 UTC – left: from up to bottom (a) σ^0_{HH}, (b) σ^0_{VV}, (c) γ_{co}, and right: σ^0_{HH} (red), σ^0_{VV} (green), γ_{co} (blue). From Geldsetzer and Yackel (2009), original synthetic aperture radar (SAR) data © 2004 European Space Agency.

Thin ice exhibits a variety of colors (see the yellow, light-brown, and mottled light-blue areas in Fig. 3.36). This reflects the variety of types and hence scattering properties of thin ice. However, authors note the difficulty of identifying specific types of thin ice, namely new, nilas, and young ice, without *in situ* observations. The open water associated with blue and striated light-blue areas in Fig. 3.36 show high values of σ^0_{VV} and γ_{co}.

3.5.6 Geology

The structure of the natural terrain is very complicated and the interaction of the electromagnetic radiation with the highly variable ground surface is not yet fully understood.

Singhroy and Molch (2004) demonstrated using polarimetric metrics for the physical properties and distribution of surficial materials mapping basing on CV-580 C-band full polarimetric data acquired over bare fields in the Ottawa test site in November 1999 (Fig. 3.37). This is essential for construction engineering and groundwater, mineral, and aggregate exploration studies. Surficial materials are unconsolidated sediments including weathered and transported materials composed of mineral, rock, organic materials, and sediments deposited by water, wind,

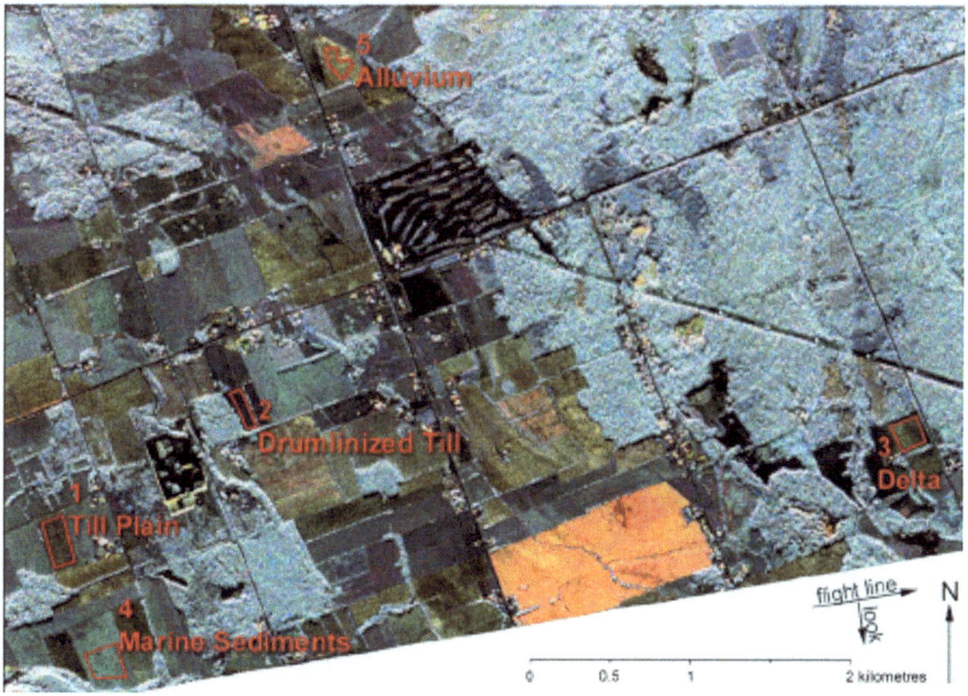

Fig. 3.37. Images HH (red), VV (green), and VH (blue) of CV-580 C-band polarimetric data acquired in November 1999 over the Ottawa study site. The arias of different surficial materials under consideration are outlined in red (1, till plain; 2, drumlinized till; 3, delta; 4, marine sediments; 5, alluvium). From Singhroy and Molch (2004).

ice, gravity, or any combination of these agents. Until now, large-scale aerial photographs and field mapping are the most common techniques used to do this.

In Fig. 3.37, the locations of the five areas of different surficial materials are shown. The polarimetric signatures averaged over each of the areas are extracted. Results are shown in Fig. 3.38.

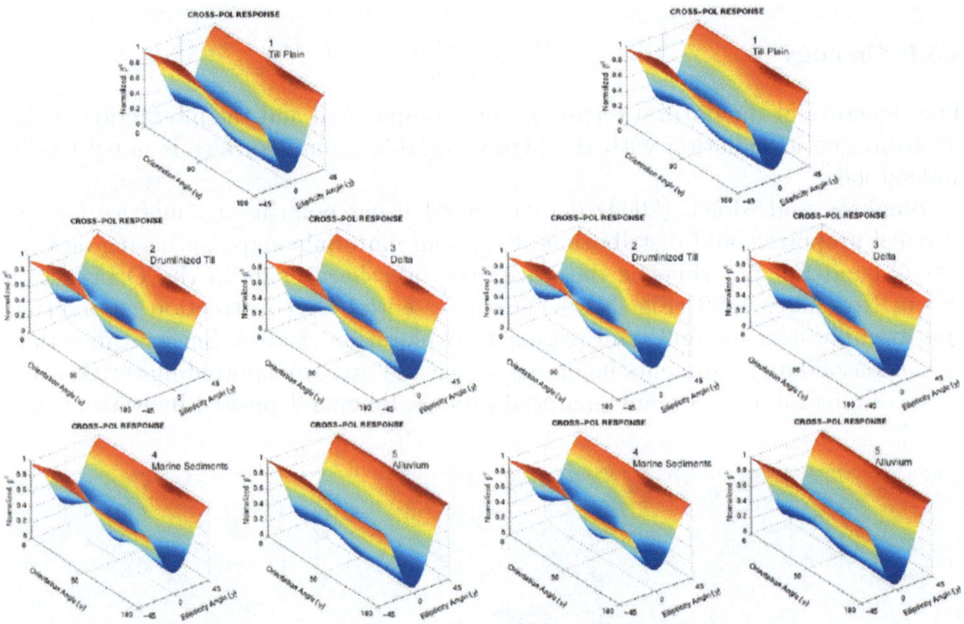

(a)

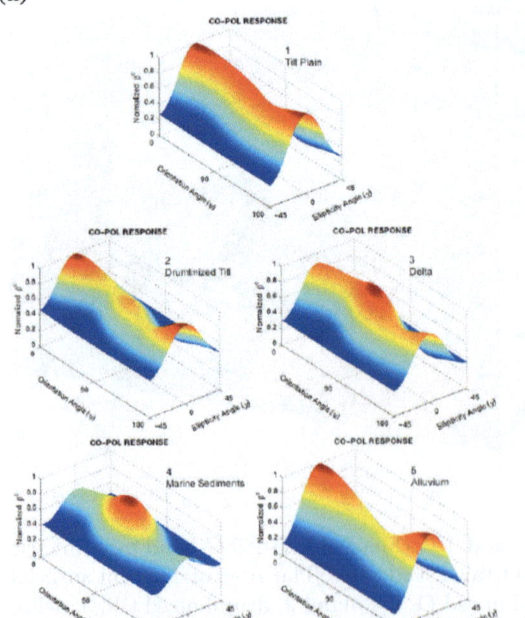

(b)

Fig. 3.38. Cross- (a) and co-polarized (b) polarimetric signatures extracted from CV-580 C-band synthetic aperture radar (SAR) data for the study sites under consideration near Ottawa. From Singhroy and Molch (2004).

The cross-polarized signatures show very little differences for areas considered, while the co-polarized signatures are different. The features of the co polarized signatures can be grouped into three types: silty material of the marine and deltaic sediments, clayey till (drumlinized and till plain), and sand–silt alluvium. The last one mainly results from residue cover, surface roughness, direction of harrowing, surface moisture, and other management practices (McNairn et al., 2002). This may potentially be useful for characterizing the engineering properties of surficial materials.

Blumberg (1998) presented a series of experiments geared toward evaluating the use of polarimetric SAR in the study of windblown morphologies in desert environments. The ability to identify dune types in remote sensing data and images is important for geomorphic mapping and climate reconstruction. This is even more interesting because two other terrestrial planets, Mars and Venus, display an abundance of wind streaks, dune fields, and even yardangs (aerodynamically shaped hills.) Hence, remote sensing is an extremely useful tool for gathering data to study such landscapes.

Sand dunes are bodies of sand particles ranging from 62 μm to 2,000 μm in diameter, spanning a variety of morphologies. The dunes are classified by wind regimes responsible for their formation and by the number of slipfaces (McKee, 1979). The linear dunes are elongated, the vegetated type have a somewhat blunt crest, and the seif type have a sharp undulating crest. Star dunes consist of multiple slip faces with a single central peak resembling a giant starfish. This dune type has at least three arms diverging from a central peak.

The study cases reported in Blumberg (1998) utilize NASA's Jet Propulsion Laboratory AIRSAR polarimetric system mounted on a DC-8 aircraft. The AIRSAR system operates three wavelengths (λ): the C-band ($\lambda = 55.6$ cm), L-band ($\lambda = 524$ cm), and the P-band ($\lambda = 567$ cm). The system is fully polarimetric.

One study case, the Stovepipe Wells dune field in Death Valley, California, utilizes SIR-C/X-SAR data (see Fig. 3.39). Depending on the operation mode, SIR-C/X-SAR is fully polarimetric in L- and C-bands, and operates only in VV in X-band. The site is located in the hottest and topographically lowest area of North America. It is especially interesting because it contains a wide variety of dune types in a small area including star, linear, transverse, and reverse transverse dunes.

The star dunes achieving heights 40 m above the surrounding surfaces are recognized by high backscatter relative to the surroundings (bright pixels in Fig. 3.39). The high backscatter is attributed by authors to corner topography between the dune arms. In Fig. 3.40, normalized polarization signature plots measured for corner reflector in L-band (a) and generated for the bright pixel of a star dune (b) are presented. Fair agreement between the two signature plots demonstrates that this technique can be used to identify star dunes in remote sensing with scanty surface information.

The effect of polarization on wind streak visibility was tested based on AIRSAR data acquired on June 1993 for a site located in the vicinity of the Cerro Quisharo volcano near the Altiplano of Bolivia at an elevation of about 4,000 m between the eastern and western Cordillera of the Andes (see Fig. 3.41). Spatial resolution is 8.27 m/pixel in azimuth and 6.66 m/pixel in range.

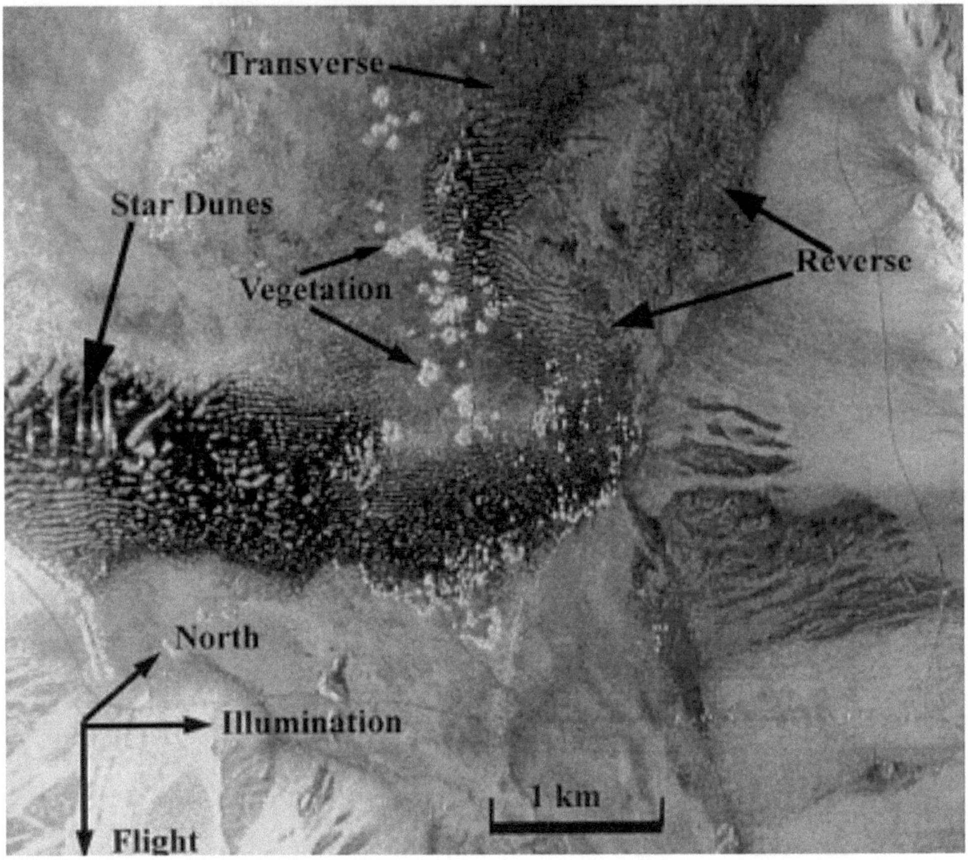

Fig. 3.39. *L*-band AIRSAR *HH* image of Stovepipe Wells Dune Field in Death Valley, California. The data were acquired in April 1994. From Blumberg (1998).

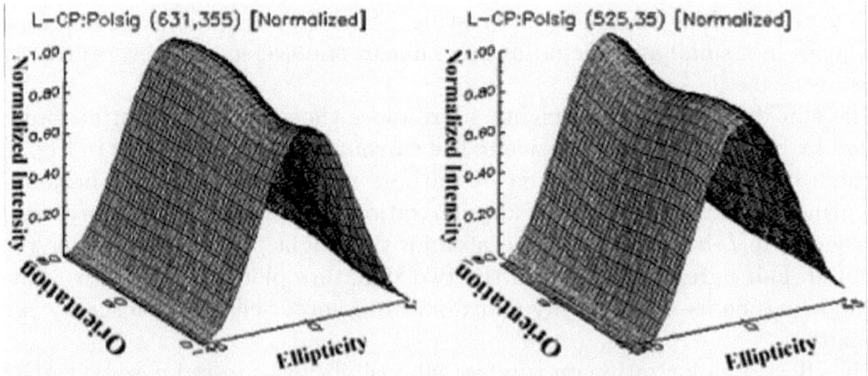

Fig. 3.40. The co-polarized polarization signatures for a trihedral corner reflector deployed at (a) Rosamond Lake and for (b) a star dune at Stovepipe Wells in Death Valley (image cm 140-2a, pixel 525, 35). From Blumberg (1998).

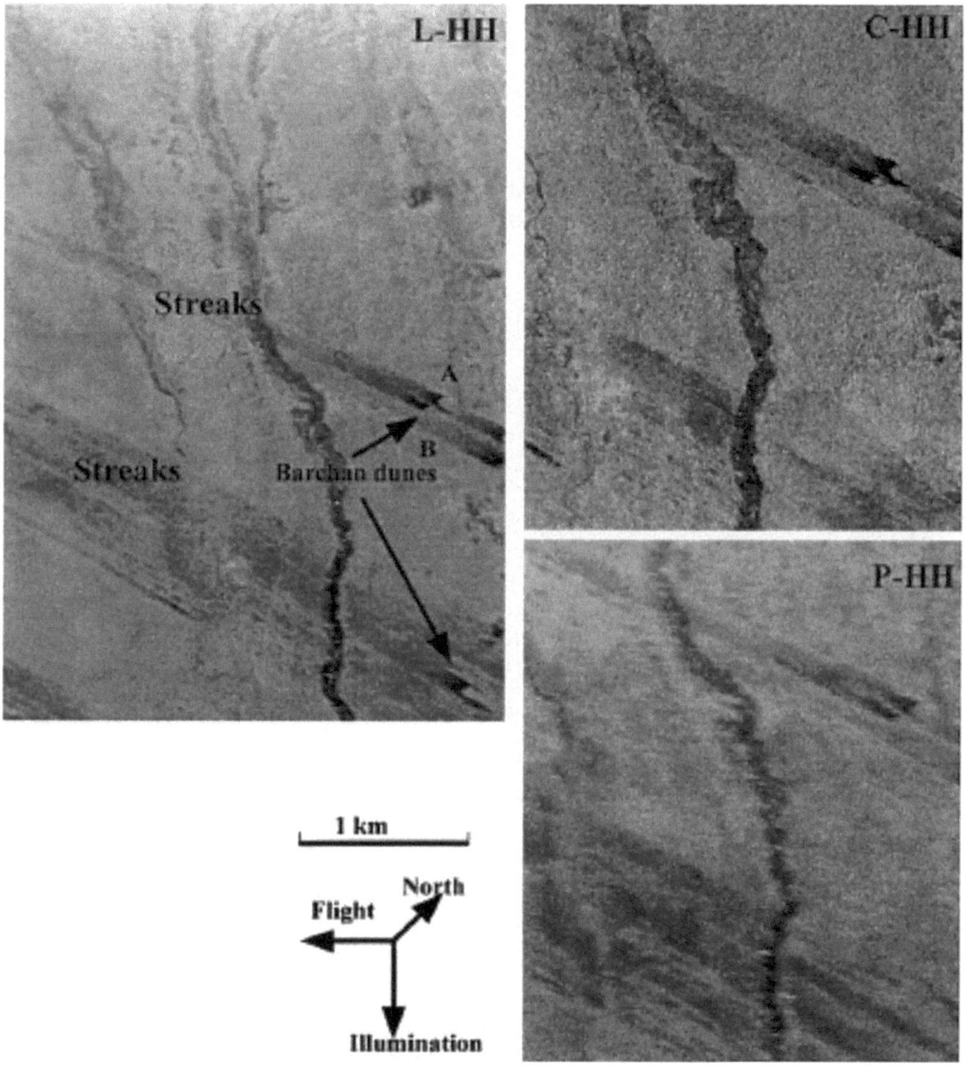

Fig. 3.41. *L*-, *C*-, and *P*-band AIRSAR *HH* images of wind streaks near the Cerro Quisharo volcano on the Altiplano of Bolivia. From Blumberg (1998).

Figure 3.41 shows cross-sections of the backscatter from A to B (see *L–HH* image) in three wavelength ranges over a wind streak.

The profiles show that *C*-band manifests much less contrast than do the *L*- and *P*-bands. Indeed, in the *C*-band data, the streaks are partly invisible. The contrast ratio is twice greater in the *P*-band than the *C*-band (0.6 versus 0.3). The results from the *C*-band are characterized by a diffuse return from ripples and other small-scale roughness on and off the streak, while, in the *L*- and *P*-bands, the streak is seen as a smooth surface (appears dark in the image). The last one is consistent with the sparse vegetation and uniform medium-sized sand sediments which were earlier observed in field studies at Cerro Quisharo (Greeley et al., 1989). From the

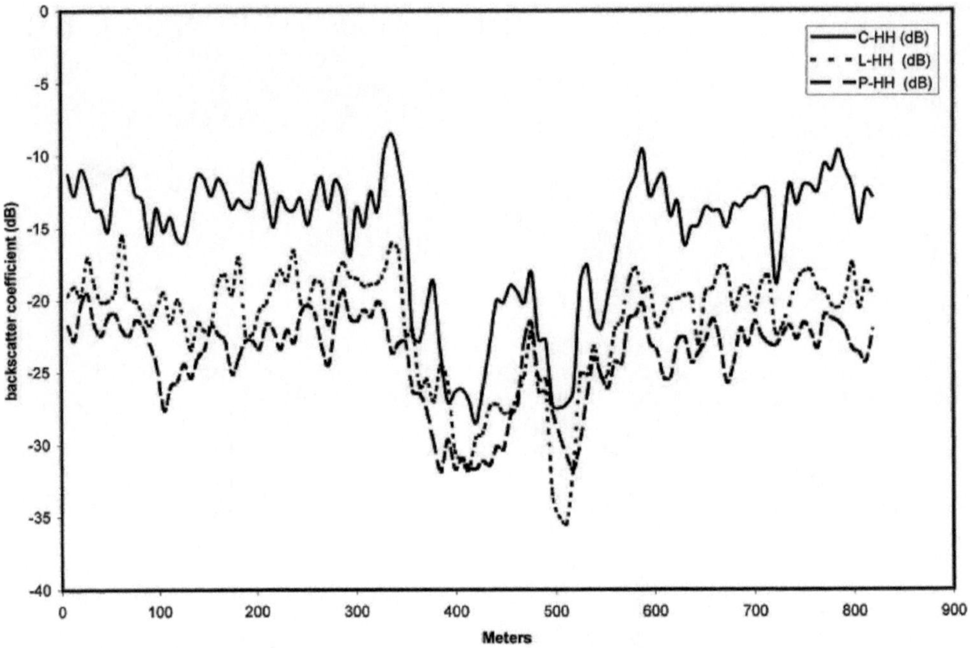

Fig. 3.42. Profile of *C–HH*, *L–HH*, and *P–HH* backscatter values over a barchan dune observed in the image shown in Fig. 3.41 from A to B. From Blumberg (1998).

results presented in Figs 3.41 and 3.42, the prevailing wind direction in this area can be deduced. In this case, it is westerly.

Another site is of interest as it demonstrates the effect of observation angle on the visibility of dunes in polarization imaging. The site, Roter Kamm, is an impact crater (Koeberl et al., 1993) in the heart of a diamond-rich area in the Namibian Desert of Southern Africa. The data were obtained during the SIR-C/X-SAR missions in April and October 1994 using the *L*-, *C*-, and *X*-band systems on board the Space Shuttle *Endeavour*. The Shuttle obtained the data in multiple passes: the first at 25° and the second at 53°. This provides the opportunity to compare the visibility of the dunes for different observation angles.

Figure 3.43 represents the images from opposing directions in *X*-band *VV* (A and B) and in *L*-band *HH* (C and D). A and C are right-looking and B and D left-looking.

The image for 25° displays specular scattering from the dunes in all three wavelengths, which can be seen in Fig. 3.43 as a dark background peppered by bright spots. The 54° image displays a dark streak. Additionally, Fig. 3.43B shows that the longer wavelength (*L*-band) is characterized by somewhat higher contrast.

Changes in snow cover and glaciers are sensitive indicators of the environment and climate changes, which are evidently related to human life on Earth (Konig et al., 2001). In the glaciology, radar polarimetry allows carrying out the research on remote areas that are hard to access, to gather data on a spatial extent not possible by fieldwork alone, and, in addition, are not influenced by clouds. This is

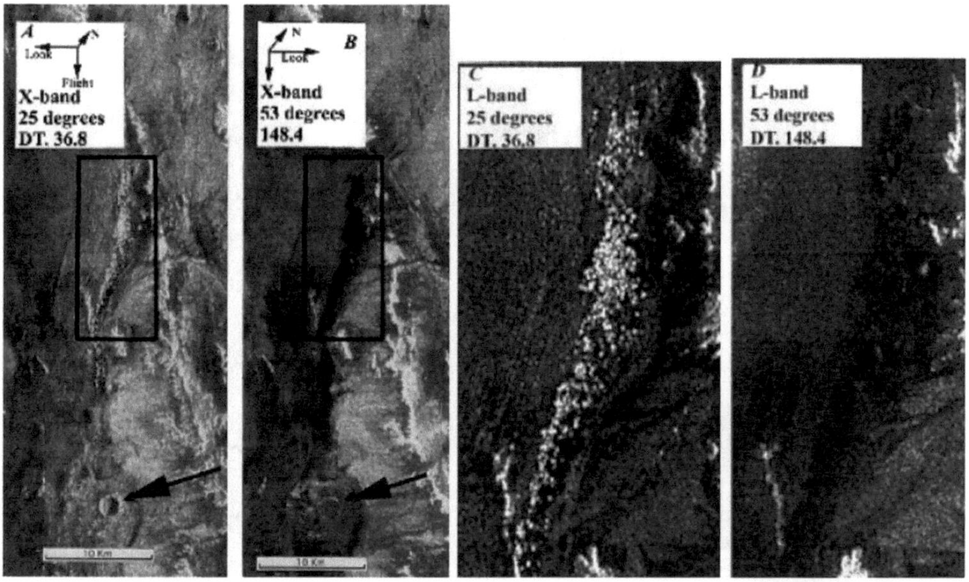

Fig. 3.43. Polarization images of the Roter Kamm impact crater area. The black boxes in A and B mark the area shown in C and D. The circular morphology shown by an arrow is the Roter Kamm impact crater. From Blumberg (1998).

extremely important because glacial areas are usually under cloud cover, especially during the summer months.

In Huang et al. (2011), the ability of the target decomposition features, namely the backscattering coefficients, the H/A/α and Pauli decompositions, for ground-cover classification in glacial areas are verified.

The study area is around the Dongkemadi glacier (as shown in Fig. 3.44), which lies in the eastern part of the Qinghai–Tibetan plateau. The Dongkemadi glacier consists of a main glacier, which is called the Da–Dongkemadi (DD) glacier, and a branch glacier, which is called the Xiao–Dongkemadi (XD) glacier. The altitude of the DD glacier ranges from 5,275 to 6,060 m and, on the XD glacier, the altitude ranges from 5,380 to 5,926 m. The lengths of the DD glacier and the XD glacier are 5.4 and 2.8 km, respectively. The terrain of the two glaciers is relatively gentle.

The land cover in the study area can be classified into five types: wet snow, ice, river outwash, rocky land, and soil land with sparse low-growing grass. These types are shown in Fig. 3.45(I) (C-band RADARSAT-2). In Fig. 3.45(I), a special, high-brightness type called a natural corner reflector formed by the structure of the glacier can be seen. Note that the natural corner reflector is not a kind of land cover object, but rather a natural structure.

To verify the possibility of differentiating the ice from the ground, the co- and cross-polarization sigtatures and metrics basing on the Pauli and $H/A/\alpha$ target decompositions (see section 3.4) along the centerline of the DD glacier shown in Fig. 3.45(II) are generated. The glacier is divided into four parts denoted by A, B, C, D, and E: the section AB corresponds to wet snow, BC to bare ice, CD to river

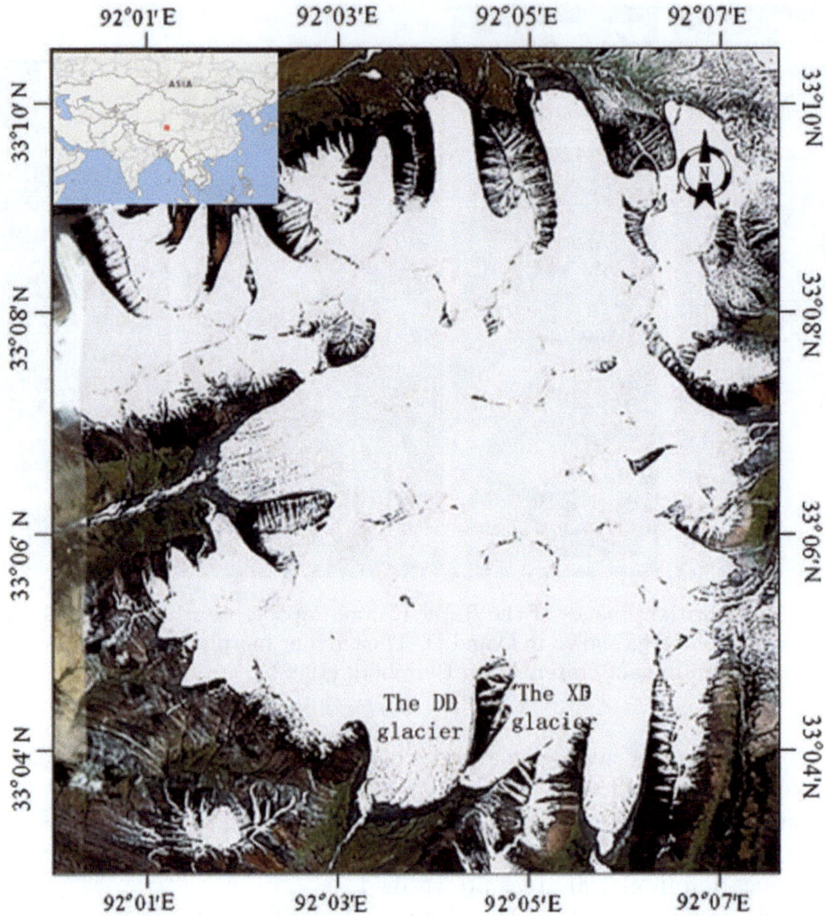

Fig. 3.44. The study area around the Dongkemadi glacier acquired from Google Earth (the original image was acquired in May 2009). The map in the top-left corner shows the geographic location of the glaciers. From Huang et al. (2011).

outwash, and DE to soil land, respectively (see Fig. 3.45II). Results are presented in Fig. 3.46.

The wet-snow areas (section AB) demonstrate lower values of the three scattering metrics of Pauli and $H/A/\alpha$ decompositions in comparison with other sections of the glacier shown in Fig. 3.45II. This results from the fact that this area has impure surface scattering but is close to volume scattering. The average α value of the river outwash area (section CD) is very close to AB, while section DE, namely the ground area, behaves differently, implying that DE is mainly characterized by surface scattering. The ice area along BC is between these two cases.

Outside the centerline, there are two different types of ground objects in the study area: the natural corner reflector and the rocky land. The natural corner reflector, a special structure formed by the ground and relief glacier edge, demonstrates strong backscatter, which distinguishes it from other objects. The rocky

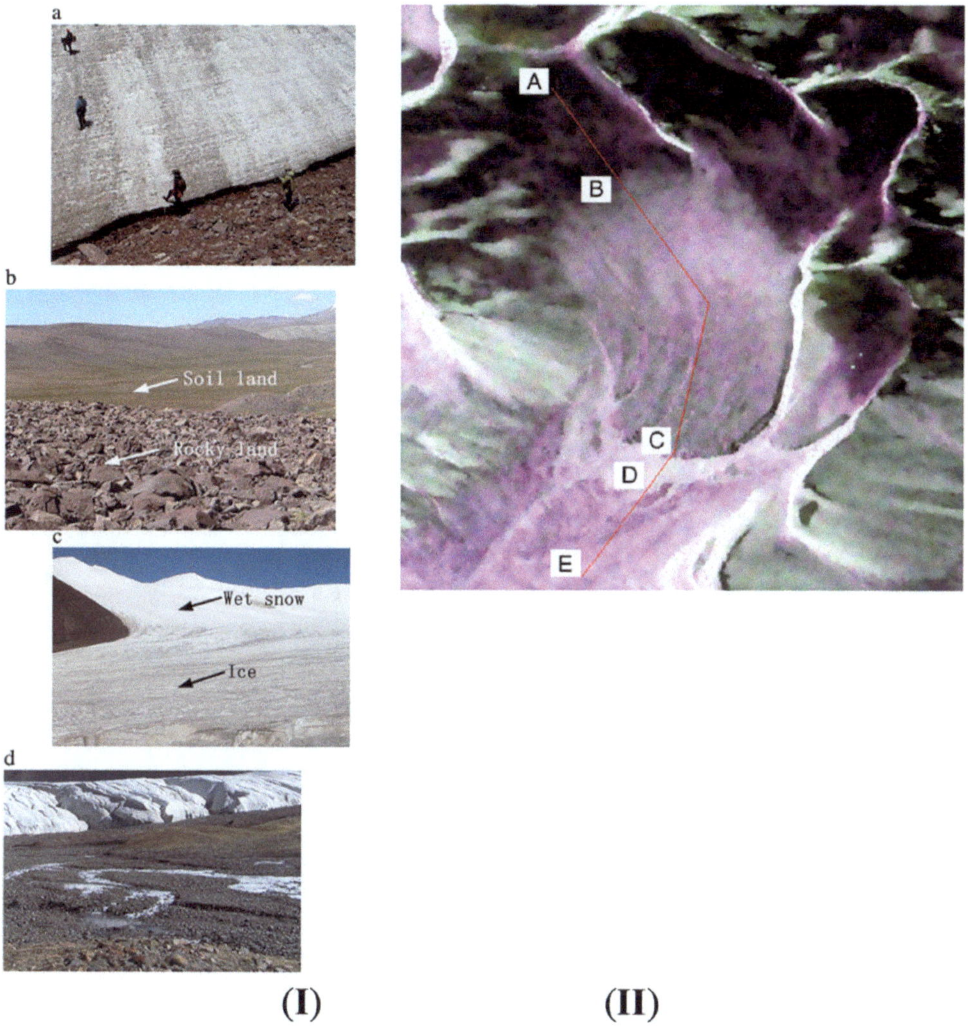

Fig. 3.45. (I) Field photos taken around the Dongkemadi glacier. (a) the steep edge of the glacier that forms the natural corner reflector; (b) soil land and rocky land; (c) wet snow and ice on the glacier surface; (d) the river outwash at the terminus of the glacier. (II) The centerline of the DD glacier (see Fig. 3.44): *HH*, red; *HV*, green; *VV*, blue. From Huang et al. (2011).

land is characterized by high backscatter; for example, the values of average entropy and α of the rocky land area are 0.829 and 36.907, respectively. However, this is lower than that for the natural corner reflector.

On the whole, the comparison of the performance in different feature spaces shows that the ice and the ground can be better distinguished in the $H/A/\alpha$ decomposition feature space than in the co- and cross-polarization signature feature spaces. Furthermore, Pauli and $H/A/\alpha$ decompositions should be accomplished simultaneously to better understand the scattering mechanisms of the study area.

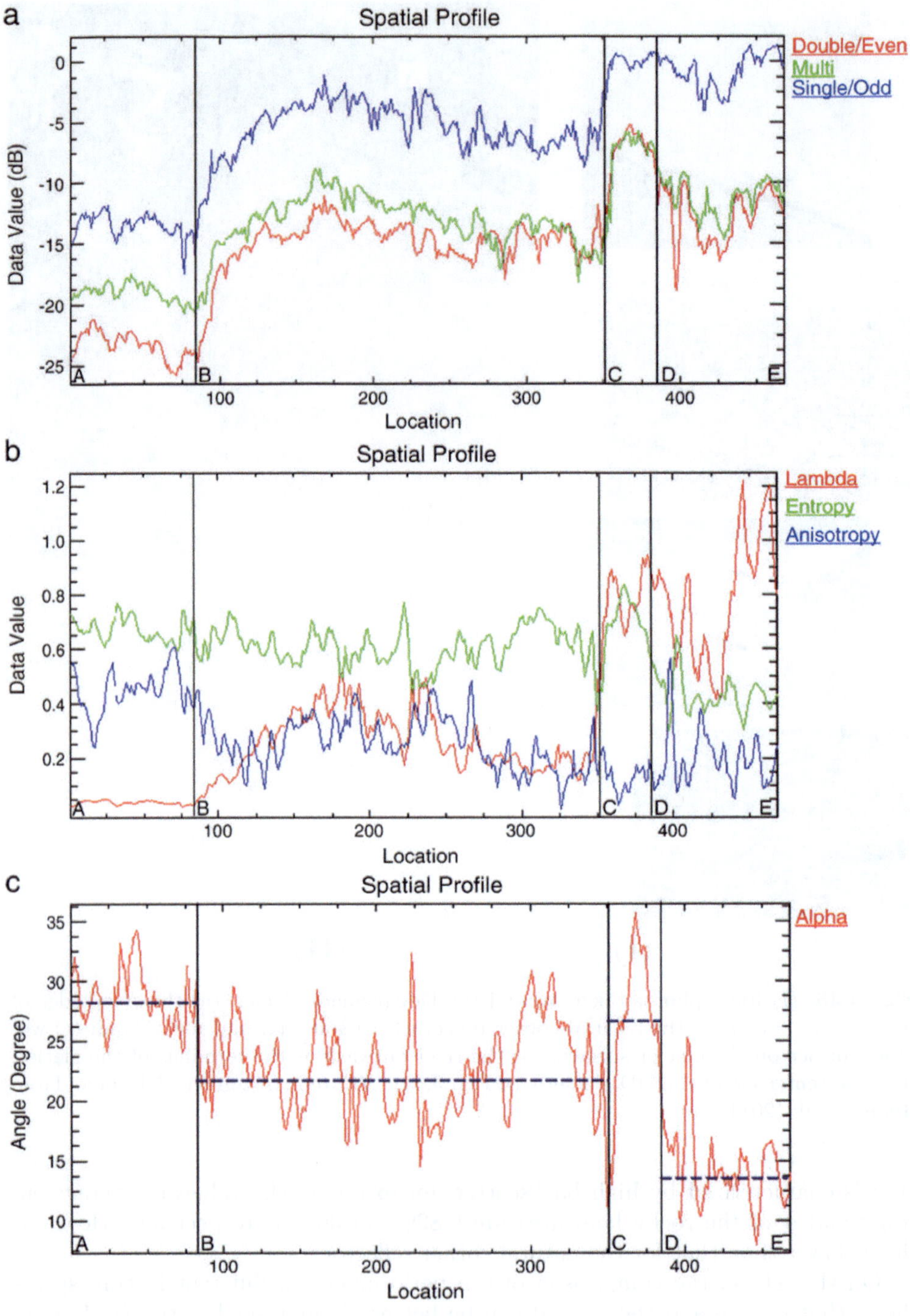
a
Spatial Profile
Double/Even
Multi
Single/Odd
Data Value (dB)
0
-5
-10
-15
-20
-25
A
B
C
D
E
100
200
300
400
Location
b
Spatial Profile
Lambda
Entropy
Anisotropy
Data Value
1.2
1.0
0.8
0.6
0.4
0.2
A
B
C
D
E
100
200
300
400
Location
c
Spatial Profile
Alpha
Angle (Degree)
35
30
25
20
15
10
A
B
C
D
E
100
200
300
400
Location

Fig. 3.46. (a)–(c)

d

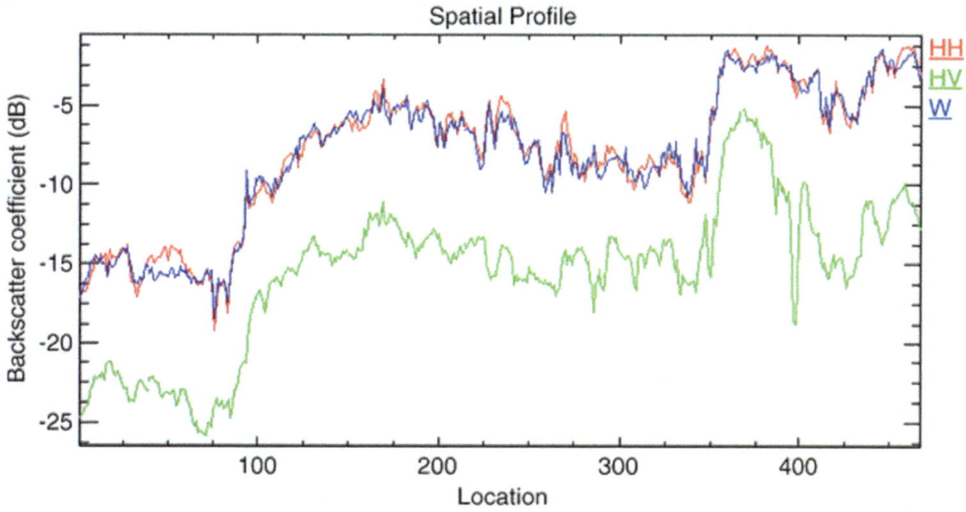

Fig. 3.46. The results of (a) Pauli decomposition; (b) and (c) $H/A/\alpha$ target decomposition; (d) the co- and cross-polarization signatures. All results are generated for geometry along the centerline of the DD glacier (Fig. 3.44). The blue dotted lines show the average values for the four zones. From Huang et al. (2011).

In addition, authors verify that the $H/A/\alpha$ decomposition is characterized by higher classification accuracy than the co- and cross-polarization signatures and the Pauli decomposition.

Acknowledgment

I would like to thank Dr. Alexander Kokhanovsky for comments and suggestions on an earlier version of this chapter that helped improve its readability and presentation.

References

Abhyankar, K.D. and Fymat, A.L. (1969) 'Relations between the elements of the phase matrix for scattering,' *J. Math. Phys.*, **10**, 1935–8.

Agrawal, A.B. and Boerner, W.M. (1989) 'Redevelopment of Kennaugh's target characteristic polarization state theory using the polarization transformation ratio formalism for the coherent case,' *IEEE Transactions on Geoscience and Remote Sensing*, **27**, 2–14.

Anderson, D.G.M. and Barakat, R. (1994) 'Necessary and sufficient conditions for a Mueller matrix to be derivable from a Jones matrix,' *J. Opt. Soc. Am. A.*, **11**, 2305–19.

Azzam, R.M. (1997) 'Mueller-matrix ellipsometry: A review,' *Proc. SPIE*, **3121**, 396–405.

Azzam, R.M. and Bashara, N.M. (1977) *Ellipsometry and Polarized Light*, North-Holland, New York.

Baghdadi, N., Holah, N., Dubois-Fernandez, P., Dupuis, X. and Garestier, F. (2006) 'Evaluation of polarimetric *L*- and *P*-bands RAMSES data for characterizing Mediterranean vineyards,' *Can. J. Remote Sensing*, **32**, 380–9.

Barakat, R. (1963) 'Theory of the coherency matrix for light of arbitrary spectral bandwidth,' *J. Opt. Soc. Am.*, **53**, 317–22.

Barakat, R. (1987) 'Conditions for the physical realizability of polarization matrices characterizing passive systems,' *J. Mod. Opt.*, **34**, 1535–44.

Bernard, A. and Rose, W.I. (1990) 'The injection of sulfuric acid aerosol in the stratosphere by El Chichon volcano and its related hazard to international air traffic,' *Nat. Hazards*, **3**, 59–67.

Bicout, D., Brosseau, Ch., Martinez, A.S. and Schmitt, J.M. (1994) 'Depolarization of multiply scattered waves by spherical diffusers: Influence of the size parameter,' *Phys. Rev. E.*, **49**, 1767–70.

Bindlish, R. and Barros, A.P. (2001) 'Parameterization of vegetation backscatter in radar-based, soil moisture estimation,' *Remote Sensing of Environment*, **76**, 130–7.

Bjerhammar, A. (1951) 'Application of calculus of matrices to method of least squares: With special references to geodetic calculations,' *Trans. Inst. Technol. (Stockholm)*, **49**, 1–86.

Blumberg, D.G. (1998) 'Remote sensing of desert dune forms by polarimetric Synthetic Aperture Radar (SAR),' *Remote Sens. Environ.*, **65**, 204–16.

Boerner, W.-M. (1992) *Direct and Inverse Methods in Radar Polarimetry*, Kluwer, Dordrecht.

Boerner, W.M., Mott, H., Lüneburg, E., Livingstone, C., Brisco, B., Brown, R.J., Paterson, J.S., Cloude, S.R., Krogager, E., Lee, J.S., Schuler, D.L., van Zyl, J.J., Randall, D., Budkewitsch, P. and Pottier, E. (1998) 'Polarimetry in radar remote sensing: Basic and applied concepts,' in Henderson, F.M. and Lewis, A.J. (eds), *Manual of Remote Sensing: Principles and Applications of Imaging Radar*, Wiley, New York.

Bohren, C.F. and Huffman, E.R. (1983) *Absorption and Scattering of Light by Small Particles*, Wiley, New York.

Borel, C.C. and McIntosh, R.E. (1990) 'Millimeter wave backscatter from deciduous trees,' *IEEE Trans. Antennas Propagat.*, **38**, 1391–8.

Borgeaud, M., Shin, R.T. and Kong, J.A. (1987) 'Theoretical models for polarimetric radar clutter,' *J. Electromagnetic Waves and Applications*, **1**, 73–89.

Brosseau, Ch. (1990) 'Analysis of experimental data for Mueller polarization matrices,' *Optik*, **85**, 83–6.

Brosseau, Ch. (1998) *Fundamentals of Polarized Light: A Statistical Optics Approach*, Wiley, New York.

Brosseau, Ch., Givens, C.R. and Kostinski, A.B. (1993) 'Generalized trace condition on the Mueller–Jones polarization matrix,' *J. Opt. Soc. Am. A.*, **10**, 2248–51.

Bueno, J.M. (2001) 'Indices of linear polarization for an optical system,' *J. Opt. A: Pure Appl. Opt.*, **3**, 470–6.

Chang, P.S., Mead, J.B., Langlois, Ph.M. and McIntosh, R.E. (1992) 'Millimeter-wave measurements of natural surfaces at 35, 95, and 225 GHz,' *Proc. ISAP*, 541–4.

Charlson, R.J., Schwartz, S.E., Hales, J.M., Cess, R.D., Coakley, J.A., Jr., Hansen, J.E. and Hofmann, D.J. (1992) 'Climate forcing by anthropogenic aerosols,' *Science*, **255**, 423–31.

Chiapello, I., Moulin, C. and Prospero, J.M. (2005) 'Understanding the long-term variability of African dust transport across the Atlantic as recorded in both Barbados surface concentrations and large-scale Total Ozone Mapping Spectrometer (TOMS) optical thickness,' *J. Geophys. Res.*, **110**, D18S10.

Chipman, R.A. (1995) 'Polarimetry,' in Bass, M. (ed.), *Handbook of Optics*, Vol. II, McGraw Hill, New York.

Chipman, R.A. (1999) 'Depolarization,' *Proc. SPIE*, **3754**, 14–20.

Chipman, R.A. (2005) 'Depolarization index and the average degree of polarization,' *Appl. Opt.*, **44**, 2490–5.
Cloude, S.R. (1986) 'Group theory and polarization algebra,' *Optik*, **7**, 26–36.
Cloude, S.R. (2010) *Polarization: Applications in Remote Sensing*, Oxford University Press, New York.
Cloude, S.R. and Pottier, E. (1995) 'Concept of polarization entropy in optical scattering,' *Opt. Eng.*, **34**, 1599–610.
Cloude, S.R. and Pottier, E. (1996) 'A review of target decomposition theorems in radar polarimetry,' *IEEE Trans. Geosci. Remote Sensing*, **34**, 498–518.
Cloude, S.R. and Pottier, E. (1997) 'An entropy based classification scheme for land applications of polarimetric SAR,' *IEEE Transactions on Geoscience and Remote Sensing*, **35**, 68–78.
Cloude, S.R., Papathanassiou, K.P. and Pottier, E. (2001) 'Radar polarimetry and polarimetric interferometry,' *IEICE Trans. Electron.*, **E84-C**, 1814–22.
Collett, E. (1993) *Polarized Light: Fundamentals and Applications*, Marcel Decker, New York.
De Martino, A., Kim, Y.-K., Garcia-Caurel, E., Laude, B. and Drevillon, B. (2003) 'Optimized Mueller polarimeter with liquid crystals,' *Opt. Lett.*, **28**, 616–18.
Deibler, L.L. and Smith, M.H. (2001) 'Measurement of the complex refractive index of isotropic materials with Mueller matrix polarimetry,' *Appl. Opt.*, **40**, 3659–67.
De Roo, R.D., Du, Y., Ulaby, F.T. and Dobson, M.C. (2001) 'A semi-empirical backscattering model at L-band and C-band for a soybean canopy with soil moisture inversion,' *IEEE Trans. Geosci. Remote Sensing*, **39**, 864–72.
Dong, Y. and Richards, J.A. (1995a) 'Forest discrimination using SAR multifrequency and multipolarization data,' *Proc. IGARSS*, **1**, 616–18.
Dong, Y. and Richards, J.A. (1995b) 'Studies of the cylinder-ground double bounce scattering mechanism in forest backscatter models,' *IEEE Trans. on Geoscience and Remote Sensing*, **33**, 229–31.
Dong, Y., Forster, B.C. and Ticehurst, C. (1996) 'Decompositions of radar polarization signatures from built and natural targets,' *Int. Archives of Photogrammetry and Remote Sensing*, **XXXI**, Part B7, 196–202.
Dubois, P.C. and Norikane, I. (1987) 'Data volume reduction for imaging radar polarimetry,' *Proc. IGARSS*, 691–6.
Dubovik, O., Holben, B.N., Lapyonok, T., Sinyuk, A., Mishchenko, M.I., Yang, P. and Slutsker, I. (2002) 'Non-spherical aerosol retrieval method employing light scattering by spheroids,' *Geophys. Res. Lett.*, **29**, 1415.
Durden, S.L., Morrissey, L.A. and Livingston, G.P. (1995) 'Microwave backscatter and attenuation dependence on leaf area index for flooded rice fields,' *IEEE Trans. Geosci. Remote Sens.*, **33**, 807–10.
Elachi, C. (1987) *Introduction to the Physics and Techniques of Remote Sensing*, Wiley, New York.
Engheta, N. and Elachi, C. (1982) 'Radar scattering from a diffuse vegetation layer over a smooth surface,' *IEEE Trans. Geosci. Remote Sens.*, **20**, 212–16.
Espinosa-Luna, R. and Bernabeu, E. (2007) 'On the Q(M) depolarization metric,' *Opt. Commun.*, **277**, 256–8.
Evans, D.L., Farr, T.G., van Zyl, J.J. and Zebker, H.A. (1988) 'Radar polarimetry: Analysis tools and applications,' *IEEE Trans. Geosci. Remote Sensing*, **26**, 774–89.
Freeman, A. and Durden, S.L. (1998) 'A three-component scattering model for polarimetric SAR data,' *IEEE Trans. Geosci. Remote Sens.*, **36**, 345–51.

Fromard, F., Puig, H., Mougin, E., Marty, G., Betoulle, J.L. and Cadamuro, L. (1998) 'Structure, above-ground biomass and dynamics of mangrove ecosystems: New data from French Guiana,' *Oecologia*, **115**, 39–53.

Fry, E.S. and Kattawar, G.W. (1981) 'Relationships between elements of the Stokes matrix,' *Appl. Opt.*, **20**, 2811–14.

Geldsetzer, T. and Yackel, J.J. (2009) 'Sea ice type and open water discrimination using dual co-polarized C-band SAR,' *Can. J. Remote Sensing*, **35**, 73–84.

Gerrard, A. and Burch, J.M. (1975) *Introduction to Matrix Methods in Optics*, Wiley, New York.

Gil, J.J. (2000) 'Characteristic properties of Mueller matrices,' *J. Opt. Soc. Am. A.*, **17**, 328–34.

Gil, J.J. (2007) 'Polarimetric characterization of light and media,' *Eur. Phys. J. Appl. Phys.*, **40**, 1–47.

Gil, J.J. and Bernabeu, E. (1985) 'A depolarization criterion in Mueller matrices,' *Opt. Acta*, **32**, 259–61.

Gil, J.J. and Bernabeu, E. (1986) 'Depolarization and polarization indexes of an optical system,' *Opt. Acta*, **33**, 185–9.

Gil, J.J. and Bernabeu, E. (1987) 'Obtainment of the polarizing and retardation parameters of a non-depolarizing optical system from the polar decomposition of its Mueller matrix,' *Optik*, **76**, 67–71.

Givens, C.R. and Kostinski, A.B. (1993) 'A simple necessary and sufficient condition on physically realizable Mueller matrices,' *J. Mod. Opt.*, **40**, 471–81.

Gopala Rao, A.V., Mallesh, K.S. and Sudha (1998a) 'On the algebraic characterization of a Mueller matrix in polarization optics. I. Identifying a Mueller matrix from its N matrix,' *J. Mod. Opt.*, **45**, 955–87.

Gopala Rao, A.V., Mallesh, K.S. and Sudha (1998b) 'On the algebraic characterization of a Mueller matrix in polarization optics. II. Necessary and sufficient conditions for Jones-derived Mueller matrices,' *J. Mod. Opt.*, **45**, 989–99.

Greeley, R., Christensen, P. and Carrasco, R. (1989) 'Shuttle radar images of wind streaks in the Altiplano, Bolivia,' *Geology*, **17**, 665–8.

Guissard, A. (1994) 'Mueller and Kennaugh matrices in radar polarimetry,' *IEEE Transactionson Geoscience and Remote Sensing*, **32**, 590–7.

Hajnsek, I. (2001) 'Inversion of surface parameters from polarimetric SAR,' *DLR Res. Rep. FB*, 2001–30.

Hajnsek, I., Pottier, E. and Cloude, S.R. (2003) 'Inversion of surface parameters from polarimetric SAR,' *IEEE Trans. Geosci. Remote Sensing*, **41**, 727–44.

Han, H., Wang, J., Wei, J. and Liu, S. (2010) 'Backwasting rate on debris-covered Koxkar glacier, Tuomuer Mountain, China,' *Journal of Glaciology*, **56**, 287–96.

Hansen, J., Lacis, A., Ruedy, R. and Sato, M. (1992) 'Potential climate impact of Mount Pinatubo eruption,' *Geophys. Res. Lett.*, **19**, 215–18.

Hauge, P.S. (1980) 'Recent developments in instrumentation in ellipsometry,' *Surface Sci.*, **96**, 108–40.

Held, D.N., Brown, W.E., Freeman, A., Klein, J.S., Zebker, H., Sato, T., Miller, T., Nguyen, Q. and Lou, Y. (1988) 'The NASAIJPL Multifrequency, Multipolarization Airborne SAR System,' *Proc. ICARSS*, 345–9.

Hess, L.L., Melack, J.M. and Simonett, D.S. (1990) 'Radar detection of flooding beneath the forest canopy: A review,' *Int. J. Remote Sens.*, **14**, 1313–25.

Hovenier, J.W. (1969) 'Symmetry relations for scattering of polarized light in a slab of randomly oriented particles,' *J. Atmos. Sci.*, **26**, 488–99.

Hovenier, J.W. (1970) 'Principles of symmetry for polarization studies of planets,' *Astron. Astrophys.*, **7**, 86–90.

Hovenier, J.W. (1994) 'Structure of a general pure Mueller matrix,' *Appl. Opt.*, **33**, 8318–24.

Hovenier, J.W. and Mackowski, D.W. (1998) 'Symmetry relations for forward and backward scattering by randomly oriented particles,' *J. Quant. Spectrosc. Radiat. Transfer*, **60**, 483–92.

Hovenier, J.W. and van der Mee, C.V.M. (2000) 'Basic relationships for matrices describing scattering by small particles,' in Mishchenko, M.I., Hovenier, J.W. and Travis, L.D. (eds), *Light Scattering by Nonspherical Particles: Theory, Measurements, and Applications*, Academic Press, San Diego, pp. 61–85.

Hovenier, J.W., van de Hulst, H.C. and van der Mee, C.V.M. (1986) 'Conditions for the elements of the scattering matrix,' *Astron. Astrophys.*, **157**, 301–10.

Hu, C., Li, D., Chen, C., Ge, J., Muller-Karger, F.E., Liu, J., Yu, F. and He, M.-X. (2010) 'On the recurrent *Ulva prolifera* blooms in the Yellow Sea and East China Sea,' *J. Geophys. Res.*, **115**, C05017.

Hu, Ch.-R., Kattawar, G.W., Parkin, M.E. and Herb, P. (1987) 'Symmetry theorems on the forward and backward scattering Mueller matrices for light scattering from a nonspherical dielectric scatterer,' *Appl. Opt.*, **26**, 4159–73.

Huang, L., Li, Zh., Tian, B.-S., Chen, Q., Liu, J.-L. and Zhang, R. (2011) 'Classification and snow line detection for glacial areas using the polarimetric SAR image,' *Remote Sensing of Environment*, **115**, 1721–32.

Huang, L., Li, Zh., Tian, B., Zhou, J. and Chen, Q. (2014) 'Recognition of supraglacial debris in the Tianshan Mountains on polarimetric SAR images,' *Remote Sensing of Environment*, **145**, 47–54.

Hurwitz, H. and Jones, C.R. (1941) 'A new calculus for the treatment of optical systems. II. Proof of three general equivalence theorems,' *J. Opt. Soc. Am.*, **31**, 493–9.

Huynen, J.R. (1970) 'Phenomenological theory of radar targets,' Ph.D. dissertation, Delft Technical University, The Netherlands.

Imhoff, M., Story, M., Vermillion, C., Khan, F. and Polcyn, F. (1986) 'Forest canopy characterization and vegetation penetration assessment with space-borne radar,' *IEEE Trans. Geosci. Remote Sens.*, **24**, 535–42.

Jaggard, D.L., Hill, C., Shorthill, R.W., Stuart, D., Glantz, M., Rosswog, F., Taggard, B. and Hammond, S. (1981) 'Light scattering from particles of regular and irregular shape,' *Atmos. Environ.*, **15**, 2511–19.

Jones, R.C. (1941) 'A new calculus for the treatment of optical systems. I. Description and discussion of the calculus,' *J. Opt. Soc. Am.*, **31**, 488–93.

Jones, R.C. (1942) 'A new calculus for the treatment of optical systems. IV,' *J. Opt. Soc. Am.*, **32**, 486–93.

Jones, R.C. (1947) 'A new calculus for the treatment of optical systems. V. A more general formulation and description of another calculus,' *J. Opt. Soc. Am.*, **37**, 107–10.

Jones, R.C. (1956) 'A new calculus for the treatment of optical systems. VIII. Electromagnetic theory,' *J. Opt. Soc. Am.*, **46**, 126–31.

Kalashnikova, O.V. and Sokolik, I.N. (2004) 'Modeling the radiative properties of nonspherical soil-delivered mineral aerosols,' *J. Quant. Spectrosc. Radiat. Transfer*, **87**, 137–66.

Kaplan, B., Ledanois, G. and Drevillon, B. (2001) 'Mueller matrix of dense polystyrene latex sphere suspensions: Measurements and Monte Carlo simulation,' *Appl. Opt.*, **40**, 2769–77.

Karam, M.A., Amar, F., Fung, A.K., Mougin, E., Lopes, A., Le Vine, D.M. and Beaudoin, A. (1995) 'A microwave polarimetric scattering model for forest canopies based on vector radiative transfer theory,' *Remote Sens. Environ.*, **53**, 16–30.

Kennaugh, K. (1951) 'Effects of type of polarization on echo characteristics,' *Tech. Rep.*, **OH 389-4**, 35, and **OH 381-9**, 39.
Kennaugh, K. (1952) 'Polarization properties of radar reflections,' M.Sc. thesis, Ohio State University, Columbus, Ohio.
Kim, K., Mandel, L. and Wolf, E. (1987) 'Relationship between Jones and Mueller matrices for random media,' *J. Opt. Soc. Am. A.*, **4**, 433–7.
Kim, Y.L., Pradhan, P., Kim, M.H. and Backman, V. (2006) 'Circular polarization memory effect in low-coherence enhanced backscattering of light,' *Opt. Lett.*, **31**, 2744–6.
Kliger, D.S., Lewis, J.W. and Randall, C.E. (1990) *Polarized Light in Optics and Spectroscopy*, Academic Press, New York.
Koeberl, C., Hartung, J.B., Kunk, M.J., Klein, J., Matsuda, J.-I., Nagao, K., Reimold, W.U. and Storzer, D. (1993) 'The age of Roter Kamm impact crater, Namibia: Constraints from ^{40}Ar-^{39}Ar, K-Ar, Rb-Sr, fission-track, and ^{10}Be-^{26}Al studies,' *Meteoritics*, **28**, 204–12.
Koepke, P., Hess, M., Shultz, I. and Shettle, E.P. (1997) 'Global aerosol data set,' Rep., 44, Max Planck Inst. for Meteorol.
Kokhanovsky, A.A. (2001) *Light Scattering Media Optics: Problems and Solutions*, Praxis, Chichester, UK.
Kokhanovsky, A.A. (2003a) 'Optical properties of irregularly shaped particles,' *J. Phys. D: Appl. Phys.*, **36**, 915–23.
Kokhanovsky, A.A. (2003b) 'Parameterization of the Mueller matrix of oceanic waters,' *J. Geophys. Res.*, **108**, 3175.
Kokhanovsky, A.A. (2003c) *Polarization Optics of Random Media*, Praxis, Chichester, UK.
Kong, J.A., Swartz, A.S., Yueh, H.A., Novak, L.M. and Shin, R.T. (1987) 'Identification of terrain cover using the optimum polarimeter classifier,' *J. Electromagnetic Waves and Applications*, **2**, 171–94.
Konig, M., Winther, J.G. and Isaksson, E. (2001) 'Measuring snow and glacier ice properties from satellite,' *Reviews of Geophysics*, **39**, 1–27.
Koren, I., Ganor, E. and Joseph, J.H. (2001) 'On the relation between size and shape of desert dust aerosol,' *J. Geophys. Res.*, **106**, 18037–45.
Kostinski, A.B. (1992) 'Depolarization criterion for incoherent scattering,' *Appl. Opt.*, **31**, 3506–8.
Kostinski, A.B., Givens, C.R. and Kwiatkowski, J.M. (1993) 'Constraints on Mueller matrices of polarization optics,' *Appl. Opt.*, **32**, 1646–51.
Krishnan, S. and Nordine, P.C. (1994) 'Mueller-matrix ellipsometry using the division-of-amplitude photopolarimeter: A study of depolarization effects,' *Appl. Opt.*, **33**, 4184–92.
Krogager, E. (1990) 'A new decomposition of the radar target scattering matrix,' *Electron. Lett.*, **26**, 1525–6.
Lancaster, P. and Tismenetsky, M. (1985) *The Theory of Matrices*, Academic Press, San Diego.
Lasaponara, R. and Masini, N. (2013) 'Satellite synthetic aperture radar in archaeology and cultural landscape: An overview,' *Archaeological Prospection Archaeol. Prospect.*, **20**, 71–8.
Lawless, R., Xie, Y., Yang, P., et al. (2006) 'Polarization and effective Mueller matrix for multiple scattering of light by nonspherical ice crystals,' *Opt. Express*, **14**, 6381–93.
Lee, J.S. and Pottier, E. (2009) *Polarimetric Radar Imaging: From Basics to Applications*, Taylor & Francis, New York.
Li, X., Maring, H., Savoie, D., Voss, K. and Prospero, J.M. (1996) 'Dominance of mineral dust in aerosol light scattering in the North Atlantic trade winds,' *Nature*, **380**, 416–

19.
Lu, S.-Y. and Chipman, R.A. (1994) 'Homogeneous and inhomogeneous Jones matrices,' *J. Opt. Soc. Am. A.*, **11**, 766–73.
Lu, S.-Y. and Chipman, R.A. 1996) 'Interpretation of Mueller matrices based on polar decomposition,' *J. Opt. Soc. Am. A.*, **13**, 1106–13.
Lüneburg, E. (1995) 'Principles of radar polarimetry: Using the directional Jones vector approach,' *IEICE Transactions on Electronics*, **E78-C**, 1339–45.
Lumme, K. (2000) 'Scattering properties of interplanetary dust particles,' in Mishchenko, M.I., Hovenier, J.W. and Travis, L.D. (eds), *Light Scattering by Nonspherical Particles*, Academic, San Diego, CA.
Mackowski, D.W. and Mishchenko, M.I. (1996) 'Calculation of the T matrix and the scattering matrix for ensembles of spheres,' *J. Opt. Soc. Am. A*, **11**, 2266–78.
Manickavasagam, S. and Menguc, M.P. (1998) 'Scattering-matrix elements of coated infinite-length cylinders,' *Appl. Opt.*, **37**, 2473–82.
Manninen, T., Stenberg, P., Rautiainen, M., Voipio, P. and Smolander, H. (2005) 'Leaf area index estimation of boreal forest using ENVISAT ASAR,' *IEEE Trans. Geosci. Remote Sens.*, **43**, 2627–35.
Mar'enko, V.V. and Savenkov, S.N. (1994) 'Representation of arbitrary Mueller matrix in the basis of matrices of circular and linear anisotropy,' *Opt. Spectrosc.*, **76**, 94–6.
Martin, W.E., Hesse, E., Hough, J.H., Sparks, W.B., Cockell, C.S., Ulanowski, Z., Germer, T.A. and Kaye, P.H. (2010) 'Polarized optical scattering signatures from biological materials,' *J. Quant. Spectrosc. Radiat. Transfer.*, **111**, 2444–59.
McKee, E.D. (1979) 'Introduction to a study of global sand seas,' in McKee, E.D. (ed.), *A Study of Global Sand Seas*, United States Department of Interior, Washington.
McNairn, H., Duguay, C., Brisco, B. and Pultz, T.J. (2002) 'The effect of soil and crop residue characteristics on polarimetric radar response,' *Remote Sensing of Environment*, **80**, 308– 20.
Mead, J.B. and McIntosh, R.E. (1990) 'A 225 GHz polarimetric radar,' *IEEE Transactions on Microwave Theory and Techniques*, **38**, 1252–7.
Mead, J.B. and McIntosh, R.E. (1991) 'Polarimetric backscatter measurements of deciduous and coniferous trees at 225 GHz,' *IEEE Trans. Geosci. Remote Sensing*, **29**, 21–8.
Mishchenko, M.I. and Hovenier, J.W. (1995) 'Depolarization of light backscattered by randomly oriented nonspherical particles,' *Opt. Lett.*, **20**, 1356–8.
Mishchenko, M.I. and Travis, L.D. (2000) 'Polarization and depolarization of light,' in Moreno, F. and González, F. (eds), *Light Scattering from Microstructures*, Springer, Berlin, pp. 159–75.
Mishchenko, M.I., Hovenier, J.W. and Travis, L.D. (eds) (2000) *Light Scattering by Nonspherical Particles: Theory, Measurements, and Applications*, Academic Press, San Diego.
Mishchenko, M.I., Travis, L.D. and Lacis, A.A. (2002) *Scattering, Absorption, and Emission of Light by Small Particles*, Cambridge University Press, Cambridge, UK.
Mishchenko, M.I., Rosenbush, V.K., Kiselev, N.N., Lupishko, D.F., Tishkovets, V.P., Kaydash, V.G., Belskaya, I.N., Efimov, Y.S. and Shakhovskoy, N.M. (2010) *Polarimetric Remote Sensing of Solar System Objects*, Akademperiodyka, Kyiv.
Mishchenko, M.I., Travis, L.D. and Lacis, A.A. (2006) *Multiple Scattering of Light by Particles: Radiative Transfer and Coherent Backscattering*, Cambridge University Press, Cambridge, UK.
Moore, E.H. (1920) 'On the reciprocal of the general algebraic matrix,' *Bull. Am. Math. Soc.*, **26**, 394–5.
Mott, H. (2007) *Remote Sensing with Polarimetric Radar*, Wiley, New York.

Mougin, E., Proisy, C., Marty, G., Fromard, F., Puig, H., Betoulle, J.L. and Rudant, J.P. (1999) 'Multifrequency and multipolarization radar backscattering from mangrove forests,' *IEEE Trans. Geosci. Remote Sens.*, **37**, 94–102.

Mueller, H. (1948) 'The foundation of optics,' *J. Opt. Soc. Am.*, **38**, 661.

Muinonen, K., J. Piironen, Shkuratov, Yu.G., Ovcharenko, A. and Clark, B.E. (2002) *Asteroid Photometric and Polarimetric Phase Effects in Asteroids III*, Univeristy of Arizona Press, Tucson.

Munoz, O., Volten, H., de Haan, J.F., Vassen, W. and Hovenier, J.W. (2001) 'Experimental determination of scattering matrices of fly ash and clay particles at 442 and 633 nm,' *J. Geophys. Res.*, **106**, 22833–44.

Munoz, O., Volten, H., de Haan, J.F., Vassen, W. and Hovenier, J.W. (2002) 'Experimental determination of the phase function and degree of linear polarization of El Chichon and Pinatubo volcanic ashes,' *J. Geophys. Res.*, **107**(D13), 4174.

Munoz, O., Volten, H., Hovenier, J.W., Nousiainen, T., Muinonen, K., Guirado, D., Moreno, F. and Waters, L.B.F.M. (2007) 'Scattering matrix of large Saharan dust particles: experiments and computations,' *J. Geophys. Res.*, **112**, D 13215.

Munoz, O., Volten, H., Hovenier, J.W., Veihelmann, B., van der Zande, W.J., Waters, L.B.F.M. and Rose, W.I. (2004) 'Scattering matrices of volcanic ash particles of Mount St. Helens, Redoubt, and Mount Spurr Volcanoes,' *J. Geophys. Res.*, **109**, D 16201.

Muttiah, R.S. (ed.) (2002) *From Laboratory Spectroscopy to Remotely Sensed Spectra of Terrestrial Ecosystems*, Kluwer, Dordrecht.

Nagirner, D.I. (1993) 'Constrains on matrices transforming Stokes vectors,' *Astron. Astrophys.*, **275**, 318–24.

Narayanan, R.M., Borel, C.C. and McIntosh, R.E. (1988) 'Radar backscatter characteristics of trees at 215 GHz,' *IEEE Trans. Geosci. Remote Sensing*, **26**, 217–28.

Nousiainen, T., Muinonen, K. and Raisanen, P. (2003) 'Scattering of light by large Saharan dust particles in a modified ray optics approximation,' *J. Geophys. Res.*, **108**(D1), 4025.

Novak, L.M. and Burl, M.C. (1990) 'Optimal speckle reduction in polarimetric SAR imagery,' *IEEE Transactions on Aerospace and Electronic Systems*, **26**, 293–305.

Oberbeck, V.R., Danielsen, E.F., Snetsinger, K.G. and Ferry, G.V. (1983) 'Effect of the eruption of El Chichon on stratospheric aerosol size and composition,' *J. Geophys. Lett.*, **10**, 1021–4.

Oberemok, E.A. and Savenkov, S.N. (2003) 'Structure of deterministic Mueller matrices and their reconstruction in the method of three input polarizations,' *J. Appl. Spectrosc.*, **70**, 224–9.

Oh, Y. (2008) 'Radar remote sensing of soil moisture and surface roughness for vegetated surfaces,' *Korean Journal of Remote Sensing*, **24**, 427–36.

Ossikovski, R. (2008) 'Interpretation of nondepolarizing Mueller matrices based on singular-value decomposition,' *J. Opt. Soc. Am. A.*, **25**, 473–82.

Ossikovski, R. (2009) 'Analysis of depolarizing Mueller matrices through a symmetric decomposition,' *J. Opt. Soc. Am. A.*, **26**, 1109–18.

Parke III, N.G. (1948) 'Matrix optics,' Ph.D. thesis, Massachusetts Institute of Technolody, Cambidge, MA.

Parke III, N.G. (1949) 'Optical algebra,' *J. Math. Phys.*, **28**, 131–9.

Patterson, E.M., Gillete, D.A. and Stockton, B.H. (1977) 'Complex index of refraction between 300 and 700 nm for Saharan aerosols,' *J. Geophys. Res.*, **82**(C21), 3153–60.

Penrose, R. (1955) 'A generalized inverse for matrices,' *Proc. Cambridge Philos. Soc.*, **151**, 406–13.

Perrin, F. (1942) 'Polarization of light scattering by isotropic opalescent media,' *J. Chem. Phys.*, **10**, 415–27.

Pospergelis, M.M. (1969) 'Spectroscopic measurements of the four stokes parameters for light scattered by natural objects,' *Soviet Physics – Astronomy*, **12**, 973–6.

Potton, R.J. (2004) 'Reciprocity in optics,' *Rep. Prog. Phys.*, **67**, 717−54.

Priezzhev, A.V., Tuchin, V.V. and Shubochkin, L.P. (1989) *Laser Diagnostics in Biology and Medicine*, Nauka, Moscow (in Russian).

Proisy, C., Mougin, E., Fromard, F. and Karam, M.A. (2000) 'Interpretation of polarimetric radar signatures of mangrove forests,' *Remote Sens. Environ.*, **71**, 56–66.

Prospero, J.M. and Lamb, P.J. (2003) 'African droughts and dust transport to the Caribbean: Climate change implications,' *Science*, **302**, 1024–7.

Pueschel, R.F., Russell, P.B., Allend, D.A., Ferry, G.V., Snetsinger, K.G., Livingston, J.M. and Verma, S. (1994) 'Physical and optical properties of the Pinatubo volcanic aerosol: Aircraft observations with impactors and a Sun-tracking photometer,' *J. Geophys. Lett.*, **99**(12), 915–22.

Ray, T.W., Farr, T.G. and van Zyl, J.J. (1992) 'Detection of land degradation with polarimetric SAR,' *Geophys. Res. Lett.*, **19**, 1587–90.

Richards, J.A., Woodgate, P.W. and Skidmore, A.K. (1987) 'An explanation of enhanced radar backscattering from flooded forests,' *Int. J. Remote Sens.*, **18**, 1319–32.

Rojas-Ochoa, L.F., Lacoste, D., Lenke, R., et al. (2004) 'Depolarization of backscattered linearly polarized light,' *J. Opt. Soc. Am. A.*, **21**, 1799−804.

Rose, W.I. (1986) 'Interaction of aircraft and explosive eruption clouds: A volcanologist's perspective,' *AIAA J.*, **25**, 52.

Salberg, A.-B., Solberg, S., Weydahl, D.J. and Astrup, R. (2009) 'Leaf area index estimation using ENVISAT ASAR and Radarsat-2,' *Note No.SAMBA/30/09*, Norwegian Computing Center, NR.

Savenkov, S.N. (2002) 'Optimization and structuring of the instrument matrix for polarimetric measurements,' *Opt. Eng.*, **41**, 965–72.

Savenkov, S.N. (2007) 'Analysis of generalized polarimetric measurement equation,' *Proc SPIE*, **6682**, 668214.

Savenkov, S.N. and Muttiah, R.S. (2004) 'Inverse polarimetry and light scattering from leaves,' in Videen, G., Yatskiv, Y. and Mischenko, M. (eds), *Photopolarimetry in Remote Sensing*, Springer, The Netherlands.

Savenkov, S.N., Marienko, V.V., Oberemok, E.A. and Sydoruk, O.I. (2006) 'Generalized matrix equivalence theorem for polarization theory,' *Phys. Rev. E.*, **74**, 056607.

Savenkov, S.N., Mishchenko, L.T., Muttiah, R.S., Oberemok, Y.A. and Mishchenko, I.A. (2004) 'Mueller polarimetry of virus-infected and healthy wheat under field and microgravity conditions,' *J. Quant. Spectrosc. Radiat. Transfer*, **88**, 327–43.

Savenkov, S.N., Muttiah, R.S. and Oberemok, Y.A. (2003) 'Transmitted and reflected scattering matrices from an English oak leaf,' *Appl. Opt.*, **42**, 4955–62.

Savenkov, S.N., Muttiah, R.S., Yushtin, K.E. and Volchkov, S.A. (2007a) 'Mueller-matrix model of an inhomogeneous, linear, birefringent medium: Single scattering case,' *J. Quant. Spectrosc. Radiat. Transfer*, **106**, 475–86.

Savenkov, S.N., Sydoruk, O.I. and Muttiah, R.S. (2005) 'The conditions for polarization elements to be dichroic and birefringent,' *J. Opt. Soc. Am. A.*, **22**, 1447−52.

Savenkov, S.N., Sydoruk, O.I. and Muttiah, R.S. (2007b) 'Eigenanalysis of dichroic, birefringent, and degenerate polarization elements: A Jones-calculus study,' *Appl. Opt.*, **46**, 6700−9.

Saxon, D.S. (1955) 'Tensor scattering matrix for electromagnetic fields,' *Phys. Rev.*, **100**, 1771−5.

Sekera, Z. (1966) 'Scattering matrices and reciprocity relationships for various representations of the state of polarization,' *J. Opt. Soc. Am.*, **56**, 1732−40.

Shen, H., Perrie, W., Liu, Q. and He, Y. (2014) 'Detection of macroalgae blooms by complex SAR imagery,' *Marine Pollution Bulletin*, **78**, 190–5.

Shindo, Y. (1995) 'Applications of polarized modulator technique in polymer science,' *Opt. Eng.*, **34**, 3369−84.

Shurcliff, W.A. (1962) *Polarized Light: Production and Use*, Harvard University Press, Harvard, MA.

Sikdar, M., MacIntosh, S., Cumming, I. and Brisco, B. (2005) 'Incorporating a vegetation index into a soil moisture retrieval model: Results from Conair-580 SAR data,' *Proc. IGARSS*, 383–6.

Simon, R. (1982) 'The connection between Mueller and Jones matrices of polarization optics,' *Opt. Commun.*, **42**, 293−7.

Simon, R. (1987) 'Mueller matrices and depolarization criteria,' *J. Mod. Opt.*, **34**, 569−75.

Sinclair, G. (1950) 'Transmission and reception of elliptically polarized waves,' *Proc. IRE*, **38**, 148–51.

Singhroy, V. and Molch, K. (2004) 'Geological case studies related to RADARSAT-2,' *Can. J. Remote Sensing*, **30**, 893–902.

Smith, M.H. (2002) 'Optimization of a dual-rotating-retarder Mueller matrix polarimeter,' *Appl. Opt.*, **41**, 2488–95.

Snetsinger, K.G., Ferry, G.V., Russell, P.B., Pueschel, R.F. and Oberbeck, V.R. (1987) 'Effects of El Chichon on stratospheric aerosols late 1982 to early 1984,' *J. Geophys. Res.*, **92**, 761–71.

Solberg, S., Brunner, A., Hanssen, K.H., Lange, H., Nusset, E., Rautiainen, M. and Stenberg, P. (2009) 'Mapping LAI in a Norway spruce forest using airborne laser scanning,' *Remote Sensing of Environment*, **113**, 2317–27.

Solleillet, P. (1929) 'Sur les paramètres caractérisant la polarisation partielle de la lumière dans les phénomènes de fluorescence,' *Ann. Phys. Biol. Med.*, **12**, 23–97.

Sridhar, R. and Simon, R. (1994) 'Normal form for Mueller matrices in polarization optics,' *J. Mod. Opt.*, **41**, 1903−15.

Swami, M.K., Manhas, S., Buddhiwant, P., Ghosh, N., Uppal, A. and Gupta, P.K. (2006) 'Polar decomposition of 3x3 Mueller matrix: A tool for quantitative tissue polarimetry,' *Opt. Express*, **14**, 9324–37.

Tang, S.T. and Kwok, H.S. (2001) '3x3 Matrix for unitary optical systems,' *J. Opt. Soc. Am. A*, **18**, 2138–45.

Touzi, R. and Lopes, A. (1994) 'The principle of speckle filtering in polarimetric SAR imagery,' *IEEE Transactions on Geoscience and Remote Sensing*, **32**, 1110–14.

Touzi, R., Boerner, W.M., Lee, J.S. and Lüneburg, E. (2004) 'A review of polarimetry in the context of synthetic aperture radar: concepts and information extraction,' *Can. J. Remote Sensing*, **30**, 380–407.

Tsang, L., Kong, J.A. and Shin, R.T. (1985) *Theory of Microwave Remote Sensing*, Wiley, New York.

Tsitas, S.R. and Yung, Y.L. (1996) 'The effect of volcanic aerosols on ultraviolet radiation in Antarctica,' *Geophys. Res. Lett.*, **23**, 157–60.

Tuchin, V.V. (ed.) (2002) *Handbook of Optical Biomedical Diagnostics*, SPIE Press, Bellingham, WA.

Tuchin, V.V., Wang, L.V. and Zimnyakov, D.A. (2006) *Optical Polarization in Biomedical Applications*, Springer, Berlin.

Twietmeyer, K. and Chipman, R.A. (2008) 'Optimization of Mueller polarimeters in the presence of error sources,' *Opt. Express*, **16**, 11589–603.

Tyo, J.S., Wang, Zh., Johnson, S.J. and Hoover, B.G. (2010) 'Design and optimization of partial Mueller matrix polarimeters,' *Appl. Opt.*, **49**, 2326–33.

Ulaby, F.T., Allen, C.T. and Eger III, G. (1984) 'Relating the microwave backscattering coefficient to leaf area index,' *Remote Sensing of Environment*, **14**, 113–33.

Ulaby, F.T., Moore, R.K. and Fung, A.K. (1986) *Microwave Remote Sensing: Radar Remote Sensing and Surface Scattering and Emission Theory*, Vol. II, Artech House Inc., Norwood.

Valenzuela, G.R. (1967) 'Depolarization of EM waves by slightly rough surfaces,' *IEEE Trans. on Antennas and Propagation*, **AP-15(4)**, 552–7.

van de Hulst, H.C. (1957) *Light Scattering by Small Particles*, Wiley, New York.

van der Mee, C.V.M. (1993) 'An eigenvalue criterion for matrices transforming Stokes parameters,' *J. Math. Phys.*, **34**, 5072–88.

van der Mee, C.V.M. and Hovenier, J.W. (1992) 'Structure of matrices transforming Stokes parameters,' *J. Math. Phys.*, **33**, 3574–84.

van Zyl, J.J. (1989) 'Unsupervised classification of scattering behavior using radar polarimetry data,' *IEEE Transactions on Geoscience and Remote Sensing*, **27**, 37–45.

van Zyl, J.J. (1992) 'Application of Cloude's target decomposition theorem to polarimetric imaging radar data,' *Proc. SPIE*, **1748**, 184–91.

van Zyl, J.J. and Ulaby, F.T. (1990) 'Scattering matrix representation for simple targets,' in Ulaby, F.T. and Elachi, C. (eds), *Radar Polarimetry for Geosciences Applications*, Artech House, Norwood, MA.

van Zyl, J.J. and Zebker, H.A. (1990) 'Imaging radar polarimetry,' in Kong, J.A. (ed.), *PIER 3, Progress in Electromagnetics Research*, Elsevier, New York.

van Zyl, J.J., Zebker, H.A. and Elachi, C. (1987) 'Imaging radar polarization signatures: Theory and observation,' *Radio Science*, **22**, 529–43.

Vansteenkiste, N., Nignolo, P. and Aspect, A. (1993) 'Optical reversibility theorems for polarization: Application to remote control of polarization,' *J.Opt.Soc.Am. A.*, **10**, 2240–5.

Volten, H., Munoz, O., Rol, E., de Haan, J.F., Vassen, W., Hovenier, J.W., Muinonen, K. and Nousianen, T. (2001) 'Scattering matrices of mineral aerosol particles at 441.6nm and 632.8nm,' *J. Geophys. Res.*, **106**(17), 375.

Voss, K.J. and Fry, E.S. (1984) 'Measurement of the Mueller matrix for ocean water,' *Appl. Opt.*, **23**, 4427–36.

Wang, Y. and Imhoff, M.L. (1993) 'Simulated and observed L-HH radar backscatter from tropical mangrove forests,' *Int. J. Remote Sens.*, **14**, 2819–28.

Wang Y., Hess, L.L., Filoso, S. and Melack, J.M. (1995) 'Understanding the radar backscattering from flooded and non flooded Amazonian forests: Results from canopy backscatter modeling,' *Remote Sens. Environ.*, **54**, 324–32.

West, R.A., Doose, L.R., Eibl, A.M., Tomasko, M.G. and Mishchenko, M.I. (1997) 'Laboratory measurements of mineral dust scattering phase function and linear polarization,' *J. Geophys. Res.*, **102**, 16871–81.

Whitney, C. (1971) 'Pauli-algebraic operators in polarization optics,' *J. Opt. Soc. Am.*, **61**, 1207–13.

Woodhouse, I.H. (2006) 'Predicting backscatter-biomass and height-biomass trends using a macroecology model,' *IEEE Transactions on Geoscience and Remote Sensing*, **GRS-44**, 871–7.

Xie, C., Ding, Y., Chen, C. and Han, T. (2007) 'Study on the change of Koxkar Glacier during the last 30 years, Mt. Tuomuer, Western China,' *Environmental Geology*, **51**, 1165–70.

Xing, Z.-F. (1992) 'On the deterministic and nondeterministic Mueller matrix,' *J. Mod. Opt.*, **39**, 461–84.

Xu, L., Ding, J. and Cheng, A.Y.S. (2002) 'Scattering matrix of infrared radiation by ice finite circular cylinders,' *Appl. Opt.*, **41**, 2333–48.

Yamaguchi, Y., Moriyama, Y., Ishido, M. and Yamada, H. (2005) 'Four component scattering model for polarimetric SAR image decomposition,' *IEEE Transaction on Geoscience and Remote Sensing*, **43**, 1699–706.

Zebker, F.A. and van Zyl, J.J. (1991) 'Imaging radar polarimetry: A review,' *Proc. IEEE*, **79**, 1583–606.

Zubko, E., Shkuratov, Y.G. and Videen, G. (2004) 'Coherent backscattering effect for non-zero elements of Mueller matrix of discrete media at different illumination-observation geometries,' *J. Quant. Spectrosc. Radiat. Transfer*, **89**, 443–52.

4 Modeling polarized solar radiation of the ocean–atmosphere system for satellite remote sensing applications

Wenbo Sun, Rosemary R. Baize, Constantine Lukashin, Gorden Videen, Yongxiang Hu, and Bing Lin

4.1 Introduction

Reflected solar radiation from Earth's ocean–atmosphere system can be significantly polarized by Earth's surface and by atmospheric components such as air molecules and aerosols. Measurements of many satellite radiometric instruments have some dependence on the polarization state of the reflected light. The radiance measurement errors associated with the sensors' polarization dependence could decrease the measurement accuracy below the requirement of a reliable climate modeling (Wielicki et al., 2013; Lukashin et al., 2013; Sun and Lukashin, 2013). In order to use NASA's Climate Absolute Radiance and Refractivity Observatory (CLARREO) mission (Wielicki et al., 2013) data to inter-calibrate solar imagers like the Moderate Resolution Imaging Spectrometer (MODIS) (King et al., 1992) or its follow-on instrument the Visible Infrared Imaging Radiometer Suite (VIIRS) and geostationary imagers, the polarization-induced measurement errors of these solar imagers must be corrected. To correct these errors, the polarization state of the reflected light at the top of the atmosphere (TOA) must be well known. At the three wavelengths of 490, 670, and 865 nm, the Polarization and Anisotropy of Reflectances for Atmospheric Science instrument coupled with observations from lidar (PARASOL) data (Tanre et al., 2011) can be used to obtain the polarization distribution models (PDMs) empirically (Lukashin et al., 2013). However, to obtain the PDMs for the solar spectra from 320 to 2,300 nm at which the CLARREO will perform the inter-calibrations with various space-borne sensors, spectrally modeling the TOA polarized solar radiation cannot be avoided (Sun and Lukashin, 2013).

Usually, sensitivity to polarization of imaging radiometers such as MODIS and VIIRS is obtained during instrument characterization before launch, and expressed in polarization factor. These factors are corrections of the baseline (unpolarized) gain, and depend on the instrument band, scan angle, and angle of polarization (Sun and Xiong, 2007). The PDM should be developed to provide information on the degree of polarization (DOP), and the angle of linear polarization (AOLP) of the reflected solar radiation at the TOA.

To briefly review the fundamentals relevant to polarization of reflected light, following Mishchenko and Travis (1997), we set a right-handed Cartesian coordinate system as shown in Fig. 4.1, with the z-axis directed vertically to the upper boundary of the atmosphere and xoz being the principal plane. The Sun is in the principal plane and over the negative x-axis of the coordinate system in this figure. The direction of the reflected light from the ocean–atmosphere system is specified by the unit vector $\mathbf{e}_r$, and θ and φ denote the viewing zenith angle (VZA) and relative azimuth angle (RAZ), respectively. In the local right-handed orthonormal coordinate system formed by the unit vectors $\mathbf{e}_r$, $\mathbf{e}_\theta$, and $\mathbf{e}_\varphi$, we have $\mathbf{e}_r = \mathbf{e}_\theta \times \mathbf{e}_\varphi$, where $\mathbf{e}_\theta$ lies in the meridian plane of the reflected light beam. The AOLP of the reflected radiance in the direction of $\mathbf{e}_r$ is the angle between the local meridian line and the electric vector of the linearly polarized light, measured anticlockwise when viewing in the reverse direction of the reflected radiance. Also, in the local right-handed orthonormal coordinate system formed by the unit vectors $\mathbf{e}_r$, $\mathbf{e}_\theta$, and $\mathbf{e}_\varphi$, the common intensity and the polarization state of any quasi-monochromatic light can be completely specified by the Stokes parameters I, Q, U, and V. Following the definition in Hansen and Travis (1974), we have

$$I = \langle E_\theta E_\theta^* + E_\varphi E_\varphi^* \rangle \,, \tag{4.1a}$$

$$Q = \langle E_\theta E_\theta^* - E_\varphi E_\varphi^* \rangle \,, \tag{4.1b}$$

$$U = \langle E_\theta E_\varphi^* + E_\varphi E_\theta^* \rangle \,, \tag{4.1c}$$

$$V = i\langle E_\varphi E_\theta^* - E_\theta E_\varphi^* \rangle \,, \tag{4.1d}$$

where E_θ and E_φ are the θ and φ components of the electric field in the local right-handed orthonormal coordinate system, respectively. The asterisk denotes the complex-conjugate value, and angular brackets denote averaging in time. It is well known that any arbitrarily polarized incoherent radiation denoted Stokes parameters I, Q, U, and V can be represented by a sum of an unpolarized part and a 100% polarized part as

$$\begin{bmatrix} I \\ Q \\ U \\ V \end{bmatrix} = \begin{bmatrix} I - \sqrt{Q^2+U^2+V^2} \\ 0 \\ 0 \\ 0 \end{bmatrix} + \begin{bmatrix} \sqrt{Q^2+U^2+V^2} \\ Q \\ U \\ V \end{bmatrix} . \tag{4.2}$$

For a sensor with polarization dependence, the measurement of the polarized portion of light $[\sqrt{Q^2+U^2+V^2},\ Q,\ U,\ V]$ is a function of the polarization angle. As an extreme example, a linearly polarized lens can transmit a linearly polarized light $[\sqrt{Q^2+U^2},\ Q,\ U,\ 0]$ from 0 to 100%, depending on the AOLP relative to the polarization direction of the lens. Since the circularly polarized radiance from the ocean–atmosphere system is negligible ($V \approx 0$) (Coulson, 1988), only the total intensity I and linearly polarized radiance Q and U need to be calculated in this study. Without considering the circularly polarized radiance ($V \approx 0$) at TOA, the DOP (which is equal to the degree of linear polarization (DOLP) when V is not considered) and AOLP are defined in terms of Stokes parameters, respectively, as (Sun and Lukashin, 2013)

$$DOP = DOLP = \frac{\sqrt{Q^2+U^2}}{I} \,, \tag{4.3}$$

and

$$AOLP = \frac{1}{2}\tan^{-1}\left(\frac{U}{Q}\right) + \alpha_0\,, \tag{4.4}$$

where $\alpha_0 = 0°$ if $Q > 0$ and $U \geq 0$; $\alpha_0 = 180°$ if $Q > 0$ and $U < 0$; $\alpha_0 = 90°$ if $Q \leq 0$. The physical meaning of AOLP is illustrated in Fig. 4.1.

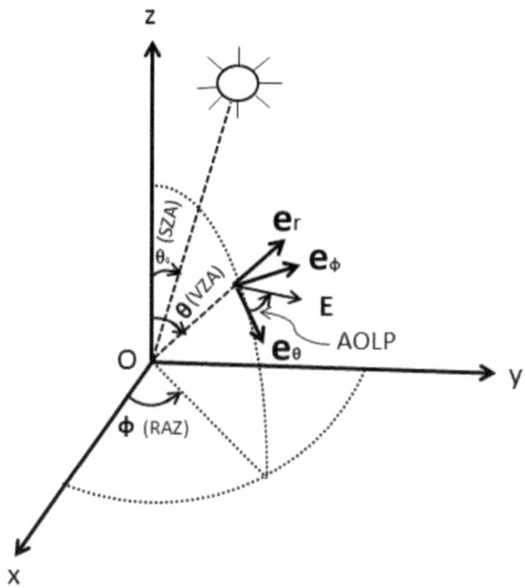

Fig. 4.1. The geometry of the system showing the scattered light from a surface located in the x–y plane (from Sun and Lukashin, 2013). The Sun is located on the principal plane (xoz) and over the negative x-axis (i.e., the solar azimuth angle is 180°). The direction of the reflected light is specified by the unit vector $\mathbf{e}_r$, and θ and ϕ denote the viewing zenith angle (VZA) and relative azimuth angle (RAZ), respectively. In the local right-handed orthonormal coordinate system, $\mathbf{e}_r = \mathbf{e}_\theta \times \mathbf{e}_\phi$, where $\mathbf{e}_\theta$ lies in the meridian plane of the reflected light beam. The angle of linear polarization (AOLP) of the reflected beam is the angle between the local meridian line and the electric vector E of the linearly polarized light, counted anticlockwise when viewing in the reverse direction of the reflected beam. The electric field E of the linearly polarized part of the reflected light oscillates in the plane of $\mathbf{e}_\theta$ and $\mathbf{e}_\phi$.

When both DOP and AOLP of the radiance are known, the sensor-measured intensity counts (which is the product of actual incidence intensity and gain factor of the sensor for unpolarized radiation) of light can be expressed as (Sun and Lukashin, 2013)

$$C_m = G_0 \cdot (1 - DOP) \cdot I + G_p(AOLP) \cdot DOP \cdot I\,, \tag{4.5}$$

where I denotes the actual intensity of the light to be measured, and G_0 and $G_p(AOLP)$ are the sensor's gain factors for unpolarized radiation and linearly polarized radiation, respectively. $G_p(AOLP)$ is the ratio of the sensor-measured intensity count to the intensity of the linearly polarized incident light used for calibrating the instrument, and G_0 can be derived as a mean value of $G_p(AOLP)$ over the AOLP. From Eq. (4.5), the actual radiance can be derived from the measured value as

$$I = \frac{C_m/G_0}{1 + \left[\frac{G_p(AOLP)-G_0}{G_0}\right] \cdot DOP}, \tag{4.6}$$

where C_m/G_0 is simply the measured intensity without polarization correction, and $\left[\frac{G_p(AOLP)-G_0}{G_0}\right]$, as a function of AOLP, is the imager's sensitivity-to-polarization factor, obtained during the prelaunch calibration of the instrument. Note here that, for a well-depolarized imager, the sensitivity-to-polarization factor, $m(AOLP) = \left[\frac{G_p(AOLP)-G_0}{G_0}\right]$, should be a quantity with $|m(AOLP)| \ll 1$. Also, the DOP is always ≤ 1. Therefore, using $\frac{1}{1+x} \approx 1 - x$ for small x, Eq. (4.6) can be expressed as

$$I \approx (C_m/G_0) \cdot [1 - m(AOLP) \cdot DOP]\,. \tag{4.7}$$

The relative error (RE) of the measured intensity due to polarization can then be calculated as

$$RE = \frac{(C_m/G_0) - I}{I} \approx \frac{m(AOLP) \cdot DOP}{1 - m(AOLP) \cdot DOP} \approx m(AOLP) \cdot DOP\,. \tag{4.8}$$

In deriving Eq. (4.8), we neglected the second-order small value $[m(AOLP) \cdot DOP]^2$. Thus, for example, for a sensor with a sensitivity-to-polarization factor of only 1%, its measurement for light with a DOP of 30% will have a RE of 0.3% solely due to the polarization.

4.2 Radiative-transfer model

A variety of techniques have been developed for computing radiative transfer including multiple scattering of light through the atmosphere. In Sun and Lukashin (2013), we have employed the adding-doubling method (Hansen and Travis, 1974; Stokes, 1682; Peebles and Plesset, 1951; van de Hulst, 1963; Twomey et al., 1966; Hansen and Hovenier, 1971; de Haan et al., 1987; Evans and Stephens, 1991), and coupled it with a rough-ocean-surface light-reflection matrix (Mishchenko and Travis, 1997), to model the reflected solar radiation from the ocean–atmosphere system. The adding-doubling method (Hansen and Travis, 1974; Evans and Stephens, 1991) separates each model layer of the medium in which the light propagates into many optically thin sub-layers. The optical thickness of each sub-layer is set to be so small that the optical properties of the sub-layer can be represented by single scattering. When reflection and transmission are known for each of the two adjacent sub-layers, the reflection and transmission from the combined layer can be obtained by computing the successive reflections back and forth between the two sub-layers. If the optical properties of the two sub-layers are identical, the results for the combined layer can be built up rapidly in a doubling manner. In the practice of adding-doubling programming, Fourier decomposition is made for the Stokes vector and scattering-phase matrix. The numerical solution is obtained for each Fourier component, and the Stokes vector of the transferred light is calculated with these Fourier components (de Haan et al., 1987; Evans and Stephens, 1991). Due to its simple essence, the adding-doubling method is a standard radiative-transfer algorithm that has a long history of applications and documentations and does not need to be further introduced here. The adding-doubling model of Sun and Lukashin (2013) follows the latest program development in this method (Hansen and Travis, 1974; Evans and Stephens, 1991), which can be applied to calculate all Stokes parameters of the radiation through a plane-parallel atmosphere composed of absorbing gas, scattering molecules, scattering particulates including various aerosols, water-cloud droplets, and ice-cloud particles.

The atmosphere is assumed to be plane-parallel and separated into 32 layers with the ocean surface as the reflecting boundary. The atmospheric profiles, which give the pressure, temperature, water vapor, and ozone as functions of altitude, are from the tables of tropical (TPC), mid-latitude summer (MLS), mid-latitude winter (MLW), subarctic summer (SAS), and subarctic winter (SAW) atmospheric profiles (McClatchey et al., 1972). The US Standard Atmosphere (STD) (National Oceanic and Atmospheric Administration, National Aeronautics and Space Administration, and United States Air Force, 1976) is also used in sensitivity studies in this chapter. We use gas absorption coefficients from the k-distribution treatment (Kato et al., 1999) of the spectral data from the line-by-line radiative-transfer model (LBLRTM) (Clough et al., 1992; Clough and Iacono, 1995) using the MODTRAN 3 data set (Kneizys et al., 1988). Ozone-absorption coefficients are also taken from the ozone cross-section table provided by the World Meteorological Organization (1985) for wavelengths smaller than 700 nm. Molecular scattering optical thickness above any pressure level, P, is from Hansen and Travis (1974):

$$\tau = 0.008569\lambda^{-4}(1 + 0.0113\lambda^{-2} + 0.00013\lambda^{-4})\left(\frac{P}{P_0}\right), \tag{4.9}$$

where λ is the wavelength of light in μm, P is the pressure in millibars (mb), and P_0 = 1,013.25 mb is the standard surface pressure. The scattering-phase matrix elements of molecular atmosphere are based on Rayleigh-scattering solution with a depolarization factor of 0.03 (Hansen and Travis, 1974). Single-scattering properties of aerosol and cloud particles are calculated differently, depending on their characteristics: for spherical droplets, including stratospheric sulfur aerosols and water-cloud particles, the Mie solution (Mie, 1908; Fu and Sun, 2001), is applied; for solid or mixed phase aerosols or small ice crystals, which are usually nonspherical particles, the finite-difference time domain (FDTD) light-scattering model (Sun et al., 1999, 2002, 2013b), the scattered-field pseudo-spectral time domain (PSTD) light-scattering model (Sun et al., 2013b), and the discrete-dipole approximation (DDA) light-scattering model (Zubko et al., 2006, 2013) are used; and for large ice crystals, we use the geometric optics approximation (GOA) (Macke, 1993; Yang and Liou, 1996). The C1 size distribution (Deirmendjian, 1969) is used for water-cloud droplets. This size distribution is based on a modified gamma (MG) particle size distribution (PSD)

$$dN/dR = N_0 R^{\nu} \exp\left(-\nu\frac{R}{R_0}\right), \tag{4.10}$$

where R denotes the droplet radius, R_0 is the modal radius, ν defines the shape of the distribution, and

$$N_0 = \frac{\nu^{\nu+1}}{\Gamma(\nu+1)R_0^{\nu+1}} N_{tot} \quad (4.11)$$

is a constant with $\Gamma(\nu+1)$ being the gamma function and N_{tot} being the total number of particles per unit volume. The commonly used C1 size distribution (Deirmendjian, 1969) is a specific case of the MG PSD with $R_0 = 4$ μm and $\nu = 6$.

The 28 measured ice-crystal size distributions used in Fu (1996) and the two other size distributions in Mitchell et al. (1996) are used to calculate the volume single-scattering properties of ice clouds. To include the effect of aerosols in the modeling, a two-mode log-normal size distribution (Davies, 1974; Whitby, 1978; Reist, 1984; Ott, 1990; Porter and Clarke, 1997) is applied in a form

$$dN/d\log D = \quad \mathrm{mod}\ e_1 + \quad \mathrm{mod}\ e_2\,, \quad (4.12a)$$

$$\mathrm{mod}\ e = \frac{M}{D\log\sigma_g\sqrt{2\pi}} \exp\left[\frac{-(\log D - \log D_g)^2}{2\log^2\sigma_g}\right]. \quad (4.12b)$$

In Eqs (4.12a) and (4.12b), D is the aerosol diameter in μm, M is a multiplier, σ_g is the geometric standard deviation, and D_g is the geometric mean diameter in μm.

Traditional radiative-transfer models generally assume independent radiation processes for molecules and aerosol or cloud particles, which means that the molecular radiation process and particulate process are assumed at different layers of the atmosphere to avoid the convolution of the light-scattering phase functions of molecules and particulates in the radiative-transfer calculations. This may involve errors due to unphysical single-scattering properties in each layer. For layers with more than one type of scattering atmospheric components, such as layers with both air molecules and aerosols, and layers composed of air molecules, aerosols, and cloud particles, Sun and Lukashin (2013) calculated the mixed single-scattering properties as the mixed optical properties of the layer. In calculations of the mixed values, single-scattering properties of individual agents are weighted by their optical thicknesses.

In the radiative-transfer calculations, the phase-matrix elements of particulate atmospheric components are input to the radiative-transfer model as Legendre polynomial series (Evans and Stephens, 1991). If the phase-matrix elements of the scattering particles have strong forward-scattering peaks, as for large cloud particles, many Legendre high-order terms are needed to approach the original phase-matrix elements, which could heavily increase the computation time and memory of the radiative-transfer calculation due to the large increment in the Legendre terms and stream number in the modeling. To avoid this problem, we conduct a delta adjustment (Hansen, 1968) to the layer's mixed single-scattering properties prior to the Legendre series expansion of the scattering-phase-matrix elements. This means a truncation of the forward-scattering peak in the phase-matrix elements for particulate atmospheric components. This is implemented in a way as exemplified by the conventional phase function (P_{11}) of a water cloud: the scattering peak of P_{11} between the scattering angle of θ_0 and the forward-scattering direction is truncated and replaced by values from a linear extrapolation algorithm

$$\frac{\log_{10} P_{11}(\theta) - \log_{10} P_{11}(\theta_1)}{(\theta - \theta_1)} = \frac{\log_{10} P_{11}(\theta_0) - \log_{10} P_{11}(\theta_1)}{(\theta_0 - \theta_1)}, \tag{4.13}$$

where θ_0 and θ_1 are two neighboring scattering angles and $\theta_1 > \theta_0$, and θ is any scattering angle between 0 and θ_0. The truncated phase function is integrated over the scattering angle to obtain the energy loss fraction f due to the truncation of the forward-scattering peak, which is the difference between the integral of the original phase function and the integral of the truncated phase function over the scattering angles. After the energy loss fraction f is obtained, the scattering optical thickness τ_s', total optical thickness τ', and single-scattering albedo α' of the cloud are adjusted as:

$$\tau_s' = (1 - f)\tau_s\,, \tag{4.14}$$

$$\tau' = \tau_a + \tau_s'\,, \tag{4.15}$$

and

$$\alpha' = \tau_s'/\tau'\,, \tag{4.16}$$

where τ_s and τ_a are the scattering and absorption optical thickness of the original clouds, respectively. Other elements of the phase matrix are also adjusted by conserving their ratio values to the conventional phase function (P_{11}) (e.g., P_{12}/P_{11} does not change after the adjustment). The adjusted phase-matrix elements are renormalized by $(1 - f)$.

The delta-adjustment treatment may cause some numerical errors in calculation of the directional irradiance from clouds due to the strong forward-scattering peak in their phase-matrix elements. However, since radiation from clouds generally has a small DOP, which we show later, the delta-adjustment approximation should not cause any significant errors in the correction of radiance measurements caused by the polarization state of light in remote-sensing inter-calibration applications.

4.3 Surface-reflection model

The major expansion to the adding-doubling method in Sun and Lukashin (2013) is the coupling of the rough-ocean-surface light-reflection matrix with the atmospheric layers. The ocean-surface-light-reflection matrix is obtained based on an empirical foam-spectral-reflectance model (Koepke, 1984), an empirical spectral-reflectance model for water volume below the surface (Morel, 1988), and the standard Kirchhoff approach under the stationary-phase approximation (Mishchenko and Travis, 1997) for foam-free waves with slope distribution as given in Cox and Munk (1954, 1956). To examine the dependence of the reflected light's polarization on the direction of wind over ocean, the wave-slope-distribution models with the Gram–Charlier series expansion (Cox and Munk, 1954), and without the Gram–Charlier series expansion (Cox and Munk, 1956), are both integrated in the adding-doubling radiative-transfer model. The surface-reflection matrix with 4×4 elements is calculated as

$$R_0(\theta_s,\theta_v,\varphi) = fR_{WC}+(1-f)R_{WL}+(1-f)\frac{\pi M(\theta_s,\theta_v,\varphi)}{4\cos^4\beta\cos\theta_s\cos\theta_v}P(Z_x,Z_y)\,, \quad (4.17)$$

where θ_s, θ_v, and φ denote solar zenith angle, VZA, and relative azimuth angle of the reflected light, respectively. The fraction of whitecap (WC or foam) is denoted as f. The fraction of WCs has a large uncertainty, which not only depends on the wind speed, but also on the fetch, history of winds, and the factors altering the mean lifetime of the WCs, such as water temperature and thermal stability of the lower atmosphere (Koepke, 1984). Sun and Lukashin (2013) use the expression by Monahan and O'Muircheartaigh (1980)

$$f = 2.95\times 10^{-6}W^{3.52}\,, \quad (4.18)$$

where the wind speed W is in the units of m/s. In Eq. (4.17), $\mathbf{R}_{\mathrm{WC}}$ is the WC reflection matrix. Since foam is generally assumed to be a Lambertian reflector, the only non-zero element of $\mathbf{R}_{\mathrm{WC}}$ is the reflectance, which is from an empirical foam spectral-reflectance model (Koepke, 1984) in this study. The water-leaving (WL) reflection is also assumed to be Lambertian. Similarly to $\mathbf{R}_{\mathrm{WC}}$, $\mathbf{R}_{\mathrm{WL}}$ has only one non-zero element, the reflectance of water volume below the surface, which is obtained from an empirical spectral-reflectance model (Morel, 1988) with an ocean-water pigment concentration of 0.01 mg/m^3. The 4×4 elements of $M(\theta_s,\theta_v,\varphi)$ for each wave-facet orientation are calculated in the same way as in Mishchenko and Travis (1997), based on the Fresnel laws. As given in Cox and Munk (1954, 1956), $P(Z_x,\,Z_y)$ is the wave-slope probability distribution as a function of the two components of the surface slope

$$Z_x = \frac{\partial Z}{\partial x} = \frac{\sin\theta_v\cos\varphi-\sin\theta_s}{\cos\theta_v+\cos\theta_s}\,, \quad (4.19)$$

and

$$Z_y = \frac{\partial Z}{\partial y} = \frac{\sin\theta_v\sin\varphi}{\cos\theta_v+\cos\theta_s}\,, \quad (4.20)$$

where Z denotes the height of the surface. In Eq. (4.17), β is the tilting angle of the wave facet, thus

$$\tan\beta = \sqrt{Z_x^2+Z_y^2}\,. \quad (4.21)$$

If wind direction is not accounted for, Cox and Munk (1956) gives $P(Z_x,\,Z_y)$ as a function of wind speed in the form

$$P(Z_x,Z_y) = \frac{1}{\pi\sigma^2}\exp\left(-\frac{Z_x^2+Z_y^2}{\sigma^2}\right)\,, \quad (4.22)$$

where σ^2 is linearly related to wind speed W (m/s) in an empirical form

$$\sigma^2 = 0.003+5.12\times 10^{-3}W. \quad (4.23)$$

Furthermore, to study the sensitivity of the reflected light's polarization state to wind direction over the ocean, we also integrate in the model a form of the wave-slope probability distribution with a Gram–Charlier series expansion (Cox and

Munk, 1954)

$$
\begin{aligned}
P(Z_c, Z_u) = & \frac{1}{2\pi\sigma_c\sigma_u}\exp\left(-\frac{\xi^2+\eta^2}{2}\right)\left[1-\frac{c_{21}}{2}(\xi^2-1)\eta-\frac{c_{03}}{6}(\eta^3-3\eta)\right. \\
& +\frac{c_{40}}{24}(\xi^4-6\xi^2+3)+\frac{c_{04}}{24}(\eta^4-6\eta^2+3) \\
& \left.+\frac{c_{22}}{4}(\xi^2-1)(\eta^2-1)+\cdots\right], \qquad (4.24)
\end{aligned}
$$

where $\xi = \frac{Z_c}{\sigma_c}$ and $\eta = \frac{Z_u}{\sigma_u}$, with Z_c and Z_u denoting the two components of the surface slope crosswind and upwind, σ_c and σ_u denoting the root-mean-square (r.m.s.) values of Z_c and Z_u, respectively. σ_c^2, σ_u^2, and the coefficients c_{21}, c_{03}, c_{40}, c_{04}, and c_{22} are all empirical linear functions of wind speed as given in Cox and Munk (1954). The geometry for the definition of wind direction is illustrated in Fig. 4.2.

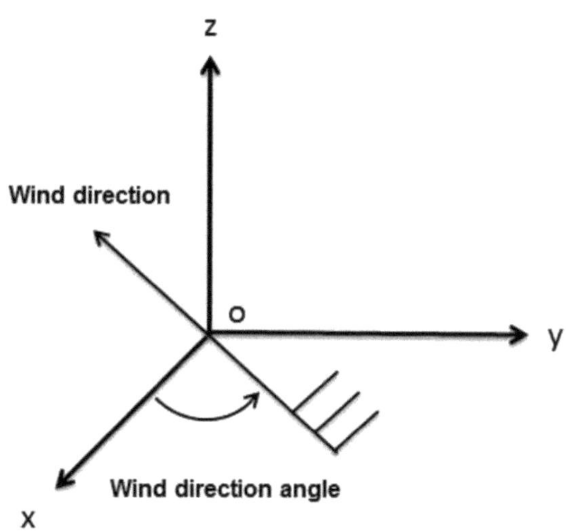

Fig. 4.2. Geometry of wind direction in the coordinate system. From Sun and Lukashin (2013).

The ocean surface is treated as the bottom boundary in the adding-doubling calculations in this study. The cosine azimuth expansion modes (coefficients) and sine azimuth expansion modes from the Fourier transformation of the ocean-reflection matrix elements are integrated into the adding-doubling process as the bottom layer optical properties functioning as boundary conditions. In this study, the 4×4 ocean-reflection matrix elements calculated in Eq. (4.17) are transformed into cosine azimuth modes or sine azimuth modes by a discrete Fourier transform (DFT) over azimuth angles with a numerical Gauss integration.

4.4 Numerical calculations

In the adding-doubling radiative-transfer model (ADRTM) (Sun and Lukashin, 2013), the adding-doubling scheme is actually conducted on the expansion modes (coefficients) of the Fourier series of the Stokes parameters over azimuth angle (de Haan et al., 1987; Evans and Stephens, 1991). After the adding-doubling calculations, the cosine expansion modes and sine expansion modes (Evans and Stephens, 1991) of the Stokes parameters are transformed back into the Stokes parameters. Therefore, the number of the cosine and sine expansion modes in the Fourier series of the Stokes parameters affects the accuracy of the radiative-transfer calculation. A larger number is required for accurate calculation of radiation with stronger anisotropy. Sensitivity of modeled Stokes parameters to expansion mode number shows that an expansion mode number of 18 is adequate for modeling most atmospheric and ocean conditions when solar zenith angle is not larger than $\sim 50°$.

Also, in adding-doubling schemes, the calculation for transferred light's sine and cosine expansion modes of the Stokes parameters is conducted only on streams over discrete VZAs. Using more streams means higher resolution in the zenith angles and better accuracy in the results, but also requires more computational resources. To keep the computing time reasonable, we have limited the number of streams to 18. In this study, the streams follow a set of discrete Gaussian quadrature angles. Integration of radiance over limited discrete VZAs at Gaussian quadrature points can result in more accurate flux than over uniformly distributed discrete angles. However, this treatment will output Stokes parameters at Gaussian quadrature points, which is not the easiest way for storing and accessing the parameters during applications. To obtain the calculated Stokes parameters and the DOP and AOLP derived from these parameters over high-resolution uniform discrete VZAs, the Stokes parameters at only the Gaussian quadrature points are extrapolated and interpolated to the uniform grid points of VZA.

4.4.1 Validation of the ADRTM

A radiative-transfer model should be validated by both model results and measurement data (Sun et al., 2015). To validate the ADRTM with other radiative-transfer models (Sun and Lukashin, 2013), the ADRTM outputs are compared with the results from the widely validated discrete-ordinate radiative-transfer (DISORT) model (Stamnes et al., 1988). Figure 4.3 shows the comparison of the directional irradiance reflectance at the wavelength of 670 nm on the principal plane from the ADRTM and from the DISORT (Sun and Lukashin, 2013). The atmosphere is assumed to have the MLS pristine profile with only molecular scattering (Rayleigh scattering) and gas absorption. The empirical models of ocean foam (Koepke, 1984), water-leaving (WL) reflectance (Morel, 1988), and wave-slope probability distribution with a Gram–Charlier series expansion (Cox and Munk, 1954) are used. The wind speed is set to be 7.5 m/s and wind direction is assumed in the reverse direction of the x-axis as illustrated in Fig. 4.2 (i.e., wind direction is 0°). The solar zenith angle (SZA) is 23.44° for Fig. 4.3a and 4.3b, and 43.16° for Fig. 4.3c and 4.3d. In the numerical simulations, we use 36 streams in the DISORT and 18 streams in the ADRTM. In the ADRTM calculation, the number of Fourier expansion modes is set to 18. We can see that the directional irradiance reflectance from

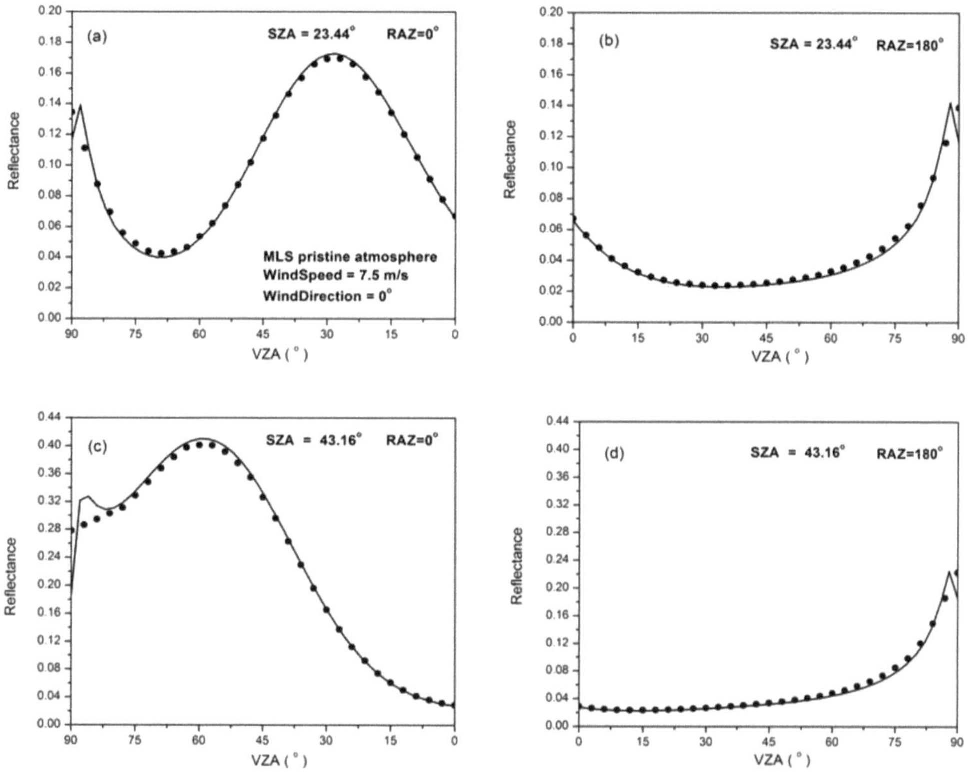

Fig. 4.3. Comparison of the directional irradiance reflectance of the ocean–atmosphere system at a wavelength of 670 nm on the principal plane (from Sun and Lukashin, 2013). Black dots and solid curves denote the results from the adding-doubling radiative-transfer model (ADRTM) and the discrete-ordinate radiative-transfer (DISORT), respectively. The atmosphere is of the mid-latitude summer (MLS) atmosphere with only Rayleigh scattering and gas absorption. The wind speed is 7.5 m/s and wind direction is 0°. The solar zenith angle (SZA) for (a) and (b) is 23.44° and for (c) and (d) is 43.16°. Thirty-six streams are used in the DISORT and 18 streams and 18 Fourier expansion modes in the ADRTM.

the ADRTM is very close to that from the DISORT, with significant differences only at VZA $> \sim 80°$, when the atmospheric optical thickness along the path of light is large. For VZA $< \sim 80°$, the relative difference in the reflectances from the ADRTM and the DISORT is smaller than ~5%. Since most in-orbit sensors do not report observations for VZA larger than ~70°, the focus of the modeling qualities is on the VZA range of 0° to 70°. As a scalar approximation to a vector radiative-transfer problem, DISORT has errors due to the negligence of polarization (Lacis et al., 1998). At small VZA, since the path-length optical thickness is small at the near-IR wavelength, the errors in the DISORT caused by the polarization of scattered light are also small. So the DISORT result is very close to the ADRTM data. At a larger SZA of 43.16°, for the same VZA, the agreement of the DISORT and ADRTM results are even better. For VZA $< \sim 80°$, the relative difference in reflectance from the ADRTM and the DISORT is smaller than ~3%. We also can see in the two cases that the directional irradiance reflectance from DISORT is generally larger than that of ADRTM in the forward-reflecting directions and smaller than that of ADRTM in the backward-reflecting directions. This is consistent with the results in Lacis et al. (1998).

For remote-sensing application, our ADRTM results are also validated with the PARASOL on-orbit polarization measurements (Sun et al., 2015). Figure 4.4 shows the directional irradiance reflectance and DOP at 670 nm from the PARASOL data for clear-sky oceans averaged in a SZA bin of 27° to 30° (black dots) and the ADRTM results at a SZA of 28.5° (solid curve). Also shown are the standard deviations of the PARASOL results, representing the potential uncertainties of the observation values. The PARASOL data are from the 24-day measurements for a wind-speed range of 6–9 m/s. The 24 days of PARASOL data are taken from the first 2 days of each month across 2006 to reflect the mean conditions of atmospheric profiles. In the modeling, the wind speed is 7 m/s, the sea-salt aerosol optical depth (AOD) is 0.06 at the wavelength of 670 nm, and the US Standard Atmosphere (STD) is used. We can see that the 24-day mean directional irradiance reflectance from the PARASOL data at a RAZ of $\sim$180° is significantly larger than the reflectance from the ADRTM model. The reflectance difference is systematically larger than the standard deviation of the PARASOL data. At a RAZ of $\sim$180°, the DOPs from the PARASOL and the ADRTM generally agree. However, at the RAZ of $\sim$0° and when the VZA is larger than $\sim$30°, the DOPs from the PARASOL and the ADRTM are significantly different: the differences in the DOP values from the ADRTM and from the PARASOL are larger than the standard deviations of the PARASOL data.

Our sensitivity studies demonstrate that changing the surface wind speed and varying AOD and water vapor amount (i.e., water vapor absorption) cannot improve the agreements of the model results and the PARASOL data in Fig. 4.4 systematically in both the DOP and the directional irradiance reflectance. Super-thin clouds undetectable for most passive remote sensing sensors may play a role in these differences. Super-thin clouds with optical depths smaller than $\sim$0.3 exist globally (McFarquhar et al., 2000; Sun et al., 2011a) and are very difficult to detect even with the 1.38 μm strong-water-vapor-absorption channel (Gao and Kaufman, 1995; Roskovensky and Liou, 2003). Without an advanced cloud detection process as introduced in Sun et al. (2014), the PARASOL clear-sky data could be contaminated by these undetected clouds. Based on this reasoning, we incorporate a thin layer of cirrus cloud with an optical depth (OD) of 0.18 between the altitudes of 7 and 9 km in the ADRTM model. The thin cirrus OD of 0.18 is derived by least squares fitting of the spatial and temporal mean of the total reflectance from the PARASOL to the modeled values. The cirrus clouds in the modeling are assumed to have an average size distribution based on all 13 size distributions analyzed by the ice particle replicator for the December 1991 FIRE II case study (Mitchell et al., 1996). The cirrus clouds are assumed to be composed of various ice-crystal habits for mid-latitude cirrus clouds as described in Baum et al. (2000). We can see from Fig. 4.5 that the DOP values from the PARASOL data and the ADRTM model are in better agreement after the incorporation of super-thin cloud in the model. Although the simulated directional irradiance reflectance may still be slightly smaller than that from PARASOL observations in the RAZ = $\sim$180° case, the reflectance

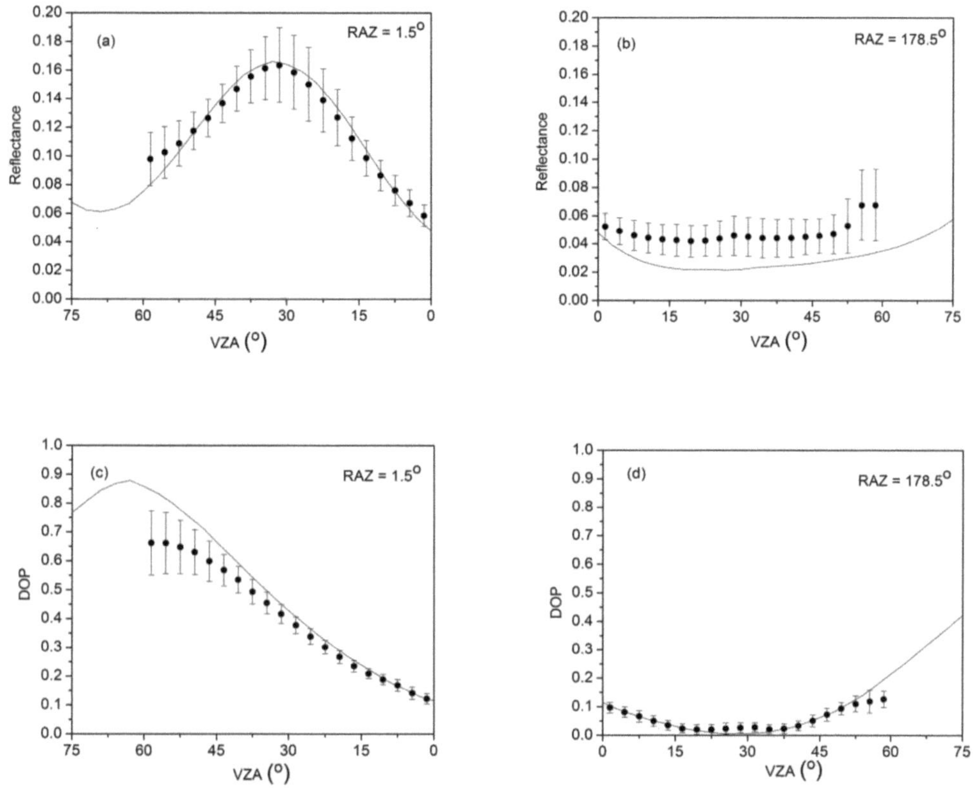

Fig. 4.4. Directional irradiance reflectance and degree of polarization (DOP) at 670 nm from PARASOL data for clear-sky oceans averaged in a SZA bin of 27° to 30° (black dots) and adding-doubling radiative-transfer model (ADRTM) results at a SZA of 28.5° (solid curve), and the standard deviations of the PARASOL data (from Sun et al., 2015). The PARASOL data are from the 24-day measurements for a wind-speed range of 6–9 m/s. In the modeling, the wind speed is 7 m/s, the sea-salt aerosol optical depth (AOD) is 0.06 at the wavelength of 670 nm, and the US Standard Atmosphere (STD) is used.

differences for all cases are within the uncertainty ranges of the PARASOL data, indicating the high consistency of simulated results with observations.

The AOLP values from the PARASOL data and the ADRTM results for Figs 4.4 and 4.5 are displayed in Fig. 4.6. Figure 4.6a and 4.6b show the mean values of AOLP and standard deviations, respectively, of the 24-day PARASOL measurements for clear-sky oceans. The 24 days of PARASOL data are collected in a SZA bin of 27° to 30° with ocean surface wind speeds 6–9 m/s, and averaged in 3° bins in both viewing zenith and relative azimuth angles. Figure 4.6c shows the AOLP results of the ADRTM modeling cases in Fig. 4.4 for clear-sky oceans, whereas Fig. 4.6d plots the AOLP values for the same cases as in Fig. 4.6c except a layer of cirrus clouds with an OD of 0.18 is added as those in Fig. 4.5. As already discussed in Sun et al. (2014), the AOLP values from the PARASOL data (Fig. 4.6a) and the ADRTM for clear-sky oceans (Fig. 4.6c) are close to each other except that at ~8° off the exact-backscatter direction (RAZ = ~ 180° and VZA = ~ 20° and

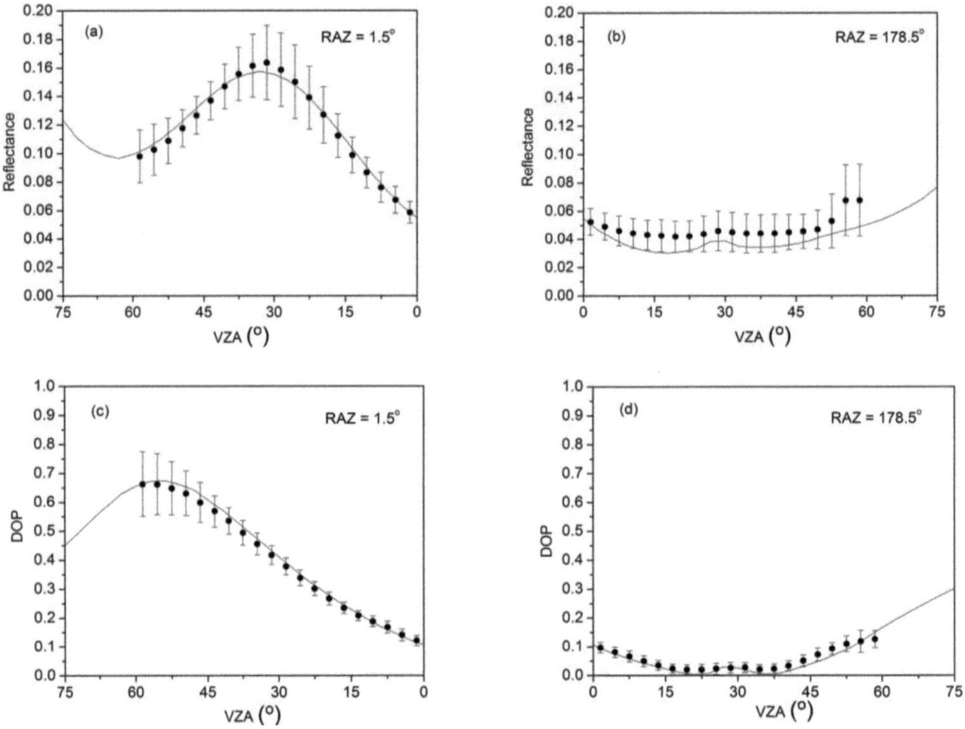

Fig. 4.5. Same as in Fig. 4.4, but a layer of cirrus cloud having an optical depth (OD) of 0.18 is added in the adding-doubling radiative-transfer model (ADRTM). From Sun et al. (2015).

~36°), the PARASOL result shows some distinct features. At these viewing angles, the backscattered linearly polarized electric field in the PARASOL data for clear-sky oceans is nearly perpendicular to Earth's surface, whereas the modeled AOLPs for clear-sky oceans are all ~90 degree around the principal plane – that is, the backscattered linearly polarized electric field is parallel to Earth's surface. Our sensitivity studies by model simulations demonstrate that altering ocean surface roughness due to winds and changing atmospheric gas or water vapor absorptions in practical ranges cannot cause a significant rotation of the linearly polarized electric field of the backscattered light. However, this rotation does appear when we incorporate a thin layer of cirrus into the atmosphere. The ADRTM result in Fig. 4.6d demonstrates that the presence of a layer of super-thin cirrus clouds can reproduce the AOLP features in Fig. 4.6a. Note here that a significant increase of AOD also can change the AOLP in the neighborhood of the principal plane. But, at the specific viewing angles near the principal plane where clouds cause ~0° AOLP characteristics, the rotation of the linearly polarized electric field due to aerosols is in a direction opposite to that due to the presence of clouds. Furthermore, Fig. 4.6b shows that the standard deviations of the AOLP values from the PARASOL clear-sky-ocean data have significant viewing-angle dependence. The AOLP of forward-reflecting light has very small variations, whereas the AOLP

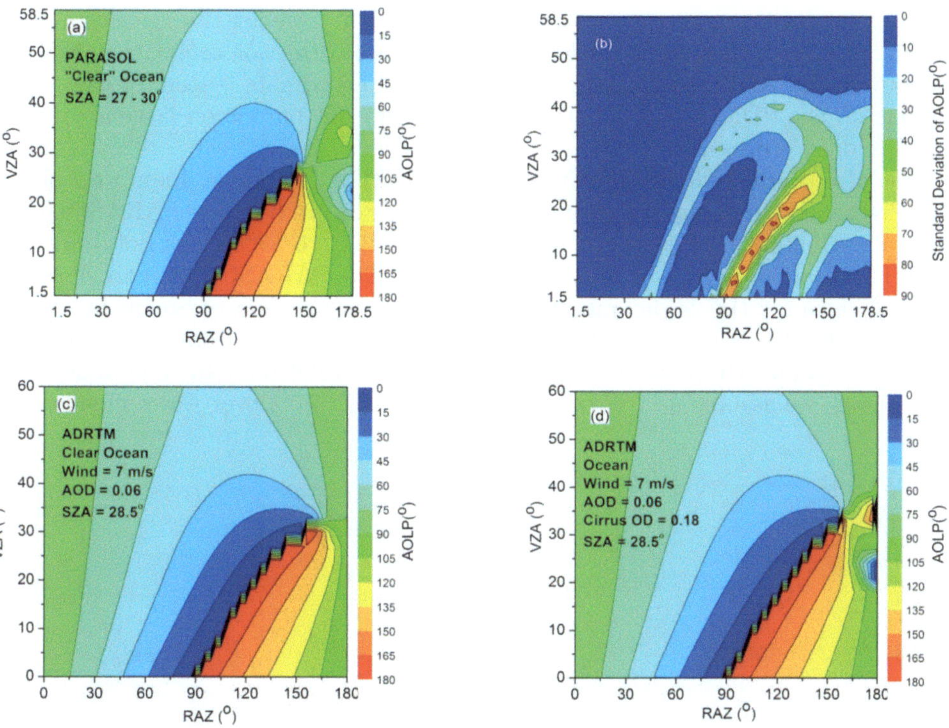

Fig. 4.6. Same as in Figs 4.4 and 4.5, but for (a) angles of linear polarization (AOLPs) from PARASOL data for clear-sky oceans, (b) the standard deviations of the AOLPs from PARASOL, (c) AOLPs from adding-doubling radiative-transfer model (ADRTM) for clear-sky oceans, and (d) AOLPs from ADRTM for oceans with a layer of cirrus cloud having an optical depth (OD) of 0.18. From Sun et al. (2015).

of side-reflecting light and back-reflecting light has large variations within most viewing-angle bins. The significant variability of AOLP in the neighborhood of backscattering angles is obviously due to the super-thin clouds undetected by the PARASOL cloud-screening algorithm. However, the large standard deviations of the AOLP at the transition area from the AOLP of 0° to the AOLP of 180° are due to the ambiguity of the AOLP values in angle bins across the transition interface between AOLP = 0° and AOLP = 180°. It is worth noting here that, except in the neighborhood of the backscatter directions where clouds or aerosols cause special patterns, most of the AOLP features as a function of viewing geometry result from the angle between the meridian plane (i.e., the reference plane in this study) and the scattering plane (i.e., the plane containing the incident light beam and the scattered light beam, frequently used by other researchers as the reference plane) (François-Marie Bréon, personal communications). When the scattering plane is used as a reference, the AOLP is very close to 90°, except in the backscatter direction. This may help to demonstrate the AOLP features of clouds or aerosols more clearly. For light-scattering physics study, using scattering plane as reference plane is convenient. However, for polarization-caused error correction of satellite data, defining parameters in the local coordinate system of the sensor is more appropriate. For satellite measurement correction application, if we use the scattering plane as reference, each time when we do the satellite data correction calculation, we have to convert the AOLP of this system to the local coordinate system of the sensor to look up the PDM tables. This may exert significant burden on large amount of satellite data processing.

We also conducted the validation of the ADRTM for clouds. Figure 4.7 shows the directional irradiance reflectance and DOP at 865 nm from the PARASOL data for water-cloud oceans averaged in a SZA bin of 54° to 57° (black dots) and the ADRTM results at a SZA of 54.74° (solid curve). The PARASOL data are obtained from the mean of the 24-day measurements for water clouds over oceans in a wind-speed range of 6–9 m/s. The 24-day mean water-cloud OD from the PARASOL data is 5.3. Since the water-cloud data of the PARASOL may include undetected thin cirrus, we assume a layer of cirrus with an OD of 0.3 over water clouds with

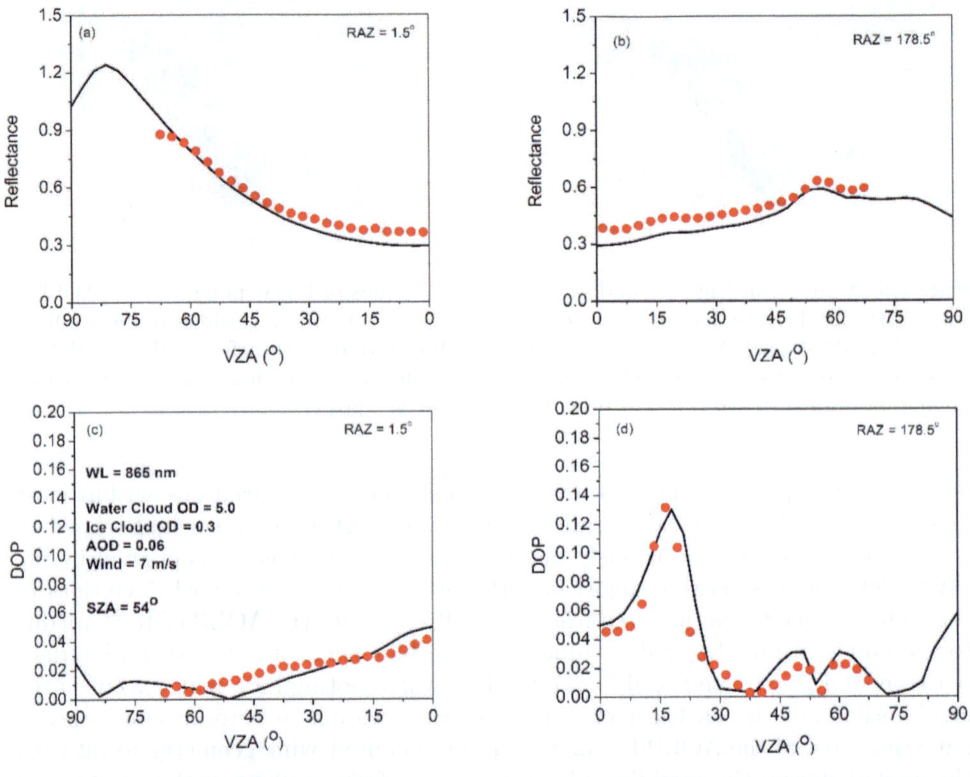

Fig. 4.7. Directional irradiance reflectance and degree of polarization (DOP) at 865 nm from the PARASOL data for water-cloud oceans averaged in a SZA bin of 54° to 57° (red dots) and the adding-doubling radiative-transfer model (ADRTM) results at a SZA of 54.74° (solid curve). The PARASOL data are from the mean of the 24-day measurements of water clouds over oceans in a wind-speed range of 6–9 m/s. In the model, the US Standard Atmosphere (STD) is used, an ocean-boundary wind speed of 7 m/s is assumed, and the sea-salt aerosol optical depth (AOD) and the water-cloud optical depth (OD) are set to be 0.06 and 5.0, respectively. A layer of cirrus with an OD of 0.3 over water clouds is assumed in the modeling.

an OD of 5.0 in the modeling. The sea-salt AOD is set to be 0.06. The STD is used with an ocean-boundary wind speed of 7 m/s in the model. It can be seen that the results from the ADRTM generally agree well with the PARASOL data.

Furthermore, as shown in Fig. 4.8, the specific angular distribution of water clouds' AOLP features in the neighborhood of backscattering angles from the ADRTM is also very similar to that from the PARASOL observation. This means that the ADRTM can simulate the AOLP values of the reflected light from water clouds very well.

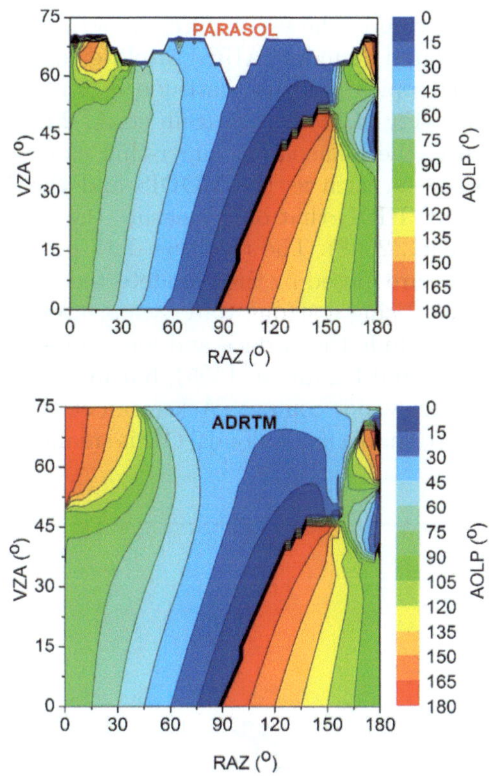

Fig. 4.8. Same as in Fig. 4.7, but for the angle of linear polarization (AOLP) values from the mean of the 24-day PARASOL data for water-cloud oceans, and the AOLP values from the adding-doubling radiative-transfer model (ADRTM) for oceans covered by a layer of cirrus having an optical depth (OD) of 0.3 over a layer of water cloud having an OD of 5.0.

4.4.2 A super-thin cloud detection method revealed by the ADRTM and PARASOL

From the AOLP pattern in Fig. 4.6a, we have concluded that the clear-sky data of PARASOL must be contaminated by undetected super-thin clouds (Sun et al., 2011a, 2014). This explains why a good agreement of the PARASOL data with the ADRTM results can and must be achieved by incorporating a thin layer of clouds in the model. This also provides a robust method to detect super-thin clouds.

As already mentioned in section 4.4.1, globally distributed super-thin clouds (Sun et al., 2011a; Gao and Kaufman, 1995) can seriously affect the remote sensing of aerosols (Sun et al., 2011; Omar et al., 2013), surface temperature (Sun et al., 2011a), and atmospheric composition gases (Christi and Stephens, 2004). These clouds are also important to the radiation energy balance of Earth's climate system (Sun et al., 2011a; Sassen and Benson, 2001). Climate models must incorporate these clouds correctly to account for Earth's radiation energy budget. For example, as shown in Fig. 4.9, the AOD retrieved from the MODIS data could be overestimated by ~100% when these clouds exist (Sun et al., 2011a). Also, Fig. 4.10 shows that failing to detect these clouds, the sea-surface temperature (SST) retrieved from NASA's Atmospheric Infrared Sounder (AIRS) satellite data (Chahine et al., 2006) could be ~5 K lower at tropical and mid-latitude regions, where the occurrence frequency of these clouds (Fig. 4.11) is high (Sun et al., 2011b). Due to uncertainties in surface reflectance, the transparent super-thin clouds generally cannot be detected by satellite imagers, such as the MODIS and the Advanced Very High Resolution Radiometer (AVHRR) that only measure the intensity of the reflected solar light (Ackerman et al., 2008). The resulting data products of many satellite and ground measurements are biased by these undetected clouds (Sun et al., 2011a, 2011b; Omar et al., 2013). Using a strong-water-vapor-absorption channel such as the 1.38 μm radiance to exclude the surface and low-atmosphere effects can be effective on high cirrus (Gao and Kaufman, 1995), but may encounter difficulties for atmospheres with low water vapor content (Ackerman et al., 1998). The reliability of this method is also questionable if the clouds' OD is smaller than ~0.5, when their

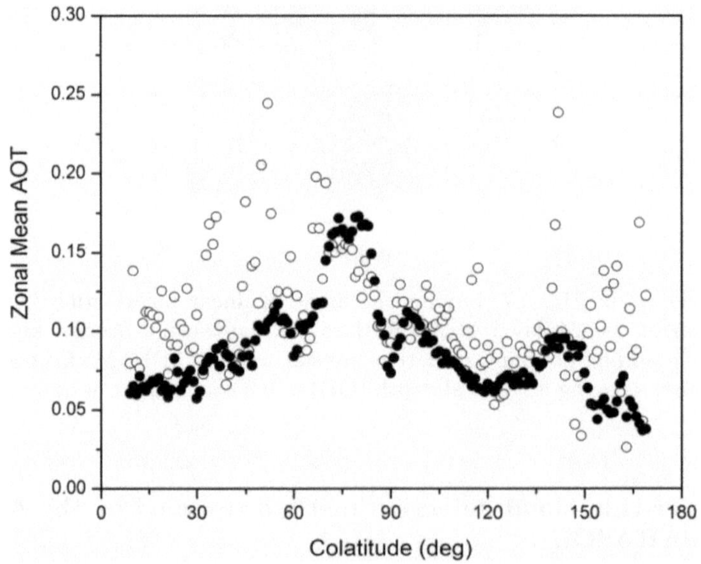

Fig. 4.9. One-year (2007) zonal mean MOD04 aerosol optical depth (AOD) at 0.55 μm for clear (filled circle) and super-thin cloud (open circle) ocean identified by Cloud-Aerosol Lidar and Infrared Pathfinder Satellite Observation (CALIPSO) Lidar. From Winker et al. (2007) and Sun et al. (2011a).

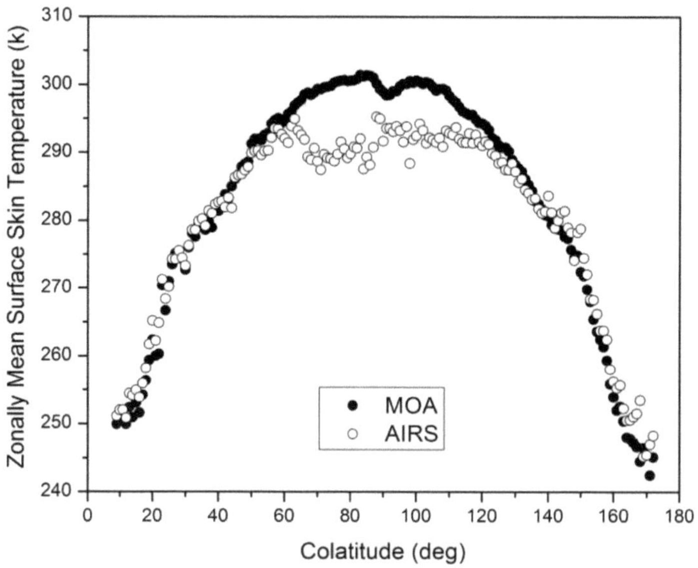

Fig. 4.10. One-year (2007) zonal mean surface skin temperatures from the AIRS and the Meteorology, Ozone, and Aerosol (MOA) data set for night-time oceans. From Sun et al. (2011b).

backscattered intensity is low (Roskovensky and Liou, 2003). In addition, super-thin clouds may also exist in the lower layers of the atmosphere where there is ample water vapor. The sensitivity of the 1.38 μm channel is weak in this region, hampering detection capabilities. The National Oceanic and Atmospheric Administration (NOAA)'s polar-orbiting High Resolution Infrared Radiation Sounder (HIRS) multispectral infrared data are usually used with the CO2-slicing method for detecting thin cirrus clouds (Wylie et al., 1995; Wylie and Menzel, 1999). However, for super-thin clouds, this requires the radiance of their background atmosphere and surface to be very close to that of the pre-defined reference clear-sky environment, which is difficult, as the terrestrial background changes on various spatial and temporal scales. In addition, this method is problematic when the difference between clear-sky and cloudy radiances for a spectral band is less than the instrument noise (Wylie et al., 1995), as for the cases of super-thin clouds.

Currently, the Cloud-Aerosol Lidar and Infrared Pathfinder Satellite Observation (CALIPSO) Lidar (Winker et al., 2007) is the only instrument in orbit that can detect super-thin clouds. However, long-term global surveys of super-thin clouds using the space-borne lidar are limited by the large operational cost and narrow field of view of this active instrument. There is a driving need for a passive technique to quantify super-thin clouds. A passive polarimetric instrument such as the aerosol polarimetry sensor (APS) (Mishchenko et al., 2007) that was a part of NASA's Glory mission may provide such a technique. Since NASA's Glory mission was failed during launch, NASA's Advanced Composition Explorer (ACE) or CNES/ESA's 3MI could be the potential mission for this measurement. However, a new satellite mission with a fast polarimetric sensor is necessary for an accurate measurement of the super-thin clouds.

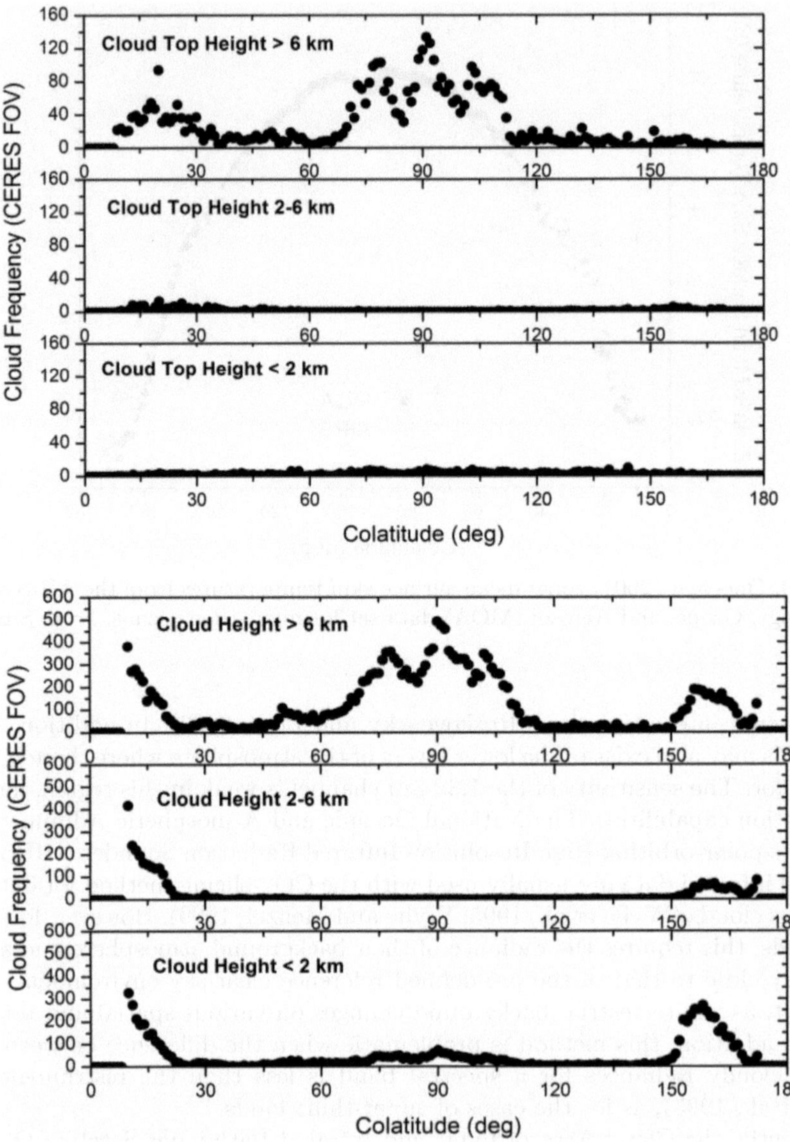

Fig. 4.11. Zonal and altitude distribution of super-thin cloud occurrence frequency in the unit of CERES (Wielicki et al., 1996) field-of-view (FOV) number for daytime (upper panel) and night-time (lower panel) ocean (from Sun et al., 2011a).

As discussed above, natural solar radiation is polarized by surface reflections as well as by scattering of atmospheric molecules and particles. When sunlight propagates through the clear atmosphere and is scattered back toward the Sun, the resulting signal is nearly unpolarized when the SZA is not larger than ~40° (Sun and Lukashin, 2013). System asymmetries, like preferentially oriented waves, can produce some residual polarization. Unlike intensity I, the DOP and AOLP are insensitive to surface roughness and absorption by atmospheric water vapor and other gases (Sun and Lukashin, 2013), thus these polarization measurements are robust to different environmental conditions. Remote sensing using polarization measurements provides a means to minimize surface, molecule, and absorbing gas interferences, and increase the sensitivity to atmospheric particulates, like super-thin clouds.

Although thin clouds can change the DOP of the reflected light (Sun and Lukashin, 2013), the DOP dynamic range is insufficient to identify super-thin clouds when the background is unknown. As shown in Figs 4.6a and 4.6d, the AOLPs contain two distinct features at ~8° off the exact-backscatter direction (RAZ = ~180° and VZA = ~20° and ~36°), where the dominant backscattered electric field rotates perpendicular to Earth's surface when a super-thin layer of cloud exists. The dominant backscattered electric field for clear sky is parallel to Earth's surface (Fig. 4.6c). This rotation appears when we incorporate a layer of cirrus as thin as OD = 0.1 for cirrus clouds as reported in Sun et al. (2014). In sensitivity studies, this rotation remains significant even down to ODs of ~0.06. The mechanism responsible for this feature is the optical glory phenomenon. The glory is an angular region of higher intensity that may extend several degrees from the exact-backscatter direction. It is accompanied by a strong p-polarization (i.e., electric field polarized in the same plane as the incident ray and the scattering surface normal) component a few degrees from the exact-backscatter direction. Spherical cloud droplets can produce an especially strong glory. Figure 4.12a shows ADRTM results for water clouds with an OD of 0.1. Also, model results in Fig. 4.12b demonstrate that subvisual water clouds having an OD of only 0.01 still display a prominent polarization feature. However, this polarization feature can be suppressed when particle absorption increases and morphology becomes irregular (Muinonen et al., 2011; Volten et al., 2001). While cirrus clouds composed solely of irregularly shaped

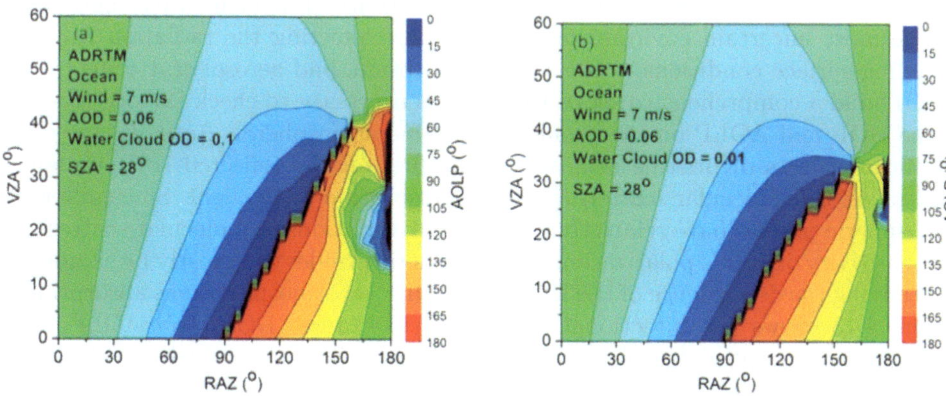

Fig. 4.12. The angle of linear polarization (AOLP) at 670 nm as a function of viewing zenith angle (VZA) and relative azimuth angle (RAZ) from the adding-doubling radiative-transfer model (ADRTM) for (a) water clouds with an optical depth (OD) of 0.1 and (b) water clouds with an OD of 0.01 at 670 nm over oceans (from Sun et al., 2014). In the modeling, the wind speed is 7 m/s, the sea-salt aerosol optical depth (AOD) is 0.06, the SZA is 28°, and the US Standard Atmosphere (STD) is used.

aggregated ice particles tend to have a weaker glory feature (Sun and Lukashin, 2013), thin cirrus clouds generally contain significant quantities of crystals having simple particle shapes such as hollow or solid columns, plates, droxtals, bullet rosettes, etc. (Baum et al., 2000), especially near cloud tops or in cold and dry regions. Additionally, the glory prominence is dependent on particle morphology and absorption, and our ADRTM simulations incorporating a dust cloud (Volten et al., 2001) do not result in a noticeable glory feature. Interestingly, such a glory feature was observed with the Hubble Space Telescope in observations during the 2003 Mars opposition. It was suggested that the p-polarization was the result of thin clouds of nucleating ice crystals. These crystals were seen at the forward edge of a prominent dust storm and were visible against a desert background. The dust storm itself did not display such p-polarization features (Shkuratov et al., 2005), suggesting that it had different composition. This demonstrates that polarization-based detection of super-thin clouds can be used over different terrain types and can differentiate such clouds from dust.

4.4.3 Effects of water vapor, surface, wavelength, and aerosol on polarization

One major objective of our numerical calculations is to study the sensitivity of the polarization state of the reflected light on various parameters, including incident solar wavelength, incident and viewing geometries, surface conditions, and atmospheric composition (Sun and Lukashin, 2013). These include SZA, VZA, RAZ, reflection surface conditions, and gas absorption, and molecular and particulate scattering to the light in the atmosphere. These sensitivity studies can help us make PDMs as functions of only necessary input parameters for quick access. Sensitivities of the polarization of reflected light to incident and viewing geometries, and molecular and cloud particulate scattering have been mentioned in previous discussions. In this section, we focus on the sensitivity of incident wavelength, and on the most uncertain environmental parameters affecting the radiation transfer including surface conditions, water vapor absorption, and aerosol scattering.

To build a comprehensive set of PDMs, it is necessary to check the dependence of the DOP and AOLP on the profiles of pristine atmospheres. Actually, the only significant effects of pristine atmospheric profiles on the reflected solar radiance spectra are the molecular scattering and gas absorption to the light. Since different atmospheric profiles have relatively insignificant variations in molecular scattering, the sensitivity of light's polarization to atmospheric profiles can mostly be examined by studying the dependence of the DOP and AOLP on atmospheric gas absorptions. Figure 4.13 shows the directional irradiance reflectance, DOP, and AOLP at a water-vapor-absorption wavelength of 1,200 nm at the TOA, which is calculated with the ADRTM at a SZA of 43.16° over a pristine clear-sky ocean with a wind speed of 7.5 m/s. The reflectance and DOP with the SAW atmosphere (solid curves) and those with the tropical atmosphere (black dots) are shown. Also shown are the AOLP with the SAW atmosphere (Fig. 4.13e) and those with the tropical atmosphere (Fig. 4.13f).

Since the tropical atmosphere has a much larger amount of water vapor than the SAW atmosphere, the total reflectance at the water-vapor-absorption wavelength

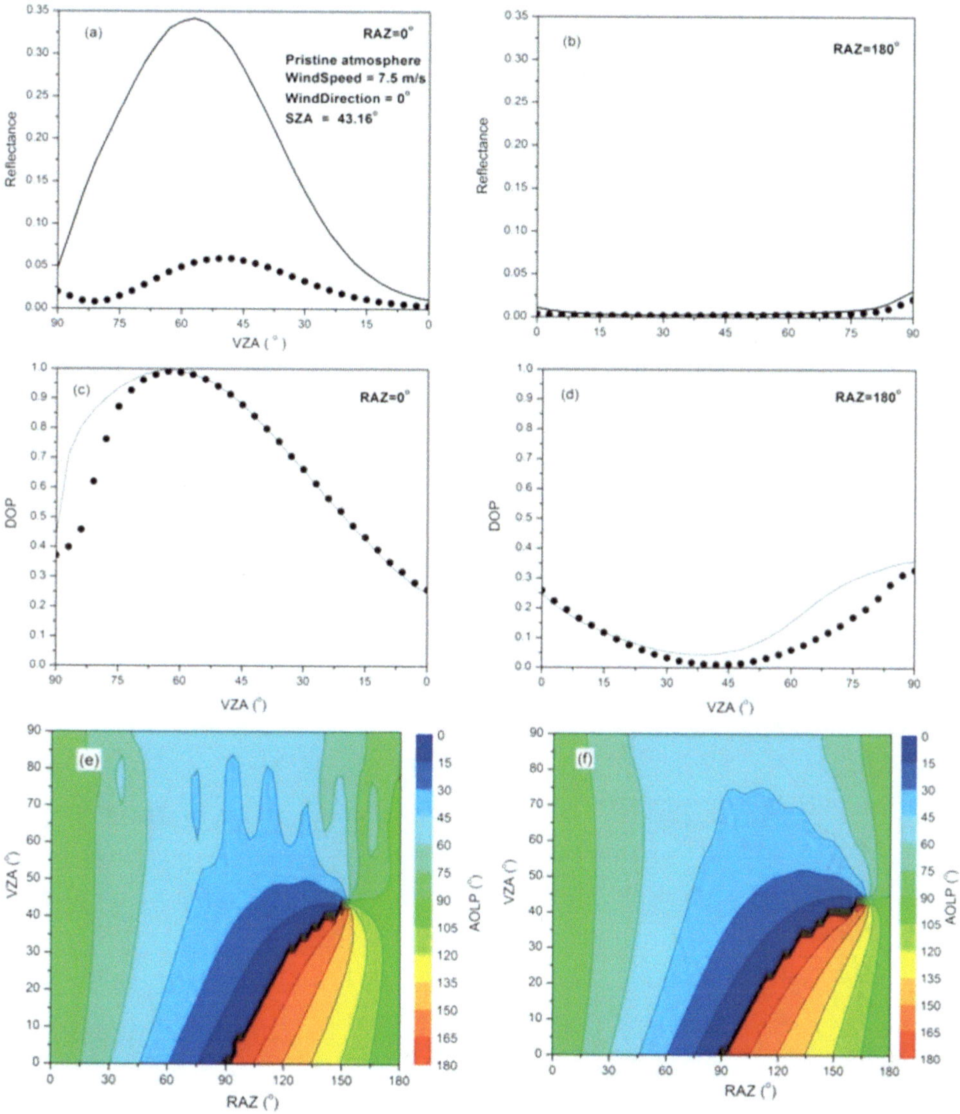

Fig. 4.13. The directional irradiance reflectance, degree of polarization (DOP), and angle of linear polarization (AOLP) at a wavelength of 1,200 nm at the top of the atmosphere (TOA), which is calculated with the adding-doubling radiative-transfer model (ADRTM) at a solar zenith angle (SZA) of 43.16° over a pristine clear sky ocean with a wind speed of 7.5 m/s and a wind direction of 0° (Sun and Lukashin, 2013). Solid curves denote the reflectance and DOP with the subarctic winter (SAW) atmosphere. Black dots denote the reflectance and DOP with the tropical (TPC) atmosphere. Panel (e) shows the AOLP with SAW atmosphere and (f) shows the AOLP with TPC atmosphere.

of 1,200 nm from the tropical atmosphere is much smaller than that from the subarctic atmosphere. However, we can see that the DOP and AOLP are much less affected by the atmospheric water vapor absorptions. This is consistent with the results for single scattering by particles (Sun et al., 2002). Therefore, gas absorption in different atmospheric profiles is important for modeling the intensity, but has relatively little effect on the DOP and AOLP of the reflected light. A stratification of the atmospheric profiles into the TPC, MLS, MLW, SAS, and SAW atmosphere should be sufficient for accurate modeling of the DOP and AOLP of the reflected solar light over the globe.

It is well known that wind-caused ocean surface roughness can significantly affect the directional reflectance of solar light. However, it is not yet known how much wind speed and direction can affect the polarization state of reflected solar light at the TOA. Figure 4.14 shows the directional irradiance reflectance and DOP of the ocean–atmosphere system on the principal plane, which is calculated with the ADRTM at wavelength 670 nm. A pristine atmosphere with the STD profile is assumed. The SZA is 33.30°. Wind direction is assumed to be at 0°. Wind speeds are given as 5.0, 7.5, 10.0, and 15.0 m/s, respectively. We can see that wind speed significantly affects the total reflectance at RAZ = 0°, but has only a small effect

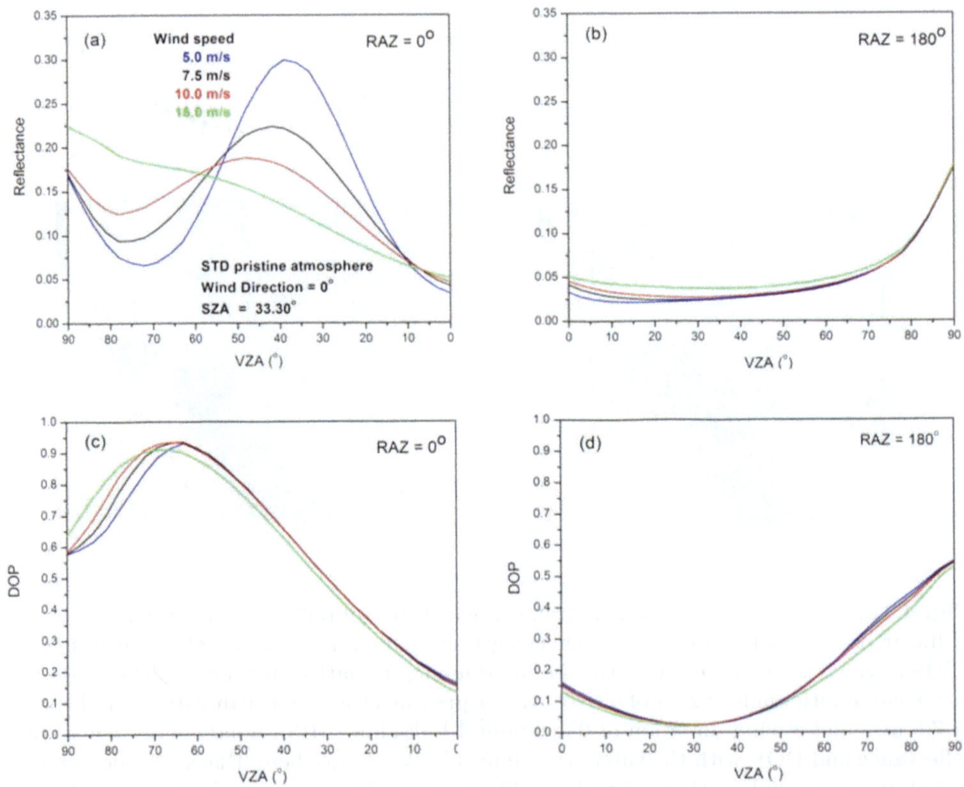

Fig. 4.14. The directional irradiance reflectance and degree of polarization (DOP) of the ocean–atmosphere system on the principal plane, which is calculated with the adding-doubling radiative-transfer model (ADRTM) at a wavelength of 670 nm (from Sun and Lukashin, 2013). A pristine US Standard Atmosphere (STD) is used. The solar zenith angle (SZA) is 33.30°. Wind direction is at 0°. Wind speeds are 5.0, 7.5, 10.0, and 15.0 m/s, respectively.

on the total reflectance at RAZ = 180°. Wind-speed effect on the DOP is also insignificant.

The AOLPs of the cases in Fig. 4.14 are shown in Fig. 4.15. We can see that the wind-speed effect on the AOLP is insignificant. In the PDM development and applications, we will be able to use wind speed assimilated from the Global Modeling and Assimilation Office (GMAO) weather data products. Our modeling results show that the uncertainty of wind-speed data will only have a small impact on the DOP and AOLP.

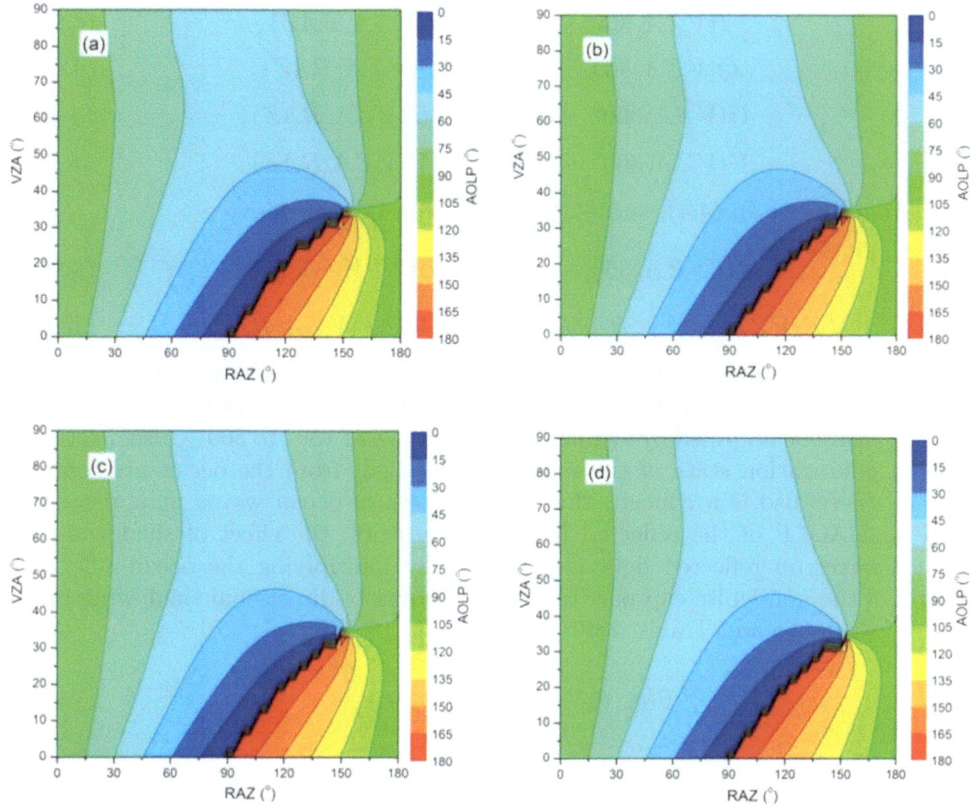

Fig. 4.15. Same as in Fig. 4.14, but for angle of linear polarization (AOLP) over relative azimuth angle (RAZ) of 0° to 180° at a wind speed of (a) 5.0 m/s, (b) 7.5 m/s, (c) 10.0 m/s, and (d) 15.0 m/s, respectively. From Sun and Lukashin (2013).

Additionally, our sensitivity studies (Sun and Lukashin, 2013) demonstrate that varying wind direction or using different ocean-wave slope probability distribution models (Cox and Munk, 1954, 1956) can significantly change the directional irradiance reflectance at RAZ = 0°, but has little impact on the reflectance at RAZ = 180°. Varying wind direction, or using different ocean-wave slope probability distribution models, has almost no effect on the DOP and the AOLP. Since wind direction is very hard to be obtained over ocean, and wind direction and wave-slope-distribution models have little effect on the polarization state of reflected light at TOA, wind direction need not to be considered in the correction of the reflected solar radiation from water bodies. Without considering wind direction over oceans, we can simply use the Cox-and-Munk ocean-wave slope probability distribution model without wind-direction dependence (Cox and Munk, 1956) for reflected radiation polarization parameter calculations. This can make the reflection field symmetric to the principal plane. However, it is worth pointing out that, among the Stokes parameters, I and Q are symmetric to the principal plane, but U and V are oddly symmetric to the principal plane – that is,

$$I(VZA, 360° - RAZ) = I(VZA, RAZ)\,, \tag{4.25a}$$

$$Q(VZA, 360° - RAZ) = Q(VZA, RAZ)\,, \tag{4.25b}$$

$$U(VZA, 360° - RAZ) = -U(VZA, RAZ)\,, \tag{4.25c}$$

$$V(VZA, 360° - RAZ) = -V(VZA, RAZ)\,. \tag{4.25d}$$

And, from Eqs (4.3), (4.4), and (4.25a–c), we can further derive

$$DOP(VZA, 360° - RAZ) = DOP(VZA, RAZ)\,, \tag{4.26a}$$

$$AOLP(VZA, 360° - RAZ) = 180° - AOLP(VZA, RAZ)\,. \tag{4.26b}$$

Therefore, in modeling PDMs, we only need to calculate and store the DOP and AOLP over the RAZ of 0° to 180°, as practiced in previous sections. These quantities can easily be obtained by symmetry for the RAZ of 180° to 360°. Also, in studying the polarization state of the reflected solar light from the ocean–atmosphere system, there also is a concern that the shadows of ocean waves may affect the DOP and AOLP of the reflected light. In this work, the effect of shadowing by surface waves on reflected light is examined by multiplying the ocean-reflection matrix of the non-white-cap part of the ocean by a bi-directional shadowing function (Mishchenko and Travis, 1997; Tsang et al., 1985)

$$S(\theta_s, \theta_v) = \frac{1}{1 + \Lambda(\theta_s) + \Lambda(\theta_v)}\,, \tag{4.27}$$

where

$$\Lambda(\theta) = \frac{1}{2}\left\{\frac{\sigma}{\cos\theta}\left[\frac{(1-\cos^2\theta)}{\pi}\right]^{1/2}\exp\left[-\frac{\cos^2\theta}{\sigma^2(1-\cos^2\theta)}\right] - erfc\left[\frac{\cos\theta}{\sigma\sqrt{1-\cos^2\theta}}\right]\right\} \tag{4.28}$$

and σ is calculated with Eq. (4.23) and $erfc(x)$ is the complementary error function. Our studies show that the ocean-wave shadows do not significantly affect the DOP and the AOLP of the reflected solar light (Sun and Lukashin, 2013).

To derive PDMs that are adequate for applications over broad solar spectra, we need to investigate the sensitivity of the polarization state of reflected light on its wavelength. Figure 4.16 shows the directional irradiance reflectance and the DOP on the principal plane, which is calculated with the ADRTM at the wavelengths of 470 and 865 nm, respectively. A pristine STD is used. The SZA is 33.30°. Wind speed is 7.5 m/s. We can see that changing the wavelength can affect the total reflectance significantly. Although changing solar wavelength also significantly affects the DOP, especially when VZA > ~45°, the DOP's sensitivity to wavelength is not as great as that of the reflectance. Though not shown here, we also find that increasing the SZA can greatly increase the difference between the DOPs of different wavelengths at all VZAs. The AOLPs for the cases in Fig. 4.16 are displayed in Fig. 4.17. The AOLP's dependence on wavelength is also noticeable. Therefore, the PDMs must be made as a function of solar wavelengths, but may not require very high spectral resolution.

One of the most uncertain components in the atmosphere is aerosols. The effect of aerosols on solar reflectance has been widely studied. Aerosols' effect on polarization of light also has attracted many efforts (Mishchenko and Travis, 1997; Chowdhary et al., 2002; Sun et al., 2013a; Mishchenko et al., 2013). In this study, to calculate the effect of aerosols on the polarization state of reflected light at the TOA, we choose an example of an STD with sea-salt aerosols. The calculation is conducted at the visible wavelength of 550 nm, where the refractive index of sea

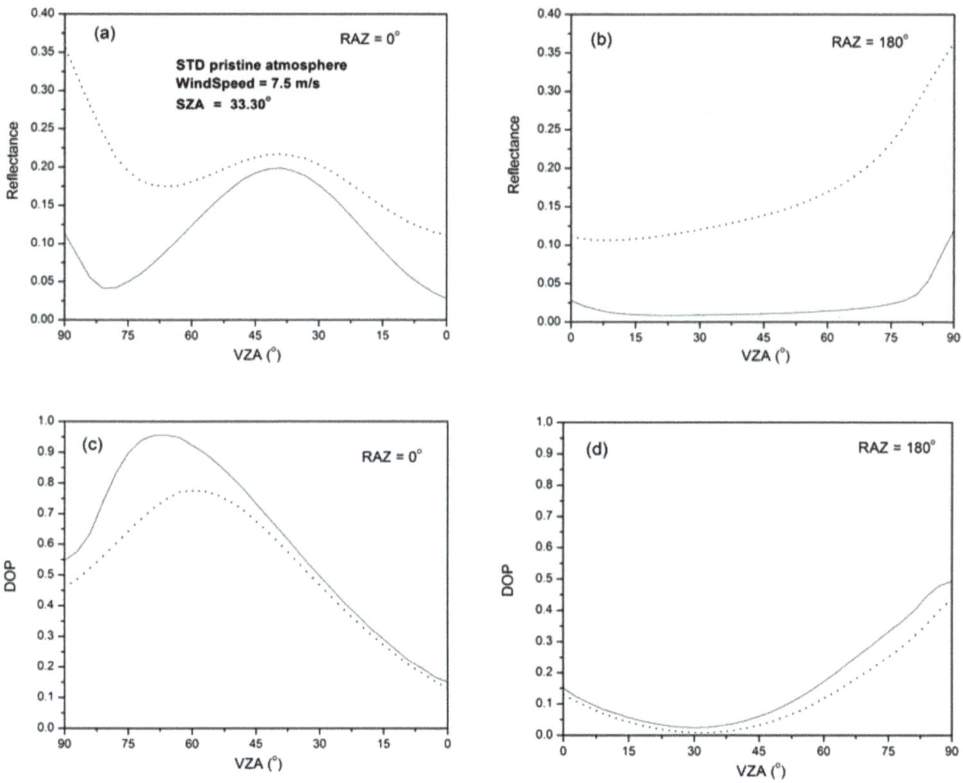

Fig. 4.16. The directional irradiance reflectance and degree of polarization (DOP) of the ocean–atmosphere system on the principal plane, which is calculated with the adding-doubling radiative-transfer model (ADRTM) at the wavelengths of 470 nm (dashed curve) and 865 nm (solid curve), respectively (from Sun and Lukashin, 2013). A pristine US Standard Atmosphere (STD) is used. The solar zenith angle (SZA) is 33.30°. Wind speed is 7.5 m/s.

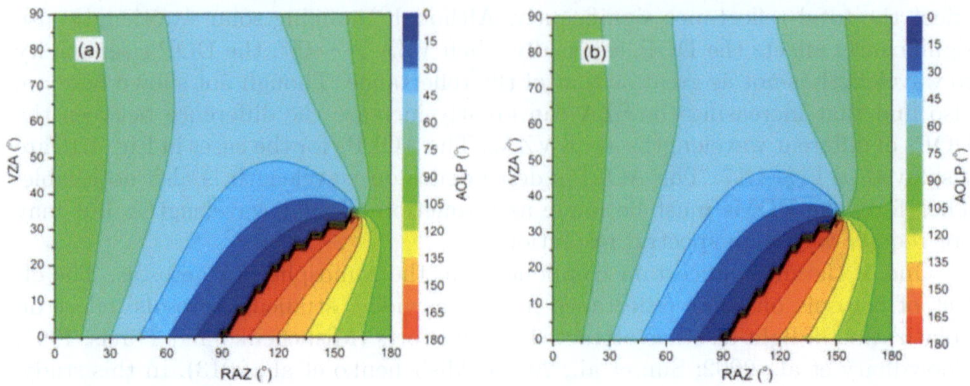

Fig. 4.17. Same as in Fig. 4.16, but for angles of linear polarization (AOLPs) at the wavelengths of (a) 470 nm and (b) 865 nm. From Sun and Lukashin (2013).

salt is given as $1.5 + i10^{-8}$ (Chamaillard et al., 2002). The sea-salt aerosol particle shapes are assumed to be the agglomerated debris, and their single-scattering properties are from DDA calculations (Zubko et al., 2006, 2013). A two-mode log-normal size distribution (Porter and Clarke, 1997) is applied for the sea-salt aerosols (see Eq. (4.12)). In this study, we chose a sea-salt aerosol size distribution for wind speeds between 5.5 and 7.9 m/s from Porter and Clarke (1997) with all the parameters for fine and coarse modes shown in Fig. 4.18. The optical thickness of the aerosol layer is given as 0.1. Figure 4.19 shows a comparison of the directional irradiance reflectance and DOP on the principal plane from the ocean–atmosphere system without aerosols and with sea-salt aerosols, respectively. The ocean wind speed is 7.5 m/s. The SZA is 33.30°. We can see that aerosols can increase the total

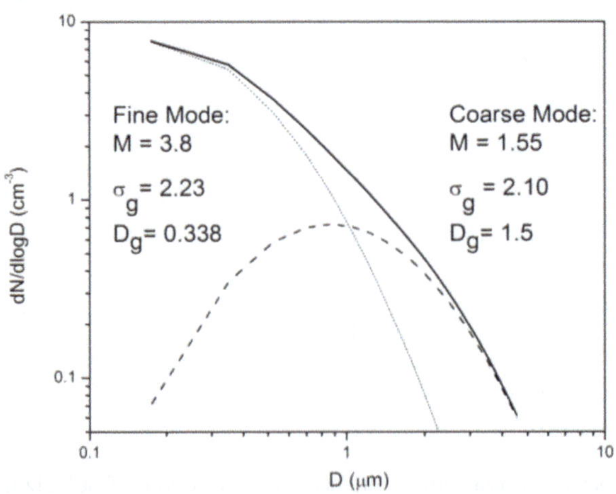

Fig. 4.18. A two-mode sea-salt aerosol size distribution from Porter and Clarke (1997) for ocean wind speed between 5.5 and 7.9 m/s (from Sun and Lukashin, 2013). The dotted curve denotes the fine-mode size distribution, and the dashed curve denotes the coarse-mode size distribution. The combined size distribution is denoted by the solid curve.

reflectance and decrease the DOP significantly. Although Fig. 4.20 shows aerosols' insignificant effect on the AOLP, aerosols play an important role in the DOP of the reflected solar light. Thus, in building the PDMs, the aerosol effect must be taken into account.

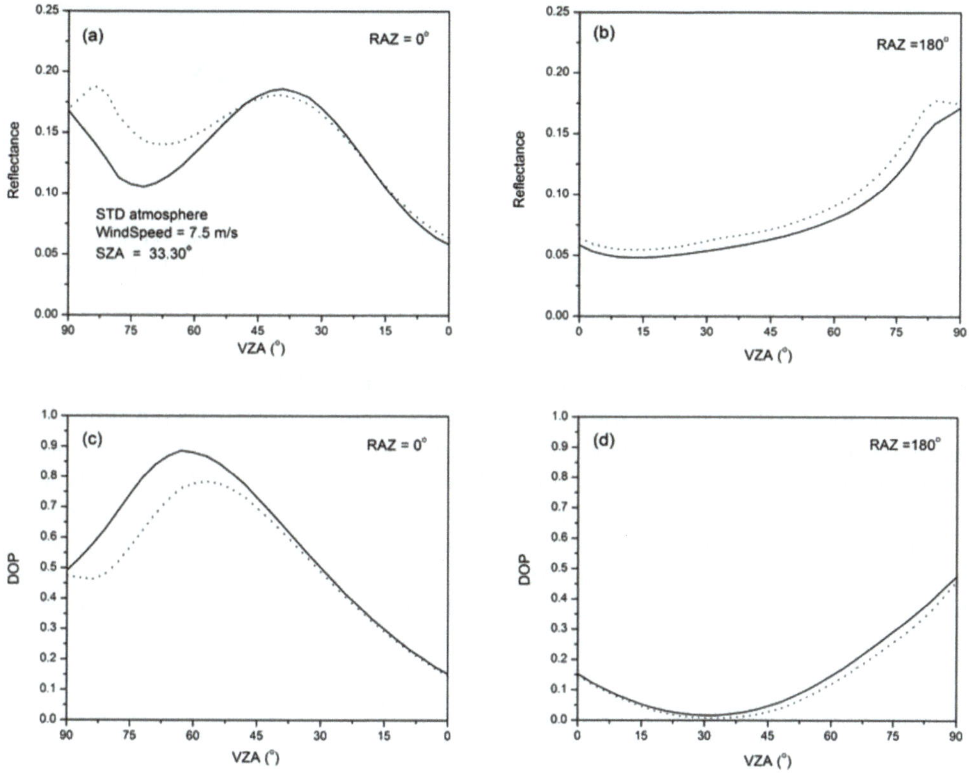

Fig. 4.19. The directional irradiance reflectance and degree of polarization (DOP) on the principal plane, which is calculated with the adding-doubling radiative-transfer model (ADRTM) at the wavelengths of 550 nm from the ocean–atmosphere system without aerosols (solid curve) and with sea-salt aerosols (dashed curve) of optical thickness 0.1 (from Sun and Lukashin, 2013). The ocean wind speed is 7.5 m/s. The SZA is 33.30°. The US Standard Atmosphere (STD) is used.

4.5 Summary and conclusion

The reflected solar radiance from the Earth–atmosphere system is polarized. Radiance measurements can be affected by the reflected light's state of polarization if the radiometric sensor is sensitive to the polarization of the observed light. To correct measurement errors due to the use of the polarization-dependent imagers, such as the MODIS, the polarization state of the reflected solar light must be known with sufficient accuracy. In this chapter, our recent studies of the polarized solar

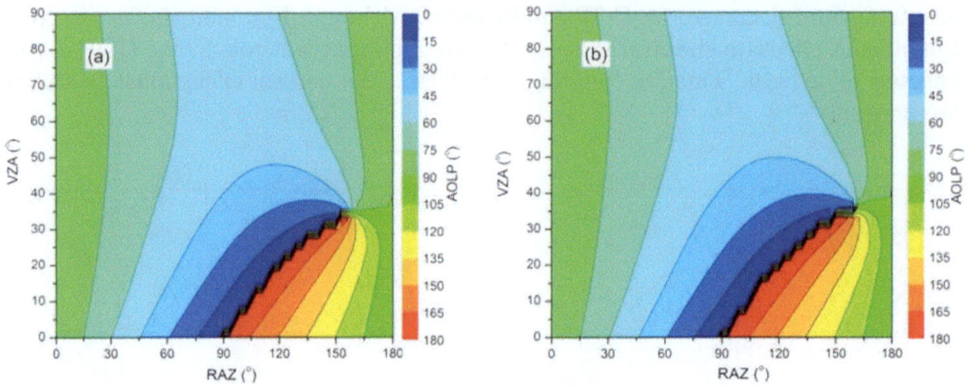

Fig. 4.20. Same as in Fig. 4.19, but for the angles of linear polarization (AOLPs) of the ocean–atmosphere system (a) without aerosols and (b) with sea-salt aerosols, respectively. From Sun and Lukashin (2013).

radiation from the ocean–atmosphere system with the ADRTM are reviewed. The modeled polarized solar radiation quantities are compared with PARASOL satellite measurements and DISORT model results. Good agreement between model results and satellite data is shown for both liquid water clouds and ice clouds. Differences between model results and satellite measurements for clear-sky oceans are due to the presence of undetected clouds in the PARASOL clear-sky scenes. We also note a distinct feature in the AOLP of solar radiation that is backscattered from clouds. The dominant backscattered electric field from the clear-sky Earth–atmosphere system is nearly parallel to Earth's surface. However, when clouds are present, this electric field can rotate significantly away from the parallel direction. Model results demonstrate that this polarization feature can be used to detect super-thin cirrus clouds having an OD of only ~0.06 and super-thin liquid water clouds having an OD of only ~0.01. Such clouds are too thin to be sensed using any current passive satellite instruments. Based on these results, a novel method for detecting cloud particles in the atmosphere is suggested. Sensitivities of reflected solar radiation's polarization to various ocean-surface and atmospheric conditions are also addressed. These studies suggest that the modeling can provide a reliable approach for making the spectral PDMs for NASA's future CLARREO mission's inter-calibration application, which cannot be achieved by empirical PDMs based on the analysis of the data from polarimetric sensors.

Acknowledgments

This work is supported by NASA's CLARREO mission. The authors thank Bruce A. Wielicki and David F. Young for this support. Wenbo Sun and Gorden Videen also thank Hal B. Maring and Michael I. Mishchenko for the support of NASA Glory fund 09-GLORY09-0027 for this work.

References

Ackerman, S.A., Holz, R.E., Frey, R., Eloranta, E.W., Maddux, B.C. and McGill, M. (2008) 'Cloud detection with MODIS. Part II: Validation,' *J. Atmos. Oceanic. Technol.*, **25**, 1073–86.

Ackerman, S.A., Strabala, K.I., Menzel, W.P., Frey, R.A., Moeller, C.C. and Gumley, L.E. (1998) 'Discriminating clear sky from clouds with MODIS,' *J. Geophys. Res.*, **103**, 32141–57.

Baum, B.A., Kratz, D.P., Yang, P., Ou, S.C., Hu, Y., Soulen, P.F. and Tsay, S.C. (2000) 'Remote sensing of cloud properties using MODIS Airborne Simulator imagery during SUCCESS. I. Data and models,' *J. Geophys. Res.*, **105**, 11767–80.

Chahine, M.T., Pagano, T.S., Aumann, H.H., Atlas, R., Barnet, C., Blaisdell, J., Chen, L., Divakarla, M., Fetzer, E.J., Goldberg, M., Gautier, C., Granger, S., Hannon, S., Irion, F.W., Kakar, R., Kalnay, E., Lambrigtsen, B.H., Lee, S.-Y., Le Marshall, J., McMillan, W.W., McMillin, L., Olsen, E.T., Revercomb, H., Rosenkranz, P., Smith, W.L., Staelin, D., Strow, L.L., Susskind, J., Tobin, D., Wolf, W. and Zhou, L.H. (2006) 'AIRS: Improving weather forecasting and providing new data on greenhouse gases,' *Bull. Amer. Meteor. Soc.*, **87**, 911–26.

Chamaillard, K., Jennings, S.G., Kleefeld, C., Ceburnis, D. and Yoon, Y.J. (2003) 'Light backscattering and scattering by nonspherical sea-salt aerosols,' *J. Quant. Spectrosc. Radiat. Trans.*, **79–80**, 577–97.

Chowdhary, J., Cairns, B. and Travis, L.D. (2002) Case studies of aerosol retrievals over the ocean from multiangle, multispectral photopolarimetric remote sensing data,' *J. Atmos. Sci.*, **59**, 383–97.

Christi, M.J. and Stephens, G.L. (2004) 'Retrieving profiles of atmospheric CO2 in clear sky and in the presence of thin cloud using spectroscopy from the near and thermal infrared: A preliminary case study,' *J. Geophys. Res.*, **109**, D04316.

Clough, S.A. and Iacono, M.J. (1995) 'Line-by-line calculations of atmospheric fluxes and cooling rates. II: Application to carbon dioxide, ozone, methane, nitrous oxide, and the halocarbons,' *J. Geophys. Res.*, **100**, 16519–35.

Clough, S.A., Iacono, M.J. and Moncet, J.-L. (1992) 'Line-by-line calculation of atmospheric fluxes and cooling rates: Application to water vapor,' *J. Geophys. Res.*, **97**, 15761–85.

Coulson, K.L. (1988) *Polarization and Intensity of Light in the Atmosphere*, A. Deepak, Hampton, VA.

Cox, C. and Munk, W. (1954) 'Measurement of the roughness of the sea surface from photographs of the sun's glitter,' *J. Opt. Soc. Amer.*, **44**, 838–50.

Cox, C. and Munk, W. (1956) 'Slopes of the sea surface deduced from photographs of sun glitter,' *Bull. Scripps Inst. Oceanogr.*, **6**, 401–88.

Davies, C. (1974) 'Size distribution of atmospheric particles,' *Aerosol Science*, **5**, 293–300.

de Haan, J.F., Bosma, P.B. and Hovenier, J.W. (1987) 'The adding method for multiple scattering calculations of polarized light,' *Astron. Astrophys.*, **183**, 371–91.

Deirmendjian, D. (1969) *Electromagnetic Scattering on Spherical Polydispersions*, American Elsevier Publishing Company, Inc., New York.

Evans, K.F. and Stephens, G.L. (1991) 'A new polarized atmospheric radiative transfer model,' *J. Quant. Spectrosc. Radiat. Trans.*, **46**, 413–23.

Fu, Q. (1996) 'An accurate parameterization of the solar radiative properties of cirrus clouds for climate models,' *J. Climate*, **9**, 2058–82.

Fu, Q. and Sun, W. (2001) 'Mie theory for light scattering by a spherical particle in an absorbing medium,' *Appl. Opt.*, **40**, 1354–61.

Gao, B.C. and Kaufman, Y.J. (1995) 'Selection of 1.375-μm MODIS channel for remote sensing of cirrus clouds and stratospheric aerosols from space,' *J. Atmos. Sci.*, **52**, 4231–7.

Hansen, J.E. (1968) 'Exact and approximation solutions for multiple scattering by cloudy and haze planetary atmospheres,' *J. Atmos. Sci.*, **26**, 478–87.

Hansen, J.E. and Hovenier, J.W. (1971) 'The doubling method applied to multiple scattering of polarized light,' *J. Quant. Spectrosc. Radiat. Trans.*, **11**, 809–12.

Hansen, J.E. and Travis, L.D. (1974) 'Light scattering in planetary atmospheres,' *Space Sci. Rev.*, **16**, 527–610.

Kato, S., Ackerman, T.P., Mather, J.H. and Clothiaux, E.E. (1999) 'The k-distribution method and correlated-k approximation for a shortwave radiative transfer model,' *J. Quant. Spectrosc. Radiat. Trans.*, **62**, 109–21.

King, M.D., Kaufman, Y.J., Menzel, W.P. and Tanré, D. (1992) 'Remote sensing of cloud, aerosol and water vapor properties from the Moderate Resolution Imaging Spectrometer (MODIS),' *IEEE Trans. Geosci. Remote Sens.*, **30**, 2–27.

Kneizys, F.X., Shettle, E.P., Abreu, L.W., Chetwynd, J.H., Anderson, G.P., Gallery, W.O., Selby, J.E.A. and Clough, S.A. (1988) *Users Guide to LOWTRAN 7*, AFGL-TR-88-0177.

Koepke, P. (1984) 'Effective reflectance of oceanic whitecaps,' *Appl. Opt.*, **23**, 1816–24.

Lacis, A.A., Chowdhary, J., Mishchenko, M.I. and Cairns, B. (1998) 'Modeling errors in diffuse-sky radiation: Vector vs. scalar treatment,' *Geophys. Res. Lett.*, **25**, 135–8.

Lukashin, C., Wielicki, B.A., Young, D.F., Thome, K.J., Jin, Z. and Sun, W. (2013) 'Uncertainty estimates for imager reference inter-calibration with CLARREO reflected solar spectrometer,' *IEEE Transac. Geosci. Remote Sens.*, **51**, 1425–36.

Macke, A. (1993) 'Scattering of light by polyhedral ice crystals,' *Appl. Opt.*, **32**, 2780–8.

McClatchey, R.A., Fenn, R.W., Selby, J.E.A., Volz, F.E. and Garing, J.S. (1972) *Optical Properties of the Atmosphere*, 3rd edn, Environ Res. Pap. 411, Air Force Cambridge Res. Lab., Bedford, MA.

McFarquhar, G.M., Heymsfield, A.J., Spinhirne, J. and Hart, B. (2000) 'Thin and subvisual tropopause tropical cirrus: Observations and radiative impacts,' *J. Atmos. Sci.*, **57**, 1841–53.

Mie, G. (1908) 'Beiträge zur optik trüber medien, speziell kolloidaler metallösungen,' *Leipzig Ann. Phys.*, **330**, 377–445.

Mishchenko, M.I. and Travis, L.D. (1997) 'Satellite retrieval of aerosol properties over the ocean using polarization as well as intensity of reflected sunlight,' *J. Geophys. Res.*, **102**, 16989–7013.

Mishchenko, M.I., Cairns, B., Kopp, G., Schueler, C.F., Fafaul, B.A., Hansen, J.E., Hooker, R.J., Itchkawich, T., Maring, H.B. and Travis, L.D. (2007) 'Accurate monitoring of terrestrial aerosols and total solar irradiance: Introducing the Glory mission,' *Bull. Amer. Meteorol. Soc.*, **88**, 677–91.

Mishchenko, M.I., Liu, L. and Mackowski, D.W. (2013) 'T-matrix modeling of linear depolarization by morphologically complex soot and soot-containing aerosols,' *J. Quant. Spectrosc. Radiat. Trans.*, **123**, 135–44.

Mitchell, D.L., Chai, S., Liu, Y., Heymsfield, A.J. and Dong, Y. (1996) 'Modeling cirrus clouds. Part I: Treatment of bimodal size spectra and case study analysis,' *J. Atmos. Sci.*, **53**, 2952–66.

Monahan, E.C. and O'Muircheartaigh, I.G. (1980) 'Optimal power-law description of oceanic whitecap coverage dependence on wind speed,' *J. Phys. Oceanogr.*, **10**, 2094–9.

Morel, A. (1988) 'Optical modeling of the upper ocean in relation to its biogenous matter content (Case I waters),' *J. Geophys. Res.*, **93**, 10749–68.

Muinonen, K., Tyynelä, J., Zubko, E., Lindqvist, H., Penttilä, A. and Videen, G. (2011) ‘Polarization of light backscattered by small particles,’ *J. Quant. Spectrosc. Radiat. Trans.*, **112**, 2193–212.

National Oceanic and Atmospheric Administration, National Aeronautics and Space Administration, and United States Air Force (1976) *country-U.S. Standard Atmosphere*, NOAA-S/T 76-1562.

Omar, A.H., Winker, D.M., Tackett, J.L., Giles, D.M., Kar, J., Liu, Z., Vaughan, M.A., Powell, K.A. and Trepte, C.R. (2013) ‘CALIOP and AERONET aerosol optical depth comparisons: One size fits none,’ *J. Geophys. Res.*, **118**, 4748–66.

Ott, W. (1990) ‘A physical explanation of the lognormality of pollutant concentrations,’ *J. Air Waste Manage. Assoc.*, **40**, 1378–83.

Peebles, G.H. and Plesset, M.S. (1951) ‘Transmission of Gamma-rays through large thicknesses of heavy materials,’ *Phys. Rev.*, **81**, 430–9.

Porter, J.N. and Clarke, A.D. (1997) ‘Aerosol size distribution models based on *in situ* measurements,’ *J. Geophys. Res.*, **102**, 6035–45.

Reist, P.C. (1984) *Introduction to Aerosol Science*, McMillan, New York.

Roskovensky, J.K. and Liou, K.N. (2003) ‘Detection of thin cirrus from 1.38 μm/0.65 μm reflectance ratio combined with 8.6–11 μm brightness temperature difference,’ *Geophys. Res. Lett.*, **30**, 1985.

Sassen, K.C. and Benson, S.A. (2001) ‘Midlatitude cirrus cloud climatology from the facility for atmospheric remote sensing. Part II: Microphysical properties derived from lidar depolarization,’ *J. Atmos. Sci.*, **58**, 2103–12.

Shkuratov, Y., Kreslavsky, M., Kaydash, V., Videen, G., Bell, J., Wolff, M., Hubbard, M., Noll, K. and Lubenow, A. (2005) ‘Hubble space telescope imaging polarimetry of Mars during the 2003 opposition,’ *ICARUS*, **176**, 1–11.

Stamnes, K., Tsay, S.-C., Wiscombe, W. and Jayaweera, K. (1988) ‘Numerically stable algorithm for discrete-ordinate-method radiative transfer in multiple scattering and emitting layered media,’ *Appl. Opt.*, **27**, 2502–9.

Stokes, G.G. (1862) ‘On the intensity of the light reflected from or transmitted through a pile of plates,’ *Proc. R. Soc. London*, **11**, 545–56.

Sun, J. and Xiong, X. (2007) ‘MODIS polarization-sensitivity analysis,’ *IEEE Trans. Geosci. Rem. Sens.*, **45**, 2875–85.

Sun, W. and Lukashin, C. (2013) ‘Modeling polarized solar radiation from ocean–atmosphere system for CLARREO inter-calibration applications,’ *Atmos. Chem. Phys.*, **13**, 10303–24.

Sun, W., Fu, Q. and Chen, Z. (1999) ‘Finite-difference time domain solution of light scattering by dielectric particles with a perfectly matched layer absorbing boundary condition,’ *Appl. Opt.*, **38**, 3141–51.

Sun, W., Kato, S., Lin, B., Lukashin, C. and Hu, Y. (2011a) ‘A study of subvisual clouds and their radiation effect with a synergy of CERES, MODIS, CALIPSO, and AIRS data,’ *J. Geophys. Res.*, **116**, D22207.

Sun, W., Lin, B., Hu, Y., Lukashin, C., Kato, S. and Liu, Z. (2011b) ‘On the consistency of CERES longwave flux and AIRS temperature and humidity profiles,’ *J. Geophys. Res.*, **116**, D17101.

Sun, W., Liu, Z., Videen, G., Fu, Q., Muinonen, K., Winker, D.M., Lukashin, C., Jin, Z., Lin, B. and Huang, J. (2013a) ‘For the depolarization of linearly polarized light by smoke particles,’ *J. Quant. Spectrosc. Radiat. Trans.*, **122**, 233–7.

Sun, W., Loeb, N.G. and Fu, Q. (2002) ‘Finite-difference time domain solution of light scattering and absorption by particles in an absorbing medium,’ *Appl. Opt.*, **41**, 5728–43.

Sun, W., Lukashin, C., Baize, R.R. and Goldin, D. (2015) 'Modeling polarized solar radiation for CLARREO inter-calibration applications: Validation with PARASOL data,' *J. Quant. Spectrosc. Radiat. Trans.*, **150**, 121–33.

Sun, W., Videen, G. and Mishchenko, M.I. (2014) 'Detecting super-thin clouds with polarized sunlight,' *Geophy. Res. Lett.*, **41**, 688–93.

Sun, W., Videen, G., Fu, Q. and Hu, Y. (2013b) 'Scattered-field FDTD and PSTD algorithms with CPML absorbing boundary conditions for light scattering by aerosols,' *J. Quant. Spectrosc. Radiat. Trans.*, **131**, 166–74.

Tanre, D., Breon, F.M., Deuze, J.L., Dubovik, O., Ducos, F., François, P., Goloub, P., Herman, M., Lifermann, A. and Waquet, F. (2011) 'Remote sensing of aerosols by using polarized, directional and spectral measurements within the A-train: The PARASOL mission,' *Atmos. Measur. Tech.*, **4**, 1383–95.

Tsang, L., Kong, J.A. and Shin, R.T. (1985) *Theory of Microwave Remote Sensing*, John Wiley, New York.

Twomey, S., Jacobowitz, N. and Howell, H.B. (1966) 'Matrix methods for multiple-scattering problems,' *J. Atmos. Sci.*, **23**, 289–96.

van de Hulst, H.C. (1963) *A New Look at Multiple Scattering*, Tech. Rep., Goddard Institute for Space Studies, New York.

Volten, H., Muñoz, O., Rol, E., de Haan, J.F. and Vassen, W. (2001) 'Scattering matrices of mineral aerosol particles at 441.6 and 632.8 nm,' *J. Geophys. Res.*, **106**, 17375–401.

Whitby, K. (1978) 'The physical characteristics of sulfur aerosols,' *Atmos. Environ.*, **12**, 135–59.

Wielicki, B.A., Barkstrom, B.R., Harrison, E.F., Lee, R.B., Smith, G.L. and Cooper, J.E. (1996) 'Clouds and the Earth's Radiant Energy System (CERES): An Earth observing system experiment,' *Bull. Amer. Meteorol. Soc.*, **77**, 853–68.

Wielicki, B.A., et al. (2013) 'Climate absolute radiance and refractivity observatory (CLARREO): Achieving climate change absolute accuracy in orbit,' *Bull. Amer. Meteor. Soc.*, **94**, 1519–39.

Winker, D.M., Hunt, W.H. and McGill, M.J. (2007) 'Initial performance assessment of CALIOP,' *Geophys. Res. Lett.*, **34**, L19803.

World Meteorological Organization (1985) *Atmospheric Ozone 1985*, Global Ozone Research and Monitoring Project, Report No. 16.

Wylie, D.P. and Menzel, M.P. (1999) 'Eight years of high cloud statistics using HIRS,' *J. Climate*, **12**, 170–84.

Wylie, D.P., Piironen, P., Wolf, W. and Eloranta, E. (1995) 'Understanding satellite cirrus cloud climatologies with calibrated lidar optical depths,' *J. Atmos. Sci.*, **52**, 4327–43.

Yang, P. and Liou, K.N. (1996) 'Geometric-optics: Integral-equation method for light scattering by nonspherical ice crystals,' *Appl. Opt.*, **35**, 6568–84.

Zubko, E., Muinonen, K., Munoz, O., Nousiainen, T., Shkuratov, Y., Sun, W. and Videen, G. (2013) 'Light scattering by feldspar particles: Comparison of model agglomerate debris particles with laboratory samples,' *J. Quant. Spectrosc. Radiat. Trans.*, **131**, 175–87.

Zubko, E., Shkuratov, Y., Kiselev, N.N. and Videen, G. (2006) 'DDA simulations of light scattering by small irregular particles with various structure,' *J. Quant. Spectrosc. Radiat. Trans.*, **101**, 416–34.

Part II

Atmospheric optics and inverse problems

5 Vertical profiles of optical and microphysical characteristics of tropospheric aerosol from aircraft measurements

Mikhail V. Panchenko and Tatiana B. Zhuravleva

5.1 Introduction

According to the International Panel on Climate Change (IPCC), the current lowest level of scientific understanding in estimates of radiative forcings of different atmospheric constituents is assigned to aerosols (Forster et al., 2007; Hansen et al., 2005), primarily due to a few factors.

The first factor is associated with the difficulty in obtaining accurate quantitative information on the required parameters and is caused by (i) strong spatiotemporal variations of the sources and sinks of aerosol particles, (ii) relatively short lifetimes (approximately 7 days) in the troposphere in contrast to greenhouse gases, requiring far more detailed knowledge of their spatiotemporal variations, and (iii) the difficulty in performing synchronous measurements of the scattering and absorbing properties of aerosols for studying their vertical and horizontal distributions.

The second complex problem is associated with the multiform impact of atmospheric aerosols on Earth's radiation budget, which is manifested by both direct and indirect aerosol effects (Haywood and Boucher, 2000).

The direct aerosol effect is the contribution to the cooling (heating) of the atmosphere by aerosol particles, which scatter (absorb) the solar radiation in the atmosphere. Thus, the net effect depends on the relationship between the scattering and absorbing properties of aerosols in the atmospheric column (Penner et al., 2003; Haywood and Ramaswamy, 1998). The indirect aerosol effects are more complex (Twomey, 1977b; Albrecht, 1989; Johnson et al., 2004). The aerosol particles, acting as condensation nuclei, determine the amount of cloud droplets and their microstructure, thus affecting both the optical properties of clouds and their lifetime in the atmosphere. Low-level clouds, as a rule, play a cooling factor role, while clouds at high altitudes can cause a cooling effect, as well as the intensification of atmospheric heating. Another problem is estimating the radiation effects caused by the presence of absorbing aerosol between cloud particles.

Still another aspect of the indirect aerosol effect is associated with the sedimentation and settling of aerosol (primarily black carbon (BC)) particles, causing a change in the reflectance (albedo) of the underlying surface. The effect of this process is most significant for the northern territories, where the supply of absorbing substances on the ice and snow cover leads to a substantial decrease in the surface albedo. A resulting consequence is an increase in absorbed radiation that promotes the melting of snow and ice. This process is challenging to accurately simulate in climate models at the present stage due to the difficulties in obtaining reliable experimental data.

An undoubted progress achieved in the development of climate models, in which the radiation code is one of the important elements, dictates that the appropriate level of the basic information on the key optical characteristics of aerosols (optical depth, single scattering albedo, and asymmetry factor of the scattering phase function) should also be reached. Clearly, regardless of the methods used for calculating the radiative characteristics, the success of prediction of climate changes will ultimately be determined by the reliability and validity of the information on the real optical parameters of the atmosphere and by the accuracy of accounting for their variations under the impact of external factors. The simulation of the optical state of the atmospheric aerosols is reduced to the generalization of their optical or microphysical properties as functions of geophysical, synoptic, and meteorological processes, which are characteristic for the different geographic regions of the planet.

In the practice of aerosol studies, we can single out two main approaches to designing the model concepts regarding the properties of the atmospheric aerosol. One is based upon information about the aerosol *microphysical* composition, obtained from experimental data, and the subsequent calculation of the necessary optical characteristics ('microphysical' approach). Another method is based on the direct measurements of the *optical* characteristics ('optical' approach). Each of these approaches has its own merits and drawbacks.

An advantage of the *microphysical* approach is the capability to calculate all of the required optical characteristics of aerosol for a given spectral range. One serious disadvantage is that any limitation or distortion of information on the microphysical parameters of particles or their shapes leads to optical parameter retrieval errors that are difficult to control.

Most problems with the microphysical approach are associated with the determination of the complex refractive index of particles for different size ranges. As a rule, optical constants of substance are retrieved from the chemical composition measurements. However, relying only upon observations of the chemical composition of atmospheric aerosol particles is almost impossible to take into account *a priori* the contribution of non-recorded small concentrations of chemical elements and their compounds to the real (n) and imaginary (κ) parts of the refractive index. It should be stressed that considerable non-controlled errors may be a consequence of the neglect of the contribution from chemical elements having large values of n and κ (such as metals). Moreover, the contribution of organic compounds to the atmospheric aerosol is still poorly understood. This is especially true for highly volatile compounds, which can either be substantially transformed in the measurement process or 'escape' from the sample in the process of its storage and preparation for analysis.

To a large extent, the approach based on measurements of the optical characteristics directly in the real atmosphere is free of the drawbacks inherent in the microphysical modeling. It almost completely eliminates the need to estimate and take into account the optical significance of particles in one or another size range or different chemical origins. The model input parameters are chosen in a natural way, depending on the degree of influence of directly measured external factors on the variations of the optical characteristics. The optical approach is disadvantageous in that the use of the obtained model is usually restricted to the list of the measured optical characteristics, spectral range, and geophysical conditions, under which the initial observational data were obtained. Expanding beyond these constraints requires extra studies and substantiation.

The difficulties arising in implementing each of these approaches can be partly overcome using the method of 'microphysical extrapolation,' suggested by G.V. Rosenberg (1967, 1968, 1976). This method is based on the idea of using information on the microstructure and complex refractive index of particles, retrieved in the course of solving the inverse 'optical' problem for a subsequent calculation of the required optical characteristics, which were not measured in experiment. Based on our experience, it should be emphasized that the retrievals of microstructure and refractive index made using measurements of a small number of optical parameters in a limited spectral interval should be used carefully. Additional studies are required to determine the boundaries for the correctly retrieved particle size spectrum, as well as the uncertainties in the estimates of the complex refractive index of the aerosol substance. At the same time, solutions to the inverse problem are most suitable for a subsequent calculation of all of the required set of optical characteristics in the same or close spectral range because they are determined to a considerable degree by the same particle size interval.

Clearly, only a complex combination of the optical and microphysical approaches to the arrangement of experiments for studying the aerosol properties will permit the most complete retrieval of information required for the radiation calculations. In particular, the measurements of aerosol microstructure in a wide particle size region can ensure the retrieval of the optical characteristics in the near-IR spectral range. A specific problem in interpreting the observation data and in creating the empirical aerosol models is the need to parameterize the effect of the *relative humidity* (RH) on either changing the size spectrum and optical constants of particles or directly influencing the optical characteristics (spectral scattering and absorption coefficients, scattering phase functions, etc.).

The relative air humidity is the most dynamic characteristic that substantially alters the aerosol optical properties. As an example, we will consider the problem of the parameterization of the optical and microphysical characteristics with RH on the basis of data from *in situ* observations. The RH varies synchronously with the change in the ambient air temperature during the day. Simultaneously with this well-pronounced process, the microstructure and vertical stratification of aerosols also change under the impact of particle generation, transport, outflow, or sedimentation. Clearly, a comprehensive analysis of all these components is required to correctly describe the effect of RH on aerosol microphysical properties because the above-mentioned processes of microstructure transformation and relative humidity variation in the atmosphere have cardinally different spatiotemporal scales.

The principles of forming model concepts and, ultimately, the principles of constructing empirical aerosol models are dictated by their aim, the nature of the simulated process, and the attained level of the knowledge.

In the second half of 20th century, the rapid development of different systems and devices operating in the optical range through the atmosphere necessitated conducting large-scale aerosol studies. In this period, an enormous amount of experimental material on practically all required characteristics of aerosol had been accumulated and became a basis for a number of models on the optical characteristics well known to specialists (see, e.g., Elterman, 1970; Gillette and Blifford, 1971; Fenn, 1979; Shettle and Fenn, 1979; Toon and Pollack, 1976; Twomey, 1977a; Jaenicke, 1980).

For various atmospheric applications, researchers are widely used the LOWTRAN model (Kneizys et al., 1996), which takes into account the stratification of particle microphysical parameters and the effect of the relative air humidity on the transformation of optical characteristics for a number of characteristic geographic zones. Similarly to most other models, this model was based on the microphysical approach; however, the authors also introduced the meteorological visibility range (optical characteristic) as an input parameter, which substantially widened the model applicability range. The vertical structure of the optical characteristics for two periods of the year, namely 'fall–winter' and 'spring–summer,' is taken into account by changing the height of the mixing layer.

Rich experimental data on the optical, microphysical, and chemical characteristics of the particles for different geographical regions, obtained using aircraft and ground-based lidar sensing of the atmosphere, were collected and generalized in aerosol models presented by scientists from the former USSR (see, e.g., Ivlev, 1969; Kondratyev and Pozdnykov, 1981; Krekov and Rakhimov, 1982; Zuev and Krekov, 1986). All advantages of the optical approach to modeling were first implemented in the model developed under the leadership of G.V. Rosenberg in the Institute of Atmospheric Physics of the Russian Academy of Sciences (Rosenberg et al., 1980). This model was based on the long-term integrated measurements and statistical analysis of the angular dependencies of components of the scattering matrices for the visible spectral range, spectral dependencies of the extinction coefficients in a wide spectral range, data on the absorbing properties of aerosol particles and their chemical composition, information on the vertical stratification of the optical parameters according to data of twilight sensing, and comprehension of the main processes of generation and transformation of aerosol particles in the atmosphere.

At this stage, the main efforts of scientists were concentrated on the *scattering* properties of atmospheric particles because the spectral ranges of instruments were chosen to lie, as a rule, in the visible spectral range and outside the absorption lines of atmospheric gases. Much less information was obtained about aerosol *absorption*, primarily due to the absence of the appropriate instrumentation. However, at the frontier of the 20th and 21st centuries, in view of the observed climate changes and ensuing requirements on the simulation of atmospheric radiative characteristics, the world scientific community had come to realize the pressing need for detailed studies on the *relationship between the scattering and absorbing properties* of atmospheric aerosol. This circumstance initiated the intense development of instrumental bases and the beginning of a new stage of large-scale comprehensive experiments and routine observations.

Among these studies, there are experiments aimed at obtaining information on the vertical structure of the aerosol characteristics, which is required to solve a whole class of problems. Haywood and Ramaswamy (1998) studied the direct radiative forcing (DRF) of sulfate and black carbon aerosols and showed that the magnitude and sign of DRF are partly determined by the vertical distribution; sulfate gives the strongest DRF at low altitudes due to the effects of RH, and BC does it at high altitudes, as the aerosol is moved above the cloudy layers of the atmosphere. Rozwadowska (2007) showed that the error assumptions about the shape of aerosol profiles can cause errors in the aerosol optical thickness retrieved from satellites. The vertical distribution of *absorbing* aerosol in the atmosphere is especially interesting. In particular, BC, in addition to the above-mentioned DRF, may also affect the clouds. Koch and Del Genio (2010) showed that the atmospheric heating through the direct aerosol effect perturbs the stability of the atmosphere and therefore cloud distributions, potentially causing warming or cooling (semi-direct effect). black carbon aerosols may also affect warm clouds by acting as cloud condensation nuclei, thus leading to an increase in cloud brightness and decrease in precipitation (the indirect effect; Kristjansson, 2002; Bauer et al., 2010).

Data on the stratification of aerosol characteristics can be obtained on the basis of lidar and aircraft sensing. The vertical profiles of aerosol are studied by the methods of *lidar* sensing in the frameworks of the Micro Pulse Lidar Network (MPLNET, *mplnet.gsfc.nasa.gov*), the European Aerosol Research Lidar Network to Establish an Aerosol Climatology (EARLINET, *www.earlinet.org*), the Asian Dust and Aerosol Lidar Observation Network (AD-Net, *www-lidar.nies.go.jp/AD-Net*), and others. Lidar observations provide data about the vertical loading and variability of aerosol particles; however, Andrews et al. (2011) note that 'deriving profiles of aerosol *absorption* or other properties (e.g., single scattering albedo, asymmetry parameter) useful for radiative forcing calculations from lidar measurements is still in its infancy'. To increase the interpretability of data obtained using Sun photometers and lidars and to provide both columnar and vertically resolved aerosol and cloud data, *joint* measurements of aerosol characteristics with these instruments were initiated in the last few years at a number of AERONET sites (*mplnet.gsfc.nasa.gov*; *aeronet.gsfc.nasa.gov*).

The need in the aerosol property studies on the basis of *aircraft* sensing data is dictated by the following circumstances. It is well known that the concentration, composition, and stratification of tropospheric aerosol are formed under the influence of peculiarities of specific air mass (for approximately 3–10 days). The aerosol state and variations are affected by the atmospheric processes in a wide range of spatiotemporal scales, from local to regional scales. To study the regional-scale atmospheric phenomena, it is necessary either to perform *multipoint* observations or use highly mobile, equipment, capable of covering vast territories within a limited time. In this sense, aircraft sensing is the only source of data on both vertical and horizontal distributions of aerosol characteristics in the atmosphere.

Contemporary aircraft can lift the numerous sensors and instruments, integrated into a single information-measurement system, to the required height and supply the measurement process by means of software, energy, and meteorological services. An advantage of this approach is the comprehensive character of observations; that is, most physical characteristics of air can be measured in a universal-time system. An important process made possible by research aircraft is accompanying the studied atmospheric phenomenon along its propagation path over large distances and, thereby, objectively estimating the spatiotemporal variations in atmospheric characteristics.

At present, aircraft measurement data are used to solve a wide range of problems, including those associated with climate, ecology, and validation of satellite data, among others. This chapter provides the brief overview of the studies devoted to the problems of retrieving vertical profiles of aerosol microphysical and optical characteristics in the context of their subsequent application in radiation calculations. We focused on the descriptions of the *approaches*, which were used to solve these tasks in the course of (i) specific comprehensive experiments and (ii) long-term observations (optical and microphysical approaches, method of microphysical extrapolation). The purpose of the specific comprehensive experiments, lasting usually for up to a few weeks, was to study the vertical structure of *certain types of aerosols* (marine, dust, smoke, etc.) in different regions of the globe. Considering that the data on vertical structure of aerosol optical characteristics during these experiments were obtained by different research groups with the use of instruments installed onboard aircrafts, ships, and satellites, as well as ground-based equipment, much attention was paid to results of *aerosol and radiation closure tests*. The second collection, namely the data of multi-year observations, is the foundation for creation of *regional* models of aerosol optical characteristics. As far as the authors of the review know, at present, the information about the stratification of extinction coefficient, single scattering albedo, and asymmetry factor, taking into account their seasonal variations for more than a 5-year period, is obtained for two regions of the globe: rural Oklahoma (USA) and western Siberia (Russia). In generalization of these data to the level of model representations of the atmospheric aerosol it was taken into account that their optical and microphysical properties were the functions of geophysical, synoptic, and meteorological processes, characteristic for these regions.

We emphasize that our aim was to review the *approaches* used at present to construct the vertical profiles of optical and microphysical characteristics of tropospheric aerosol according to data of aircraft observations. The particular *retrieval results* of aerosol stratification were presented in detail in publications cited below.

5.2 Specific comprehensive experiments

Aerosol properties have been intensely measured over several regions of the globe in major international field campaigns conducted over the past two decades (see, e.g., Yu et al., 2006; Bates et al., 2006; McNaughton et al., 2011). During each of these comprehensive missions, aerosols were studied in great detail, using combinations of *in situ* and remote sensing observations of physical and chemical properties from various platforms (e.g., aircraft, ship, satellite, and ground-network) and numerical modeling.

This section describes the aspects of some of these large-scale experiments, one of the main purposes of which was to study the *vertical structure* of certain types of aerosols in different regions of the globe; these experiments included the Tropospheric Aerosol Radiative Forcing Observational Experiment (TARFOX), Aerosol

Characterization experiments (ACE-2), the Southern African Regional Science Initiative (SAFARI 2000), and the SaHAran Dust Experiment (SHADE). As far as possible, the aircraft measurements, performed by different methods, were compared with each other (aerosol and radiative closures), as well as with data obtained from ground-based and satellite observations.

5.2.1 Instrumentation

The observation campaigns used different sets of instruments for retrieving the aerosol characteristics. The main devices were (i) the set of instruments for measuring the size distributions of aerosol particles, (ii) the three-wavelength nephelometer, and (iii) the absorption photometer for measuring light absorption of the aerosol. The aircraft were also equipped with instrumentation for determining the chemical composition of aerosol, Sun photometers for measuring the aerosol optical depth, radiometers for measuring the broadband and spectral fluxes of solar radiation, etc. The key instruments mentioned in the review are briefly described below.

Aerosol size distributions for fine particles were measured by the Particle Measuring System (PMS) Passive Cavity Aerosol Spectrometer Probe (PCASP). With the heaters switched *on*, it has been demonstrated that the size distributions are measured in a dehydrated state regardless of the ambient conditions (see, e.g., Strapp et al., 1992). In conditions of low RH and for particles that are not greatly hygroscopic, it can be expected that the measurements with the *non-deiced* PCASP will be representative of ambient aerosol. This observation was confirmed, in particular, by comparing the size distributions of biomass-burning aerosol obtained with the heater switched off and from measurements made by an AERONET Sun photometer during SAFARI-2000 (Haywood et al., 2003a). However, with the non-deiced PCASP, the degree of drying of the aerosol particles within conditions of high RH is uncertain and difficult to quantify, and errors in the determination of the true particle size may occur (Strapp et al., 1992; Hignett et al., 1999).

The PMS Forward Scattering Spectrometer Probe (FSSP) and Small Ice Detector (SID; Hirst et al., 2001) were used to measure larger-sized (radii between 1 and 20 μm) particles. Despite the fact that both devices had sizing problems (Osborne et al., 2004; Reid et al., 2003), it was shown that the effect which they exert on the mid-visible optical properties was negligibly small (see, e.g., Osborne et al., 2004). However, particle size determination errors should be considered if the microphysical data are to be applied in calculations at longer wavelengths.

In most cases, the instrument used for aerosol particle *scattering coefficient* measurements was a TSI 3563 integrating nephelometer (Bodhaine et al., 1991; Heintzenberg and Charlson, 1996). It measured the aerosol light-scattering coefficient at three wavelengths (450, 550, and 700 nm) in two ranges of angular integration (7° to 170°, total scattering; 90° to 170°, hemispheric backscattering). The scattering data have been corrected for angular nonidealities (Anderson and Ogren, 1998). The *absorption coefficient* at 567 nm was measured with a Radiance Research Particle Soot Absorption Photometer (PSAP). The data from this photometric device are corrected for inaccuracies in the flow rate, area of exposure of the filter, and absorption artifacts, according to the method of Bond et al. (1999).

Aerosol *chemistry* was determined from isokinetic sampling onto filter substrates (Andreae et al., 2000; Formenti et al., 2003a, 2003b, among others).

The six-channel NASA Ames Airborne Tracking Sunphotometer (AATS-6) (Matsumoto et al., 1987) and its extended version, the 14-channel AATS-14 flown by aircraft, were used in a number of experiments for *aerosol optical depth* (AOD) determination. The center wavelengths of the aerosol channels were 380, 451, 525, and 1021 nm for the AATS-6 and 380, 448, 453, 499, 525, 605, 667, 779, 864, 1019, and 1558 nm for the AATS-14. Under cloudless conditions, the Sun photometers reported the spectral AOD (τ_λ) above the altitude of the aircraft. The extinction coefficient values can be derived by differentiating τ_λ at two vertically separated points, and uncertainties in these derived values are approximately 15% to 20% (Schmid et al., 2003, 2006).

Solar *radiative fluxes* were measured by upward- and downward-facing Eppley broadband radiometers (BBRs), which covered the 0.3–3.0 and 0.7–3.0 μm ranges, respectively (see, e.g., Hignett et al., 1999; Haywood et al., 2003b, 2003c). In addition to broadband fluxes, the Meteorological Office Scanning Airborne Filter Radiometer (SAFIRE) was designed to measure *radiances* in 16 bands across the visible and near-infrared regions of the spectrum (Francis et al., 1999).

5.2.2 Specific experiments

Consider in more detail approaches to the retrieval of the aerosol characteristics, which have been implemented during some aerosol experiments.

The *Tropospheric Aerosol Radiative Forcing Observational Experiment* (TARFOX) was conducted in July and August of 1996 and focused on the plume of *pollutant* haze that moves off the US East Coast over the Atlantic Ocean (Russell et al., 1999a). The overall goal of TARFOX was to reduce uncertainties in the effects of *anthropogenic* aerosols on climate by determining the direct radiative impacts, as well as the chemical, physical, and optical properties of the aerosols carried over the western Atlantic Ocean from the US.

The measurements of the vertical profiles of particle size distributions, scattering and absorption coefficients, individual chemical components, layer AODs, and radiative characteristics were performed from three aircraft (UK Meteorological Research Flight C-130, University of Washington (UW) C-131A, and the Pelican of CIRPAS (Center for Interdisciplinary Remotely Piloted Aircraft Studies)) (Russell et al., 1999a; Hignett et al., 1999; Hobbs, 1999).

The results of some *aerosol closure* tests were presented by Hegg et al. (1997). In particular, AOD, which is a key parameter in the determination of direct aerosol radiative forcing, was measured with the six-channel NASA AATS-6 (τ^{sp}) and derived from the vertical profile measurements of the dried aerosol light-scattering ($\sigma_{sc}^{dry}(z)$) and absorption ($\sigma_{abs}^{dry}(z)$) coefficients and the hygroscopic growth factor ($\tau^{in\ situ}$). The changes in aerosol light scattering with RH increasing from 30% to 85% were analyzed using scanning humidographs (Kotchenruther et al., 1999) and were taken into consideration in the measurement-based calculations of the AOD:

$$\tau^{in\ situ}(z_1, z_2) = \int_{z_1}^{z_2} \left(\sigma_{sc}^{dry}(z)\, f_{sc}(z, RH) + \sigma_{abs}^{dry}(z)\, f_{abs}(z, RH) \right) dz\,, \tag{5.1}$$

where $f_{sc}(z, RH)$ and $f_{abs}(z, RH)$ are the hygroscopic factors as a function of altitude z for scattering and absorption, respectively. The agreement between these two independent methodologies for determining AOD was very good; the slope of the regression line of $\tau^{in\ situ}$ onto τ^{sp} is 0.86 at wavelength 450 nm, and the *in situ* values were systematically lower than the Sun photometer measurements.

Hignett et al. (1999), Francis et al. (1999), and Russell et al. (1999b) compared the measured and simulated radiative characteristics, which were calculated using the experimental data (*radiative closures*).

The extinction coefficient (σ_{ext}), single scattering albedo (ω), and asymmetry parameter (g) were calculated using Mie theory and the particle size distribution measurements collected by PSASP and FSSP. PCASP was operated with the heaters *off* to give the best estimate of the ambient particle size distributions (Hignett et al., 1999; Francis et al., 1999). The chemical components (50% ammonium sulfate, 45% organic carbon, and 5% elemental (black) carbon by mass) (Hegg et al., 1997; Novakov et al., 1997) were assumed to be *uniformly mixed internally* within each particle and across the particle size spectrum. The particle refractive index was calculated as the volume-weighted average of the refractive indices of the individual components, including water:

$$n = \frac{1}{\sum_j V_j} \sum_j n_j V_j, \quad \kappa = \frac{1}{\sum_j V_j} \sum_j \kappa_j V_j, \tag{5.2}$$

where n_j and κ_j are the real and imaginary parts of the refractive index of species j, taken from literature data, and V_j is the volume occupied by species j within the internally mixed aerosol.

The simulations of diurnally averaged direct aerosol forcing were performed for different values (0%, 80%, and 90%) of the RH (Hignett et al., 1999). The modeled and observed values best agreed for high relative humidities, and this matches with the observation during TARFOX that much of the aerosol loading was present at a high RH and aerosol had a significant absorbing component (see also Hegg et al., 1997). The distributions of the *sky radiance fields* were calculated at several wavelengths in the visible and near-infrared regions of the spectrum from these optical data (Francis et al., 1999). Reasonable agreement with the observed radiance distributions was obtained for two of the three cases considered.

Unlike Hignett et al. (1999) and Francis et al. (1999), Russell et al. (1999b) applied the method of microphysical extrapolation and calculated the optical characteristics using particle size spectrum, which was retrieved on the basis of spectral measurements of AOD performed in the AATS-6 (C-131A) and AATS-14 (Pelican) and the constrained linear inversions (King et al., 1978). The refractive index model was determined from information on the chemical composition (Novakov et al., 1997; Hegg et al., 1997) and data of Palmer and Williams (1975) (volume-mixing rule, Eq. (5.2)). Because the accuracy of size distributions is less important for radiation simulation than the accuracy of the τ, ω, and g spectra (see, e.g., Gonzales and Orgen, 1996), the calculated flux changes were compared to measured flux changes only as a function of AOD. Russell et al. (1999b) confirmed the agreement between the overall dependency of downward and upward flux change on optical depth obtained from radiative measurements and calculations based on measured aerosol properties.

A unified approach used to retrieve the optical and microphysical characteristics of aerosol on the basis of aircraft observations during TARFOX, ACE-2, SAFARI, and SHADE was presented by Osborne and Haywood (2005). The observed *in situ* aerosol size distributions (PCASP, FSSP, SID) were fitted with two (maritime aerosol) or three (industrial, biomass burning, dust) log-normal curves. In deriving the aerosol optical properties, the authors used the common assumption that each log-normal mode is composed of an internal mixture. These data were input parameters in the Mie scattering code for calculating the σ_{ext}, ω, and g for a sum of all modes ('TOT') and for modes 1 and 2 ('1+2'). In contrast to SHADE and SAFARI, the condensed water during TARFOX accounted for a large fraction (up to 40%). A resulting consequence had been significant changes in the refractive index and, hence, the differences in the optical characteristics between dehydrated and hydrated (ambient) aerosols. Except for TARFOX, all experiments compared the columnar single scattering albedo of submicron aerosol at wavelength 550 nm, obtained with the use of the microphysical ω^{1+2} and optical approaches $\omega^{in\ situ}$ ($\omega^{in\ situ}$ is the measured value from the nephelometer and PSAP). Comparisons of ω^{1+2} and $\omega^{in\ situ}$ showed reasonable agreement. At the same time, the g^{1+2} and g^{TOT}, σ_{ext}^{1+2} and σ_{ext}^{TOT} values may differ significantly from each other, depending on the type of aerosol which was dominant during the specific experiment. This finding is consistent with results obtained by other authors (see details below, SHADE description) and should be taken into account in the radiative calculations with the obtained experimental data.

The second *Aerosol Characterization Experiment* (ACE-2) was conducted in June and July of 1997 in the north-eastern Atlantic in a region bounded roughly by the Canary Islands of Spain to the south and Sagres, Portugal, to the north. The purpose of ACE-2 was to study the properties, processes, and effects of various aerosol types in this region, including background marine and anthropogenic pollution aerosol in the marine boundary layer (MBL) and background aerosol and mineral dust in the overlaying free troposphere (FT) (Raes et al., 2000).

One of six focused ACE-2 activities was to conduct a clear-sky column closure experiment CLEARCOLUMN, including a variety of *aerosol closure tests*. The set of instruments installed onboard the research aircrafts (Pelican, C-130) to measure the aerosol properties was similar to the equipment of TARFOX (Johnson et al., 2000; Collins et al., 2000; Russel and Heintzenberg, 2000).

The results of the ground-based measurements of Swietlicki et al. (2000) indicated that, during ACE-2, 80% or more of the particles were hygroscopic in nature, having wet diameters (RH $\sim$ 80%–90%) typically 1.5 times larger than their corresponding dry diameters (RH $\sim$ 10%). Aerosol characteristics depending on the relative humidity in the atmosphere were estimated using the data of two nephelometers operated at different RH (see Öström and Noone, 2000; Gassó et al., 2000, for details).

Using the additional nephelometers, Gassó et al. (2000) implemented a methodology based on measuring the scattering coefficient at known humidities below and above ambient conditions to derive airborne ambient scattering coefficients and aerosol hygroscopic properties with high temporal and spatial resolution. Particle scattering coefficients σ_{sc} (530 nm) from the two nephelometers are used to solve for the exponent γ in the equation

$$\sigma_{sc}(\mathrm{RH}) \sim (1 - \mathrm{RH}/100)^{-\gamma} . \tag{5.3}$$

From all ACE-2 Pelican flights, the authors obtained γ values (mean and standard deviation) for dust, polluted, and clean marine aerosols and used these γ values to estimate σ_{sc} at the ambient RH and also to estimate $\sigma_{sc}(RH = 80\%)/\sigma_{sc}(RH = 30\%)$. Öström and Noone (2000) used the data of Gassó et al. (2000) and the measurements at the ground-based site on Tenerife during ACE-2 (Swietlicki et al., 2000) to estimate the effect of the RH not only on the scattering coefficient, but also on the single scattering albedo ω, according to the relationship derived by Hänel (1976).

Collins et al. (2000) presented comparisons between the optical properties determined through the measured aerosol size distributions and those measured directly by an airborne 14-wavelength Sun photometer and nephelometers. Aerosol size distributions adjusted to ambient RH were used in conjunction with the size-resolved chemical compositions as inputs into Mie theory, from which σ_{ext} and σ_{sc} were calculated. The required refractive indices for elemental, organic carbon, and dust particles were chosen according to literature data or were based upon the previous analysis of samples taken on Tenerife, and the parameters n and κ for each of the salts were calculated on the basis of the partial molar refraction approach described by Moelwyn-Hughes (1961). Agreement between the extinction and scattering coefficients obtained using optical and microphysical approaches varied for different measurements and for different cases. Nonetheless, the differences between the model optical characteristics and those averaged over the four case studies did not exceed the uncertainties caused by the errors from measurements of the entire set of characteristics and by assumptions used in the calculations.

A similar aerosol test and analogous results were obtained by Schmid et al. (2000). Schmid et al. (2000) also estimated the aerosol size distributions by inverting spectral AOD using the constrained linear inversion method of King et al. (1978). The consent between aerosol size distributions based on *in situ* measurements and inverted AATS-14 extinction spectra was achieved in the MBL in contrast to the dust layer (due to the assumption on the wavelength-independent refractive index in the inversion of the AATS-14 spectra, according to the authors).

The *Southern African Regional Science Initiative* (SAFARI 2000) was a major surface, airborne, and space-borne field campaign carried out in southern Africa in 2000–01 (Swap et al., 2002). The time frame of the campaign coincided with the period of anomalously high rainfall and prolific vegetative growth; therefore, *biomass-burning* emissions became the overwhelming influence on the southern African atmosphere during the following dry season. During SAFARI, the coordinated, coincident, and spatially extensive observations of the troposphere were ensured by a few research aircraft and, in particular, by the University of Washington Convair-580 and the UK Met Office C-130 (Schmid et al., 2003).

Using the instrumentation flown by the C-130, the aerosol size distributions (PCASP-100X, SID), absorption, and scattering coefficients (PSAP and TSI 3563 nephelometer) were measured (Johnson et al., 2000; Haywood et al., 2003b, 2003c). Aerosol chemistry was determined from isokinetic sampling onto filter substrates (Formenti et al., 2003b). The radiation equipment consisted of upward- and downward-facing Eppley broadband radiometers, SAFIRI, and the Short-Wave Spectrometer (SWS) which uses two Carl Zeiss spectrometer modules operating in the spectral ranges of 0.30–0.95 and 0.95–1.70 μm. The Convair-580, in addition to the analogous TSI nephelometer and PSAP, also carried the AATS-14 Sun photometer (Magi et al., 2003).

The measurement results were used for comparing the single scattering albedo and the optical depth obtained by different methods.

Haywood et al. (2003c) showed that ω derived using measured aerosol size distributions are consistent with those from the nephelometer and PSAP. Data on the chemical composition of aerosol were used to calculate the aerosol refractive index on the basis of two rules: volume weighting (Eq. (5.2)) and Maxwell-Garnett mixing rule (Chýlek et al., 1988; Martins et al., 1998). This test took into account that the mean relative humidity was mainly $\sim$30% and only once RH exceeded 55%; therefore, the influence of RH on the aerosol properties was ignored. At the same time, the effects of RH on aerosols in southern Africa during the biomass-burning season were discussed in more detail by Magi and Hobbs (2003). The values of σ_{sc} were obtained from σ_{sc}^{dry} and appropriate humidographs on the basis of an empirical formula (Magi and Hobbs, 2003; Magi et al. 2003):

$$\sigma_{sc} = \sigma_{sc}^{dry} \left[1 + a_1 \left(RH/100\right)^{a_2}\right] , \tag{5.4}$$

where a_1 and a_2 are fitting parameters, which were representative of the ambient air in southern Africa during SAFARI 2000.

Haywood et al. (2003a) and Magi and Hobbs (2003) presented the results of aerosol closure tests, regarding the optical depths. Haywood et al. (2003a) estimated the value of AOD from measurements of (i) the scattering and absorption coefficients, (ii) the aerosol size distribution with the subsequent use of the Mie theory combined with suitable refractive indices to determine the aerosol extinction, (iii) the direct and diffuse components of the downwelling irradiance (Hignett et al., 1999), (iv) the radiance as a function of scattering angle (Francis et al., 1999), and (v) the magnitude and spectral dependence of the upwelling radiance from above the aerosol. It was shown that the estimates of optical depth from the various approaches were in good agreement. One exception occurred when τ was derived from the PCASP because this method was shown to be extremely sensitive to the pitch angle of the aircraft, and AOD differs for profile ascents and profile descents. At the same time, the aerosol size distribution measured by the PCASP and derived from the AERONET site agreed well over the 0.05–1.0 μm radius range, as did the derived refractive indices and single scattering albedo.

The *in situ* measurements of the absorption and scattering coefficients were used to calculate the layer AOD between heights z_1 and z_2 ($\tau^{in\ situ}$, Eq. (5.1)), and this value was compared with the τ^{sp} data derived from the AATS-14 Sun photometer measurements (Magi et al., 2003). The obtained results ensured the most successful closure; at the wavelength of 550 nm, the slope of regression line of $\tau^{in\ situ}$ onto τ^{sp} was equal to 0.96 ($r^2 = 0.98$). The reason for the excellent agreement between $\tau^{in\ situ}$ and τ^{sp} may be the fact that the aerosols in SAFARI 2000 were dominated by small, relatively nonhygroscopic, biomass-burning aerosols, which are easier to sample than hygroscopic and/or larger particles, such as mineral dust or sea salt (see the results obtained by Hegg et al. (1997) and Hartley et al. (2000) during TARFOX).

Based on SAFARI-2000 data, a method suggested by Magi et al. (2007) was tested in addition to the use of the traditional optical and microphysical approaches. (This technique was close in concept to the above-mentioned method of microphysical extrapolation.) The straightforward retrieval algorithm searches look-up tables constructed using Mie theory to find a size distribution and refractive indices that most closely reproduce the *in situ* and remote sensing measurements of aerosol optical properties. The first step of the algorithm is to retrieve the *optically equivalent* (*oe*) unimodal log-normal submicron size distribution (described by geometrical mean diameter $D_{g,oe}(z)$, standard deviation $\sigma_{g,oe}(z)$, and aerosol number concentration $N_{a,oe}(z)$) and the optically equivalent refractive indices $(n_{oe,\lambda}(z) - i \times \kappa_{oe\lambda}(z))$ at $\lambda = 450$, 550, and 700 nm at all vertical levels z. These wavelengths together most closely reproduce the measured values of the extinction coefficient $\sigma_{ext,\lambda}(z)$, single scattering albedo $\omega_\lambda(z)$, and backscattering coefficient $b_\lambda(z)$ at $\lambda = 450$, 550, and 700 nm. When measurements of the extinction coefficient and single scattering albedo are available at other wavelengths, this information, along with the optically equivalent size distribution, can be used at the next algorithm step to retrieve $n_{oe,\lambda}$ and $\kappa_{oe,\lambda}$ at the other wavelengths. Evidently, the optically equivalent size distribution and refractive indices are not necessarily representative of the real aerosol size distribution and refractive indices, especially if the aerosol is not composed of spherical particles.

Magi et al. (2007) showed that, during SAFARI-2000, the *optically equivalent* size distributions are similar to the submicron modes of the size distributions in the works of Haywood et al. (2003a, 2003c) and retrieved from AERONET (Eck et al., 2003). Because smaller-diameter particles were included in their consideration, the concentration of aerosol particles, retrieved in accordance with Magi et al. (2007), was higher than the data presented in Haywood et al. (2003c). At the same time, the optically equivalent real and imaginary refractive indices were ~14% and ~50% greater than those derived from AERONET retrievals for three study cases and, in some cases, there was a very different wavelength dependency.

The Magi et al. (2007) approach has limited application because the aerosol size distribution is assumed to be unimodal with a dominating contribution of submicron particles. In addition, to properly constrain the retrieval, the information on $\sigma_{ext,\lambda}(z)$, $\omega_\lambda(z)$, and $b_\lambda(z)$ is needed at least at three wavelengths. This information is widely available for $\sigma_{sc,\lambda}(z)$ and $b_\lambda(z)$ from nephelometry (Anderson and Ogren, 1998) and to a lesser degree for the absorption coefficient $\sigma_{abs,\lambda}(z)$, although multi-wavelength measurements of $\sigma_{abs,\lambda}(z)$ are becoming more common (Ganguly et al., 2005; Sheridan et al., 2005; Virkkula et al., 2005; Schmid et al., 2006).

The largest dust source in the world is the Sahara (Shao et al., 2011). Dust plays an important role in climate through a variety of mechanisms, thus necessitating an accurate understanding of the optical properties of dust for use in the weather and climate models.

Since 2000, several aircraft campaigns have attempted to measure the properties of mineral dust over the Sahara: the African Monsoon Multidisciplinary Analysis (AMMA, 2006; Redelsperger et al., 2006); the Dust And Biomass burning Experiment (DABEX, 2006; Haywood et al., 2008); the Saharan Mineral Dust Experiment 1 (SAMUM-1, 2006; Heintzenberg, 2009); and others. Moreover, the majority of experiments were conducted over the eastern and western Atlantic Ocean: the SaHAran Dust Experiment (SHADE, 2000; Haywood et al., 2003b); the Puerto Rico Dust Experiment (PRIDE, 2000; Reid and Maring, 2003); SAMUM-2 (2008; Ansmann et al., 2011); and others.

One problem of the result analysis obtained in the course of the above-mentioned campaigns is that coarse particles are difficult to measure on aircraft. In addition, it was not always clear whether the differences in descriptions of the coarse mode were due to instrumental and sampling differences or due to the different geographical location and dust event types (see, e.g., bibliography in Ryder et al., 2013). Direct measurements of scattering and absorption also suffered from the difficulty of measuring the coarse mode on aircraft because such measurements are limited to representing the aerosol accumulation mode (Haywood et al., 2003b; McConnell et al., 2008; Petzold et al., 2011).

SHADE was one of the first experiments when the physical and optical properties of Saharan dust aerosols were studied. It took place in September 2000 off the coast of West Africa near the Cape Verde Islands. The aerosol and radiation instrumentation onboard the C-130, as well as the main approaches to retrieving and analyzing the radiative parameters of aerosol, were identical to those used in the SAFARI campaign (Tanré et al., 2003; Haywood et al., 2003b). Aerosol size distributions were available only from PCASP because, like throughout SAFARI 2000, Fast FSSP did not operate correctly. The mean aerosol size distributions were fitted with five log-normal size distributions (Haywood et al., 2003b); however, mode 5 was poorly constrained by measurements with PCASP which detects particles with radii up to approximately 1.5 μm. The size distributions retrieved from surface-based Sun photometers sited in Sal and Dakar were used to describe the largest mode (>1.5 μm) (Nakajima et al., 1996; Dubovik and King, 2000). Haywood et al. (2003b) showed that including mode 5 leads to a significant reduction in $\sigma_{ext,550}$, causes a slight decrease in g_{550}, and has little effect on ω_{550}. In this regard, the method of integrating the scattering and absorption coefficients from the nephelometer and PSAP underestimated τ_λ, which might be the reason for the errors in the estimates of radiation effects of dust aerosol (Myhre et al., 2003). Radiometric measurements of τ_{550} showed good consistency with each other and agreement with the AERONET measurements of AOD made at Sal and Dakar (Haywood et al., 2003b).

New *in situ* aircraft measurements of Saharan dust originating from Mali, Mauritania, and Algeria were performed in the course of the Fennec 2011 aircraft campaign (Washington et al., 2012). A brief description of flight patterns, performed on the UK's BAe-146-301 Large Atmospheric Research Aircraft operated by the Facility for Airborne Atmospheric Measurements (FAAM), and instrumentation during Fennec was presented in Ryder et al. (2013).

The new size distribution instrumentation operated on the FAAM aircraft has allowed measurements of dust coarse mode particles to be extended to larger sizes (300 μm in diameter) than previous experiments have achieved. Similarly to SAMUM-1, Fennec was characterized by the presence of a larger number of coarse particles; these campaigns had in common that they measured dust closer to source regions, with shorter transport times and distances than other campaigns. The single scattering albedo from direct measurements (nephelometer, PSAP) ranged from 0.91 to 0.99; these results were similar to those previously measured for dust using inlets which restrict the measurement of coarse particles (Haywood et al., 2003b; McConnell et al., 2008; Petzold et al., 2011).

Ryder et al. (2013) also presented the model calculations of the ω value, using the measured particle size distributions and an iterative process of retrieving the imaginary part of the refractive index, as described by McConnell et al. (2010). Ambient size distributions are input into a Mie scattering code, along with a refractive index at $\lambda = 550$ nm, where the real part is fixed ($n_{550} = 1.53$) and imaginary part is varied ($0.001 \leq \kappa_{550} \leq 0.008$). At each iteration step, the calculated value of ω_{550} is compared to that measured by the nephelometer and PSAP; the value of κ_{550} which gave the closest agreement was chosen as the best estimate.

Because the size distribution used as input to the Mie code should represent the size distribution reaching the nephelometer and PSAP, the authors used measurements adjusted for inlet and pipe losses and enhancements, namely those with reduced coarse mode, in their calculations. The calculated ω_{550} values covered the interval 0.95–0.99 in satisfactory agreement with the *in situ* measurements of single scattering albedo. Simulation with the use of the *full*-sized distributions revealed high absorption ranging from 0.70 to 0.97, depending on the refractive index. The ω value was found to be strongly dependent on the effective particle diameter; when D_{eff} is greater than 2 μm, direct measurements, behind Rosemount inlets, overestimate ω_{550} by up to ~0.1. (For this reason, previous measurements of ω from aircraft measurements may also have been overestimated.) This result should be taken into consideration when the radiation effects of mineral dust are estimated; neglecting the coarse mode may lead to underestimates in shortwave atmospheric heating rates by two to three times.

5.3 Long-term observations of vertical profiles of tropospheric aerosol characteristics: Usage in models

The above-described approaches made it possible to obtain information on the vertical profiles of aerosol characteristics in different regions of the globe. A specific feature of these data is that they characterize the state of aerosol characteristics immediately in the period of measurements and, in this sense, they predominately reflect information on the optical/microphysical characteristics of separate types of aerosol (marine, arid, smoke, anthropogenic, etc.). As a consequence, the specific features of different *regions* of the globe cannot be described completely because the real aerosol represents a far more complex and variable mixture.

Recently, there appeared the possibility of aircraft measurements of aerosol properties within one or another region of Earth over a longer term than separate short periods of observation. The accumulated results of *multi-year* aircraft sensing are a basis for creating the statistically representative models, characterizing the typical values of aerosol optical characteristics for different seasons and regions.

Examples of these studies are the results obtained over Oklahoma, US (Andrews et al., 2011), and western Siberia, Russia (Panchenko et al., 2012).

In this section, we describe the specific features of the vertical profiles over the territory of Oklahoma and present in more detail the approach, which we developed to create the empirical model of radiative characteristics of the tropospheric aerosol over western Siberia.

Andrews et al. (2011) summarized the studies of the vertical profiles of the aerosol optical characteristics over *rural Oklahoma*. The main objectives of a multi-year study included the following issues:

- to identify the climatically significant features of formation of vertical profiles of the aerosol scattering and absorption characteristics for different seasons of the year, based on regular multi-year aircraft measurements;
- to compare the obtained measurements with data of ground-based complex equipment, considering that the world's unique contemporary instrumentation complex for atmospheric radiation measurements is available at the Department of Energy (DOE)'s Southern Great Plains (SGP) site (*www.arm.gov*); and
- to compare the results and conclusions, obtained with the help of ground-based observations and aircraft sensing.

A unique experimental material was obtained on the basis of measurements between March 2000 and December 2007 over the DOE's SGP site near Lamont, Oklahoma (~600 flights). The research airplane (Cessna 172XP) was originally equipped with an integrating nephelometer TSI-3563 and a filter-based light absorption instrument PSAP (Andrews et al., 2004). After the summer of 2005, the instrumentation was updated: (i) the aerosol inlet was changed to the shrouded inlet design, and the cut size has been estimated to be ~ 7 μm; and (ii) PSAP was upgraded to a three-wavelength instrument (467, 530, and 660 nm). A heater upstream of the instrumentation ensured that measurements were made under low relative humidity conditions (RH $<$ 40%). Ambient temperature and RH were also measured utilizing a Vaisala Humicap sensor mounted on the outside of the aircraft.

From the measured aerosol absorption, scattering, and backscattering coefficients, several climatically important aerosol optical parameters were calculated. These characteristics included the single scattering albedo, scattering (α_{sc}) and absorption (α_{abs}) Ångström exponents, and asymmetry parameter, estimated with the empirical relationship $g = -7.1439b^3 + 7.4644b^2 - 3.9636b + 0.9893$, where b is the ratio of hemispheric backscattering to total backscattering (Andrews et al., 2006).

In analysis of the vertical distribution, the authors focused attention on the results of their measurements obtained for dry aerosol (RH $<$ 40%). This approach is reasonable because the main formation processes of microphysical composition and vertical distribution of particles, on one hand, and the 'modulating' effect of humidity on the diurnal or seasonal behavior, on the other hand, have substantially different spatiotemporal scales. In addition, the 2005–07 period appeared to be drier for all seasons, which was consistent with the ongoing drought conditions in Oklahoma; the median RH values were less than 60%.

Andrews et al. (2011) compared the AOD and other column-average aerosol optical properties (α_{sc}, ω, g) derived from the 2005–07 airplane profiles with the same aerosol optical properties determined from AERONET climatologies over the 1994–2011 period. The adjustment of the *in situ* data to ambient RH was performed using seasonal median-fit parameters derived from hygroscopic growth measurements at the SGP surface site (Sheridan et al., 2001).

It was shown that the median AOD derived from *in situ* measurements tends to be lower than the AERONET AOD for all seasons except spring, when the median values are quite similar. In the authors' opinion, the hygroscopic growth correction alone is not responsible for the difference observed between the AERONET and *in situ* measurements. In addition to the RH effect, possible reasons for the AOD discrepancies could have been the loss of large particles due to sample line content, the neglect of change in hygroscopic growth with height, the difference in the time intervals within which the averaging was performed, etc. These same factors could also partly influence the differences between α_{sc} and g derived from the *in situ* and AERONET measurements. Regarding the single scattering albedo, it is important to emphasize that the ω retrievals from AERONET are severely limited by the constraint that τ_{440} must be greater than 0.4. At the same time, the revealed characteristic features in the seasonal behavior of the optical state of the atmosphere at a given geographic site obtained by different methods agree with each other well.

The scrupulous analysis of the experimental data allowed correctly estimating the unavoidable uncertainties of retrieval on the basis of relatively limited experimental information and discussing the influence of atmospheric processes, which determined the seasonal differences for the entire complex of the optical characteristics. The obtained results represent an excellent basis for constructing an empirical model of the vertical profiles for the entire complex of the climatically significant optical characteristics required for the radiation calculations.

Based on aircraft sensing data, an empirical model of the optical and radiative characteristics of the tropospheric aerosol *over western Siberia* was developed in two stages. The first stage (1985–88) had the purpose to develop a model for supplying information to systems and devices, operating in the optical range through the atmosphere. Solving this problem primarily required information on the scattering properties of aerosol for a *wide range of atmospheric situations*. In this regard, the aircraft was flown in regimes of maximum possible regularity in all seasons of the year and at different times of the day, except relatively rare cases of dense fogs, strong thunderstorms, and squall wind.

The second model development stage was associated with the need to obtain the *complete* data set for estimating the radiation effects, which primarily involved studying the *absorbing* properties of aerosol. From 1999 to the present, the flights were performed once monthly in midday under the stable conditions of clear-sky and low-cloud atmosphere.

Two main aspects underlie in the model. *The first aspect* consists of the necessity to parameterize the relationship between aerosol characteristics and relative air humidity, which strongly affects the transformation of the microphysical parameters of aerosol particles. The approach we have developed is based on the principle of separate investigation of variations in the aerosol particle dry matter and aerosol condensation activity under the external factors (Panchenko et al., 2004). *The second aspect* of our approach is that an empirical model is constructed by combining the microphysical and optical approaches (Panchenko et al., 1998; Panchenko and Terpugova, 2002).

The measurement instrumentation and the calibration techniques used here were described in detail in earlier publications (Panchenko et al., 1996, 2000). Here, we only give a brief instrument characterization.

The polarized nephelometer was used for measuring the aerosol angular-scattering coefficient at the angle of 45° and at the wavelength 510 nm. The nephelometer was calibrated in absolute units in each flight using the values of the molecular scattering coefficient of pure air at the heights of 0.5, 1.0, 2.0, 3.0, 4.0, and 5.0 km (at different pressures). The sensitivity of the instrument was 1 Mm^{-1}.

The particle size distribution function was recorded by a photoelectric particle counter in 12 channels in the radius range of 0.2–5 μm.

The mass concentration of absorbing substance was measured by an aethalometer, which implemented the method for measuring the diffuse absorption by a layer of particles during their deposition from airflow onto a filter. The instrument is capable of measuring the mass concentration of BC starting from the values of 0.01 $\mu\mathrm{g/m}^3$ when 10–20 L of air pass through it. The aethalometer was calibrated in absolute units under laboratory conditions using a special pyrolysis generator of BC particles and by comparing the data of synchronous optical and gravimetric measurements of the BC content.

The meteorological parameters (temperature and RH) of air outside the aircraft and immediately in the scattering volume were recorded simultaneously with the nephelometric measurements. The true values of the *in situ* scattering coefficient were reconstructed on this basis.

When developing the empirical model, we used the data obtained in two measurement campaigns. The model of the vertical profiles of the *scattering coefficient* and the *particle size distribution function* at heights up to 5 km (Panchenko et al., 1998; Panchenko and Terpugova, 2002) was constructed on the basis of the airborne measurement data that were collected in the 1986–88 period. The data array consists of more than 600 vertical profiles. The input parameters of the model are the geophysical factors (season, type of air mass, and time of day) and the measured parameters (near-ground value of the angular-scattering coefficient, AOD, vertical profiles of temperature, and RH of the air). In this period of measurements, the airborne instrumentation had no devices for absorber ('soot,' or BC) measurements.

Comparison of the synoptic and meteorological conditions during 2.5 years (1986–88) with aerological sensing data over the stations located in the west Siberian region (Aleksandrovskoe, Barabinsk, Novosibirsk, and Omsk) showed that the bulk of data obtained on the optical parameters of aerosol and the factors of its variability reflect the most general regional features of western Siberia. Measurements in the second campaign (1999–2007) were not as numerous; flights were performed once a month only in the afternoon. Nonetheless, the results of the analysis presented in Panchenko et al. (2012) indicate the possibility of using the data of these two different periods; according to the data shown in Fig. 5.1, the shapes of the vertical profiles of the aerosol angular-scattering coefficient $\sigma_{sc}^{dry,1}(45°)$ and $\sigma_{sc}^{dry,2}(45°)$ obtained in the measurement periods 1986–88 and 1999–2007 agree well with each other. The possibility of combining the results of airborne data is also confirmed by the fact that a time series with more than 10 years of avail-

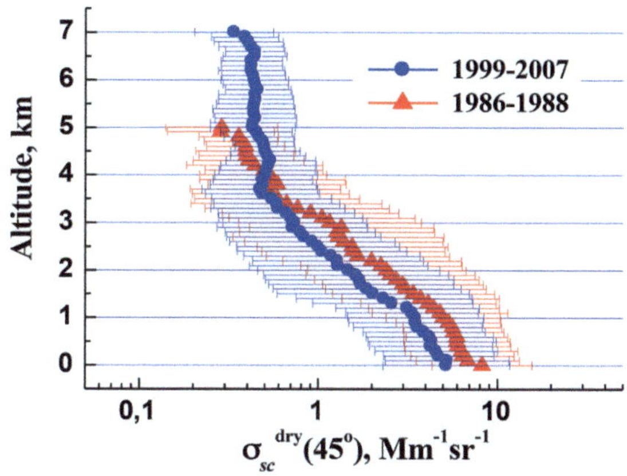

Fig. 5.1. The summer average vertical profiles and standard deviation of the aerosol angular-scattering coefficient according to data of airborne measurements in 1986–88 and 1999–2007 over Western Siberia.

able continuous measurements of the mass concentrations of aerosol and BC in the near-ground layer of the atmosphere shows no significant trend in these parameters (Kozlov et al., 2007).

The proposed procedure for calculating the optical and microphysical aerosol parameters is described below. The block-diagram of the generalized model is shown in Fig. 5.2. The first block includes the earlier developed model of the vertical profile of the scattering coefficient and the size distribution function of dry aerosol (Panchenko et al., 1998; Panchenko and Terpugova, 2002). The second block of the model is related to the incorporation of the aerosol absorption properties into the calculations, namely the seasonal average values of the mass concentration of BC $M_{BC}(z)$ at different heights (Kozlov et al., 2009). Then, the optical constants (real and imaginary parts of the refractive index) of the dry aerosol fractions are calculated by the simple mixture rule. At the next stage of calculations, the seasonal average vertical profile of relative humidity $RH(z)$ and the parameter of condensation activity $\gamma(z)$, which determines the dependence of aerosol parameters on RH (Hänel, 1976), are included in the calculations. Finally, applying the Mie theory, the radiative parameters (the spectral extinction coefficient, single scattering albedo, and scattering phase function) are calculated at all prescribed altitudes z.

The measured *aerosol size distribution functions* at each altitude (0, 0.5, 1.0, 1.5, 2.0, 3.0, 4.0, and 5.0 km) were fitted by a sum of two log-normal functions (Panchenko et al., 1998). Then, the angular-scattering coefficient of the dry fraction $\sigma_{sc}^{dry}(45°)$ was calculated with Mie formulae using the parameters of submicrometer (s) and coarse (c) fractions (median radius Ri, standard deviation of the radius logarithm ν_i and volume concentration V_i, $i = s, c$) at each altitude. Aerosol particles were assumed to be non-absorbing, with the real part of the refractive index being $n_{aer} = 1.5$. Then, the volume concentrations V_i, $i = s, c$, were recalculated so that the obtained value $\sigma_{sc}^{dry}(45°)$ coincides with the average seasonal value $\sigma_{sc}^{dry,1}(45°)$, and the ratio between V_s and V_c remains constant (Panchenko et al., 1996).

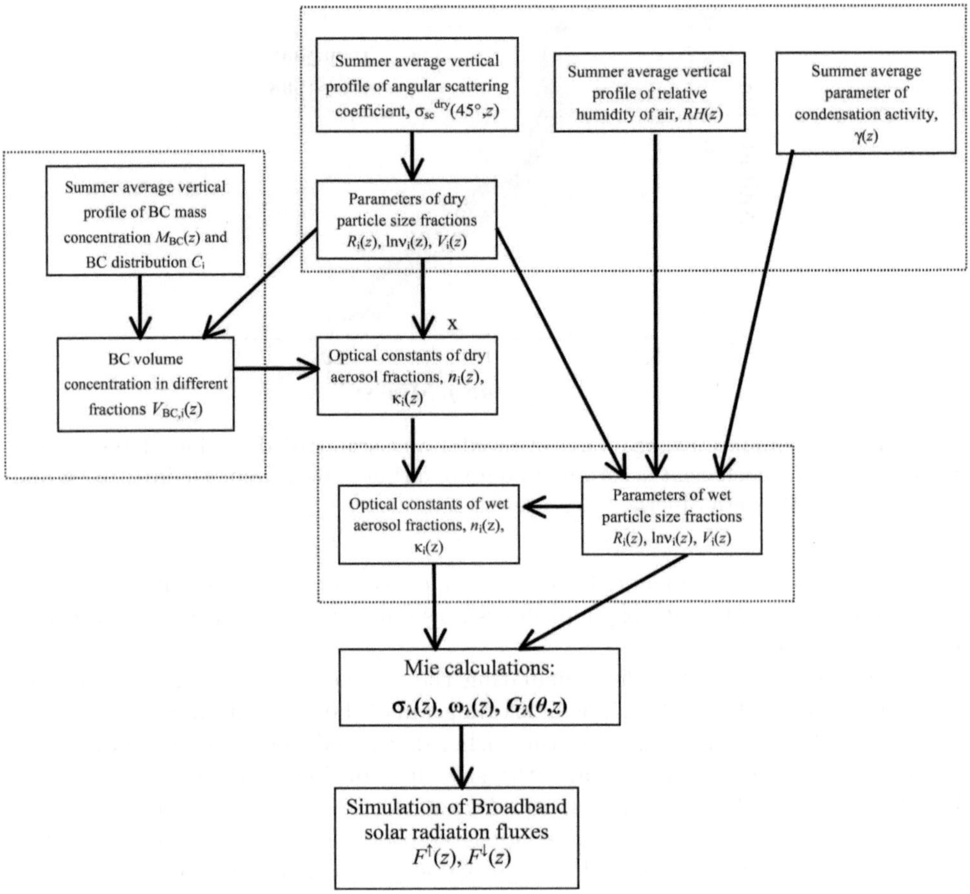

Fig. 5.2. Block-diagram of the model.

Table 5.1 presents the average scattering coefficients of the aerosol dry matter and RH measured at these altitudes, as well as the distribution parameters R_i, ν_i, and V_i adjusted for both particle fractions, $i = s, c$.

In the following simulations, we assume BC to be the only absorbing substance. Data from particular experiments dealing with BC particle size distributions are available in the literature (Hitzenberger and Tohno, 2001; Kozlov et al., 2002; Höler et al., 2002; Delene and Ogren, 2002). Their results indicate that particles with radii of 0.1–0.12 μm contain the largest amount of BC and that the relative BC content decreases with particle size. An analysis of BC impactor measurements made in 1998–99 in Vienna and Uji (Japan) (Höler et al., 2002) shows that large particles contain approximately 10% of the total BC mass, with the remaining 90% distributed in approximately equal parts as microdispersed (less than 0.1 μm) and submicron particles (0.1–1 μm). Delene and Ogren (2002) reported that the average contribution of coarse particles to the total absorption coefficient varies from 7% to 20%, depending on the anthropogenic loading in the region of observation.

Table 5.1. Parameters of the log-normal distributions (median radius R, standard deviation of the radius logarithm ν, and volume concentration of fraction V) for the submicrometer and coarsely dispersed fractions, angular-scattering coefficients for $\sigma_{sc}^{dry}(45°)$, and the mass concentrations of the absorbing substance M_{BC}^{*} at different altitudes z.

z (km)	M_{BC}^{*} (μg/m^3)	$\sigma_{sc}^{dry}(45°)$ (Mm^{-1}sr^{-1})	RH (%)	Submicrometer fraction ($i = s$)			Coarsely dispersed fraction ($i = c$)		
				Lnν_s	R_s (μm)	V_s	Lnν_c	R_c (μm)	V_c
0	1.16	9.77	72	0.8	0.114	1.548E-11	0.65	2.078	1.145E-11
0.5	0.634	7.07	70	0.8	0.096	1.414E-11	0.65	1.804	1.573E-11
1	0.556	6.18	75	0.8	0.098	1.236E-11	0.65	1.797	1.321E-11
1.5	0.458	5.09	74	0.8	0.098	1.018E-11	0.65	1.791	1.044E-11
2	0.417	3.48	71	0.8	0.096	6.956E-12	0.65	1.784	6.841E-12
3	0.248	2.08	69	0.8	0.095	4.156E-12	0.65	1.771	3.738E-12
4	0.091	0.61	62	0.8	0.092	1.213E-12	0.65	1.758	9.914E-13
5	0.045	0.30	53	0.8	0.089	6.022E-13	0.65	1.745	4.435E-13

The values M_{BC}^{*} were calculated based on Eq. (5.5).

Here, it should be noted that measurements of the BC mass concentration were carried out in the second campaign (1999–2007), whereas the model of the vertical profiles of the scattering coefficient was created from the data from the first campaign (1986–88). Thus, to match the data used for estimating the volume concentration of BC in the model calculations, the coefficient was introduced as equal to the ratio of the average aerosol angular-scattering coefficients in the first and second campaigns:

$$M_{BC}^{*} = M_{BC}\sigma_{sc}^{dry,1}(45°)/\sigma_{sc}^{dry,2}(45°)\,, \quad V_{BC} = M_{BC}^{*}/\rho_{BC}\,, \tag{5.5}$$

where M_{BC}^{*} is the mass concentration of BC used for calculations of the complex refractive index (Table 5.1) and ρ_{BC} is the density of BC matter, $\rho_{BC} = 1.85$ g/cm^3 (Gelencser, 2004; Bond and Bergstrom, 2006).

The volume concentration of BC in every ith fraction, $i = s, c$, was calculated using the formula

$$V_{BC,i} = C_i V_{BC}\,, \tag{5.6}$$

where C_i is the portion of BC in the ith aerosol fraction ($C_s = 0.9$, $C_c = 0.1$, in accordance with the accepted hypothesis).

The refractive index of the mixture in each fraction ($n_{dry,i}$; $\kappa_{dry,i}$), $i = s, c$, was determined at each altitude z, according to the internal mixture rule (Eq. (5.2)):

$$\begin{aligned} n_{dry,i} &= (n_{aer} \times V_{aer,i} + n_{BC} \times V_{BC,i})\,/\,(V_{aer,i} + V_{BC,i})\,, \\ \kappa_{dry,i} &= \kappa_{BC} \times V_{BC,i}/(V_{aer,i} + V_{BC,i})\,, \end{aligned} \tag{5.2a}$$

where $V_{aer,i}$ is the volume concentration of the non-absorbing aerosol in the ith fraction. The complex refractive index of BC was set to ($n_{BC} = 1.8$, $\kappa_{BC} = 0.74$) (Gelencser, 2004; Bond and Bergstrom, 2006).

In the next stage, we recalculated the refractive index of dry aerosol in accordance with the mean seasonal relative humidities $RH(z)$ obtained during airborne sensing.

The dependencies of the aerosol parameters (radius, volume concentration, and scattering coefficient) on RH were approximated by the Hänel formula (Hänel, 1976). In particular, for the angular-scattering coefficient, the formula can be written as (see Eq. (5.3)):

$$\sigma_{sc}^{wet}(45°) = \sigma_{sc}^{dry}(45°) \times (1 - RH\,(z)\,/100)^{-\gamma(z)}\ , \tag{5.7}$$

where σ_{sc}^{wet} is the scattering coefficient under ambient conditions, σ_{sc}^{dry} is the dry scattering coefficient at relative humidity <30%, and $RH(z)$ and $\gamma(z)$ are the mean seasonal relative humidity and parameter of the condensation activity at a height of z, respectively.

The γ values at different altitudes, except for those at the ground level $z = 0$, are among the input parameters of the empirical model; they were obtained in the first measurement campaign (Panchenko et al., 1996). The mean seasonal data from multi-year measurements in Tomsk were taken as the near-ground γ value (Panchenko et al., 2005). For summer, the condensation activity parameter was $\gamma = 0.3$ at $z = 0$, with $\gamma = 0.5$ at all other altitudes. The results of processing more than 200 realizations of hygrograms at heights of up to 5 km showed that the differences between the γ values at $z = 0$ and other heights are more than 95% statistically significant (Panchenko et al., 1996).

For the submicron aerosol fraction, $\sigma_{sc}^{dry}(45°)$ is linearly related to the particle volume concentration; therefore, for the wet aerosol, the following formula is valid:

$$V_{wet,s} = V_{dry,s} \times (1 - RH(z)/100)^{-\gamma(z)}\ , \tag{5.8}$$

where $V_{dry,s}$ and $V_{wet,s}$ are the volume concentrations of the respective 'dry' and 'wet' (ambient) aerosols.

To decide whether to include the effect of humidity on *coarse* particles, we compared data from earlier experiments with the aerosol optical parameters calculated on the condition that only the submicron fraction is hygroscopic. Figure 5.3 presents the calculated values of (a) the degree of linear polarization of the scattered

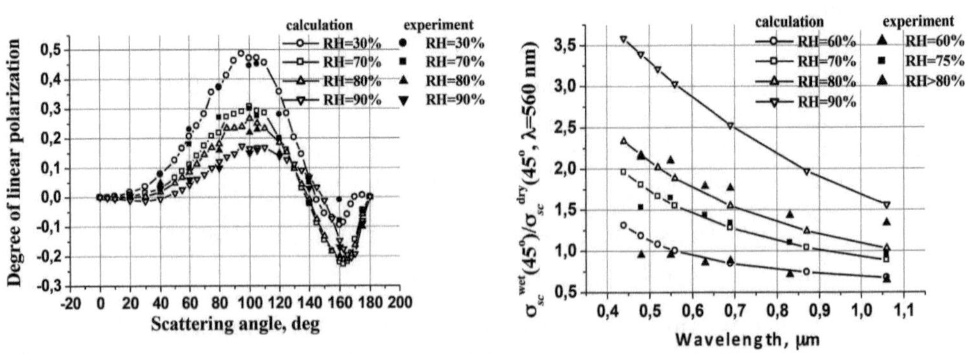

Fig. 5.3. Comparison of calculated and measured ground-level aerosol parameters: (a) the degree of linear polarization of the scattered radiation and (b) spectral dependence of the ratio of the scattering coefficient at several values of relative humidity (RH) to the scattering coefficient of the aerosol dry matter.

radiation and (b) the spectral dependency of the ratio of the $\sigma_{sc}^{wet}(45°)/\sigma_{sc}^{dry}(45°)$ values in comparison with experimental data (Kabanov et al., 1988). The comparison showed a good agreement between the model and experimental characteristics; therefore, in further calculations, we considered only variations in the wetting of the submicron fraction.

The refractive index ($n_{wet,s}$; $\kappa_{wet,s}$) of the submicron aerosol fraction was estimated under the assumption that, during wetting, the volume of particles increases due to water uptake only:

$$n_{wet,s} = (n_{dry,s}V_s + n_w V_w)/V_{wet,s}, \quad \kappa_{wet,s} = \kappa_{dry,s}V_s/V_{wet,s}\,. \tag{5.2b}$$

Here, $n_w = 1.33$ is the refractive index of the liquid water, and $V_w = V_{wet,s} - V_{dry,s}$ is the condensed water volume, namely the increment in the volume of the aerosol particles.

The values of the complex refractive index of the submicron and coarse fractions used for further calculations are presented in Table 5.2.

Table 5.2. Real and imaginary parts of the aerosol complex refractive indices of the submicrometer and coarse fractions.

z (km)	0	0.5	1	1.5	2	3	4	5
n_s	1.4486	1.4266	1.4182	1.4200	1.4261	1.4294	1.4413	1.4538
n_c	1.50	1.50	1.50	1.50	1.50	1.50	1.50	1.50
κ_s	0.0171	0.0087	0.0079	0.0081	0.0113	0.0117	0.0161	0.0179
κ_c	0.0040	0.0016	0.0017	0.0018	0.0024	0.0027	0.0037	0.0041

Mie calculations of $\sigma_\lambda(z)$, $\omega_\lambda(z)$, and scattering phase function $G_\lambda(\theta, z)$ (θ is the scattering angle) were performed at the prescribed atmospheric levels in the altitude range of 0–5 km. The optical depth τ_λ was calculated as the integral of the aerosol extinction coefficient in the altitude range of 0–5 km.

The normalized spectral behavior of τ_λ/τ_{550} in the wavelength range of 370–1020 nm and the Ångström exponent α(440–870 nm) are in good agreement with multi-year Sun photometer measurements collected on the territory of Siberia (Sakerin and Kabanov, 2007; Sakerin et al., 2009) (Fig. 5.4a). The α(440–870 nm) value is also close to the Ångström exponent for the OPAC model of continental aerosol (continental average, RH = 70%; Hess et al., 1998) and is equal to approximately 1.4.

The aerosol single scattering albedo $\omega_\lambda(z)$ varies quite substantially with altitude; at all wavelengths, the maximum values of $\omega_\lambda(z)$ (~0.92–0.93) are observed in the mixing layer at 0.5–1.5 km and, outside this layer, they decrease toward Earth's surface and with increasing altitude (Fig. 5.4b). At the same time, the altitude variations in the asymmetry parameter $g_\lambda(z)$ are much smaller, not exceeding ~0.02 throughout the entire wavelength range of 440–870 nm (Fig. 5.4c).

We use the obtained values for comparison with the *columnar* single scattering albedo ω_λ^{atm} and the asymmetry factor g_λ^{atm}, retrieved according to ground-based photometric observations on the AERONET site located in the suburb of Tomsk

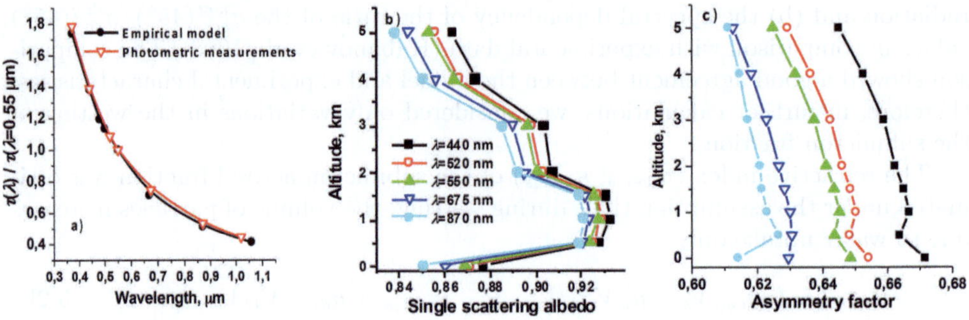

Fig. 5.4. (a) The spectral behavior of the normalized aerosol optical depth (AOD) according to data of multi-year airborne and ground-based photometric measurements and vertical profiles of (b) single scattering albedo and (c) asymmetry factor in the altitude range of 0–5 km in the summer period.

during the summers (June–August) of 2004–09 (*aeronet.gsfc.nasa.gov*). Table 5.3 presents the *average* ω_λ^{atm} and g_λ^{atm} values, which were obtained using the algorithm (Dubovik and King, 2000) under the conditions of $\tau_{440} \geq 0.4$ and the algorithm (Bedareva and Zhuravleva, 2011, 2012) at $\tau_{440} \leq 0.4$. In addition, Table 5.3 gives the optical characteristics of the continental average aerosol (RH = 70%) from the OPAC model (Hess et al., 1998).

Table 5.3. The columnar single scattering albedo and the asymmetry factor obtained with the empirical model in comparison with the methods of Dubovik and King (2000) and Bedareva and Zhuravleva (2011, 2012), and the continental aerosol model (RH = 70%) presented in the model OPAC (Hess et al., 1998).

	λ (nm)	Panchenko et al. (2012)	Dubovik and King (2000)	Bedareva and Zhuravleva (2011, 2012)	Hess et al. (1998)
$\omega^{atm}(\lambda)$	440	0.92	0.94	0.92	0.92
	675	0.90	0.93	0.90	0.9
	870	0.90	0.92	–	0.87
$g^{atm}(\lambda)$	440	0.67	0.69	0.68	0.72
	675	0.63	0.61	0.61	0.69
	870	0.62	0.57	–	0.66

A comparison of the data shows satisfactory agreement between the ω_λ^{atm} and g_λ^{atm} values, obtained according to the aircraft sensing and the method presented by Bedareva and Zhuravleva (2011, 2012). The single scattering albedo obtained from ground-based photometric measurements using the Dubovik and King algorithm (2000) under high-turbidity conditions is slightly greater than that obtained with the model presented here. In our opinion, this is not only due to the difference in retrieval methods of aerosol radiative parameters but to the fact that data under the conditions of $\tau_{440} \geq 0.4$ corresponded to situations influenced by the Siberian forest fires. As shown in Kozlov et al. (2008), the intrusion of forest fire smoke in the region of observation leads to an increased single scattering albedo in comparison with background conditions. As for the model of continental aerosol OPAC, its single scattering albedo data are close to the values we obtained for the territory of western Siberia, while the qualitative differences in the asymmetry factor values are much more significant.

An example of testing the empirical aerosol model for simulation of broadband fluxes is presented in this subsection. The broadband solar radiative fluxes in the molecular-aerosol, plane-parallel atmosphere were calculated using the Monte Carlo algorithm developed earlier (Zhuravleva, 2008; Zhuravleva et al., 2009). The shortwave spectral interval 0.2–5.0 μm was divided into 31 subdivisions (Slingo, 1989).

To obtain a complete set of the input parameters for the simulation, the empirical model was complemented as follows:

- the optical characteristics in the altitude interval 5–35 km were taken from the OPAC model of continental average RH = 70%;
- the extinction coefficients were extrapolated to a wider range of 370–1020 nm using the Ångström exponent α(440–870 nm) and, outside the interval 370–1,020 nm, the extinction coefficients were assumed to be constant and equal to the respective boundary values for $\lambda = 370$ nm and $\lambda = 1020$ nm;
- the scattering phase functions and single scattering albedo of aerosol outside the interval 440–870 nm were assumed to be constant and equal to the respective boundary values for $\lambda = 440$ nm and $\lambda = 870$ nm.

We used this approach because the Ångström exponents α(370–1020 nm) according to the data of airborne and AERONET measurements are comparable. In addition, the neglect of the spectral dependences of the scattering phase function and single scattering albedo does not significantly affect the solar radiative fluxes (see Panchenko et al. (2012) for more detail).

The vertical profiles of upward and downward radiative fluxes in the altitude range $0 \leq z \leq 5$ km at the solar zenith angle SZA $= 75.62°$ (07:00 LT, local time) are presented in Fig. 5.5a. The instantaneous values of the downward radiative fluxes at the underlying surface level as functions of the solar zenith angle are shown in Fig. 5.5b. The daily average values of the shortwave component of the aerosol direct radiative effect (DRE) at the above-mentioned input parameters are (-15.1) W/m^2 and (-6.8) W/m^2 at the bottom and top of the atmosphere, respectively.

We compared the obtained estimates with the DRE simulation results using three OPAC models of continental aerosol: clean, average, and polluted (Hess et al., 1998). The wide range of variations of the single scattering albedo and the asymmetry factor presented in these models make it possible to describe the sufficiently large number of situations which can be observed in the atmosphere. Calculations of daily average values of DRE performed for $\tau_{500} = 0.16$ have shown that the range of DRE with the use of three above-mentioned models is (-16.5) – (-14.6) W/m^2 and (-7.8) – (-5.0) W/m^2 at the bottom and top of the atmosphere, respectively.

The estimates of DRE obtained using our empirical model and OPAC continental aerosol model are in good agreement. At the same time, one should note that we compared the aerosol radiative effects only at the boundaries of the atmosphere. The majority of aerosol models, and the data obtained by AERONET, in contrast

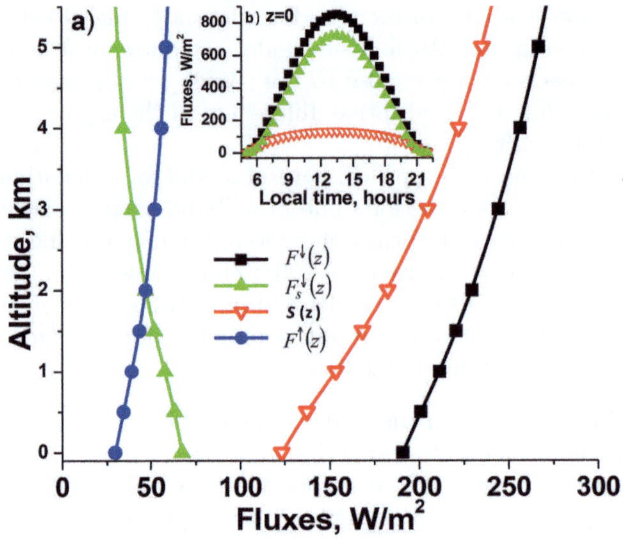

Fig. 5.5. (a) Vertical profiles of the solar radiative fluxes calculated using empirical model (Panchenko et al., 2012) at 07:00 LT on July 15 (SZA = 75.618°) and (b) time variations in the downward radiative fluxes at the surface level; $\tau_{500} = 0.16$; $F^{\downarrow(\uparrow)}$, downward (upward) fluxes; $F_s^{\downarrow}$ and S, diffuse and direct fluxes.

to the model of the vertical profiles of the aerosol characteristics, do not enable one to obtain the data on the change of the upward and downward radiative fluxes inside the atmosphere and to more adequately estimate, for example, the change in the cooling rates.

In the conclusion, we will briefly discuss some aspects of our approach. First, the model (Panchenko et al., 1998; Panchenko and Terpugova, 2002) developed on the basis of airborne sensing provides for the retrieval of vertical profiles of aerosol parameters under different conditions. In particular, knowledge of the air mass type and time of day (night, morning, day, or evening) and the availability of data on near-ground values of the scattering coefficient, temperature, and RH make it possible to more reliably estimate the sought aerosol parameters. In the case when data on the AOD and/or lidar vertical profiles are available, one can essentially reduce the retrieval errors, which are characteristic of all statistical models. However, in Panchenko et al. (2012), we considered only the *average* summer values of the measured parameters, and the data were not divided into subsets corresponding to absolutely clear-sky and low-cloud situations.

Second, when calculating the refractive indices, we used the mixture rule; hence, we initially assumed that the absorbing substance exists as an 'internal mixture' in the aerosol particles. In our opinion, this assumption is justified because, according to literature data (see, e.g., Bond and Bergstrom, 2006), there are reasons to assume that BC exists in aerosol as an 'external mixture' only in the early phase of its emission – that is, in the regions close to the actual sources of hydrocarbon material burning. The state of an internal mixture of aerosol and BC is realized in the process of the long-range transport of air masses during aerosol 'aging.' It is important to note that all of our measurements in the atmosphere of western Siberia were carried out in background regions far from large industrial centers.

The next problem under discussion is related to our choice of the BC concentration ratio for submicron and coarse particles. Our photoacoustic measurements of the spectral dependencies of the aerosol light absorption in the near-ground layer of the atmosphere (Tikhomirov et al., 2005) showed that, with high accuracy, this dependency is close to λ^{-1}, suggesting that the main portion of absorbing substance is localized in particles smaller than visible wavelengths. At the same time, by assuming the ratio of BC content to be 90% in the submicron fraction and 10% in the coarse one, we followed the data from previous studies (Gelencser, 2004; Hitzenberger and Tohno, 2001; Kozlov et al., 2002; Höler et al., 2002; Delene and Ogren, 2002). Taking into account the methodical remarks, we note that the main assumptions introduced into the scheme of the model construction do not contradict the data from numerous measurements of atmospheric aerosol properties.

We assume that the development of our approach can be achieved by more thoroughly accounting for the size distribution of absorbing materials, as well as by comparing the retrieved optical properties for specific atmospheric situations and the results of lidar and photometric measurements.

5.4 Conclusion

In conclusion, it should be noted that large volume of information on the optical characteristics of the tropospheric aerosol in different geographic regions of the planet is already obtained to date on the basis of aircraft measurements. Without pretending to present a complete review of the results, we discussed the basic approaches that are used to retrieve the vertical profiles of the microphysical and optical aerosol characteristics needed for the radiative calculations.

Most of the data were collected during comprehensive aerosol experiments. During the *intensive observation periods*, measurements were performed by instruments installed on board aircraft, ships, and satellites, as well as ground-based equipment. The comprehensive nature of these studies ensured the implementation of several closure tests to examine the magnitude of the uncertainties associated with the various techniques used to measure and estimate aerosol microphysical and optical properties. Most closure tests were intended to compare extinction coefficients/optical depths and single scattering albedo calculated in several ways: (i) microphysical properties such as size distribution and chemical composition determined by *in situ* measurements; (ii) *in situ* optical measurements of aerosol scattering and absorption coefficients; and (iii) airborne and ground-based radiometric measurements. A review of published data showed that the agreement of aerosol characteristics obtained by the different methods is essentially determined by the specific features of the aerosol (dust, biomass burning, etc.) and geophysical conditions during the concrete experiment.

Multi-year aircraft observations, which are held in a pre-determined area of the planet, can be the basis for models of vertical profiles of climatically significant aerosol parameters specific for this region. Using an *optical* approach for measurements that are regularly performed over the territory of Oklahoma (2000–07) has revealed seasonal differences of the vertical profiles of the single scattering albedo, asymmetry factor and absorption, and scattering Ångström exponent of dry matter aerosol in the visible range (Andrews et al., 2011). The combination of *optical and microphysical* approaches constitutes the basis of the empirical model of the vertical profiles of the extinction coefficient, albedo, and phase scattering function developed for the territory of western Siberia from measurements collected during two aircraft campaigns (1986–88 and 1999–2007) (Panchenko et al., 2012). The synthesis of different approaches to create model concepts on the optical aerosol characteristics not only makes it possible to appreciably advance the solution of radiation problem, but also ensures the possibility of optimal selection of instruments, which is required to arrange long-term aircraft studies. We stress that only data of *regular* observations for a specific geographic region make it possible to identify the effects of local, regional, and global processes on variations in aerosol optical characteristics and to estimate the relationship between the natural and anthropogenic factors for different seasons of the year and for different synoptic situations, and, on this basis, to ensure the creation of empirical models.

References

Albrecht, B. (1989) 'Aerosols, cloud microphysics, and fractional cloudiness,' *Science*, **245**(4923), 1227–30.

Anderson, T.L. and Ogren, J.A. (1998) 'Determining aerosol radiative properties using the TSI 3563 Integrating Nephelometer,' *Aerosol Sci. Technol.*, **29**, 57–69.

Andreae, M.O., Elbert, W., Gabriel, R., Johnson, D.W., Osborne, S. and Wood, R. (2000) 'Soluble ion chemistry of the atmospheric aerosol and SO_2 concentrations over the eastern North Atlantic during ACE-2,' *Tellus*, **52B**, 1066–87.

Andrews, E., Ogren, J.A., Sheridan, P.J. and Ferrare, R. (2004) 'Vertical properties of aerosol optical properties at the ARM Southern Great Plains CART site,' *J. Geophys. Res.*, **109**, D06208.

Andrews, E., Sheridan, P.J. and Ogren, J.A. (2011) 'Seasonal differences in the vertical profiles of aerosol optical properties over rural Oklahoma,' *Atmos. Chem. Phys.*, **11**, 10661–76.

Andrews, E., Sheridan, P.J., Fiebig, M., McComiskey, M., Ogren, J.A., Arnott, P., Covert, D., Elleman, R., Gasparini, R., Collins, D., Jonsson, H., Schmid, B. and Wang, J. (2006) 'Comparison of methods for deriving aerosol asymmetry parameter,' *J. Geophys. Res.*, **111**, D05S04.

Ansmann, A., Petzold, A., Kandler, K., Tegen, I., Wendisch, M., Müller, D., Weinzierl, B., Müller, T. and Heintzenberg, J. (2011) 'Saharan Mineral Dust experiments SAMUM-1 and SAMUM-2: What have we learned?' *Tellus B*, **63**, 403–29.

Bates, T.S., Anderson, T.L., Baynard, T., Bond, T., Boucher, O., Carmichael, G., Clarke, A., Erlick, C., Guo, H., Horowitz, L., Howell, S., Kulkarni, S., Maring, H., McComiskey, A., Middlebrook, A., Noone, K., O'Dowd, C.D., Ogren, J., Penner, J., Quinn, P.K., Ravishankara, A.R., Savoie, D.L., Schwartz, S.E., Shinozuka, Y., Tang, Y., Weber, R.J. and Wu, Y. (2006) 'Aerosol direct radiative effects over the northwest Atlantic, northwest Pacific, and North Indian Oceans: Estimates based on in-situ chemical and optical measurements and chemical transport modeling,' *Atmos. Chem. Phys.*, **6**, 1657–732.

Bauer, S.E., Menon, S., Koch, D., Bond, T.C. and Tsigaridis, K. (2010) 'A global modeling study on carbonaceous aerosol microphysical characteristics and radiative effects,' *Atmos. Chem. Phys.*, **10**, 7439–56.

Bedareva, T.V. and Zhuravleva, T.B. (2011) 'Retrieval of aerosol scattering phase function and single scattering albedo according to data of radiation measurements in solar almucantar: Numerical simulation,' *Atmos. Ocean. Opt.*, **24**, 373–85.

Bedareva, T.V. and Zhuravleva, T.B. (2012) 'Estimation of aerosol absorption under summer conditions of western Siberia from Sun photometer data,' *Atmos. Ocean. Opt.*, **25**, 216–23.

Bodhaine, B.A., Ahlquist, N.C. and Schnell, R.C. (1991) 'Three-wavelength nephelometer suitable for aircraft measurement of background aerosol scattering coefficient,' *Atmos. Environ.*, **25A**, 2267–76.

Bond, T.C. and Bergstrom, R.W. (2006) 'Light absorption by carbonaceous particles: An investigative review,' *Aer. Sci. and Technol.*, **40**, 27–67.

Bond, T.C., Anderson, T.L. and Campbell, D. (1999) 'Calibration and intercomparison of filter-based measurements of visible light absorption by aerosols,' *Aerosol Sci. Technol.*, **30**, 582–600.

Chýlek, P., Srivastava, V., Pinnick, R.G. and Wang, R.T. (1988) 'Scattering of electromagnetic waves by composite spherical particles: Experiment and effective medium approximations,' *Appl. Opt.*, **27**, 2396–404.

Collins, D.R., Jonsson, H.H., Seinfeld, J.H., Flagan, R.C., Gassó, S., Hegg, D.A., Russell, P.B., Schmid, B., Livingston, J.M., Ostrom, E., Noone, K.J., Russell, L.M. and Putaud, J.P. (2000) 'In-situ aerosol size distributions and clear column radiative closure during ACE-2,' *Tellus*, **52B**, 498–525.

Delene, D.J. and Ogren, J.A. (2002) 'Variability of aerosol optical properties at four North American surface monitoring sites,' *J. Atmos. Sci.*, **59**, 1135–50.

Dubovik, O. and King, M. (2000) 'A flexible inversion algorithm for retrieval of aerosol optical properties from Sun and sky radiance measurements,' *J. Geophys. Res.*, **105**, 20673–96.

Eck, T.F., Holben, B.N., Ward, D.E., Mukelabai, M.M., Dubovik, O., Smirnov, A., Schafer, J.S., Hsu, N.C., Piketh, S.J., Queface, A., Le Roux, J., Swap, R.J. and Slutsker, I. (2003) 'Variability of biomass burning aerosol optical characteristics in southern Africa during the SAFARI 2000 dry season campaign and a comparison of single scattering albedo estimates from radiometric measurements,' *J. Geophys. Res.*, **108**(D13), 8477.

Elterman, L. (1970) *Vertical Attenuation Model with Eight Surface Meteorological Ranges 2 to 13 Kilometers*, Report AFCRL-70-0200, AFCRL, Bedford, StateMA.

Fenn, R.W. (1979) *Models of the Lower Atmosphere and the Effects of Humidity Variations on Their Optical Properties*, Report AFCRL-TR-79-0214, AFCRL, Bedford, StateMA.

Formenti, P., Elbert, W., Maenhaut, W., Haywood, J. and Andreae, M.O. (2003a) 'The chemical composition of mineral dust during the SHADE airborne campaign over Cape Verde, September 2000,' *J. Geophys. Res.*, **108**(D18), 8576.

Formenti, P., Elbert, W., Maenhaut, W., Haywood, J., Osborne, S. and Andreae, M.O. (2003b) 'Inorganic and carbonaceous aerosols during the Southern African Regional Science Initiative (SAFARI 2000) experiment: Chemical characteristics, physical properties, and emission data for smoke from African biomass burning,' *J. Geophys. Res.*, **108**(D13), doi:10.1029/2002JD002408.

Forster, P., Ramaswamy, V., Artaxo, P., Berntsen, T., Betts, R., Fahey, D.W., Haywood, J., Lean, J., Lowe, D.C., Myhre, G., Nganga, J., Prinn, R., Raga, G., Schulz, M. and Van Dorland, R. (2007) 'Changes in atmospheric constituents and in radiative forcing,' in Solomon, S., Qin, D., Manning, M., Chen, Z., Marquis, M., Averyt, K.B.,

Tignor, M. and Miller, H.L. (eds), *Climate Change 2007: The Physical Science Basis: Contribution of Working Group I to the Fourth Assessment Report of the Intergovernmental Panel on Climate Change*, Cambridge University Press, Cambridge, UK, and New York, USA, pp. 129–234.

Francis, P.N., Hignett, P. and Taylor, J.P. (1999) 'Aircraft observations and modeling of sky radiance distributions from aerosol during TARFOX,' *J. Geophys. Res.*, **104**(D2), 2309–19.

Ganguly, D., Jayaraman, A., Gadhavi, H. and Rajesh, T.A. (2005) 'Features in wavelength dependence of aerosol absorption observed over central India,' *Geophys. Res. Lett.*, **32**, L13821.

Gassó, S., Hegg, D.A., Covert, D.S., Noone, K.J., Öström, E., Schmid, B., Russell, P.B., Livingston, J.M., Durkee, P.A. and Jonsson, H.H. (2000) 'Influence of humidity on the aerosol scattering coefficient and its effect on the upwelling radiance during ACE-2,' *Tellus*, **52B**, 546–67.

Gelencser, A. (2004) *Carbonaceous Aerosol*, Springer, Dordrecht, The Netherlands.

Gillette, D.E. and Blifford, I.N. (1971) 'Composition of tropospheric aerosol as function of altitude,' *J. Atmos. Sci.*, **28**, 1199–210.

Gonzales, J.H. and Ogren, J.A. (1996) 'Sensitivity of retrieved aerosol properties to assumptions in the inversion of spectral optical depths,' *J. Atmos. Sci.*, **53**, 3669–83.

Hänel, G. (1976) 'The properties of atmospheric aerosol particles as functions of the relative humidity at thermodynamic equilibrium with the surrounding moist air,' in Landsberg, H.E. and Van Mieghen, J. (eds), *Advances in Geophysics*, Academic Press, New York, pp. 73–188.

Hansen, J., Sato, M., Ruedy, R., Nazarenko, L., Lacis, A., Schmidt, G.A., Russell, G., Aleinov, I., Bauer, M., Bauer, S., Bell, N., Cairns, B., Canuto, V., Chandler, M., Cheng, Y., Del Genio, A., Faluvegi, G., Fleming, E., Friend, A., Hall, T., Jackman, C., Kelley, M., Kiang, N.Y., Koch, D., Lean, J., Lerner, J., Lo, K., Menon, S., Miller, R.L., Minnis, P., Novakov, T., Oinas, V., Perlwitz, Ja, Perlwitz, Ju, Rind, D., Romanou, A., Shindell, D., Stone, P., Sun, S., Tausnev, N., Thresher, D., Wielicki, B., Wong, T., Yao, M. and Zhan, S. (2005) 'Efficacy of climate forcings,' *J. Geophys. Res.*, **110**, D18104.

Hartley, S.W., Hobbs, P.V., Ross, J.L., Russell, P.B. and Livingston, J.M. (2000) 'Properties of aerosols aloft relevant to direct radiative forcing off the mid-Atlantic coast of the United States,' *J. Geophys. Res.*, **105**, 9859– 85.

Haywood, J. and Boucher, O. (2000) 'Estimates of the direct and indirect radiative forcing due to tropospheric aerosols: A review,' *Rev. Geophys.*, **38**(4), 513–43.

Haywood, J., Francis, P., Dubovik, O., Glew, M. and Holben, B. (2003a) 'Comparison of aerosol size distributions, radiative properties, and optical depths determined by aircraft observations and Sun photometers during SAFARI 2000,' *J. Geophys. Res.*, **108**(D13), 8471.

Haywood, J., Francis, P., Osborne, S., Glew, M., Loeb, N., Highwood, E., Tanré, D., Myhre, G., Formenti, P. and Hirst, E. (2003b) 'Radiative properties and direct radiative effect of Saharan dust measured by the C-130 aircraft during SHADE: 1. Solar spectrum,' *J. Geophys. Res.*, **108**(D18), 8577.

Haywood, J.M. and Ramaswamy, V. (1998) 'Global sensitivity studies of the direct radiative forcing due to anthropogenic sulphate and black carbon aerosols,' *J. Geophys. Res.*, **103**(D6), 6043–58.

Haywood, J.M., Osborne, S.R., Francis, P.N., Keil, A., Formenti, P., Andreae, M.O. and Kaye, P.H. (2003c) 'The mean physical and optical properties of regional haze dominated biomass burning aerosol measured from the C-130 aircraft during SAFARI 2000,' *J. Geophys. Res.*, **108**(D13), 8473.

Haywood, J.M., Pelon, J., Formenti, P., Bharmal, N., Brooks, M., Capes, G., Chazette, P., Chou, C., Christopher, S., Coe, H., Cuesta, J., Derimian, Y., Desboeufs, K., Greed, G., Harrison, M., Heese, B., Highwood, E.J., Johnson, B., Mallet, M., Marticorena, B., Marsham, J., Milton, S., Myhre, G., Osborne, S.R., Parker, D.J., Rajot, J.L., Schulz, M., Slingo, A., Tanré, D. and Tulet, P. (2008) 'Overview of the Dust and Biomass-burning Experiment and African Monsoon Multidisciplinary Analysis Special Observing Period-0,' *J. Geophys. Res.*, **113**(D23), D00C17.

Hegg, D.A., Livingston, J., Hobbs, P.V., Novakov, T. and Russell, P.B. (1997) 'Chemical apportionment of aerosol column optical depth off the mid-Atlantic coast of the United States,' *J. Geophys. Res.*, **102**, 25293–303.

Heintzenberg, J. (2009) 'The SAMUM-1 experiment over Southern Morocco: Overview and introduction,' *Tellus B*, **61**, 2–11.

Heintzenberg, J. and Charlson, R.J. (1996) 'Design and applications of the integrating nephelometer: A review,' *J. Atmos. Oceanic Technol.*, **13**, 987–1000.

Hess, M., Koepke, P. and Shult, I. (1998) 'Optical properties of aerosols and clouds: The software package OPAC,' *Bull. Am. Meteorol. Soc.*, **79**, 831–44.

Hignett, P., Taylor, J.P., Francis, P.N. and Glew, M.D. (1999) 'Comparison of observed and modeled direct aerosol forcing during TARFOX,' *J. Geophys. Res.*, **104**(D2), 2279–87.

Hirst, E., Kaye, P.H., Greenaway, R.S., Field, P. and Johnson, D.W. (2001) 'Discrimination of micrometre-sized ice and supercooled droplets in mixed-phase cloud,' *Atmos. Environ.*, **35**, 33–47.

Hitzenberger, R. and Tohno, S. (2001) 'Comparison of black carbon (BC) aerosols in two urban areas – concentrations and size distributions,' *Atmos. Environ.*, **35**, 2153–67.

Hobbs, P.V. (1999) 'An overview of the University of Washington airborne measurements and results from the Tropospheric Aerosol Radiative Forcing Observational Experiment (TARFOX),' *J. Geophys. Res.*, **104**(D2), 2233–8.

Höler, R., Tohno, S., Kasahara, M. and Hitzenberger, R. (2002) 'Long-term characterization of carboneous aerosol in Uji, Japan,' *Atmos. Environ.*, **36**, 1267–75.

Ivlev, L.S. (1969) 'Aerosol model of the atmosphere,' in *Problems of Atmospheric Physics*, Issue 8, Publishing House of Leningrad State University, Leningrad, pp. 125–60.

Jaenicke, R. (1980) 'Atmospheric aerosol and global climate,' *J. Aerosol Sci.*, **11**, 577–88.

Johnson, D.W., Osborne, S., Wood, R., Suhre, K., Johnson, R., Businger, S., Quinn, P.K., Raes, F., Bates, T., McGovern, F. and Liedekerke, M. (2000) 'The 2nd Aerosol Characterisation Experiment (ACE-2): General overview and main results,' *Tellus*, **52B**, 111–25.

Johnson, B., Shine, K. and Forster, P. (2004) 'The semi-direct aerosol effect: Impact of absorbing aerosols on marine stratocumulus,' *Quart. J. Roy. Meteor. Soc.*, **130**(599), 1407–22.

Kabanov, M.V., Panchenko, M.V., Pkhalagov, Yu.A., Veretennikov, V.V., Uzhegov, V.N. and Fadeev, V.Ya. (1988) *Optical Properties of Coastal Atmospheric Hazes*, Novosibirsk, Nauka.

King, M.D., Byrne, D.M., Herman, B.M. and Reagan, J.A. (1978) 'Aerosol size distributions obtained by inversion of spectral optical depth measurements,' *J. Atmos. Sci.*, **35**, 2153–67.

Kneizys, F.X., Robertson, D.S., Abreu, L.W., Acharya, P., Anderson, G.P., Rothman, L.S., Chetwynd, J.H., Selby, J.E.A., Shetle, E.P., Gallery, W.O., Berk, A., Clough, S.A. and Bernstein, L.S. (1996) 'The MODTRAN 2/3 report and LOWTRAN 7 model,' Phillips Laboratory, Geophysics Directorate, Hanscom AFB, MA 01731-3010.

Koch, D. and Del Genio, A.D. (2010) 'black carbon semi-direct effects on cloud cover: Review and synthesis,' *Atmos. Chem. Phys.*, **10**, 7685–96.

Kondratyev, K.Ya. and Pozdnykov, D.V. (1981) *Aerosol Models of the Atmosphere*, Nauka, Moscow.

Kotchenruther, R.A., Hobbs, P.V. and Hegg, D.A. (1999) 'Humidification factors for atmospheric aerosols off the mid-Atlantic coast of the United States,' *J. Geophys. Res.*, **104**(D2), 2239–51.

Kozlov, V.S., Panchenko, M.V. and Yausheva, E.P. (2007) 'Time content variations of submicrometer aerosol and soot in the near-ground layer of the West Siberia atmosphere,' *Atmos. Ocean Opt.*, **20**, 987–90.

Kozlov, V.S., Panchenko, M.V. and Yausheva, E.P. (2008) 'Mass fraction of black carbon in submicrometer aerosol as an indicator of influence of smokes from remote forest fires in Siberia,' *Atmos. Environ.*, **42**, 2611–20.

Kozlov, V.S., Panchenko, M.V., Kozlov, A.S., Ankilov, A.N., Baklanov, A.M. and Malyshkin, S.B. (2002) 'Instrumental and technical approach to prompt field measurements of size distributions of aerosol absorbing and scattering characteristics,' Proceedings of the Twelfth Atmospheric Radiation Measurement (ARM) Science Team Meeting, St. Petersburg, Florida, *www.arm.gov/docs/documents/technical/conf_0204/kozlov(2)-vs.pdf*.

Kozlov, V.S., Shmargunov, V.P. and Panchenko, M.V. (2009) 'Seasonal variability of the vertical profiles of absorption parameters of submicrometer aerosol in the troposphere,' *Atmos. Ocean Opt.*, **22**, 635–42.

Krekov, G.M. and Rakhimov, R.F. (1982) *Optical-Radar Model of Continental Aerosol*, Nauka, Novosibirsk.

Kristjansson, J.E. (2002) 'Studies of the aerosol indirect effect from sulfate and black carbon aerosols,' *J. Geophys. Res.*, **107**, 4246.

Magi, B. and Hobbs, P. (2003) 'Effects of humidity on aerosols in southern Africa during the biomass burning season,' *J. Geophys. Res.*, **108**(D13), 8495.

Magi, B.I., Fu, Q. and Redemann, J. (2007) 'A methodology to retrieve self-consistent aerosol optical properties using common aircraft measurements,' *J. Geophys. Res.*, **112**(D24), doi:10.1029/2006JD008312.

Magi, B.I., Hobbs, P.V., Schmid, B. and Redemann, J. (2003) 'Vertical profiles of light scattering, light absorption and single scattering albedo during the dry, biomass burning season in southern Africa and comparisons of in-situ and remote sensing measurements of aerosol optical depth,' *J. Geophys. Res.*, **108**(D13), 8504.

Martins, J.V., Artaxo, P., Liousse, C., Reid, J.S., Hobbs, P.V. and Kaufman, Y. (1998) 'Effects of black carbon content, particle size, and mixing on light absorption by aerosols from biomass burning in Brazil,' *J. Geophys. Res.*, **103**, 32041–50.

Matsumoto, T., Russell, P.B., Mina, C. and Van Ark, W. (1987) 'Airborne tracking sunphotometer,' *J. Atmos. Oceanic Technol.*, **4**, 336–9.

McConnell, C.L., Formenti, P., Highwood, E.J. and Harrison, M.A.J. (2010) 'Using aircraft measurements to determine the refractive index of Saharan dust during the DODO experiments,' *Atmos. Chem. Phys.*, **10**, 3081–98.

McConnell, C.L., Highwood, E.J., Coe, H., Formenti, P., Anderson, B., Osborne, S., Nava, S., Desboeufs, K., Chen, G. and Harrison, M.A.J. (2008) 'Seasonal variations of the physical and optical characteristics of Saharan dust: Results from the Dust Outflow and Deposition to the Ocean (DODO) experiment,' *J. Geophys. Res.*, **113**, D14S05.

McNaughton, C.S., Clarke, A.D., Freitag, S., Kapustin, V.N., Kondo, Y., Moteki, N., Sahu, L., Takegawa, N., Schwarz, J.P., Spackman, J.R., Watts, L., Diskin, G., Podolske, J., Holloway, J.S., et al. (2011) 'Absorbing aerosol in the troposphere of the Western Arctic during the 2008 ARCTAS/ARCPAC airborne field campaigns,' *Atmos. Chem. Phys.*, **11**, 7561–82.

Moelwyn-Hughes, E.A. (1961) *Physical Chemistry*, Pergamon, Tarrytown, StateNew York.

Myhre, G., Grini, A., Haywood, J.M., Stordal, F., Chatenet, B., Tanré, D., Sundet, J.K. and Isaksen, I.S.A. (2003) 'Modeling the radiative impact of mineral dust during the Saharan Dust Experiment (SHADE) campaign,' *J. Geophys. Res.*, **108**(D18), 8579.

Nakajima, T., Tonna, G., Rao, R., Kaufman, Y. and Holben, B. (1996) 'Use of sky brightness measurements from ground for remote sensing of particulate polydispersions,' *Appl. Opt.*, **35**, 2672–86.

Novakov, T., Hegg, D.A. and Hobbs, P.V. (1997) 'Airborne measurements of carbonaceous aerosols on the east coast of the United States,' *J. Geophys. Res.*, **102**(D25), 30023–30.

Osborne, S.R. and Haywood, J.M. (2005) 'Aircraft observations of the microphysical and optical properties of major aerosol species,' *Atmos. Res.*, **73**, 173–201.

Osborne, S.R., Haywood, J.M., Francis, P.N. and Dubovik, O. (2004) 'Short-wave radiative effects of biomass burning aerosol during SAFARI2000,' *Q. J. Royal Meteorol. Soc.*, **130**, 1423–47.

Öström, E. and Noone, K.J. (2000) 'Vertical profiles of aerosol scattering and absorption measured in situ during the North Atlantic Aerosol Characterization Experiment,' *Tellus*, **52B**, 526–45.

Palmer, K.F. and Williams, D. (1975) 'Optical constants of sulfuric acid: Application to the clouds of Venus?' *Appl. Opt.*, **14**, 208–19.

Panchenko, M.V. and Terpugova, S.A. (2002) 'Reconstruction of the scattering coefficient in the lower troposphere using ground-based measurement,' *J. Atmos. Sci.*, **59**, 581–9.

Panchenko, M.V., Kozlov, V.S., Terpugova, S.A., Shmargunov, V.P. and Burkov, V.V. (2000) 'Simultaneous measurements of submicrometer aerosol and absorbing substance in the altitude range up to 7 km,' Proceedings of Tenth ARM Science Team Meeting, San-Antonio, Texas, *www.arm.gov/docs/documents/technical/conf_0003/panchenko–mv.pdf*.

Panchenko, M.V., Sviridenkov, M.A., Terpugova, S.A. and Kozlov, V.S. (2004) 'Active spectral nephelometry as a method for the study of submicrometer atmospheric aerosols,' *Atmos. Ocean Opt.*, **17**, 378–86.

Panchenko, M.V., Terpugova, S.A. and Pol'kin, V.V. (1998) 'empirical model of the aerosol optical properties in the troposphere over West Siberia,' *Atmos. Ocean Opt.*, **11**, 532–9.

Panchenko, M.V., Terpugova, S.A. and Tumakov, A.G. (1996) 'Annual variations of submicrometer aerosol fraction as assessed from the data of airborne nephelometric measurements,' *Atmos. Res.*, **41**, 203–15.

Panchenko, M.V., Terpugova, S.A., Kozlov, V.S., Pol'kin, V.V. and Yausheva, E.P. (2005) 'Annual behavior of the condensation activity of submicrometer aerosol in the atmospheric surface layer of Western Siberia,' *Atmos. Ocean Opt.*, **18**, 607–11.

Panchenko, M.V., Zhuravleva, T.B., Terpugova, S.A., Polkin, V.V. and Kozlov, V.S. (2012) 'An empirical model of optical and radiative characteristics of the tropospheric aerosol over West Siberia in summer,' *Atmos. Meas. Tech.*, **5**, 1513–27.

Penner, J.E., Zhang, S.Y. and Chuang, C.C. (2003) 'Soot and smoke aerosol may not warm climate,' *J. Geophys. Res.*, **108**, 4657.

Petzold, A., Veira, A., Mund, S., Esselborn, M., Kiemle, C., Weinzierl, B., Hamburger, T., Ehret, G., Lieke, K. and Kandle, K. (2011) 'Mixing of mineral dust with urban pollution aerosol over Dakar (Senegal): Impact on dust physico-chemical and radiative properties,' *Tellus B*, **63**, 619–34.

Raes, F., Bates, T., McGovern, F. and Liedekerke, M. (2000) 'The 2nd Aerosol Characterisation Experiment (ACE-2): General overview and main results,' *Tellus B*, **52**, 111–25.

Redelsperger, J.L., Thorncroft, C.D., Diedhiou, A., Lebel, T., Parker, D.J. and Polcher, J. (2006) 'African monsoon multidisciplinary analysis: An international research project and field campaign,' *Bull. Am. Meteorol. Soc.*, **87**, 1739–46.

Reid, J.S. and Maring, H.B. (2003) 'Foreword to special section on the Puerto Rico Dust Experiment (PRIDE),' *J. Geophys. Res.*, **108**(D19), 8585.

Reid, J.S., Jonsson, H.H., Maring, H.B., Smirnov, A., Savoie, D.L., Cliff, S.S., Reid, E.A., Livingston, J.M., Meier, M.M., Dubovik, O. and Tsay, S.-C. (2003) 'Comparison of size and morphological measurements of coarse mode dust particles from Africa,' *J. Geophys. Res.*, **108**(D19), doi:10.1029/2002JD002485.

Rosenberg, G.V. (1967) 'Properties of atmospheric aerosol according to data of optical sensing,' *Izv. Akad. Nauk SSSR, Fiz. Atmos. Okeana*, **3**(9), 936–49.

Rosenberg, G.V. (1968) 'Optical studies of the atmospheric aerosol,' *Uspekhi Fizicheskikh Nauk*, **95**(1), 159–208.

Rosenberg, G.V. (1976) 'Determination of microphysical parameters of aerosol according to data of integrated optical measurements,' *Izv. Akad. Nauk SSSR, Fiz. Atmos. Okeana*, **12**(11), 1159–67.

Rosenberg, G.V., Gorchakov, G.I., Georgievsky, Yu.S. and Lyubovtseva, Yu.S. (1980) 'Optical parameters of the atmospheric aerosol,' in *Atmospheric Physics and Climate Problem*, Nauka, Moscow.

Rozwadowska, A. (2007) 'Influence of aerosol vertical profile variability on retrievals of aerosol optical thickness optical thickness from NOAA AVHRR measurements in the Baltic region,' *Oceanologia*, **49**, 165–84.

Russell, P.B. and Heintzenberg, J. (2000) 'An overview of the ACE-2 clear sky column closure experiment (CLEARCOLUMN),' *Tellus*, **52B**, 463–83.

Russell, P.B., Hobbs, P.V. and Stowe, L.L. (1999a) 'Aerosol properties and radiative effects in the United States East Coast haze plume: An overview of the Tropospheric Aerosol Radiative Forcing Experiment (TARFOX),' *J. Geophys. Res.*, **104**, 2213–22.

Russell, P.B., Livingston, J.M., Hignett, P., Kinne, S., Wrong, J., Chien, A., Bergstrom, R., Durkee, P. and Hobbs, P.V. (1999b) 'Aerosol-induced radiative flux changes off the United States mid-Atlantic coast: Comparison of values calculated from sunphotometer and in situ data with those measured by airborne pyranometer,' *J. Geophys. Res.*, **104**(D2), 2289–307.

Ryder C.L., Highwood, E.J., Rosenberg, P.D., Trembath, J., Brooke, J.K., Bart, M., Dean, A., Crosier, J., Dorsey, J., Brindley, H., Banks, J., Marsham, J.H., McQuaid, J.B., Sodemann, H. and Washington, R. (2013) 'Optical properties of Saharan dust aerosol and contribution from the coarse mode as measured during the Fennec 2011 aircraft campaign,' *Atmos. Chem. Phys.*, **13**, 303–25.

Sakerin, S.M. and Kabanov, D.M. (2007) 'Spectral dependence of the atmosphere aerosol optical depth near 0.37–4 μm,' *Atmos. Ocean Opt.*, **20**, 141–9.

Sakerin, S.M., Beresnev, S.A., Gorda, S.Yu., Kabanov, D.M., Kornienko, G.I., Markelov, Yu.I., Mikhalev, A.V., Nikolashkin, S.V., Panchenko, M.V., Poddubnyi, V.A., Pol'kin, V.V., Smirnov, A.V., Tashilin, M.A., Turchinovich, S.A., Turchinovich, Yu.S., Holben, B. and Eremina, T.A. (2009) 'Characteristics of annual behavior of spectral aerosol optical depth of the atmosphere under conditions of Siberia,' *Atmos. Ocean Opt.*, **22**, 446–56.

Schmid, B., Ferrare, R., Flynn, C., Elleman, R., Covert, D., Strawa, A., Welton, E., Turner, D. and Jonsson, H. (2006) 'How well do state-of-the-art techniques measuring the vertical profile of tropospheric aerosol extinction compare?' *J. Geophys. Res.*, **111**(D5), doi:10.1029/2005JD005837.

Schmid, B., Livingston, J.M., Russell, P.B., Durkee, P.A., Jonsson, H.H., Collins, D.R., Flagan, R.C., Seinfeld, J.H., Gassó, S., Hegg, D.A., Öström, E., Noone, K.J., Welton, E.J., Voss, K., Gordon, H.R., Formenti, P. and Andreae, M.O. (2000) 'Clear sky closure studies of lower tropospheric aerosol and water vapor during ACE-2 using airborne sunphotometer, airborne in-situ, space-borne, and ground based measurements,' *Tellus*, **52B**, 568–93.

Schmid, B., Redemann, J., Russell, P.B., Hobbs, P.V., Hlavka, D.L. and McGill, M.J. (2003) 'Coordinated airborne, spaceborne, and groundbased measurements of massive, thick aerosol layers during the dry season in southern Africa,' *J. Geophys. Res.*, **108**(D13), 8496.

Shao, Y.P., Wyrwoll, K.H., Chappell, A., Huang, J.P., Lin, Z.H., McTainsh, G.H., Mikami, M., Tanaka, Y.Y., Wang, X.L. and Yoon, S. (2011) 'Dust cycle: An emerging core theme in Earth system science,' *Aeolian Res.*, **2**, 181–204.

Sheridan, P.J., Arnott, W.P., Ogren, J.A., Andrews, E., Atkinson, D.B., Covert, D.S. and Moosmüller, H. (2005) 'The Reno Aerosol Optics Study: An evaluation of aerosol absorption measurement methods,' *Aerosol Sci. Technol.*, **39**, 1–16.

Sheridan, P.J., Delene, D.J. and Ogren, J.A. (2001) 'Four years of continuous surface aerosol measurements from the Department of Energy's Atmospheric Radiation Measurement Program Southern Great Plains Cloud and Radiation Testbed site,' *J. Geophys. Res.*, **106**, 20735–47.

Shettle, E.P. and Fenn, R.W. (1979) *Models for the Aerosols of the Lower Atmosphere and the Effects of Humidity Variations on Their Optical Properties*, AFCRL-TR-79-0214, AFCRL, Bedford, StateMA.

Slingo, A. (1989) 'A GCM parameterization for shortwave radiative properties of water clouds,' *J. Atmos. Sci.*, **46**, 1419–27.

Strapp, J.W., Leaitch, W.R. and Liu, P.S.K. (1992) 'Hydrated and dried aerosol size distribution measurements from the particle measuring systems FSSP-300 probe and the deiced PCASP-100X probe,' *J. Atmos. Ocean. Technol.*, **9**, 548–55.

Swap, R.J., Annegarn, H.J., Suttles, T., Haywood, J. and Helminger, M.C. (2002) 'The Southern African Regional Science Initiative (SAFARI 2000): Dry-season field campaign. An overview,' *South African Journal of Science*, **98**, 125–30.

Swietlicki, E., Zhou, J., Covert, D.S., Hameri, K., Busch, B., Vakeva, M., Dusek, U., Berg, O.H., Wiedensohler, A., Aalto, P., Makela, J., Marinsson, B.G., Papaspiropoulos, G., Mentes, Frank, G. and Stratmann, F. (2000) 'Hygroscopic properties of aerosol particles in the Northeastern Atlantic during ACE-2,' *Tellus*, **52B**, 201–27.

Tanré, D., Haywood, J., Pelon, J., Léon, J.F., Chatenet, B., Formenti, P., Francis, P., Goloub, P., Highwood, E.J. and Myhre, G. (2003) 'Measurement and modeling of the Saharan dust radiative impact: Overview of the Saharan Dust Experiment (SHADE),' *J. Geophys. Res.*, **108**(D18), 8574.

Tikhomirov, A.B., Firsov, K.M., Kozlov, V.S., Panchenko, M.V., Ponomarev, Yu.N. and Tikhomirov, B.A. (2005) 'Investigation of spectral dependence of shortwave radiation absorption by ambient aerosol using time-resolved photoacoustic technique,' *Optical Engineering*, **4**(7), 071203-1–11.

Toon, O.B. and Pollack, J.B. (1976) 'A global average model of atmospheric aerosols for radiative transfer calculations,' *J. Appl. Met.*, **15**(3), 225–46.

Twomey, S. (1977a) *Atmospheric Aerosols*, Elsevier, StateNew York.

Twomey, S. (1977b) 'The influence of pollution on the shortwave albedo of clouds,' *J. Atmos. Sci.*, **34**(7), 1149–52.

Virkkula, A., Ahlquist, StateN.C., Covert, D.S., Arnott, W.P., Sheridan, P.J., Quinn, P.K. and Coffman, D.J. (2005) 'Modification, calibration and a field test of an instrument for measuring light absorption by particles,' *Aerosol Sci. Technol.*, **36**, 68–83.

Washington, R., Flamant, C., Parker, D.J., Marsham, J., McQuaid, J.B., Brindley, H., Todd, M., Highwood, E.J., Chaboureau, J.-P., Kocha, C. , Bechir, M., Saci, A. and Ryder, C.L. (2012) 'Fennec: The Saharan Climate System,' *CLIVAR Exchanges*, **17**(3), 31–2.

Yu, H., Kaufman, Y., Chin, M., Feingold, G., Remer, L., Anderson, T., Balkanski, Y., Bellouin, N., Boucher, O., Christopher, S., DeCola, P., Kahn, R., Koch, D., Loeb, N., Reddy, M., Schulz, M., Takemura, T. and Zhou, M. (2006) 'A review of measurement-based assessments of the aerosol direct radiative effect and forcing,' *Atmos. Chem. Phys.*, **6**, 613–66.

Zhuravleva, T.B. (2008) 'Simulation of solar radiative transfer under different atmospheric conditions. Part I: Deterministic atmosphere,' *Atmos. Ocean Opt.*, **21**, 99–114.

Zhuravleva, T.B., Kabanov, D.M., Sakerin, S.M. and Firsov, K.M. (2009) 'Simulation of direct aerosol radiative forcing for typical summer conditions of Siberia. Part 1: Method of calculation and choice of the input parameters,' *Atmos. Ocean Opt.*, **22**, 63–73.

Zuev, V.E. and Krekov, G.M. (1986) *Optical Models of the Atmosphere*, Gidrometeoizdat, Leningrad.

6 Light absorption in the atmosphere

Helmuth Horvath

6.1 Introduction

The atmosphere consists mainly of nitrogen, oxygen, as well as argon, water vapor, carbon dioxide, and other trace gases and particles. Whereas the fractions of nitrogen, oxygen, and the noble gases are practically constant, the others undergo temporal and spatial variations. An overview is given in Table 6.1. The limiting values are given as an indication, since the strong variability of the concentration calls for the regulations to allow a few excesses on the one hand and/or limit the upper percentiles. The variability of the atmospheric aerosol is caused by the short lifetime, often less than a week, which does not permit homogeneous mixing around the globe, as is the case for CO_2, for example (lifetime more than 30 years).

Mainly the trace gases CO_2, H_2O, O_3, SO_2, NO_2, N_2O, and the particles in the atmosphere contribute to light absorption; in total, they amount to a few percent of the constituents of the atmosphere.

Both particles and gas molecules consist of atoms, having a positive nucleus and electrons, and, upon interaction with electromagnetic radiation, the atoms become vibrating dipoles, emitting light; this process is called light scattering. At the same time, some light energy is transferred into internal energy of the molecules and finally into heat; this is called absorption.

6.2 Definitions

The scattering and absorptive properties of an aerosol/gas are characterized by magnitudes, which are described in the following (a graphical overview is given in Fig. 6.1).

6.2.1 Absorption coefficient, extinction coefficient, scattering coefficient

A parallel monochromatic beam of light with flux Φ_0 passing a distance x through a medium containing molecules, particles, bacteria, etc., is attenuated to a flux

Table 6.1. Constituents of the atmosphere.

Substance	Mole fraction	Density	Remark	Limit in EU daily mean	
N_2	0.7809	0.976 kg/m^3	Constant		
O_2	0.2095	0.299kg/m^3	Constant		
Ar	0.0093	16.6 g/m^3	Constant		
CO_2	400 ppm	785 mg/m^3	Slowly rising		
Ne	18 ppm	16 mg/m^3	Constant		
He	5.24 ppm	0.94 g/m^3	Constant		
H_2O vapor	0.0012–0.048	1–40 g/m^3	Variable		
O_3 stratosphere	163 ppb	350 μg/m^3	Constant		Slowly increasing
O_3 ground level	5–75 ppb	10–150 μg/m^3	Variable	125 μg/m^3	
SO_2	1–20 ppb	2–50 μg/m^3	Variable	125 μg/m^3	Decreasing in Europe since 1992
NO_2	6–250 ppb	5–200 μg/m^3	Variable	200 μg/m^3	Hourly mean
N_2O	320 ppb	630 μg/m^3	Rising continuously		
Particles, PM 10	–	5–200 μg/m^3	Variable	50 μg/m^3	
Particles, PM 2.5	–	2–100 μg/m^3	Variable	25 μg/m^3	
black carbon/soot		0.3–15 μg/m^3	Variable		

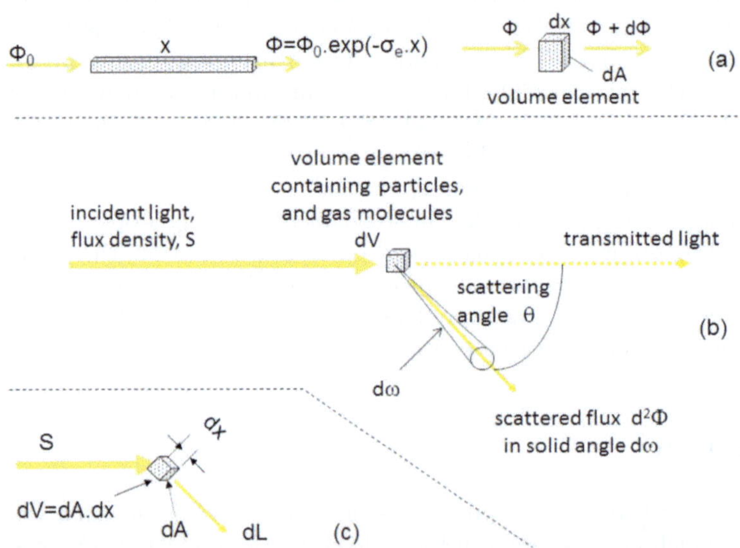

Fig. 6.1. Optical characteristics of an aerosol/gas: (a) definition of the extinction coefficient; (b) definition of the volume-scattering function; (c) radiance of a layer of illuminated aerosol.

$\Phi = \Phi_0 \exp(-\sigma_e \cdot x)$ (see Fig. 6.1a). The constant σ_e is called the extinction coefficient and is a measure for the attenuation of light, $\Phi/\Phi_0 = T = \exp(-\sigma_e \cdot x)$ is called the transmission through the medium. The negative natural logarithm of the transmission $\ln(1/T) = \sigma_e \cdot x = \delta$ is called the optical depth. In case $\sigma_e(x)$ depends on x, the optical depth is obtained by integration: $\delta = \int_0^x \sigma_e(x')\,dx'$.

Using the differential formulation of the exponential attenuation law $\frac{d\Phi}{\Phi} = -\sigma_e \cdot dx$, it is immediately evident that the extinction coefficient σ_e is the fraction of light attenuated per unit length. Its SI unit is 1/m. In atmospheric optics, frequently used units are: $\mathrm{km}^{-1} = 0.001\ \mathrm{m}^{-1}$ and $\mathrm{Mm}^{-1} = 10^{-6}\ \mathrm{m}^{-1}$. The exponential attenuation law is only valid for a parallel beam of strictly monochromatic light.

The attenuation is caused by the scattering and absorption of light by the gas molecules and particles. If only absorption takes place, which would, for example, be the case of a dye solution, then σ_e coincides with the absorption coefficient σ_a. If only scattering takes place – that is, the light energy is taken out of the parallel beam by deflecting part of it in other directions without loss of light energy – then σ_e coincides with the scattering coefficient σ_s. Usually, scattering and absorption take place simultaneously, so it is a habit to write the extinction coefficient as the sum of the scattering coefficient and the absorption coefficient, $\sigma_e = \sigma_a + \sigma_s$, where σ_a signifies the fraction of light attenuated per unit length due to absorption and σ_s is the fraction attenuated due to scattering. It must be mentioned that it is impossible to separate attenuation by scattering and attenuation by absorption from each other, which is one of the problems with the optical measurement of the absorption coefficient. All coefficients depend on the wavelength of light; furthermore, σ_e, σ_a, and σ_s are extensive properties. The wavelength dependence of the atmospheric attenuation has been studied in detail by Ångström (1929) and an empirical formula for the wavelength dependence had been found: $\sigma_{e,s,a} \propto \left(\frac{\lambda}{\lambda_0}\right)^{-\alpha}$, with λ_0 a reference wavelength (e.g., 550 nm or 1 μm). The exponent α (sometimes $-\alpha$) is called the Ångström exponent. This relationship frequently holds.

The ratio of the scattering coefficient to the extinction coefficient is called the single-scattering albedo $\varpi = \frac{\sigma_s}{\sigma_e} = 1 - \frac{\sigma_a}{\sigma_e}$. The single-scattering albedo is an intensive property.

For an aerosol consisting of particles suspended in air, as well as for gases, the extinction (scattering-, absorption-) coefficient varies with pressure. This is obvious, since, at higher pressure, the closer the scattering and/or absorbing molecules are, so more interaction with light is possible per unit length. If σ_0 is the extinction/scattering/absorption coefficient at pressure p_0, the value at pressure p will be $\sigma = \sigma_0 \cdot \left(\frac{p}{p_0}\right)$. For example, the Rayleigh-scattering coefficient of air is $11.62\ \mathrm{Mm}^{-1} \cdot \left(\frac{p}{100000\ \mathrm{Pa}}\right) \cdot \left(\frac{\lambda}{550\ \mathrm{nm}}\right)^{-4.08}$. Due to the short lifetime of the aerosol and its modification in the atmosphere, actually all optical characteristics of the aerosol depend on location and time.

6.2.2 Scattering function, phase function

The light scattered by the particles or a gas obviously depends on the properties of the scattering medium and can be described by the volume-scattering function, (see Fig. 6.1b). Consider a volume element dV of a scattering medium illuminated by a light of flux density S. The light flux $d^2\Phi$ scattered into direction θ (which is the angle between the considered direction of the scattered light and the direction of the transmitted light) is obviously proportional to the illuminating flux density, the volume, and the solid angular extent $d\omega$ into which the light is scattered, so we write $d^2\Phi = S \cdot \gamma(\theta) \cdot dV \cdot d\omega$ or $\gamma(\theta) = \frac{d^2\Phi}{S \cdot dV \cdot d\omega}$.

Integrating the volume-scattering function over the whole solid angle, the scattering coefficient is obtained, namely $\sigma_s = \int_{4\pi} \gamma(\theta) \cdot d\omega$ or, for rotational symmetry around the direction of the illuminating beam of light, we obtain $\sigma_s = 2\pi \int_0^\pi \gamma(\theta) \cdot \sin d\theta \cdot d\theta$.

Similarly to the scattering coefficient, $\gamma(\theta)$ depends on wavelength, which we always assume implicitly.

Dividing $\gamma(\theta)$ by the scattering coefficient, σ_s, produces an intensive property, the phase function $P(\theta)$. The usual definition is $P(\theta) = 4\pi \cdot \frac{\gamma(\theta)}{\sigma_s}$. The factor 4π is merely for appearance, since, then, an isotropic medium has the phase function $\gamma(\theta) = 1\ sr^{-1}$. In some publications, the factor 4π is omitted, in which case the phase function is smaller by a factor of 4π. The phase function for spherical particles can be calculated, using Mie theory (see below); the Rayleigh-scattering phase function (e.g., for air) is $P(\theta) = \frac{3}{4} \cdot (1 + \cos^2\theta)$.

6.2.3 Asymmetry parameter

The light flux scattered in the various directions depends on the type of scattering medium. An isotropic scattering medium would have a phase function of $P(\theta) \equiv 1\ sr^{-1}$ independently of the scattering angle. The Rayleigh phase (scattering) function has equal values in the forward (θ) and backward $\left(\frac{\pi}{2} - \theta\right)$ directions (i.e., the scattering is symmetric with respect to $\theta = \frac{\pi}{2}$), whereas, for the majority of aerosols, the forward scatter by far exceeds the backscatter. The asymmetry of the scattering can be characterized by the asymmetry parameter $g = \frac{\int_0^\pi \gamma(\theta)\sin(\theta)\cos(\theta)d\theta}{\int_0^\pi \gamma(\theta)\sin(\theta)d\theta}$. For symmetric scattering, it is zero; for scattering only in the forward direction, it is +1; for atmospheric aerosols, usual values are between 0.5 and 0.8.

6.2.4 Model phase functions

For computations, it may be convenient to use a model function to describe the phase function. Van de Hulst (1980, pp. 306–7) lists 29 varieties; a frequently used model phase function has been published by Henyey and Greenstein (1941):

$P = \dfrac{1-g^2}{(1+g^2-2g\cos\theta)^{1.5}}$, with g being the asymmetry parameter. This model phase function has the advantage of containing only one parameter; a disadvantage is that the phase function does not increase in the backscattering, as is the case for most aerosols (see section 6.7.4).

6.2.5 Radiance of a layer of gas/aerosol

An illuminated aerosol/gas scatters light in all directions, also toward an observer, who thus sees a more or less bright volume. The radiance L, being the radiometric equivalent to the physiologic unit brightness, can be derived easily (see Fig. 6.1c): the volume element dV is a cuboid with base dA perpendicular to the direction of observation and height dx; using the definition of the scattering function, we obtain $d^3\Phi = S \cdot \gamma(\theta) \cdot dA \cdot dx \cdot d\omega$. Rearranging this yields $\dfrac{d^3\Phi}{dA \cdot d\omega} = S \cdot \gamma(\theta) \cdot dx$, which is the definition of the radiance. Thus, the contribution to the radiance of an illuminated aerosol layer with thickness dx is $dL = S \cdot \gamma(\theta) \cdot dx$.

The radiance of an infinitesimal layer of scattering material is determined by its phase function $P(\theta)$, the single-scattering albedo ϖ, and the extinction coefficient σ_e, and amounts to $dL = S \cdot \gamma(\theta) \cdot dx = \dfrac{1}{4\pi} \cdot S \cdot \varpi \cdot \sigma_e \cdot P(\theta) \cdot dx$. When observed at a distance x, the attenuation $\exp(-\sigma_e \cdot x)$ has to be included, so $dL = \dfrac{1}{4\pi} \cdot S \cdot \varpi \cdot \sigma_e \cdot P(\theta) \cdot \exp(-\sigma_e \cdot s) \cdot dx$. For a layer extending from 0 to R, the radiance can be obtained by integration as $L = \dfrac{1}{4\pi} \cdot S \cdot \varpi \cdot P(\theta) \cdot [1 - \exp(\sigma_e \cdot R)]$. This radiance mainly depends on the illuminating flux density, the single-scattering albedo, and the phase function.

6.3 Light absorption by gases

An overview of the absorptive properties of atmospheric gases is given in Fig. 6.2. It is immediately evident that the absorption spectra are complicated, especially at high resolution, which is not shown in this figure. Since photons interact with the electrons in the gas molecules, only energy differences permitted by quantum mechanics can be transferred to the molecules, allowing only discrete wavelengths being absorbed. The energy levels in an atom are shown schematically in Fig. 6.3. The black lines represent the different electronic states. An atom/molecule can change its state from one electronic state to the other, by absorbing or emitting a photon having exactly the energy corresponding to the difference between two states. The molecule can also acquire or lose energy of rotation and vibration; thus, in addition to the electronic levels, many more energy levels exist, which are represented by the many horizontal green and red lines. The largest energy difference is between the electronic states, adjacent vibrational levels have less energy, and the least energy difference is between rotational levels. The molecule thus can absorb or emit a photon corresponding to the energy differences between two levels. One example in Fig. 6.3 is between a lower electronic level and an upper electronic level with additional vibrational and rotational energy. This energy difference usually is large, so UV absorption or emission is necessary to supply the needed energy. Transition between vibrational and/or rotational levels within the same electronic level usually is due to infrared absorption/emission. Transition between the various energy levels has to fulfill the quantum rules, but still a large number of transitions is possible, causing the complicated spectra in the visible and infrared. The HITRAN project (Rothman et al., 2013) characterizes all the absorption lines of 47 atmospheric gas molecules, which amount to a total of seven million. A verbal description of the absorption features of the various gases is given in the following (see also Fig. 6.1).

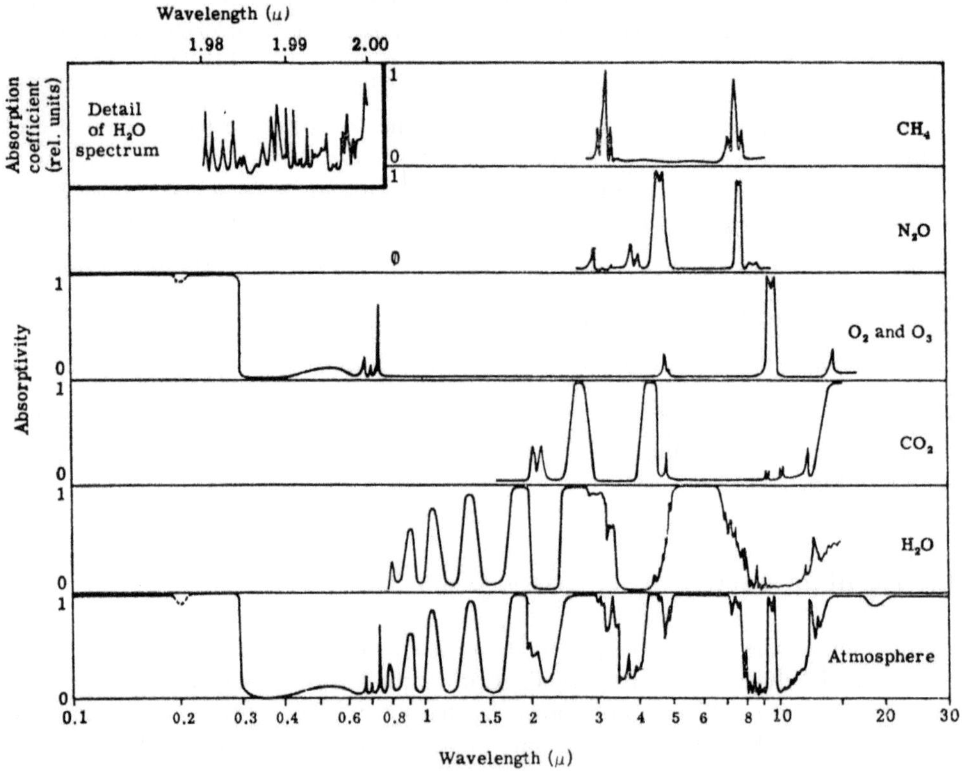

Fig. 6.2. Absorption spectra of the atmospheric gases. From Fleagle and Businger (1963).

Ozone is practically opaque for wavelengths below 300 nm: 1 cm of ozone at ground-level pressure has a transmission of 1.9×10^{-19} at 300 nm and 10^{-34} at 250 nm. The strong absorption of short-wavelength sunlight is essential for life on Earth, since all essential bio-molecules absorb UV light below a wavelength of 280 nm and are destroyed or at least lose their functionality without it. The absorption of solar UV light takes place in the stratosphere and, without this UV shield, life on Earth would not be possible. The strong absorption is caused by a continuous destruction of ozone molecules to O_2 and O by UV absorption and subsequent

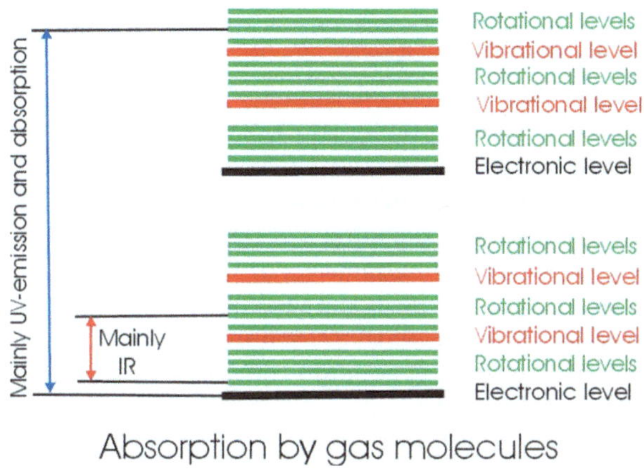

Fig. 6.3. Energy levels in an atom/molecule (schematic). The black lines represent the different electronic states, the red lines the vibrational levels, and the green lines the rotational levels.

recombination (Stolarski, 1988). The absorbed energy is transferred as heat to the stratosphere, which reaches temperatures of 270 K in 50 km. Reactive chlorine, produced in the stratosphere by UV irradiation of the extremely stable flouro-chloro-hydrocarbons, blocks the recombination of atomic and molecular oxygen during the ozone UV absorption cycle and thus destroys ozone (Stolarski, 1988). The ozone in the stratosphere is distributed between 10 and 40 km. Its integral concentration is measured in Dobson Units, with 100 Dobson Units corresponding to a layer of pure ozone of 1 mm at a pressure of 1 Bar. Presently, the integrated O3 concentration is ≈300 Dobson Units, corresponding to an ozone layer 3 mm thick under ground-level conditions. A ground-level ozone concentration of 50 ppb extending up to 2,000 m would correspond to a pure ozone layer of 0.1 mm.

At 400 nm, ozone is practically transparent, some absorption of which is between 500 and 700 nm; for a concentration of 100 ppb (polluted atmosphere), the absorption coefficient at a wavelength of 600 nm amounts to 1 Mm^{-1}.

Nitrogen dioxide can be found in polluted air, and is in equilibrium with N_2O_4 but, under the usual atmospheric conditions, the equilibrium is shifted to NO_2. It has strong absorption in the blue part of the visible spectrum, decreasing toward the red. Thus, in transmission, this gas appears reddish to yellowish or brown. It can be the cause of the appearance of brown haze (Horvath, 1971).

Oxygen absorbs mostly in the UV and near and far infrared. Below 243.2 nm, it is strongly absorbing and the interaction with a photon splits the atom, enabling the formation of ozone. In the lower atmosphere, sunlight does not contain this wavelength range due to absorption in the upper atmosphere. In the visible, oxygen has a narrow absorption band (A-band) centered at 762 nm, and two further narrow bands (B and γ) at 688 and 628 nm, whose band strengths are factors of 0.08 and 0.003 smaller than the A-band (Kiehl and Yamanouchi, 1985).

Carbon dioxide has strong absorption bands at 2.5, 4.5, and >12 μm. It is in part responsible for the greenhouse effect.

Water vapor has many absorption bands in the infrared and is the most important naturally occurring greenhouse gas.

Methane is transparent in the visible, but has absorption bands in the infrared around 3 and 7 μm. It is, for example, produced by bacteria during digestion in the stomachs of ruminants and during rice cultivation, emitted by landfills, biomass burning, and natural gas distribution, so its concentration increases with population growth. It has a 100-year global warming potential of 29; that is, in 100 years, it traps 29 times the heat of the same quantity of CO_2.

Nitrous oxide also is transparent in the visible, but has absorption bands in the infrared and it is a greenhouse gas with a warming potential of 298, due to its long atmospheric lifetime of 114 years. The main sources are nitrogen fertilizers used in agriculture which are added to the soil.

6.3.1 Attenuation law for line spectra

Most absorption by atmospheric gases is in absorption bands, being an agglomeration of many lines, which can only be resolved with a high-quality spectrometer. At a wavelength centered on a line, strong light absorption takes place. Between adjacent lines, there is a gap where either no absorption takes place or, due to the width of the line, much smaller attenuation occurs. Collision broadening and Doppler broadening determine the width of the line. Let us consider a tunable laser centered exactly at the absorption line. With this light source, the exponential attenuation law can easily be measured; for example, for a given concentration and distance, the transmission is T, so the incident flux Φ_0 is attenuated to $\Phi_1 = \Phi_0 \cdot T$. Doubling the distance or concentration, a flux of $\Phi_2 = \Phi_0 \cdot T^2$ is observed, according to the exponential attenuation law. On the other hand, for a laser having a wavelength outside the absorption line, which can be very close to the line, no attenuation takes place at all. Let us now consider a light source consisting of two lasers with a total flux of $2\Phi_0$. For the first laser, the transmitted light is $\Phi_1 = \Phi_0 \cdot T$; for the second, it is Φ_0, since no attenuation takes place, so in total it is $\Phi_1' = \Phi_0 \cdot T + \Phi_0$. Therefore, the transmission for the combined light source is $T' = \dfrac{\Phi_1'}{2\Phi_0} = \dfrac{T+1}{2}$. Doubling the concentration or distance, the total transmitted flux is $\Phi_2' = \Phi_0 \cdot T^2 + \Phi_0$, so the transmission amounts to $T'' = \dfrac{\Phi_2'}{2\Phi_0} = \dfrac{T^2+1}{2}$. If the exponential attenuation law were valid, T'' should be $(T')^2$, but it is immediately evident that $T'' > (T')^2$, thus the exponential extinction law cannot be applied. This is the case because the requirement of a monochromatic light source is not fulfilled. Therefore, for the absorption of polychromatic light in the absorption bands of atmospheric gases, a modification of the Lambert Beer law is needed. One empirical relation for water vapor is $T = \exp\left(-\sigma' \left[\dfrac{c}{c'}\right]^a \cdot x\right)$, where c' is a reference concentration and σ' the absorption coefficient at this reference concentration; a is an empirical factor between 0.5 and 1, depending on the spacing and width of the absorption lines (see, e.g., Guzzi and Rizzi, 1984). Therefore, the indication of an absorption coefficient of an absorbing gas also requires the concentration and the distance.

A few examples of absorption coefficients of atmospheric gases in the visible are indicated in Table 6.2. It is evident that, in the visible, except for a narrow range of wavelengths, atmospheric gases have a negligible contribution to light absorption of the atmosphere; light absorption in the atmosphere is dominated by the particles. The only exception might be NO_2 in high concentrations.

Table 6.2. Absorption coefficients of selected atmospheric trace gases (Mm^{-1}).

Ozone	0.2 at 500 nm	1.0 at 600 nm	0.3 at 700 nm	$c = 100$ ppb, $x = 1$ km
NO_2	86 at 400 nm	17 at 550 nm	0.8 at 700 nm	$c = 50$ ppb, $x = 1$ km
O_2	1,000	Narrow band at 761 nm		$c = 21\%$, $x = 1$ km
H_2O	100	Narrow band at 740 nm		$c = 10$ g/m^3, $x = 1$ km
Aerosol, polluted atmosphere	90 at 400 nm	66 at 550 nm	50 at 700nm	$dv = 0.5$ μm, $\sigma_g = 1.7$ $c = 100$ μg/m^3, $m = 1.56 - 0.03i$
		extinction coefficient (Mm^{-1})		
Aerosol, polluted atmosphere	679 at 400 nm	608 at 550 nm	446 at 700nm	$dv = 0.5$ μm, $\sigma_g = 1.7$ $c = 100$ μg/m^3 $m = 1.56 - 0.03i$
		Scattering coefficient of pure air (Mm^{-1})		
	42.6 at 400 nm	11.6 at 550 nm	4.3 at 700nm	$p = 100{,}000$ Pa

For the absorption coefficient of gases a distance of $x = 1$ km is considered. The concentration c, given in the last column is by volume. For the aerosol, spherical particles are assumed, having a lognormal size distribution with a volume median diameter d_v of 0.5 μm and a geometric standard deviation of σ_g of 1.7. The mass concentration of the particles is 100 μgm^{-3}, The particles are assumed to consist of a homogeneous material, having a refractive index of $m = 1.56 - 0.03i$.

Data on absorption coefficients of atmospheric gases can be found in Hodkinson (1966) for NO_2, Gast (1960) for O_3, and Balkan and Hinzpeter (1988) for O_2, CO_2, N_2O, CH_4, H_2O, and others.

6.4 Light absorption by solids

Solid materials consist of many atoms, and therefore many electrons. Due to their proximity, there is almost a continuum state of energy, but the states are grouped into bands due to the regular structure. Two cases can be distinguished:

1. In a non-conductor (insulator), the conduction band is empty and the band gap to the valence band typically is 10 eV, corresponding to a wavelength of 120 nm, so light absorption only is possible in the far UV, the material is transparent in the visible, and reflects some light on a polished surface due to its difference in refractive index to the surrounding medium. A graphical representation of the energy levels is shown in Fig. 6.4.
2. In a conductor, the conduction band either is partially filled or the valence band overlaps with the conduction band. In both cases, many 'free' electrons are

available to interact with low-energy photons, causing a strong absorption at all wavelengths. Conductors have considerable specular reflection when polished. This is the case for metals and graphite (Fig. 6.5).

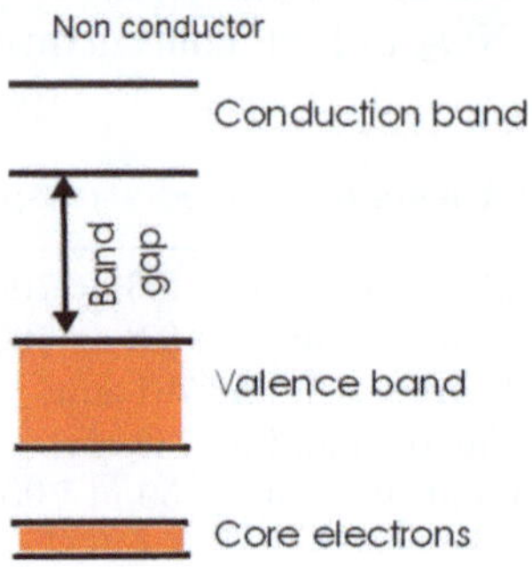

Fig. 6.4. Energy band diagram for an insulator. The energy difference between the empty conduction band and the filled valence band is about 10 eV, so photons with a wavelength less than 120 nm can be absorbed.

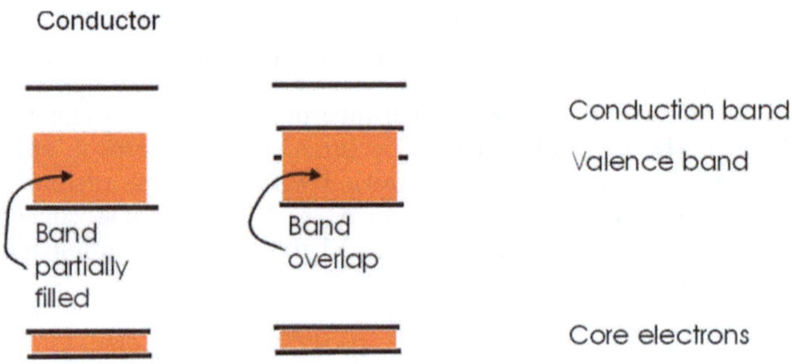

Fig. 6.5. Energy band diagram for a conductor.

6.5 Refractive index

From the point of view of optics, a solid can be characterized by its refractive index m and the bulk absorption coefficient α. For investigations of the propagation of electromagnetic waves, the material usually is characterized by its relative permittivity ε, with $m = \sqrt{\varepsilon}$. The refractive index m can be complex if the material is absorbing, usually written as $m = n - i \cdot k$; in this case, obviously also the permittivity is complex: $\varepsilon = \varepsilon' - i \cdot \varepsilon''$. The real part n of the refractive index is used for calculating the deflection of light beams upon refraction, applying Snell's law. The bulk absorption coefficient α is used to determine the transmission trough a layer of material of thickness x by $T = \exp(-\alpha \cdot x)$. The imaginary part, $-k$, of the complex refractive index $m = n - i \cdot k$ is related to the bulk absorption coefficient by $k = \dfrac{\alpha \cdot \lambda}{4\pi}$. In aerosol optics, materials which are considered light-absorbing have imaginary parts of the refractive index between −0.1 and −10, the bulk material is completely black and nontransparent, and deposits of filters look dark.[1]

Materials which are completely black from our macroscopic experience may be practically transparent from the aerosol point of view, as the following example shall demonstrate. A welding glass filter has a typical transmission of 10^{-7} at 550 nm and a thickness of 3 mm, so it appears black. The bulk absorption coefficient then is 5373 m^{-1}, which normally is thought to be considerable. The imaginary part of the refractive index using the above formula amounts to $k = 2.3 \times 10^{-4}$. This is considerably lower compared to $k = 0.1 \ldots 1$ for the material forming particles considered to be absorbing. But an imaginary part of $k > 0.1$ is needed for absorption in aerosol optics, and this can be immediately seen when going to nanometer dimensions (see Table 6.2). Let us do the following thought experiment: consider two slabs of thickness 100 nm, one consisting of welding glass and one consisting of graphite. The welding glass slab has a transmission due to absorption of 0.9994 (when including the reflection losses at the front and back surfaces, we obtain 0.9210) – that is, it is completely transparent, whereas the graphite slab only transmits 10% so it is still black at these dimensions. Let us now consider 1 m^2 of the slab atomized to particles with a diameter of 100 nm and distributed in 1 m^3 of air. The extinction coefficient of the aerosol was calculated using Mie theory (see below) and the result can also be seen in Table 6.3. The dark glass particles have slight absorption and mainly scatter light, whereas the graphite particles show considerable absorption and have some scattering. The transmission through the suspended particles is less than that of the slab, since particles have random arrangement and there always is the possibility of a photon passing through without interception. This is similar to a dense forest where also light makes it to the ground, although the leaves would overlap several times, if arranged regularly.

Table 6.3. Strongly and weakly absorbing materials.

Substance	Specification	Bulk absorption coefficient	Refractive index	Transmission of 100 nm layer	Transmission of 100 nm particles of equivalent volume	Fraction of scattering in extinction
Welding glass 3 mm thick	$T = 10^{-7}$ at 550 nm	5373 m^{-1}	$1.5 - 0.00023i$	0.9994	0.9635	0.9877
Graphite	–	22,847,947 m^{-1}	$2 - i$	0.1018	0.2255	0.142

[1] The sign for the imaginary part of the refractive index can be positive or negative, depending on the mathematical formulation of the time dependence of a wave: if $\exp(-i\omega t)$ is chosen, then the imaginary part is positive; if $\exp(i\omega t)$ is chosen, $Im(m)$ is negative. Depending on the choice of the sign of the imaginary part of the refractive index, a computer program may fail if the wrong sign is used. Bohren and Huffman (1983) use a positive imaginary part of the refractive index.

6.5.1 Refractive index of mixtures of two light-absorbing substances

Homogeneous mixtures on the atomic scale do not exist, since a solid is needed to have continuous strong light absorption. So particles of at least a few nanometers in size are needed. These small particles can be considered as giant 'molecules', but still small enough with respect to the wavelength of the interacting light, to assign an average dielectric function to the mixture. Obviously, approximations and simplifications have to be made; the Maxwell Garnett theory (1904) is the most accepted effective medium method, which was used to explain many unusual optical phenomena. Consider small spherical particles with permittivity ε embedded in a medium of permittivity ε_m. The effective macroscopic permittivity ε_{MG} of the medium containing a volume fraction f of the small spheres is obtained as

$$\varepsilon_{MG} = \varepsilon_m \frac{\varepsilon(1+2f) + 2\varepsilon_m(1-f)}{\varepsilon(1-f) + \varepsilon_m(2+f)} . \tag{6.1}$$

Bruggeman-Wassenaar (1935) has suggested another mixing rule for obtaining the effective permittivity ε_{Br} by solving the equation: $f\frac{\varepsilon - \varepsilon_{Br}}{\varepsilon + 2\varepsilon_{Br}} + (1-f)\frac{\varepsilon_m - \varepsilon_{Br}}{\varepsilon_m + 2\varepsilon_{Br}} = 0$.

The most simple rule mathematically, but without theoretical foundation, is volume mixing. Consider two substances: material 1 has the refractive index m_1 and material 2 has the refractive index m_2. The volume fractions of the two substances are f_1 and f_2, with $f_1 + f_2 = 1$. The refractive index of the mixture is $m = f_1 m_1 + f_2 m_2$. Obviously, this mixing rule can be generalized for more than two substances. It must be emphasized again that there is no theoretical foundation for this rule (see, e.g., Chylek et al., 1988).

As an example, let us consider a quasi-homogeneous mixture of small spherical carbon ($m = 1.7 - i \cdot 0.7$) particles embedded in a matrix of ammonium sulfate ($m = 1.528$). For a few volume fractions of the small carbon spheres, the effective refractive index of the quasi-homogeneous medium is listed in Table 6.4.

Table 6.4. Effective refractive index of mixed media using various rules.

Volume fraction	Maxwell Garnett		Bruggeman		Volume mixing	
F	n	k	n	k	n	k
0.00	1.528	0.000	1.528	0.000	1.528	0.000
0.02	1.533	−0.014	1.533	−0.014	1.531	−0.014
0.04	1.538	−0.028	1.538	−0.027	1.535	−0.028
0.06	1.543	−0.041	1.543	−0.041	1.538	−0.042
0.08	1.548	−0.055	1.547	−0.055	1.542	−0.056
0.10	1.553	−0.069	1.552	−0.068	1.545	−0.070
0.12	1.558	−0.083	1.556	−0.082	1.549	−0.084
0.14	1.562	−0.097	1.560	−0.096	1.552	−0.098
0.16	1.567	−0.111	1.564	−0.109	1.556	−0.112
0.18	1.572	−0.125	1.569	−0.123	1.559	−0.126
1.00	1.700	−0.700	1.700	−0.700	1.700	−0.700

It is evident that, for this case, the different mixing rules approximately give the same results. For very special cases such as metal-insulator-(semi-conductor) systems, the Maxwell Garnett theory is superior (Abeles and Gittleman, 1976); for the 'simple' case of strongly absorbing and little absorbing substances forming a particle, all methods will give useful results, as long as the size of the grains is far below the wavelength. Unfortunately, this frequently is not the case in the atmosphere.

Often, an effective refractive index is used for particles, being composed of several substances forming sections not very small compared to the wavelength of the illuminating light. In this case, the above rules cannot be applied. But, still hypothetically, the refractive is in use, which a material needs to have, in order to best describe the optical properties of the substance under investigation. But it may not always be possible to find the desired refractive index.

6.5.2 Data for the refractive index of materials forming absorbing particles

Bearing in mind that insulators have a large band gap, so no light absorption in the visible takes place, their refractive index must be a real number. On the other hand, in conductors, 'free' electrons can interact with low-energy photons, leading to considerable absorption, characterized by an imaginary part of the refractive index whose absolute value is $|k| > 0.1$.

The direct determination of the complex refractive index of a material is not possible so, besides measurements of angles, direction of polarization, flux, etc., an underlying theory is necessary. The measurement of the refractive index of transparent bulk material simply applies Snell's law, such as by finding the angle of minimum deflection of a prism. For the determination of the complex refractive index, a variety of measurement methods can be applied (see, e.g., Bohren and Huffman, 1983, p. 41): transmittance and reflectance measurement of an ultrathin slab of bulk material; ellipsometric amplitude ratio and phase shift determination for the reflected light; reflectance measurement for polarized light at different angles of incidence and application of the Fresnel formulae. For all methods, a large sample of bulk material is needed, which has to be polished and, for transmission measurements, an ultrathin section has to be cut, such as by microtomy with a diamond knife, or enough liquid has to be available to be put in an extremely thin cuvette. Although this is possible and has been done successfully, the question remains whether the material forming an aerosol particle is identical with the chunk of bulk material used, such as for ellipsometric reflectance investigations. Therefore, indirect methods are also in use, which will be described below.

The determination of the complex index of refraction has been done for various reasons: radiative properties of carbon electrodes, radiative properties of glowing soot in flames (enhancing the heat transfer), or optical data for aerosol investigations.

Senftleben and Benedict (1918) investigated the radiative properties of carbon arc lamps, so used highly polished carbon arc electrodes for reflectance measurements in order to obtain the complex refractive index.

Janzen (1979) produced a colloidal suspension of carbon black particles sized around 70 nm and measured its transmittance and, by comparing with Mie theory calculations, the refractive index is obtained.

Pluchino et al. (1980) used single spherical micrometer-sized particles, levitated them, measured their angular scattering, and applied Mie theory to determine the best-fitting index of refraction.

Kirchstetter et al. (2004) sampled black carbon particles on quartz fiber filters and measured the attenuation of light passing through. The deposited particles are considered as a thin slab, whose thickness is determined by the mass and density of the particles and the area of the deposit on the filter. An empirical correction has to be made for multiple passes of light due to the scattering or light on the fibers of the filter. The attenuation coefficient is related to the imaginary part of the refractive index by $k = \dfrac{\alpha \cdot \lambda}{4\pi}$. The same method is applied for determining the imaginary part of the organic component of soot (see Fig. 6.6).

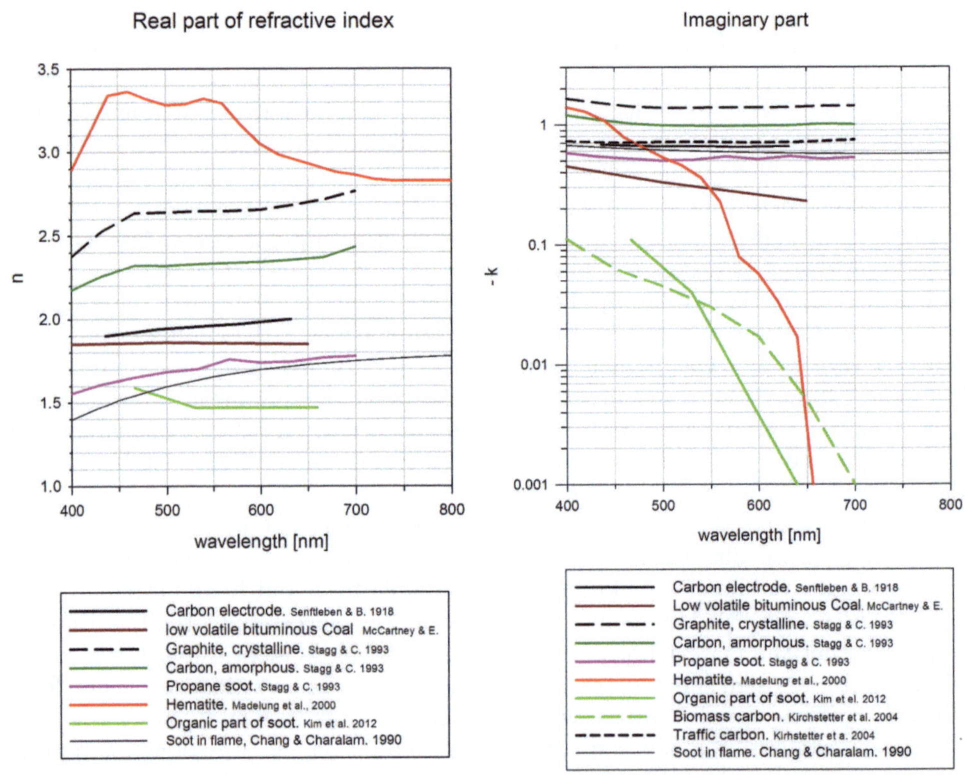

Fig. 6.6. Wavelength dependence of the real and imaginary parts of the refractive index.

Shaddix and Williams (2009) have produced soot in gas flames and, after quenching and dilution, measured the light attenuation and scattering as well as the mass of the particles. With these data, a possible range of real and imaginary parts of the refractive index could be derived.

Chang and Charalampopoulos (1990) have measured the optical properties of a flame with glowing soot: at a wavelength of 488 nm, scattering, extinction, and photon correlation are used to determine the particle size and number. The diameter of the soot particles was between 20 and 60 nm, depending on the location in the flame. Spectral extinction measurements ($\lambda = 200$ nm to 6.4 μm) in combination with Mie theory and Kramers-Krönig relations yielded spectral refractive indices.

Soot formed during combustion processes is a complex mixture of small spherical graphitic particles and organic material, condensed hydrocarbons (tar), which consist of partly oxidized and transformed fuel. The refractive index of the organic carbon can also be been determined by measuring the scattering, backscattering, and absorption of the soot aerosol, using electron microscope analysis to determine size, shape, and number of the graphitic particles, Rayleigh Gans theory for the scattering and absorption of the embedded particles, finally leading to the complex refractive index of the organic carbon (Kim et al., 2014).

Frequently, the light-absorbing particles are available as powder. A standard procedure in powder science is mixing the particles under investigation with a white powder, pressing it in a solid and determining reflection (and transmission), and applying Kubelka-Munk theory to infer the refractive index (Kortüm, 1969; Patterson and Marshall, 1982).

For Rayleigh spheres, which are only slightly absorbing, the extinction coefficient is proportional to the volume fraction of the particles and the imaginary part of the refractive index. Since the contribution of scattering to extinction is very small, an extinction measurement can simply yield the imaginary part of m (Bohren and Huffman, 1983, p. 432).

Values for the refractive index of black carbon found in the literature are listed in Table 6.5. The first part of the table lists measured values and the method to obtain the value; the latter part lists values frequently used in the literature.

It is evident that the refractive index of crystalline graphite has the largest values both for n and k, but it is unlikely for this to be found in atmospheric samples. Amorphous graphite has smaller values for n and k, close to 2 and 1, as used in early textbooks. An analysis of various types of coals showed that both n and k steadily decrease with increasing hydrogen-to-carbon content (H/C ratio) (see McCartney and Ergun, 1967). This simply explains why soot, being produced by combustion of hydrogen-containing fuels, has lower values for n and k and, depending on the combustion process, the refractive index can vary. The inter-planar spacing of graphitic layers in combustion-derived soot is 3.4–3.6 Å, compared to 3.35 Å for graphite (Chen and Dobbins, 2000). Furthermore, Janzen (1979) argues that compressing soot pellets even with a pressure of 1,600 MPa still preserves the particulate nature and the minimum void fraction corresponding to the random close-packed configuration is $(1 - 2/\pi) = 0.363$.

Let us look at a simple calculation. The refractive index of propane soot has an ellipsomentically determined refractive index of $1.701 - 0.508 \cdot i$ (Stagg and Charalampopoulos, 1993). Applying the Maxwell Garnett mixing rule, a matrix consisting of black carbon with refractive index of $2.102 - 0.755 \cdot i$ and a void fraction of $f = 0.363$ with refractive index $1.0 - 0 \cdot i$ would exactly have the measured refractive index. Thus, one can conclude – with all the reservation about

Table 6.5. Values for the refractive index of black carbon (at $\lambda = 550$ nm).

Re(m)	Im(m)	Substance	Method	Reference
1.97	−0.67	Carbon arc electrode, 0.7% impurity	Reflection measurement of polished surface	Senftleben and Benedict (1918)
1.56 1.57	−0.46 −0.50	Acetylene soot Propane soot	Polarization ratio of reflection on compressed pellets	Dalzell and Sarofin (1969)
2.15	−0.66	Graphite	Reflection	McCartney and Ergun (1967)
1.86	−0.33	Low volatile bituminous coal	Reflection	McCartney and Ergun (1967)
2	−1	Carbon black	Transmission of liquid suspension of nano-spheres and Mie theory	Janzen (1979)
1.7	−0.7	Carbon black	Angular scattering of levitated micrometer-sized particles	Pluchino et al. (1980)
1.28 1.35 1.36	−0.22 −0.19 −0.113	Kerosene soot Acetylene soot Propane soot	Polarization ratio, angular reflectance method of compressed samples	Batten 1985
2.649	−1.395	Graphite, crystalline	Ellipsometry	Stagg and Charalampopoulos (1993)
2.337	−0.997	Graphite, amorphous	Ellipsometry	Stagg and Charalampopoulos (1993)
1.701	−0.508	Propane soot	Ellipsometry of compressed pellets	Stagg and Charalampopoulos (1993)
–	−0.72	black carbon from vehicle emissions	Transmission of filter deposit	Kirchstetter et al. (2004)
1.4 … 1.9	−0.77 … −1.25	Methane, ethylene and kerosene soot	Transmission and scattering of suspended soot particles	Shaddix and Williams (2009)
1.66	−0.59	Soot in burning propane flame	Photon correlation and extinction/scattering measurement	Chang and Charalampopoulos (1990)
1.46	−0.45		Value used in model	Shettle and Fenn (1979)
2.0	−0.66		Value used in model	Bergstrom (1973)
1.95	−0.66		Value used for modeling	Kattawar and Hood (1976)
1.57	−0.56	Soot	Suggestion for a representative value	Smyth and Shaddix (1996)

mixing rules – that the black material forming the soot has a refractive index of $2.102 - 0.755 \cdot i$, which is within the range of possibilities for graphite.

Another argument for caution with ellipsometric measurements of compressed samples is formulated by Janzen (1979): a surface of compressed small spheres is not a flat surface, although it reflects light, and reflectance measurements may not give the correct value of the refractive index.

The discrepancies in the values of the refractive index not only reflect possible deficiencies in the measuring method, but also the different processes which lead to the formation the black carbon. Important factors are the C/H ratio and the extent of agglomeration. But this still does not explain the large variations in refractive index. Smyth and Shaddix (1996) analyzed a large number of available data and suggest using a refractive index of $1.57 - 0.56i$ as a compromise, since then various studies will be comparable. Shaddix and Williams (2009) measured the extinction of soot produced by various flames and found a range of possible values for the refractive index (see Table 6.5), and they consider $m = 1.7 - 1.0i$ as a good compromise. Nevertheless, one may conclude that there is no reference value for the refractive index of black carbon, and that the production process also influences its value.

6.5.3 Wavelength dependence of the refractive index

Obviously, all optical parameters are wavelength-dependent. An overview of the wavelength dependence of the refractive index in the visible for black carbon and other materials is given in Fig. 6.6.

For all curves shown in this figure, the real part of the refractive index has a slight variation, with a maximum deviation from the mean by a factor of 1.2 or less. Similarly to the black carbon data, the imaginary part is constant with respect to wavelength, although the spread between the different materials amounts to a factor of 2. A very strong wavelength dependence can be seen for hematite and, to a lesser degree, for the organic part of soot and biomass carbon. For both, the light absorption in the short-wavelength visible range is by far larger than for the long-wavelength range (see also below for examples).

6.6 Mie theory

The theory of absorption and scattering of light by small spherical particles is attributed to G. Mie (1908).[2] His aim was to explain the brilliant colors of gold colloids. He strictly applied Maxwell's theory to the propagation of electromagnetic waves in a medium containing particles and found an exact solution, which was represented by a series expansion. Since computers were not available, only few terms could be added manually, so results could be obtained only for nanoparticles, but this was sufficient to explain the color of colloids. Mie's solution has proven many times to be exact. Nowadays, computer programs are available and a Mie calculation takes milliseconds; they usually use the mathematical procedure of Debye

[2] A solution for the scattering of light by small dielectric spheres using the ether theory was published by Lorenz in 1880, unfortunately in Danish, so it has been known only to a few persons. For historical remarks, see Horvath (2009).

(1909). Computer programs can be found in Bohren and Huffman (1983) and Barber and Hill (1990). Among many others, the University of Bremen offers the possibility to download a variety of programs calculating scattering and absorption by small particles (*http://diogenes.iwt.uni-bremen.de/vt/laser/wriedt/index_ns.html*).

Using geometric optics, it is simple to determine the extinction coefficient of an ensemble of N particles per cubic meter, which we assume to be mono-dispersed with a radius r. For the moment, we assume that all light incident on the cross-section of a particle is lost due to interception. Let us consider a lamina of particles of thickness dx. One square meter of the lamina contains $N \cdot 1\ \mathrm{m}^2 \cdot dx$ particles, which have a total cross-section of $Nr^2\pi \cdot 1\ \mathrm{m}^2 \cdot dx$. The fraction of 1 m^2 of the lamina, which blocks light passing through is $N \cdot r^2\pi \cdot dx$. Compared with the infinitesimal definition of the extinction coefficient, it is immediately evident that $\sigma_e = N \cdot r^2\pi$. This is strictly valid for geometric optics, such as the shade produced by leaves of a tree. For aerosol particles, the diffraction, scattering and absorption of the waves by the particles have to be considered, which can amount to both more or less than $N \cdot r^2\pi$. This is accounted for by introducing the efficiency factor Q_e, so the extinction coefficient is obtained as $\sigma_e = N \cdot Q_e \cdot r^2\pi$. Similarly, the absorption and scattering coefficients are $\sigma_a = N \cdot Q_a \cdot r^2\pi$ and $\sigma_s = N \cdot Q_s \cdot r^2\pi$. The expression $N \cdot Q_e \cdot r^2\pi$ is the effective cross-section of the particles per unit volume, which obviously is identical to the extinction coefficient. To stress the cross-section per volume, σ_e is more precisely called the volume extinction coefficient. But it is also possible to use another reference parameter for the cross-section, such as per molecule, so it would be the molecular extinction coefficient, or per mass of the suspended particles per unit volume, namely by dividing the extinction coefficient by the density ρ_{ae} of the aerosol; this is called the mass extinction coefficient σ_e/ρ_{ae}. Since σ_e is the cross-section per volume and ρ_{ae} the mass per volume, then the mass extinction coefficient is the effective cross-section of the particles divided by the mass of the particles; the more systematic name for the mass extinction coefficient is specific cross-section.

The efficiency factors depend on the complex refractive index, and the ratio of a particle's radius and the wavelength, usually characterized by the size parameter $x = \dfrac{2\pi \cdot r}{\lambda}$ – that is, only the ratio of particle size to wavelength is important. The size parameter can be also considered as the product of the radius and wave number.

The Q factors have been calculated for two refractive indices and are shown in Fig. 6.7. For the transparent particles ($m = 1.5 - 0.0i$), the Q_e-factor (identical to Q_s) rapidly increases for size parameters between 1 and 4, and has oscillations due to interferences of the waves in and around the particle, finally approaching $Q \to 2$ for $x \to \infty$. From the point of view of geometric optics, one would expect $Q \to 1$ for $x \to \infty$. But, besides the geometric shadow, the presence of the particle also influences the wave going around the particles, so the diffraction along the edges of the particles has to be considered, which, according to Babinet's principle, amounts to the same light flux as the shadow, so $Q \to 2$ is justified (Brillouin, 1949).

For strongly absorbing particles, having the same real part of the refractive index, the Q_e curve also approaches $Q_e \to 2$ for $x \to \infty$, but without oscillations, due to absorption of part of the waves. The absorption efficiency approaches

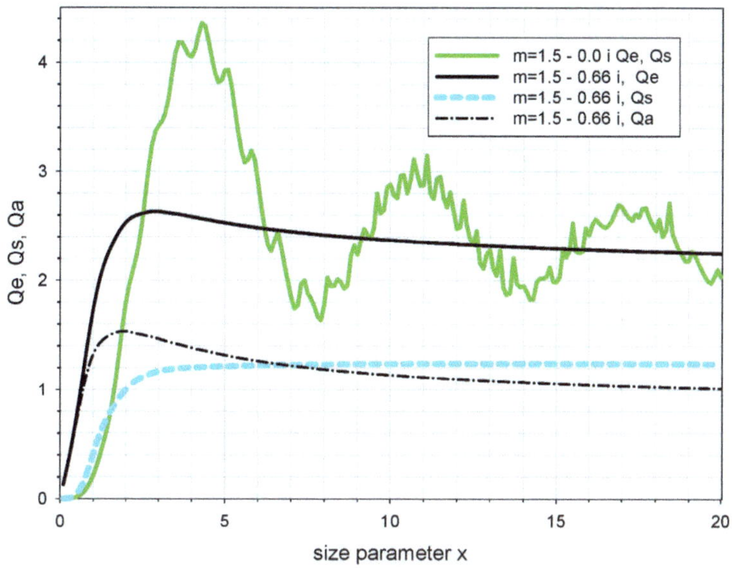

Fig. 6.7. Efficiency factors for extinction scattering and absorption as a function of the size parameter. The efficiency factor Q_e, Q_s, Q_a is the fraction of attenuated, scattered, or absorbed light intercepted by the particles relative to the light incident on the cross-section of the particle. The dimensionless size parameter $x = \frac{2r\pi}{\lambda}$ is a measure of the size of the particle relative to the wavelength, and can be understood as the ratio of the circumference of the spherical particle divided by the wavelength or the radius times wave number.

$Q_a \to 1 - R$ for $x \to \infty$ as expected from geometric optics and the scattering efficiency also approaches $Q_s \to 1 + R$ for $x \to \infty$ because of Babinet's principle. Here, R is the Fresnel reflectance coefficient integrated over the incidence angle (Kokhanovsky and Zege, 1997). For size parameters below $x = 2$ (radius smaller than $\frac{\lambda}{6}$), extinction is by far larger than for its transparent counterpart, since absorption takes place simultaneously. This is very evident when plotting on a log-log scale (Fig. 6.8). For a size parameter smaller than 1, the scattering increases $\propto x^4$, which means that the scattered light flux, being proportional to $Q_s \cdot r^2$, is proportional to the sixth power of the radius and, since $x = \frac{2\pi \cdot r}{\lambda}$, it is proportional to λ^{-4}, namely Rayleigh scattering. Light absorption dominates in the range $x < 1$ and $Q_a \propto x^1$, and, since $\sigma_a \propto Q_a \cdot r^2 \propto r^3$, it is proportional to the volume.

There is a simple physical explanation for the completely different dependences of the absorption and the scattering on particle radius (Moosmüller et al., 2009). A particle consists of many dipoles, which are vibrating due to the incident electric field and emit radiation themselves. For $x \ll 1$, the particle is small compared to the wavelength, so all dipoles vibrate in phase, and thus the emitted dipole radiation is in phase – that is, coherent (Shifrin (1951) gives a more restrictive range, $|m \cdot x| \ll 1$). The far field of the dipoles simply can be obtained by adding the field of all dipoles. The number of dipoles is proportional to the volume of the

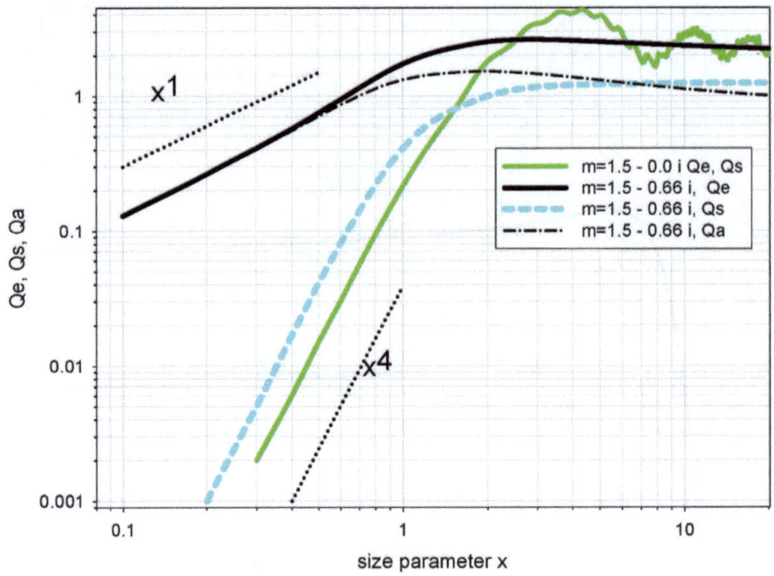

Fig. 6.8. Efficiency factors for extinction scattering and absorption as a function of the size parameter. In the log-log plot, the proportionality to x for absorption and to x^4 for scattering is clearly visible.

particle containing the dipoles; this is the electric field strength and is proportional to the volume. The light flux is proportional to the square of the electric field strength, so the scattering coefficient must be proportional to r^6. In the case of light absorption, each dipole removes a fraction of the incident light flux and, since the particle is small compared to the wavelength, all dipoles are exposed to the same field strength. Thus, the absorbed light flux is proportional to the number of dipoles, which is proportional to the volume of the particles, namely $\sigma_a \propto r^3$.

6.7 Influence of the refractive index on the absorption coefficient and related properties

6.7.1 Absorption coefficient

In Table 6.5, we have seen that a variety of refractive indices for black carbon have been measured, the real part ranging from 1.3 to 2.4 and the imaginary part ranging from 0.1 to 1.4. Bulk material with these refractive indices all would look completely black, but still the absorption coefficient of spherical 30 nm spheres can vary by a factor of 25 (see Fig. 6.9). With increasing value of k, the absorption increases; on the other hand, with increasing value for the real part n, the absorption coefficient decreases. The strong dependence of the absorption coefficient on the refractive index makes all derived properties, such as specific absorption cross-section or single-scattering albedo, dependent on the choice of the refractive index, which usually is not well known.

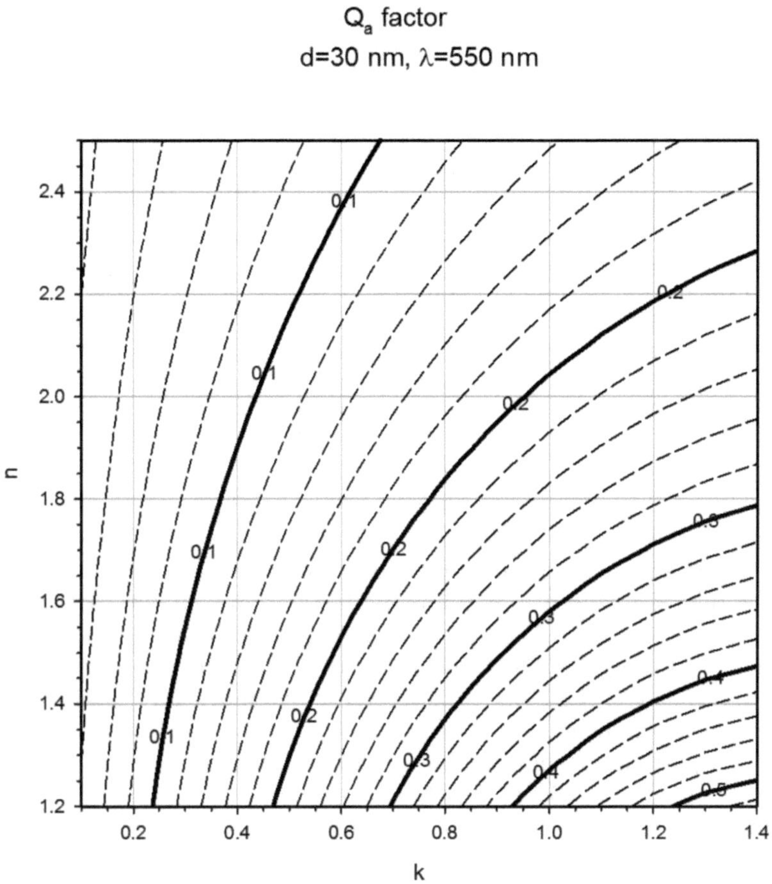

Fig. 6.9. Dependence of the absorption efficiency on the refractive index. Spherical particles with a diameter of 30 nm were assumed.

Let us now consider some hypothetic mixtures of black carbon with a transparent substance: black carbon with void; black carbon with organic carbon, such as partly burnt fuel; ammonium sulfate mixed with black carbon. For black carbon, we use the refractive index $2.102 - 0.755 \cdot i$ – a value obtained for soot carbon (Stagg and Charalampopoulos, 1993; see considerations above). We assume a homogeneous mixture and the Maxwell Garnett mixing rule, with all the reservations about homogeneous mixtures, which may not apply in practice. The characteristics of the mixtures are listed in Table 6.6.

In Fig. 6.10, the absorption efficiency factors are shown for the various mixtures. The curves for black carbon with 30% and 50% void, filled either by air or organic carbon, are almost identical. For black carbon with 50% organic carbon, the curve is below the other curves, although the organic carbon is also absorbing. This is because of the larger real part of the refractive index compared to air. As can be seen in Fig. 6.9, an increase in the real part causes a decrease in the absorption. The mixtures of black carbon with ammonium sulfate behave almost as expected for size parameters below 1: a 1% per volume mixture of black carbon has about

Table 6.6. Characteristics of mixtures used for Figures 6.10, 6.11, and 6.14.

Mixture	Refractive index at 550 nm	Density (kg/m^3)
black carbon	$2.102 - 0.775i$	1,850 (Janzen 1979)
70% BC + 30% void	$1.769 - 0.565i$	1,295
50% BC + 50% void	$1.551 - 0.426i$	926
70% BC + 30% organic C	$1.903 - 0.554i$	1,580
50% BC + 50% organic C	$1.773 - 0.410i$	1,400
100% organic C (Kim et al., 2014)	$1.47 - 0.04i$	950 (assumed)
Ammonium sulfate	$1.559i$	1,726
99% $(NH_4)_2SO_4$ + 1% BC	$1.534 - 0.0063i$	1,727
98% $(NH_4)_2SO_4$ + 2% BC	$1.541 - 0.0127i$	1,728
95% $(NH_4)_2SO_4$ + 5% BC	$1.560 - 0.0319i$	1,732
90% $(NH_4)_2SO_4$ + 10% BC	$1.592 - 0.0643i$	1,738

1.4% light absorption of the pure carbon. This is an indication of the absorption enhancement for an internal mixture. For large size parameters, the mixtures with the transparent substance have the same absorption efficiency as black carbon. This means that, for example, an ammonium sulfate mixed with 5% black carbon of size parameter 20 ($d = 3.5$ μm at $\lambda = 550$ nm) absorbs as much light as a black carbon particle of the same size. Considering the high bulk absorption coefficient of black carbon ($\alpha = 1.77 \times 10^7$ m^{-1} for the refractive index $2.1 - 0.775i$) and, using geometric optics argumentation, 90% of the light is absorbed in a layer of 0.13 μm of the 3.5 μm particle. For a 5% homogeneous mixture of ammonium sulfate and black carbon, the distance is 3.1 μm. So, using simple geometric optics arguments, one can argue that the black carbon particles absorb most of the light in the first 0.13 μm; the remaining part of the particle does not contribute much to absorption. The mixed particle, due to the lower bulk absorption coefficient, absorbs the light as well, but within the whole particle.

The effectiveness of the absorption by particles can be described by the mass absorption coefficient, which is the ratio of the absorption coefficient σ_a and the aerosol density $\rho_{aerosol}$, namely mass of particles per volume of gas (also called mass concentration). More systematically, the mass absorption coefficient is called the specific cross-section. It unit is $\left[\frac{\sigma_a}{\rho_{aerosol}}\right] = \frac{m^2}{g}$. For an aerosol consisting of monodisperse particles, it is obtained by $\frac{\sigma_a}{\rho_{aerosol}} = \frac{3Q_a}{4r\rho_{aerosol}}$. An overview of the specific cross-section of the considered mixtures is shown in Fig. 6.11 (left-hand part). The strong dependence on the particle size is easily recognized. The proportionality between mass and absorption coefficient is immediately evident by the horizontal line for particles less than $\approx$0.1 μm. For sizes above 0.3 μm, the specific absorption cross-section rapidly decreases, since most of the absorption takes place in the front part of the particle. black carbon with voids filled with air has a higher specific absorption cross-section; this is not the case for voids filled with organic carbon, since the transparent organic carbon contributes to the mass of the particles compared to voids filled with air. The absorption in the particles is caused by the black carbon, so it is of interest to investigate the

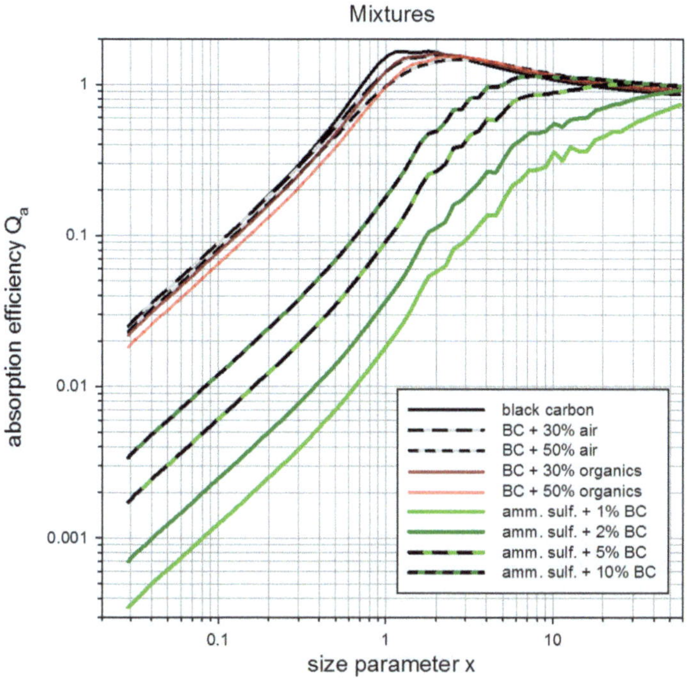

Fig. 6.10. Absorption efficiency of spheres consisting of a homogeneous mixture of absorbing and non-absorbing substances. The refractive indices of the mixtures are listed in Table 6.6.

specific absorption coefficient based on the mass of black carbon. This can be seen in Fig. 6.11 (right-hand part). Now the mixtures of the transparent ammonium sulfate have an even higher absorption per gram of black carbon compared to pure black carbon. This absorption enhancement is pronounced in particle sizes between 0.2 and 2 μm. A primitive explanation for this effect is that, since the black carbon is distributed quasi-homogeneously in a large particle, the light passing through the particle encounters many possibilities to be absorbed.

The trend shown in the figures is the same for the various values of the refractive indices available in the literature; the absolute values are strongly influenced by the choice of refractive index. One example shall demonstrate this. The specific absorption cross-section for 30 nm particles using the refractive index $(2.102 - 0.775i)$ and density (1,850 kg/m^3) as proposed by Janzen (1979) is obtained as 4.2 m^2/g. Using instead $m = 1.7 - 1.0i$ and $\rho = 1{,}740$ kg/m^3 (Shaddix and Williams, 2009) results in 7.7 m^2/g. This demonstrates clearly that the proper choice of the (usually not well known) refractive index is crucial.

For the particles smaller than 100 nm, an average specific absorption cross-section of 5–6 m^2/g seems to be an average value. Values of this magnitude are reported in the literature, although sometimes values as high as 15 m^2/g are reported (for a list see, e.g., Horvath, 1993, Table 2), which may be due to the exaggeration of the measured absorption coefficient by the method used (see below). Since the absorption coefficient and thus the specific absorption depend on wavelength,

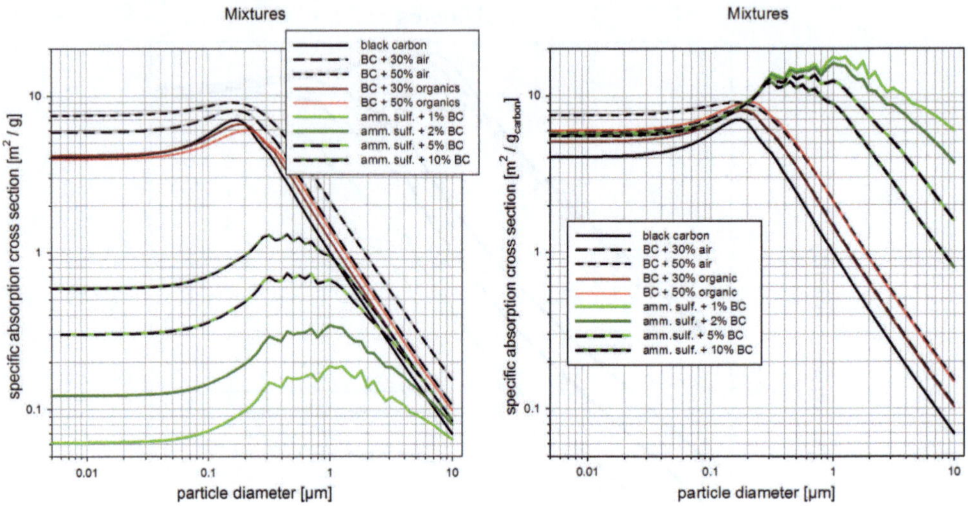

Fig. 6.11. Specific absorption cross-section of spheres consisting of a homogeneous mixture of absorbing and non-absorbing substances. In the left-hand graph, the absorption coefficient is divided by the mass of the particles. On the right-hand side, only the mass of the light-absorbing carbon is taken. The refractive indices of the mixtures are listed in Table 6.6. The wavelength is 550 nm.

also the wavelength used should be specified. Shaddix and Williams (2009) have used a dimensionless absorption (and extinction) coefficient, K_a, which is defined as $K_a = wavelength \cdot density_{soot} \cdot specific\ absorption\ coefficient$ or $K_a = \dfrac{\lambda \cdot \sigma_a}{\nu_{soot}}$ with ν_{soot} the volume of the suspended soot particles per unit volume of gas. Since $\sigma_a \propto \lambda^{-1}$, the product $\lambda \cdot \sigma_a$ is independent of wavelength, at least for particles below 30 nm. The soot particles consist of agglomerates of particles $\leq$30 nm but, since the absorption of soot aggregates is additive (Sorensen, 2000), this also holds for larger sizes of agglomerates. Using the value of $K_a = 7$ and $\rho_{soot} = 1{,}740$ kg/m^3, a specific absorption cross-section is obtained as 7.31 g/m^2, which is in agreement with our theoretical considerations. Also, this is in almost perfect agreement with the suggestion of 7.5 ± 1.2 m^2/g by Bond and Bergstrom (2006).

6.7.2 Coated sphere

A particle consisting, for example, of an absorbing core surrounded by transparent material can easily be formed in the atmosphere by condensation of a low-volatility vapor (such as α-pinene) on existing particles; cloud drops with an absorbing core are another example. The Mie solution for spherical particles can be expanded to a compound sphere consisting of spherical core, surrounded by a spherical shell, having a different refractive index (Bohren and Huffman, 1983, pp. 181–3, computer program pp. 483–9). The coated sphere absorbs more light than the absorbing center would without coating. This increase in absorption can ad hoc be explained by the 'lensing effect' (Rubinowitz, 1920; Bond et al., 2006). The curved transparent shell acts as a convex lens, concentrating the light onto the absorbing core, which thus absorbs more light than without transparent shell. But, for a lens, geometric optics requires that its diameter is considerably larger than the wavelength of light; for geometric optics (e.g., Born and Wolf, 1999), the requirement is $\lambda_0 \to 0$, thus the wavelength must be much smaller than the particle size. For example, for a 500 nm particle, this would require light with $\lambda < 100$ nm, which rarely is the case in the atmosphere, although the requirement $\lambda \ll d$ is fulfilled for the absorption enhancement of dirty clouds, having droplets with a diameter of 10 μm around an absorbing core. But still the enhancement of absorption exists for particles much smaller than 500 nm, so the lensing effect is a useful oversimplification, but not a profound argument.

As an example, calculations have been performed for a spherical particle with a diameter of 78 nm, consisting of soot and surrounded by a transparent shell. The increase in absorption coefficient as a function of shell thickness is shown in Fig. 6.12. With a shell thickness of 100 nm, the absorption coefficient doubles. An even thicker shell increases the absorption coefficient by a factor of up to four. This is important, for example, for cloud droplets formed on a light-absorbing cloud condensation nucleus.

The absorption increase due to coating was experimentally studied first by Sabbagh (2005). Carbon particles were produced by a low-voltage spark generator (Horvath and Gangl, 2003). For coating with paraffin and other waxes, a method is used which was first published by LaMer et al. (1943) and optimized by Prodi

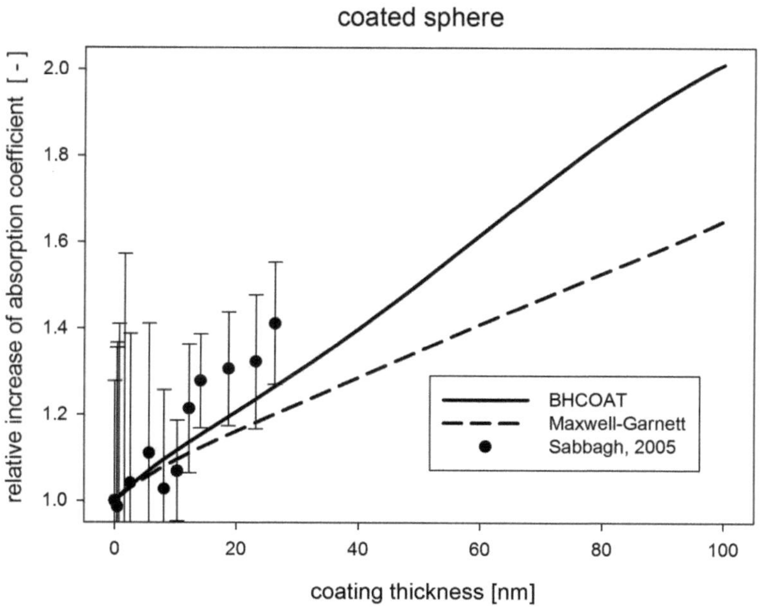

Fig. 6.12. Increase in the absorption coefficient by coating. A spherical soot particle is coated by a transparent substance. The calculated increase in the absorption coefficient is given by the solid line. Assuming the particle to be a homogeneous mixture of core and coating yields an increase given by the dashed line. The experimental data points are from Sabbagh (2005).

(1972). The carbon aerosol is mixed with vapor of wax obtained by bubbling air through liquid wax, and heated to a temperature above 400°C, making sure that there are no wax particles, and only wax vapor is present. Upon slow cooling, the wax condenses on the existing carbon particles. The coating thickness can be varied by adjusting the temperature of the liquid wax. The carbon particles of the spark generator have a fractal structure but, when coated with paraffin, the surface tension compacts the particle. In order to determine the size of the compacted primary particles, in a separate experiment, the paraffin is removed by a thermodenuder, leaving a spherical compact carbon particle. The size of the coated and uncoated particles is measured with a differential mobility analyzer (Reischl, 1991). The absorption coefficient of the particles is determined by the difference method. The extinction coefficient is measured with a White Cell (White, 1976) adapted for aerosols (Poss and Metzig, 1987); the scattering coefficient is obtained by an integrating nephelometer, which is calibrated with transparent particles, the scattering coefficient of which is obtained with the White Cell. The absorption coefficient of the aerosol simply is the difference of the measured extinction coefficient and scattering coefficient. Data points are shown in Fig. 6.12, and experimentally confirm the increase in absorption by coating with a transparent substance.

For curiosity, the increase in absorption was also calculated by assuming a homogeneous mixture of core and shell and using the refractive index obtained by the Maxwell Garnett method. This is also shown in Fig. 6.12. Obviously, the conditions for homogeneous mixing are not fulfilled, but still an absorption increase is obtained, but it is smaller compared to the exact calculation.

An instructive calculation can be done for the following case: assume a transparent particle with an absorbing shell. Obviously, the thicker the shell, the more light will be absorbed by the particle. In Fig. 6.13, we have assumed a spherical particle having a fixed outer diameter of 1 μm and a variable shell – that is, a shell

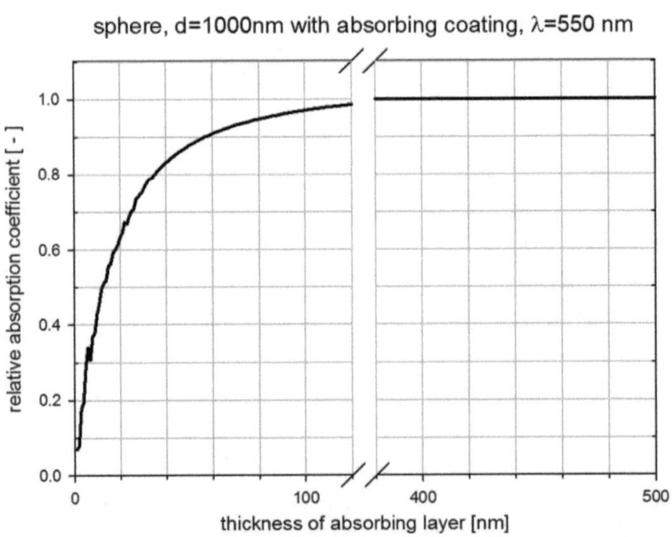

Fig. 6.13. Increase in absorption coefficient of a transparent sphere covered by a layer of absorbing material of variable thickness.

thickness of 0 nm means a transparent particle of 1 μm and a shell thickness of 500 nm is a 1 μm particle completely made of the absorbing material. The absorption coefficient of the coated particle is given relative to the value obtained for the particle made of completely absorbing material. It is evident that already a layer of thickness of 13 nm accounts for half of the absorption of a particle completely consisting of absorbing material; for a layer of 100 nm, the particle absorbs practically as much as the homogeneous particle. An atmospheric particle having small soot particles deposited on the surface can be thus considered a strongly light-absorbing particle.

6.7.3 Single-scattering albedo

Single-scattering albedo is the ratio of the scattering coefficient, σ_s, to the extinction coefficient, σ_e, or $\varpi = \dfrac{\sigma_s}{\sigma_e} = 1 - \dfrac{\sigma_a}{\sigma_e}$. It is a measure of the importance of light absorption. Non-absorbing particles have $\varpi = 1$ and, for strongly absorbing nanoparticles, $\varpi \ll 1$. We have needed the single scattering albedo for the determination of the radiance of a layer of aerosol and it plays an important role in atmospheric climate modeling. For particle sizes below 0.1 μm, the scattering coefficient is $\propto d^6$ and the absorption is $\propto d^3$, thus the single-scattering albedo is size-dependent and obviously the material is important, too. Figure 6.14 shows the dependence on both size and material. For sizes below 0.1 μm, the single-scattering

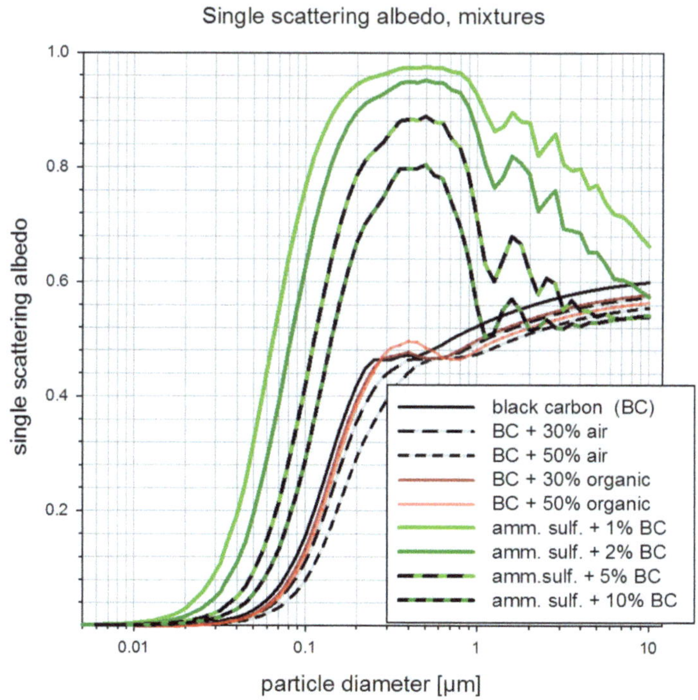

Fig. 6.14. Single-scattering albedo of light-absorbing particles.

albedo goes steadily toward zero for all substances. The mixtures with transparent substances have a high ϖ between 0.2 and 1 μm, because scattering is most efficient in this range; for larger sizes, the scattering is less efficient and, in the larger particles, there is more 'room' for absorption. The atmospheric aerosol usually is a mixture of light-absorbing and transparent particles. The absorption coefficient is determined by the absorbing particles, whereas the extinction coefficient usually is dominated by the transparent particles. Obviously, the more light-absorbing particles there are, the smaller the single-scattering albedo. In Table 6.7, the single-scattering albedo of an aerosol consisting of $d = 30$ nm black carbon particles and $d = 500$ nm ammonium sulfate particles is given.

Table 6.7. Single-scattering albedo of an aerosol consisting of absorbing and transparent particles.

Mass BC	0	5	5	5	5	5	$\mu g/m^3$
Mass ammonium sulfate	5	5	10	50	100	0	$\mu g/m^3$
	1.00	0.50	0.67	0.91	0.95	0.01	

6.7.4 Volume-scattering function

The influence of light absorption on the volume-scattering function shall be demonstrated in a few examples (Fig. 6.15). For these calculations, particles with a log-normal mass size distribution with a mean diameter as indicated were assumed. For all curves, the particles had a mass concentration of 100 $\mu g/m^3$. For particles with diameters of 30 nm, the scattering function for black carbon and transparent ammonium sulfate are small and very similar to the $(1 + \cos^2 \theta)$ dependence of the Rayleigh scatter. Surprisingly, the light-absorbing particles scatter more light than the transparent particles. But this is in accordance with Fig. 6.8, where Q_s for black carbon is about a factor of five larger than for the transparent substance. In that case, the particles are small enough to scatter light efficiently. For a size of 600 nm, black carbon particles scatter less compared to the transparent counterpart, especially in the backwards direction, since considerable absorption takes place in the volume of the particle. For the homogeneous mixture of transparent material with 5% absorbing substance, the scattering at 100° to 140° is similar to the black carbon, but the backscattering strongly increased. The external mixture scatters light almost identically to the transparent material.

6.8 Wavelength dependence of the optical properties

Generally, it is not explicitly mentioned that all optical properties depend on wavelength; the λ^{-4} dependence of the light scattered by particles in the Rayleigh size range and for air is the best known. The wavelength dependence of an optical property can be caused by the refractive index varying with wavelength and/or the size of the particle.

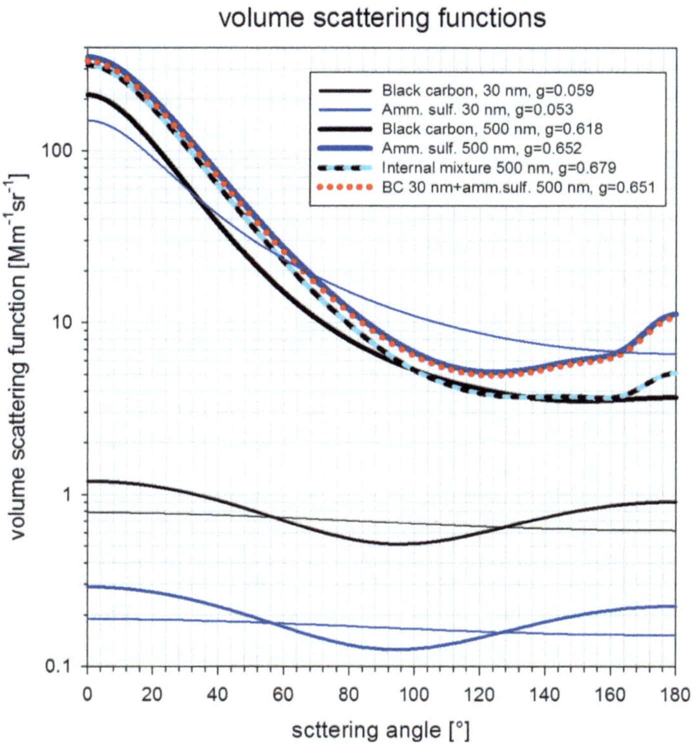

Fig. 6.15. Volume-scattering function of aerosols consisting of spherical particles. The asymmetry parameter is given as a parameter. The thin lines are derived from the Henyey–Greenstein Phase function, having the same asymmetry parameter and the same scattering coefficient.

6.8.1 Absorption coefficient

The refractive index of black carbon is almost independent of wavelength, so any wavelength dependence is a size effect. For a size parameter <0.3 – that is, a particle diameter less than 50 nm for a wavelength of 550 nm – the absorption efficiency $Q_a \propto x = \dfrac{2r \cdot \pi}{\lambda}$ and, since $\sigma_a = N \cdot r^2\pi \cdot Q_a$, it follows that $\sigma_a \propto \dfrac{r^3}{\lambda}$ – that is, the absorption coefficient has a wavelength dependence proportional to λ^{-1}. For particles larger than 50 nm, namely $x > 0.3$, the curve in the log-log presentation of Fig. 6.8 first becomes steeper and then flattens, eventually becoming horizontal, thus having little or no wavelength dependence. This is shown in detail in Fig. 6.16: a log-normal size distribution for the particles has been used. As the simplest case, the particles were assumed to consist of a material having a complex refractive index independent of wavelength (black line). For $d = 30$ and 100 nm, the absorption coefficient depends on wavelength close to λ^{-1}. For particles having a complex refractive index with some variation of refractive index, such as propane soot (Stagg and Charalampopoulos, 1993, Fig. 3), the same is true; for the wavelength of 400 nm, the absorption coefficient is slightly higher, since the real

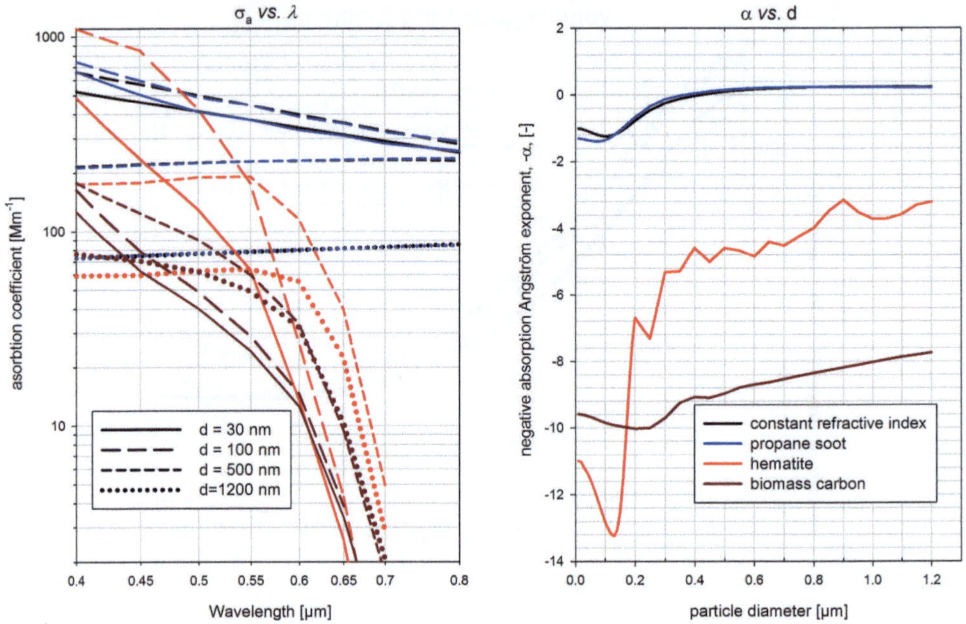

Fig. 6.16. Wavelength dependence of the absorption coefficient for particles of various materials and sizes.

part of the refractive index is lower. The absorption Ångström exponent is close to +1.

black carbon particles emitted by diesel engines are a major source of urban soot particles and are in this size range. A wavelength dependence of the light absorption close to λ^{-1} was first been observed, for example, in Santiago de Chile (Horvath et al., 1997). For larger particle sizes (500 and 1,200 nm), the wavelength dependence is weak as expected and the absorption Ångström exponent approaches zero.

A strong wavelength dependence of real and or imaginary parts of the refractive index overrides the size effect at least in part; this is shown for hematite (for refractive index, see Fig. 6.6). For particles less than 100 nm, the absorption coefficient varies by more than a factor of 500 between 0.4 and 0.65 μm, caused by the strong decrease of the imaginary part of the refractive index. For particle sizes of 500 and 1,200 nm, the curve is almost flat up to $\lambda = 550$ nm. This is caused by the larger size of the particles: as we have seen above (coated sphere), only the outer part of the shell absorbs the light and, if thick enough, even for a smaller imaginary part of the refractive index, all light is absorbed, thus there is no dependence on wavelength. Obviously, this is a particle size effect.

For biomass carbon (imaginary part by Kirchstetter et al. (2004), real part assumed as 1.7), the situation is similar. The strongest wavelength dependence is observed for 30 nm particles (but an unlikely size for biomass particles); once the particles reach a size of 500 nm or more, the dependence on the wavelength is weak. This can also be seen in the right part of the figure, which depicts the

Ångström absorption exponent. For the black carbon particles, with a refractive index almost independent of wavelength, it is around $\alpha = +1$ for sizes below 30 nm, with $\alpha \rightarrow 0$ for larger sizes. For particles consisting of hematite, the wavelength-dependent refractive index causes a wavelength dependence with α around +12 for particle sizes below 100 nm, rapidly increasing to +4 for larger sizes. The same is also true for the biomass carbon. Therefore, the wavelength dependence of the absorption coefficient is influenced by both the refractive index and the size.

A mixture of black carbon and black carbon and possibly transparent carbonaceous substances is possible, such as in biomass fire aerosol particles. The absorption of a small spherical black carbon particle having a transparent coating is enhanced; the absorption Ångström exponent can be increased due to coating from 1 to 1.6 (Lack and Cappa, 2010). When the coating consists of black carbon, the enhancement of the absorption of the core is less. On the other hand, the Ångström exponent is influenced by the wavelength dependence of the refractive index of the coating.

The single-scattering albedo is defined as $\varpi = \dfrac{\sigma_s}{\sigma_e} = \dfrac{\sigma_s}{\sigma_s + \sigma_a}$ and, for a fixed particle size, this can be also written as $\varpi = \dfrac{Q_s}{Q_s + Q_a}$.

For small light-absorbing particles (size parameter <0.5; see Fig. 6.8), Q_s is much smaller than Q_a and thus can be neglected in the denominator, so $\varpi = Q_s/Q_a$. Since $Q_a \propto x$ and $Q_s \propto x^4$, it follows that $\varpi \propto x^3$, with $x = \dfrac{2r\pi}{\lambda}$, for a fixed particle size $\varpi \propto \lambda^{-3}$; that is, for small strongly absorbing particles, the (small) single-scattering albedo decreases with wavelength. This is clearly evident in Fig. 6.17 (left part).

Usually, the atmosphere contains both small light-absorbing particles and purely scattering particles. The absorption coefficient of the light-absorbing particles is proportional to λ^{-1}, and the wavelength dependence of the extinction coefficient in many cases can be described by the Ångström relation $\sigma_e \propto \lambda^{-\alpha}$ with α frequently between 1 and 2, so $\dfrac{\sigma_a}{\sigma_e} \propto \lambda^{\alpha-1}$. Using $\varpi = 1 - \dfrac{\sigma_a}{\sigma_e}$, the single-scattering albedo is independent of wavelength for $\alpha = 1$ or decreasing with wavelength for $\alpha > 1$. The wavelength dependence of the single-scattering albedo for a 5 $\mu g/m^3$ light-absorbing aerosol with diameter of 30 nm externally mixed with an aerosol consisting of 95 mg/m^3 transparent particles with mass mean diameters of 300, 500, and 800 nm is shown in Fig. 6.17. As expected, there is little dependence on wavelength, and it is mainly determined by the wavelength dependence of the extinction coefficient of the ammonium sulfate particles. The 300 nm $(NH_4)_2SO_4$ particles have a stronger decrease in the scattering coefficient with wavelength as the absorption coefficient, so the single scattering albedo decreases with wavelength, whereas the scattering coefficient of the 800 nm particles is rather flat, so the single-scattering albedo increases. For completeness, an internally mixed aerosol, consisting of homogeneously mixed particles of the same quantities of light-absorbing and transparent substance, is also shown. Besides the increase in absorption due to internal mixing (thus lower single-scattering albedo), the trend is the same.

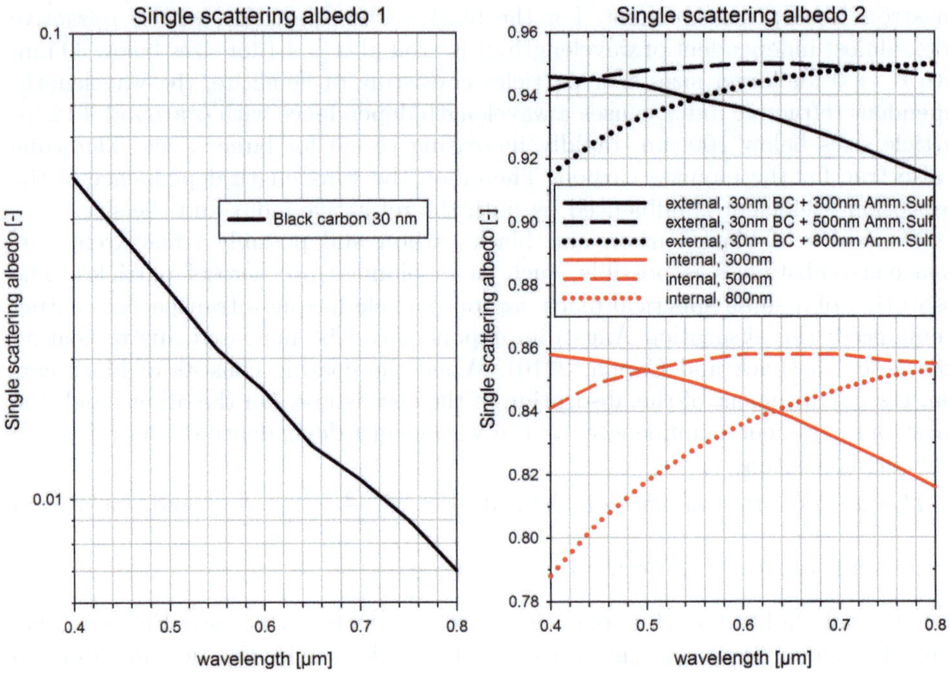

Fig. 6.17. Wavelength dependence of the single-scattering albedo. Left: black carbon spheres, with a diameter of 30 nm; right: external mixture of 30 nm black carbon and ammonium sulfate.

6.9 Multiple scattering

So far, we have considered the radiation scattered in one volume element reaching an instrument or an observer. But it is highly likely that a photon scattered in one volume element is scattered in another volume element or even many times before reaching the observer. This multiple scattering can be modeled and various techniques solving the radiative-transfer equation are in use. Here, a few examples for the effect of multiple scattering will be shown; the tables, from which the graphs are derived, as well as multiple-scattering theory and computational methods can be found in Van de Hulst (1980).

6.9.1 Radiance of the horizon

An aerosol layer, infinitely extended in the horizontal direction, of given optical depth is illuminated by the Sun having a zenith distance of 0°. The radiance of this layer is plotted in Fig. 6.18 as a function of the optical depth of the layer of aerosols having various phase functions and absorption.

Let us consider first an aerosol layer consisting of non-absorbing particles. For a vertical optical depth of 0.05, the multiple scattering amounts to 10% of the totally scattered light. For $\tau = 0.25$, multiple scattering amounts to 24% and 32%

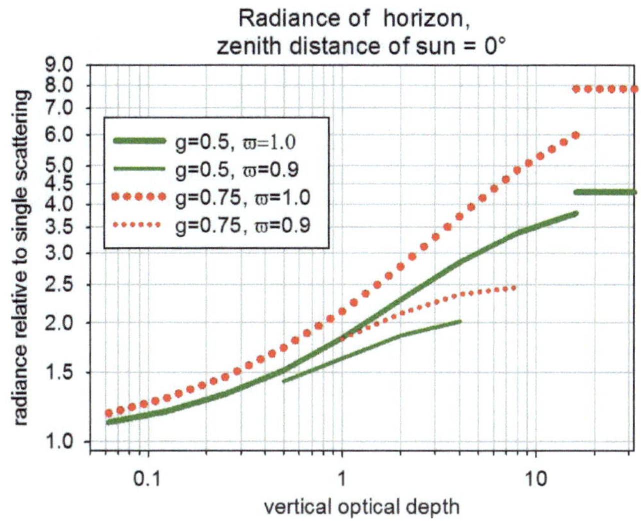

Fig. 6.18. Radiance of the horizon for an aerosol layer. The aerosol consists of particles having a Henyey–Greenstein phase function of asymmetry parameter g and a single-scattering albedo ϖ. The radiance for a layer with infinite optical depth is given as a short horizontal line on the left-hand side of the figure.

for the asymmetry parameters of $g = 0.5$ and 0.75, respectively. An atmosphere with $\tau = 0.25$ would have an extinction coefficient of 166 Mm^{-1} for a mixing height of 1.5 km, corresponding to 23.5 km visibility. The contribution of multiple scattering increases dramatically with optical depth; for example, 74% and 83% (for $g = 0.5$ and 0.75) of the observed radiance are due to multiple scattering for $\tau = 16$, in which case the visibility would be 366 m.

For a light-absorbing aerosol and single scattering, the scattered light is reduced by the factor ϖ compared to its non-absorbing counterpart having the same phase function. That is, a contribution of 10% light absorption reduces the radiance by 10%. For multiple scattering, it must be considered that the 10% scattering loss has to be applied several times, so the radiance is considerably smaller, as is evident in Fig. 6.18 from the curves for $\varpi = 0.9$ and optical depths larger than 1.

6.9.2 Reflected and transmitted light

Two examples shall demonstrate the enhancement of light absorption by multiple scattering. Let us consider a layer of aerosol of given optical depth, illuminated by the Sun in the zenith and observed at 60° to the vertical. Example 1: The light scattered upwards from the layer is shown in the left-hand part of Fig. 6.19. An increase in the optical depth increases the radiation scattered upwards, for particles both without and with absorption. But it is also evident that absorption effectively reduces the scattered light. For example, for an optical depth of 8, the light scattered upwards at 60° zenith distance for an asymmetry parameter of 0.75 is reduced by a factor of 5.5 by only 20% light absorption. Example 2: In the right-hand part of Fig. 6.19, the light transmitted by the layer downwards is shown again at 60° zenith distance. The diffuse transmittance coefficient of the layer at first increases with increasing optical depth, but decreases after reaching a maximum. This is caused by the increased upward scattering, which leaves less light for downward propagation. This is a common phenomenon: thick cumulus clouds look dark, when viewed from below, although they look snow-white when viewed from above, such as from an airplane.

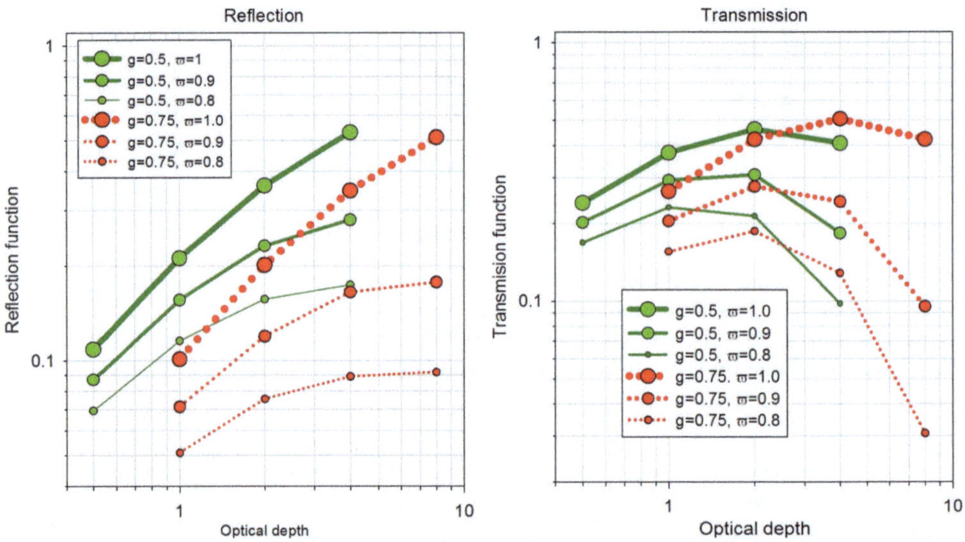

Fig. 6.19. Reflection and transmission function for a layer of aerosol. Reflection and transmission functions are proportional to the observed radiance and taken from Table 35 of Van de Hulst (1980). For both cases, the light source is located in the zenith. Left: observation from above at 60° zenith angle; right: observation from below at 60° zenith angle. The asymmetry parameter g of the Henyey–Greenstein Phase function and the single-scattering albedo ϖ is given as a parameter.

These examples clearly show that multiple scattering can greatly increase the effect of absorption. Typically, this is the case for clouds, where a small amount of light-absorbing material has enhanced absorption not only due to coating, but also due to multiple scattering. An aerosol layer above snow receives more reflected light from the ground, thus absorption effects are larger. Old historic buildings have been exposed to pollution for a long time and the depositing of black carbon causes their dark appearance. Additionally, the soot deposit in combination trace metals acts a catalyst for SO_2 or NO_2 oxidation, aiding the production of gypsum from the limestone surface, causing black crusts and eventually considerable material loss. Previously, coal combustion was the major cause, but now diesel engine emissions are most responsible (Mansfield et al., 1991).

6.10 Light-absorbing particles in the atmosphere

Three main groups of particles which absorb light can be found in the atmosphere: (i) mineral dust particles containing iron oxides, (ii) black carbon, and (iii) black carbon.

6.10.1 Mineral dust particles

On a global scale, mineral dust particles are the most abundant particles, with an estimated source strength of 600–2,400 Tg/y, with less than 20% of the mass at <2 μm (Lee et al., 2009). The main sources are deserts; more than half of the dust particles originate from northern Africa, followed by Asia and the Arabian peninsula. But also as a consequence of human activities, dry and bare soils considerably contribute to mineral dust. Mobilization of the particles is by wind and is due to a process similar to sand blasting, where saltating particles mobilize or generate smaller particles. Many substances are contained in mineral particles; most of them do not absorb light. Iron-containing minerals are the reason for the higher imaginary part of some mineral dust. Particles larger than 10 μm are more of local importance, due to rapid settling, but smaller mineral particles can even be found in other continents. For example, observations at El Arenosillo (37.10°N, 6.70°W) at least 700 km distant from the Sahara recorded periodic Sahara dust intrusions, with high absorption coefficients (Mogo et al., 2005).

6.10.2 black carbon

black carbon is a carbonaceous material with black appearance; other expressions are in use, with sometimes slightly different meaning: soot, elemental, or graphitic carbon. The black color is caused by graphitic platelets mainly forming the particles (Wentzel et al., 2003). The imaginary part of the refractive index is significantly non-zero and almost independent on wavelength. black carbon is formed in all high-temperature combustion processes of gaseous hydrocarbon fuels, such as flaming biomass fires, fires in furnaces, or internal combustion engines. Global emissions are estimated as ≈8 Tg/y, with 38% from fossil fuel, 20% from biofuel, and 42% from open burning (Bond et al., 2004). Energy-related emissions in 2000 are estimated as ≈4.5 Tg/y, with 33% from biomass and 77% from fossil fuel combustion (Bond et al., 2007). In pre-industrial times, the source of black carbon was biofuel; after 1880, coal was the main source; and, since 1995, diesel emissions have been the major source of black carbon globally.

Formation of light-absorbing particles during combustion

Soot is a product of any combustion of carbon-containing fuels and is a multi-stage process involving nucleation, coagulation, surface growth, aggregation, agglomeration, and obviously also oxidation. A detailed description of the soot-formation model can be found, for example, in Smooke et al. (2004). As a first step in combustion, oxidation and pyrolysis of the hydrocarbon fuel take place at high temperatures; in the lower part of the flame, CO, H_2, CO_2, $H_2O \cdot C_2H_2$, and radicals are present. A small fraction of the hydrocarbon fragments forms aromatic rings, forming PAHs. Nucleation, which is the collision of two PAHs, eventually evolves spherical nano-organic particles (D'Alessio et al., 2009) of a size between 1 and 5 nm with a polymer-like structure. These particles are liquid-like, almost transparent in the electron microscope, and upon coalescence remain spherical. The C/H ratio is 0.5 ... 1. The coagulation rate is low due to low sticking probability, so they survive for a long time even in the atmosphere. The density of the nano-organic particles is $\approx$1,200 kg/m^3. Further combustion, adding compounds from the gas phase in the flame by condensation and loss of hydrogen by oxidation, leads to a higher condensed ring aromatic or graphitic structure, but still organic functionalities are present on the surface. These particles have a density of $\approx$1,800 kg/m^3 and a H/C ratio of 0.05 ... 0.1. These particles agglomerate, forming the well-known chain-like cluster soot particles with sizes between 10 and 100 nm, and a fractal dimension of 1.8 ... 1.85 (Dobbins, 2009). This route of soot formation has been observed for all kinds of flames ranging from laminar diffusion flames to bituminous coal flames and diesel engines. In the terminology of combustion research, the nano-organic particles are called the condensation mode, whereas the soot particles are called accumulation-mode particles. As expected, the nano-organic particles considerably decrease in size upon thermo-desorption (from 6 nm at 20°C to 3.7 nm at 600°C), whereas the graphitic accumulation particles only exhibit a slight thermally induced shrinking in diameter (e.g., 39 nm $\rightarrow$ 37 nm; Maricq, 2009).

During atmospheric transformation and aging, such as with uptake of water, the chain-like cluster soot particles become more compact (Abel et al., 2003). Containing 85% to 95% carbon, 3% to 8% oxygen, and 1% to 3% hydrogen by mass, an atmospheric soot particle can be regarded as a complex polymer, with the ability to transfer electrons, and thus also absorb light (Andreae and Gelencsér, 2006). Most of the oxygen is on the particle's surface in various functional groups, increasing its chemical reactivity and also the wettability in contrast to pure graphite particles.

6.10.3 Brown carbon

Brown carbon is a light-absorbing carbonaceous material; its imaginary part of the refractive index strongly varies with wavelength (see, e.g., biomass carbon; Fig. 6.6). The stronger absorption in the blue makes it appear yellowish or brownish. Brown carbon is produced by smoldering (low-temperature, non-flaming) biomass fires but also in the initial stage of coal combustion, and it is mainly water-soluble. These pyrolysis products escape the flames and condense in the plume, forming secondary particles. The color is yellow to brown, and not black. But also humic/fulvic-like substances (HULIS) have similar optical properties. Water-soluble organic carbons consist of polycyclic-aromatic, acidic, and phenolic groups (Andreae and Gelencsér, 2006). Natural humic/fulvic substances are formed by polymerization from plant-degradation products and have a high degree of aromaticity. The light absorption causes the dark color of soil or swamps. Colored polymeric products may also form dienes in the presence of sulfuric acid (Limbeck et al., 2003), or the reaction of organic compounds in sulfuric acid particles (Hegglin et al., 2002). Thus, there are many routes to black carbon particles in the atmosphere.

The strong wavelength dependence of the absorption coefficient of black carbon (see Fig. 6.15, biomass carbon) and consequently the Ångström exponent well above 1 make it possible to attribute black and black carbon, using the spectral dependence of the absorption coefficient (Kirchstetter et al., 2004; Lack and Langridge, 2013). No estimates of global emissions of black carbon exist so far.

6.11 Measurement of light absorption

Light-absorbing particles have a distinct influence on the radiation in the atmosphere, so the exact knowledge of the absorption coefficient is desirable. Unfortunately, absorption of aerosol particles and scattering always occur simultaneously, so no direct method for measuring the light absorption by particles optically, using first principles, exists; therefore, only indirect methods can be used. A further complication is the usually low value of the absorption coefficient, which may be in the order of 100 Mm^{-1} or even considerably less. For attenuation measurements, a large distance (at least 100 m) is needed to be able to distinguish Φ from Φ_0. The various methods for measurement of the light-absorption coefficient can be divided into three groups, which will be discussed in the following: (i) subtraction method, (ii) integration of scattering, and (iii) use of effects of light absorption.

6.11.1 Absorption by the subtraction method

This is in principle a straightforward method: using the relation $\sigma_e = \sigma_a + \sigma_s$, the absorption coefficient is obtained by $\sigma_a = \sigma_e - \sigma_s$; thus, simultaneously measuring the extinction coefficient and the scattering coefficient of the aerosol gives the absorption coefficient by subtraction. The extinction coefficient can be determined, for example, with a White Cell (Fig. 6.20) using a multiply folded path of >100 m and adapted for aerosol measurements (Poss and Metzig, 1987), whereas the scattering coefficient is determined with an integrating nephelometer (Charlson et al., 1967; see Fig. 6.20). The difference of the two measured coefficients is the absorption coefficient. Since the absorption coefficient usually is an order of magnitude smaller than the extinction coefficient, σ_e and σ_s have a small difference, so the requirements for the accuracy of the determination σ_e and σ_s are high. Furthermore, care must be taken to do the appropriate correction for the truncation errors of the integrating nephelometer, which can be up to 10%.

The subtraction method has been successfully used for a laboratory calibration of the integrating plate technique (Horvath and Metzig, 1990). Since the light-absorbing aerosol has been generated in the laboratory, the measurement of both the extinction coefficient and the scattering coefficient could be done with very little noise of the data, giving a high accuracy of the absorption coefficient. This method has also been widely used in the AIDA soot campaign (Saathof et al., 2003; Schnaiter et al., 2005).

The White Cell (White, 1976) was originally designed for the spectroscopy of gases, so contamination of the mirrors by particles, and thus bias of the transmission measurement, was of minor importance. For atmospheric measurements, this is a point of concern. It is necessary either to frequently fill the cell with clean air

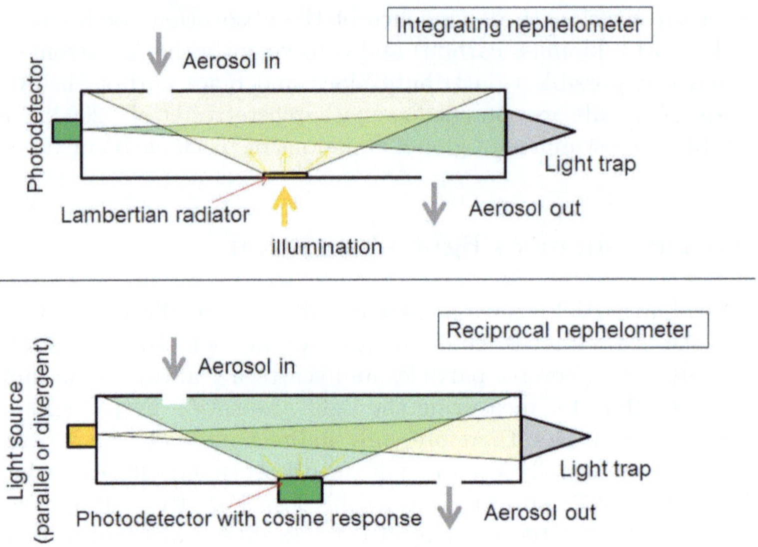

Fig. 6.20. Integrating and reciprocal nephelometer. The integrating nephelometer uses a Lambertian radiator as illumination of the measuring volume (yellow). By a series of stops (not shown), the photodetector only accepts light from a conical volume (green). The light scattered in the illuminated volume (bright green) is measured by the photodetector. On the left-hand side, forward scattering takes place; backscattering is on the right-hand side. Due to the cosine characteristic of the Lambertian radiator, the signal of the photodetector is proportional to the scattering coefficient (excluding extreme forward- and backscattering). In the reciprocal nephelometer arrangement, the light source and the detector are exchanged.

or to use filtered purge air to avoid deposits on the mirrors. Obviously, the bias due to particle deposits increases with the number of reflections. Virkkula et al. (2005) have designed an extinction cell, having a 3.3 m sampling volume with one reflection (optical path length of 6.6 m) using blue, green, and red LEDs as light sources, with a detection limit for the extinction coefficient of 5×10^{-6} m^{-1}.

An alternate method to measure the extinction coefficient is the cavity ring down principle: a high-quality optical cavity is formed by two mirrors with reflectivities larger or equal 0.99995, causing the light beam to be reflected thousands of times and thus achieving an optical path length of several kilometers for mirrors only a meter apart (Fig. 6.21) (Moosmüller et al., 2005). Light from a sub-nanosecond pulsed laser enters the cavity. The amount of light exiting the cavity is strongly influenced by the attenuation in the cavity. The decay in the light after the end of the pulse is used to determine the extinction coefficient of the aerosol present in the cavity. A detection limit of the extinction coefficient below 1 Mm^{-1} is possible.

In order to avoid artifacts due to sampling losses or different temperatures in the instruments measuring extinction and scattering, it would be desirable to measure the extinction and the scattering coefficient in the same volume. This can be achieved by equipping the extinction cell with a cosine sensor in the center of the housing, allowing determination of the scattering coefficient by the principle of the reciprocal nephelometer (Fig. 6.20) (Gerber, 1979). Unfortunately, the number of reflections in the folded extinction path could not be as high as in a single-use instrument, so the sensitivity was limited to an extinction coefficient of 100 Mm^{-1}.

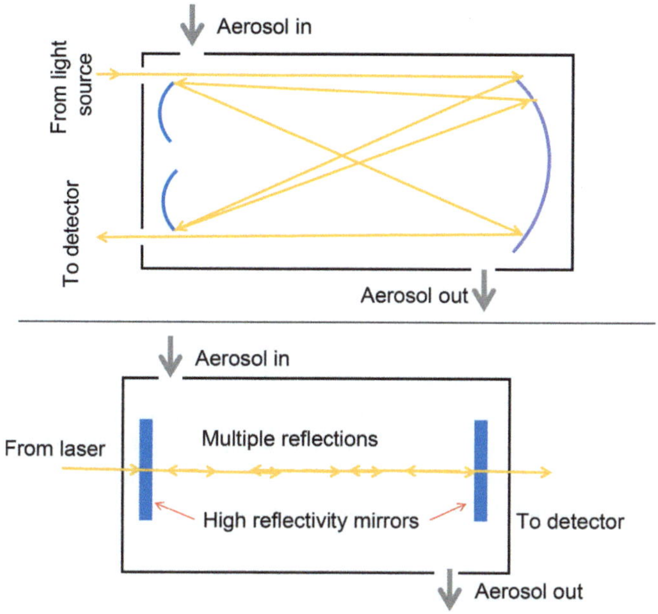

Fig. 6.21. *In situ* measurement of the extinction coefficient of the atmospheric aerosol. Top: White Cell. The mirrors are adjusted such that hundreds or more reflections take place, before the light leaves the cell, thus having a path of > 100 m. Bottom: cavity ring down principle. A light pulse, shorter than the distance between the high reflectivity, slightly concave mirrors, is reflected many times. The decay of the light transmitted through the exit mirror decreases after each reflection due to extinction. The time decay of the exiting light flux is a measure for the extinction coefficient of the aerosol in the cell.

6.11.2 Absorption by elimination of scattering

The loss of light flux through an aerosol is due to both absorption and scattering. With an extinction measurement, the sum of the two is measured. The question can be raised as to whether it is possible to design an experimental arrangement such that the transmitted light is only influenced by the absorption of the aerosol. This has first successfully been done by Fischer (1973) using an integrating sphere. The aerosol is sampled on a filter, and implicitly it is assumed that the absorptive properties of the aerosol are the same as in the airborne state. The filter is placed in the center of an integrating sphere (see Fig. 6.22).

The integrating sphere was originally designed to measure the light flux emitted by light sources, which have a variety of angular emission patterns. The interior of the sphere is coated with magnesium oxide or another diffusely reflecting material. Light is diffusely reflected on the walls of the sphere without loss and a diffuse

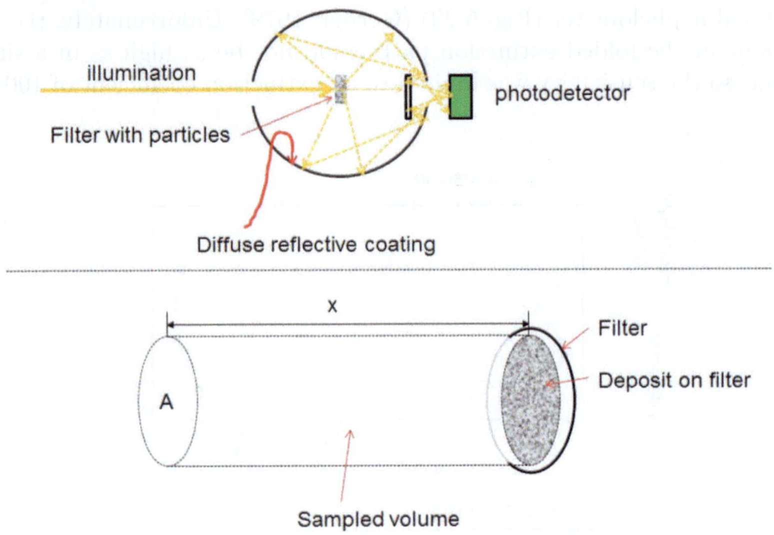

Fig. 6.22. Principle of the integrating sphere. Top: Light entering the sphere passes the filter and both transmitted and scattered light is reflected many times on the coating of the sphere, before reaching the photodetector. Bottom: Determination of the distance x for the exponential attenuation law.

light field is established. The photodetector is shielded by a plate also covered with diffuse reflecting material but, eventually, after many reflections, the light reaches the photodetector. The signal of the photodetector is proportional to the diffuse light field in the sphere. Let us assume a perfectly diffusing (non-absorbing) object put in the center of the sphere. All light incident on the diffusing object is scattered and establishes the diffuse light field in the sphere. Since all light is scattered by the diffusing object, thus no light is lost, the angular pattern of the diffusing object is not important. Let us now assume the diffusing object to be a filter (e.g., capillary disk filter) which is loaded with particles which do not absorb light. This may change the angular pattern of the scattered light but, again, no light is lost, so the same light flux is measured by the photo detector. If the particles on the filter absorb light, this obviously causes less flux reaching the photo detector, and this is exclusively due to light absorption. The light scattered by the particles and normally also causing attenuation is integrated in this optical arrangement so it is 'added' to the signal of the photo detector by the integrating sphere. Thus the ratio of the signal, Φ, measured with a loaded filter in the integrating sphere and the signal, Φ_0, with a clean filter can be interpreted as the transmission $T = \dfrac{\Phi}{\Phi_0}$ solely caused by absorption, namely $\dfrac{\Phi}{\Phi_0} = \exp(-\sigma_a \cdot x)$. The argument $\sigma_a \cdot x$ in the exponential function can be interpreted in two ways: (i) as the thickness of the deposit on the filter times the absorption coefficient of the material forming the deposit; it is difficult to introduce a thickness of the layer, since the particles can be deposited in a monolayer, with spaces in between, and then a thickness of the deposit makes no sense; and (ii) assuming the absorptive properties of the particles are the same as in the airborne state, one can treat the particles on the filter as if in the airborne state. The distance x then is obtained as the height of a cylinder with base A, which is the area of the deposit, and its volume is the volume of air passing through the filter. Therefore, if the deposit on the filter has an area of A and a volume V is sampled, then $x = \frac{V}{A}$, thus the absorption coefficient of the aerosol is $\sigma_a = -\frac{A}{V} \cdot \ln \frac{\Phi}{\Phi_0}$. Due to the diffusive light field in the integrating sphere, the scattered light is integrated. But a perfectly diffusing surface means many reflections, and also several passes of the light through the absorbing medium, so there is more absorption and using the simple formula above will result in a light-absorption coefficient which is too high. Therefore, corrections or empirical calibrations are needed.

Many other arrangements with the integrating sphere are possible. The filter can be placed at the entrance opening of the sphere, in which case the transmission through the filter plus the integrated forward scattering is determined (Hitzenberger et al., 1996). It is also possible to use a second integrating sphere to determine the backscattered light simultaneously and parallel or diffuse illumination of the filter (Pickering et al., 1992). It is also possible to dissolve the filter and position the solution in the center of the sphere (Hitzenberger et al., 1996).

A simplification of the integrating sphere is the integrating plate method (Lin et al., 1973) (see Fig. 6.23). Parallel light hits a capillary pore filter, with the particles on the opposite side of the light source. Part of the light is absorbed by the particles. Light scattered in the forward direction and the transmitted illumination reaches the opal glass, which has many scattering centers (thus its milky appearance), and integrates the forward-scattered light and the transmitted light. Part of the backscattered light is scattered by the opaque particle filter, which is caused by the capillary pores (diameter typically 0.2–0.5 μm), and in part also reaches the opal glass. Finally, light scattered by the opal glass in the backwards direction passes through the particle deposit and in part is scattered back to the opal glass. Thus, an integration similar to the integrating sphere takes place, with less emphasis on the backscattered light. The inventors of the method estimate an accuracy of the light-absorption measurement by this method by a factor of two; a laboratory calibration using the difference method (Horvath, 1997) showed that the absorption coefficient is determined as too high, and depends on the single-scattering albedo of the aerosol. For example, for $\varpi = 0.1$, 0.5, 0.8, 0.95, the absorption coefficient is measured too large by a factor of 1.25, 1.25, 1.4, and 2.8, respectively.

A variation in the integrating plate is the integrating sandwich (Fig. 6.23) (Clarke, 1982). The light source shines on an opal glass, causing several passes of the light through the loaded filter, thus increasing the sensitivity.

Fibrous filters (quartz, glass, or cellulose) also are very efficient in filtering and, due to the in-depth deposition, can be loaded by far more than the capillary pore filters. Since the fibers efficiently scatter light, a kind of integration of the scattered light takes place within the filter. Therefore, fibrous filters are optimal for continuous, automated measurement of the aerosol absorption. One example is the particle soot absorption photometer (PSAP; Bond et al., 1999) (Fig. 6.24). A filter band is clamped by the filter holder and air is sucked through one part, whereas the reference section remains unloaded. Light from an illuminated opal glass plate passes

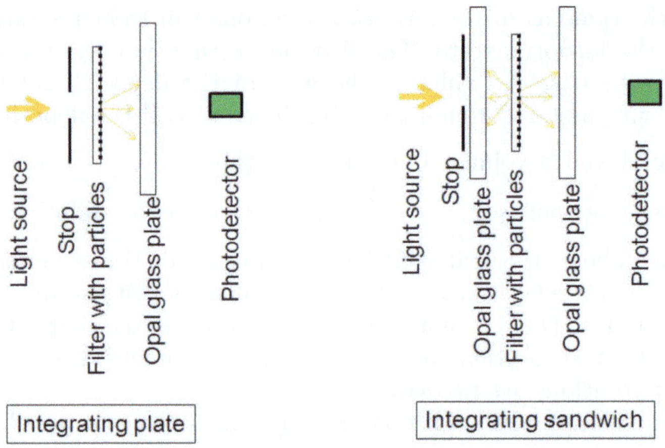

Fig. 6.23. Integrating plate and integrating sandwich. Left: integrating plate; the filter is illuminated by parallel light. The light leaving the filter is the transmitted light as well as the light scattered by the filter and the particles, and reaches the opal glass plate, which integrates most of the arriving light due to its many scattering centers. Right: Integrating sandwich; the filter is located between two opal glass plates.

through both the loaded and the reference sections. The decrease in the measured signal in relation to the reference signal is a measure of the light absorption of the deposited aerosol particles. When the signal is below a pre-determined limit, the filter holder is opened, the filter band is advanced, and a new measurement is started. The aethalometer (Hansen et al., 1984) operates in a similar way.

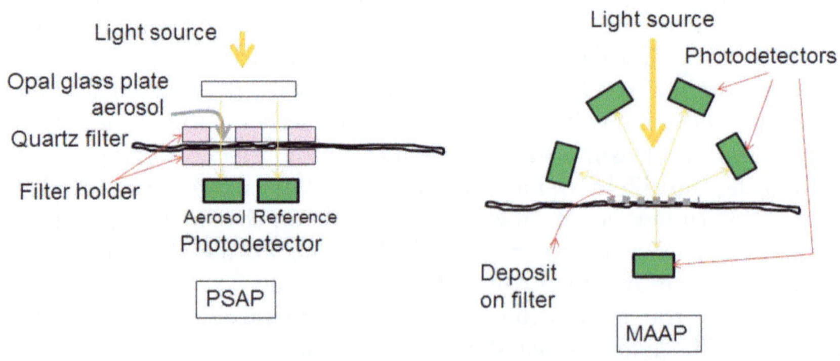

Fig. 6.24. Particles Soot Absorption Photometer and Multi Angle Absorption Photometer (schematic). A continuous fiber filter is clamped into the filter holder. Air passes through one part of the filter; a part not loaded with particles is used as reference. The scattering of light within the filter integrates the light scattered by the particles which is measured by a photodetector. When the measured signal falls below a given value, the filter is advanced. The multi-angle absorption photometer additionally measures the backscattered light.

All methods of light-absorption measurement using diffuse scattering have the problem of non-perfect integration and multiple passes of light through the deposit. Hänel (1987) has done the integration by scanning the radiation with a photodetector in all directions and adding up the signals. A combination of the Hänel method and the integration by scattering is used in the multi-angle absorption photometer (MAAP; Petzold and Schönlinner, 2004) (Fig. 6.24). The light transmission through the deposit on a filter band and the backscattering at several angles are measured and a two-stream radiative-transfer calculation is performed to estimate the absorption coefficient of the aerosol. This avoids many of the artifacts prone to the methods using integration by diffuse scattering. Vecchi et al. (2014) have successfully used a polar nephelometer to measure the scattered light at intervals of 5°, as well as the transmitted light to derive the absorption coefficient of the aerosol.

It should be mentioned that the British smoke shade or soiling index (BSI, 1969; Walter, 1980) actually is some kind of absorption measurement. The particles are sampled on a filter, which is put on a reflective base, and the filter reflectivity is determined with a special head containing light source and photocell. Originally, this method was developed to replace the tedious gravimetric mass determination of filters and the results of this method are in particle mass per volume of air. But, for a constant ratio of light-absorbing particles to total suspended particles, the light absorption will be proportional to the mass of particles per volume of air. A vast number of data using this method exists, so a conversion of black smoke to black carbon has been developed (Heal and Quincey, 2012).

6.11.3 Use of various physical effects for determining light absorption

Light absorption means that part of the light energy, interacting with the particle, remains in the particle and causes heating of the particle. Eventually, the particle transfers the heat to the surrounding gas. Absorbing particles in a very intense laser beam can heat up and emit thermal or visible radiation. This incandescence can be used to size single particles (Michelson, 2003). López-Iglesias et al. (2014) used laser-induced incandescence (LII) to determine the ratios of the absorption cross-sections at 532 and 1,064 nm. For mature soot, this value is 1.8, and increasing with decreasing maturity of the soot, namely increasing hydrogen content.

With moderate energy of the illumination, the particle transfers heat to the surrounding air, which makes the gas expand slightly and creates a small pressure increase. If the light is switched on and off periodically, the pressure change also is periodical, so the particles produce a sound wave (photoacoustic effect). The amplitude of the sound wave depends on the absorption coefficient, the laser power, and the thermal relaxation time, namely the time needed for the heat transfer to the surrounding air. The relaxation time should be small compared to the period of the modulation of the illumination. This only is a problem for particles larger than a few micrometers and can be avoided by proper choice of the modulation frequency. A schematic of a photoacoustic cell is shown in Fig. 6.25. The chopped laser light enters and leaves the cell through windows. In the center of the cell, a tube is located, the length of which is half of the wavelength of the acoustic wave (in order to achieve a high sound amplitude by resonance). On the open ends of the

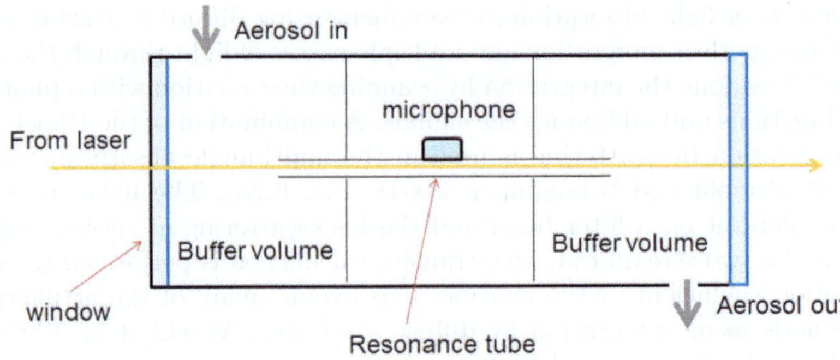

Fig. 6.25. Photoacoustic light-absorption measurement. The aerosol, passing through a tube, is illuminated by a chopped light beam. The absorbing particles periodically heat the surrounding air, and thus produce a sound wave, the amplitude of which is measured with the microphone. The microphone signal is proportional to the light-absorption coefficient of the aerosol. The frequency is chosen such that half a standing wave is formed in the tube.

tube is a pressure minimum; the maximum is in the center, where the microphone is located. If the thermal relaxation is much shorter than the period, the sound pressure measured with the microphone is proportional to the absorption coefficient of the aerosol. For more information on photoacoustic absorption measurements, see, for example, Moosmüller et al. (2009).

The heat transferred to the air by the particles absorbing light can also be measured by the slight change in the refractive index of the air. Interferometers are very sensitive and have long been used for this purpose. For an application of a folded Jamin interferometer for aerosol light-absorption measurement, see Sedlacek and Lee (2007).

Calibration for photo acoustic instruments can be done with absorbing gases such as diluted NO_2. The absorption coefficient of this gas is well known and the dilution can be measured easily, thus a reliable reference value is available.

In principle, it would also be possible to measure black carbon directly, such as by the thermochemical analysis method (Malissa et al., 1976; Chow et al., 1993) and convert the measured mass to light absorption, using an approximate value of the specific cross-section, e.g. 7.5 m^2/g. Besides problems inherent to the thermochemical method, like the right cut between organic and elemental carbon, namely the protocol used, one must be aware that the specific cross-section of 7.5 m^2/g is for black carbon and does not consider enhancement due to coating or mixing.

6.12 Data on atmospheric light absorption

At many places in the world, either black carbon or light absorption is measured. All data show a considerable variability, since the lifetime of the atmospheric particles is about 1 week and thus a strong dependence on local sources, vertical exchange, and air mass history exists. In towns, traffic is a major source of light-absorbing aerosol particles and a daily pattern resembling the traffic density with some delay is frequently measured. The lifetime of 1 week is sufficient to transport at least a small fraction of particles to remote areas, such as the South Pole. Table 6.8 gives a few examples of atmospheric light-absorption coefficients. The values were obtained from black carbon data (European Environment Agency, 2013; EPA, 2010) and converted, using the specific absorption cross-section of 7.5 g/m^2.

Table 6.8. Light-absorption coefficients in Mm^{-1}: overview.

Location	Urban	Rural
Europe	10–20	1.5–5
North America	2–18	1–5
China	70–110	2–40
Mauna Loa	Hawaii, USA, remote	0.07–0.15
South Pole	Remote	0.015–0.03
Annual averages		
Barcelona	Spain, urban	11
Huelva	Spain, rural	4
Lugano	Switzerland, urban	12
Bern	Switzerland, urban	26
London, North Kensington	Urban	15
London, Marylebone Road	Urban, heavy traffic	50

It is evident that the proximity to the sources increases the absorption coefficient of the atmospheric aerosol.

6.13 Influence of black carbon on cloud cover and global warming

6.13.1 The influence on clouds

A light-absorbing aerosol heats the surrounding atmosphere, due to absorption of sunlight. Also, the black carbon particles can act as cloud condensation and ice nuclei, so the effect of black carbon on clouds has many facets including the decrease in cloud albedo and increase in cloud lifetime. Light-absorbing particles in or near the cloud heat the layer and favor cloud evaporation (semi-indirect effect; Hansen et al., 1997). Absorbing particles below the cloud enhance convection and increase cloud cover. Light absorption above a cloud normally stabilizes the atmosphere below. This will increase stratocumulus clouds, but reduce cumulus clouds. For a review, see Koch and Del Genio (2010).

6.13.2 Light-absorbing particles and global warming

It is well known that the greenhouse effect is needed for a reasonable climate on Earth: without absorption and re-emission of the outgoing terrestrial thermal infrared radiation by the various greenhouse gases (H_2O, CO_2, CH_4, etc.), we would have an average temperature far below 0°C, which would make it uninhabitable. Our present climate is an unstable equilibrium between the two extremes, a mainly snow-covered Earth (snowball Earth; Harland, 2007) or a hot wet climate. Obviously, it is desirable to preserve the present state, but it has been disturbed by the increased CO_2 emission since the industrial revolution, initiating global warming. Non-absorbing aerosol particles counteract this. Here, we will concentrate on the effect of light-absorbing particles, which have been underestimated for a long time. The influence on climate usually is characterized by the climate forcing. This is a hypothetical value of the change in net radiation if one parameter in the climate system is altered, such as an assumed removal of sulfate particles, which scatter sunlight in the atmosphere and back to space. With no sulfate particles, we would receive more radiation from the Sun, and the climate system would adjust to it most likely with an increase in temperature.

The radiative forcing, although hypothetical, is an objective way to compare the effects of constituents of the atmosphere on the radiative balance. The radiative forcing of CO_2 and (CH_4 + N_2O + Halocarbons) is well known and amounts to 1.7 and 0.9 W/m^2, respectively.

The effect of black carbon aerosol on radiative balance is multifaceted (see Bond et al., 2013); all values of forcing, listed below, are compared to the pre-industrial era.

The direct radiative effect is due to absorption and scattering of sunlight: the power of the sunlight absorbed by the particles is transferred to the atmosphere, causing a slight warming and thus increased infrared radiation to the ground. Without absorbing particles, the sunlight would reach the ground and be reflected to space. Thus, a positive forcing results; a small correction is needed due to scattering by the light-absorbing particles. The 2013 estimate for this effect is +0.71 W/m^2, which includes contributions from fossil fuel, biofuel, and open burning sources of +0.29, +0.22, and +0.20 W/m^2.

The black carbon cloud effect comprises temperature changes by absorbing particles above, within, and below the clouds, albedo change due to incorporation of particles in cloud droplets, change in the number of cloud condensation nuclei and thus droplets, alteration of precipitation in mixed clouds, change in ice nuclei, and extent of clouds. The estimate for this effect is 0.23 W/m^2, with large uncertainties.

The black carbon snow and ice effect considers refection changes of snow cover and ice cover, which would reflect more light without black carbon deposits. The estimate for this effect is 0.13 W/m^2.

Combustion of fuels emits not only black carbon, but also transparent substances, such as sulfates, when burning coal or bio matter in open burning. These co-emitted particles need to be considered in forcing considerations because, if there were no black carbon, then there would be no combustion and thus no transparent particles. For carbon-rich sources (i.e., excluding open burning), an estimate is possible, resulting in a net forcing of 0.22 W/m^2. Including open burning, the net forcing could be −0.02 W/m^2. This may sound like no net effect, but it must be borne in mind that the black carbon forcing will be positive in the mid-latitude northern hemisphere which could change the pattern of the Asian monsoon.

6.14 Photophoresis

An absorbing particle exposed to sunlight will be warmer than the surrounding air; let us call the temperature of the particle's surface T_s. The particle will transfer its heat energy to the surrounding air by accommodation: an air molecule hitting the particle's surface with a velocity, corresponding to the temperature, T_0, of the surrounding gas, can pick up some energy from the warmer particle surface and leave the surface with a larger velocity, corresponding to a temperature T, with $T_s \geq T \geq T_0$. The probability of taking up additional thermal energy is characterized by the accommodation coefficient $\alpha = \dfrac{T - T_0}{T_s - T_0}$, with $\alpha = 0$ for no accommodation and $\alpha = 1$ for full accommodation (Knudsen, 1911), with typical values $0.3 \leq \alpha \leq 0.8$. The value of the accommodation coefficient depends on both the material and the surface structure. Since $\alpha > 0$, a molecule hitting the particle's surface will be reflected with a larger velocity corresponding to a temperature higher than the surrounding air.

The change in velocity of the air molecule upon reflection exerts a force on the particle. For a symmetric impact of gas molecules on a spherical particle, all forces caused by velocity change during reflection will compensate and the net force will be zero. Since the motion of the air molecules is random, no complete compensation occurs, and a random net force acts on the particle, which causes a random motion of the particle, called the Brownian motion. An absorbing particle, which is illuminated from one direction, is heated inhomogeneously, so the reflected gas molecules on the hotter side are faster than those on the colder side of the particle, causing a net force (and obviously a motion) in the direction of the illumination. Brownian rotation is of no importance, since the part of the particle facing the light is always heated.

Particles formed of different materials or having a complex surface structure usually show a variation in the accommodation coefficient along the surface, causing a force from the surface of higher accommodation to the surface of lower accommodation. Due to Brownian rotation, this force rapidly changes its direction, causing an enhanced Brownian motion. But this enhanced random motion is not caused by the random impact of the gas molecules, but by the random orientation of the particles and the force caused by variable accommodation. This has been observed experimentally and called the trembling effect (Ehrenhaft, 1907; Steipe, 1952). The rotation of the particles can be impeded, such as if the particle has a magnetic moment which causes partial orientation in a magnetic field; similarly, an inhomogeneous mass distribution can orient the particle by gravity. With a preferred orientation of the particle, the photophoretic force has also a preferred orientation, which can be opposite to gravity and thus cause a levitation of the particles.

Influencing factors for the photophoretic force are the flux density of the illumination, the absorption efficiency of the particle, the accommodation coefficient of the particle's surface and its variation, the size of the particle, and the gas pressure and the orienting torque. The photophoretic force is maximal, when the size of the particle is comparable to the mean free path of the gas molecules. For an overview, see Horvath (2014). For particles in the stratosphere, photophoresis can be responsible for the formation of particle layers (Cheremisin et al., 2011).

6.15 Health effects of black carbon particles

It is evident that particles smaller than a few hundred nanometers, when inhaled, have adverse health effects, such as pulmonary or cardiac problems, which can lead to early death or increased morbidity (Dockery et al., 1993); long-term effects are cancer, and lung and heart disease. Most of the increase in mortality in the six-city study was associated with fine particles from mobile combustion sources; particles from coal combustion were of minor importance (Laden et al., 2000). Potentially relevant properties of the particles are surface reactivity and adsorbed polycyclic-aromatic hydrocarbons, which the cells can extract from deposited soot particles. Upon exposure to particles, inflammation of the cells is the first reaction which is thought to be caused by oxidative stress on cells as a response to reactive oxygen species. Various *in vitro* and *in vivo* studies of exposure have shown that the age of the soot, as well as the combustion conditions, is important. The reactivity of the particles' surface depends on the combustion-related nano-structure: greater disorder on the surface creates more active sites, enhancing health effects. Freshly produced diffusion flame soot has the biggest health impact. For diesel particles, the operating load of the engine is an important factor for the response of human epithelial cells to soot. For details, see Kumfer and Kennedy (2005).

An estimate of the effect of diesel soot particles on health was performed during the Vienna Diesel Study (Horvath et al., 1988). In 1984, Vienna's automotive fuel was exclusively supplied by one refinery and negligible through traffic occurred; therefore, it was possible to determine the amount of diesel particles in the city by adding a tracer in the fuel at the refinery and analyzing the samples with the collected particles for this tracer. The diesel particles had a median diameter of 87 nm and thus could spread in the town, and they are also present indoors. The 'background', negligible traffic nearby was 10 $\mu g/m^3$ and, near a high-capacity highway, it amounted to 23 $\mu g/m^3$.

Cancer risk factors for diesel emissions are available from the literature (Cuddihy et al., 1984). The range is 0.007–0.3, due to the biological variability, with a typical value of 0.1 annual lung cancer deaths per 100,000 people per 1 $\mu g/m^3$ of diesel particles inhaled in the lifetime. Combining the measured mass of diesel particles value with the cancer risk factor, the additional deaths due to lung associated with light-absorbing diesel particles are: 1 additional cancer death when living at safe distance from a highway and 2.6 when living the whole life near a high-capacity highway. For comparison, the naturally occurring annual cancer deaths are 7 per year for non-smokers and about 80 for smokers. Although the additional risk is small, it cannot be neglected and huge efforts have been undertaken to reduce diesel emissions.

References

Abel, S.J., Haywood, J.M., Highwood, E.J., Li, J. and Busek, P.R. (2003) 'Evolution of aerosol properties from biomass burning measured during SAFARI 2000,' *Geophys. Res. Lett.*, **30**, 1783.

Abeles, B. and Gittleman, J.I. (1976) 'Composite material films: Optical properties and applications,' *Applied Optics*, **15**, 2328–32.

Andreae, M.O. and Gelencsér, A. (2006) 'black carbon or black carbon? The nature of light-absorbing carbonaceous aerosols,' *Atmospheric Chemistry and Physics*, **6**, 3131–48.

Ångström, A.A. (1929) 'On the atmospheric transmission of sun radiation and on dust in the air,' *Geografiska Annaler*, **II**, 130–59.

Balkan, S. and Hinzpeter, H. (1988) 'Atmospheric radiation,' in *Landolt-Börnstein's Numerical Data and Functional Relationships in Science and Technology*, Vol. 4: *Meteorology, Subvolume b, Physical and Chemical Properties of the Air*, Springer, Berlin, p. 140.

Barber, P.W. and Hill, S.C. (1990) *Light Scattering by Particles: Computational Methods*, World Scientific, Singapore.

Batten, C.E. (1985) 'Spectral optical constants of soots from polarized angular reflectance measurements,' *Applied Optics*, **24**, 1193–9.

Bergstrom, R.W. (1973) 'Extinction and absorption coefficients of the atmospheric aerosol as a function of particle size,' *Contr. Atmos. Phys.*, **46**, 223–34.

Bohren, C.F. and Huffman, D.R. (1983) *Absorption and Scattering of Light by Small Particles*, Wiley Interscience, New York.

Bond, T.C. and Bergstrom, R.W. (2006) 'Light absorption by carbonaceous particles: An investigative review,' *Aerosol Science and Technology*, **40**, 27–67.

Bond, T.C., Anderson, T.L. and Campbell, D. (1999) 'Calibration and intercomparison of filter-based measurements of visible light absorption by aerosols,' *Aerosol Science and Technology*, **30**, 582–600.

Bond, T.C., Bhardwaj, E., Dong, R., Jogani, R., Jung, S.K., Roden, C., Streets, D.G. and Trautmann, N.M. (2007) 'Historical emissions of black and organic carbon aerosol from energy-related combustion, 1850–2000,' *Global Biogeochemical Cycles*, **21**, doi:10.1029/2006GB002840.

Bond, T.C., Doherty, S.J., Fahey, D.W., Forster, P.M., Berntsen, T., DeAngelo, B.J., Flanner, M.G., Ghan, S., Kärcher, B., Koch, D., Kinne, S., Kondo, Y., Quinn, P.K., Sarofim, M.C., Schultz, M.G., Schulz, M., Venkataraman, C., Zhang, H., Zhang, S., Bellouin, N., Guttikunda, S.K., Hopke, P.K., Jacobson, M.Z., Kaiser, J.W., Klimont, Z., Lohmann, U., Schwarz, J.P., Shindell, D., Storelvmo, T. Warren, S.G. and Zender, C.S. (2013) 'Bounding the role of black carbon in the climate system: A scientific assessment,' *Journal of Geophysical Research*, **118**, 5380–552.

Bond, T.C., Gazala, H. and Bergstrom, R.W. (2006) 'Limitations in the enhancement of visible light absorption due to mixing state,' *Journal of Geophysical Research*, **111**, D20211.

Bond, T.C., Streets, D.G., Yarber, K.F., Nelson, S.M., Woo, J.-H. and Klimont, Z. (2004) 'A technology-based global inventory of black and organic carbon emissions from combustion,' *J. Geophysical Research*, **109**, doi:10.1029/2003JD003697.

Born, M. and Wolf, F. (1999) 'Foundation of geometric optics,' in Born, M. and Wolf, E. (eds), *Principles of Optics*, 7th edn, Cambridge University Press, Cambridge, pp. 116–17.

Brillouin, L. (1949) 'The scattering cross section of spheres for electromagnetic waves,' *Journal of Applied Physics*, **20**(11), 1110–25.

Bruggeman-Wassenaar, D.A.G. (1935) 'Berechnung verschiedener physikalischer Konstanten von heterogenen Substanzen I: Dielektrizitätskonstanten und Leitfähigkeiten der Mischkörper aus isotropen Substanzen' ['The calculation of physical constants of heterogeneous substances I: Dielectric constants and conductivities of mixtures of isotropic substances'], *Annalen der Physik*, **24**, 636–79.

BSI (1969) *Methods for the Measurement of Air Pollution. Part 2: Determination of Concentration of Suspended Matter*, British Standard 1747: Part 2: 1969, British Standards Institution, London.

Chang, H. and Charalampopoulos, T.T. (1990) 'Determination of the wavelength dependence of refractive indices of flame soot,' *Proceedings Royal Society London A: Mathematical and Physical Sciences*, **430**(1880), 577–91.

Charlson, R.J., Horvath, H. and Pueschel, R.F. (1967) 'The direct measurement of atmospheric light scattering coefficient for studies of visibility and pollution,' *Atm. Environment*, **1**, 469–78.

Chen, H.X. and Dobbins, R.A. (2000) 'Crystallogenesis of particles formed in hydrocarbon combustion,' *Combustion Science and Technology*, **159**, 109–28.

Cheremisin, A.A., Shnipkov, I.S., Horvath, H. and Rohatschek, H. (2011) 'The global picture of aerosol formation in the stratosphere and in the mesosphere under the influence of gravito-photophoretic and magneto-photophoretic forces,' *Journal of Geophysical Research: Atmospheres*, **116**, D19204.

Chow, J.C., Watson, J.G., Pritchett, L.C., Pierson, W.R., Frazier, C.A. and Purcell, R.G. (1993) 'The DRI Thermal Optical Reflectance Carbon Analysis System: Description, evaluation and applications in United-States air-quality studies,' *Atmospheric Environment Part A – General Topics*, **27**(8), 1185–201.

Chylek, P., Srivastava, V., Pinnick, R.G. and Wang, R.T. (1988) 'Scattering of electromagnetic waves by composite spherical particles: experiments and effective medium approximations,' *Applied Optics*, **27**, 2396–404.

Clarke, A.D. (1982) 'Integrating sandwich: A new method of measurement for the light absorption coefficient for atmospheric particles,' *Applied Optics*, **21**, 3011–20.

Cuddihy, R.C., Griffith, W.C. and McClellan, R.O. (1984) 'Health risk from light-duty diesel vehicles,' *Environmental Science and Technology*, **18**, 14–21.

D'Alessio, A., D'Anna, A., Minutolo, P. and Sgro, L.A. (2009) 'Nano Organic Carbon (NOC) formed in flames and their effects in urban atmospheres,' in Bockhorn, H., D'Anna, A., Sarofim, A.F. and Wang, H. (eds), *Combustion Generated Fine Carbonaceous Particles*, KIT Scientific Publishing, Karlsruhe, Germany, pp. 205–30.

Dalzell, W.H. and Sarofim, A.F. (1969) 'Optical constants of soot and their application to heat flux calculations,' *J. Heat Transf.*, **91**, 100–4.

Debye, P. (1909) 'Der Lichtdruck auf Kugeln von beliebigem Material' ['Light pressure on spheres of arbitrary material'], *Annalen der Physik*, **30**, pp. 57–136.

Dobbins, R.A. (2009) 'Precursor nanoparticles in flames and diesel engines: A review and status report,' in Bockhorn, H., D'Anna, A., Sarofim, A.F. and Wang, H. (eds), *Combustion Generated Fine Carbonaceous Particles*, KIT Scientific Publishing, Karlsruhe, Germany, pp. 179–203.

Dockery, D.W., Pope, C.A., Xu, X., Spengler, J.D., Ware, J.H., Fay, M.E., Ferris, B.G. and Speizer, F.E. (1993) 'An association between and mortality in six U.S. cities,' *New England Journal of Medicine*, **329**, 1753–9.

Ehrenhaft, F. (1907) 'Über eine der Brown'schen Molekularbewegung in Flüssigkeiten gleichartige Bewegung in den Gasen und deren molekuarkinetischer Erklärungversuch' ['On a motion in gases, being similar to the Brownian motion liquids, and an attempt of explanation based on molecular kinetics'], *Wiener Berichte*, **116**(1907), 1139–49.

EPA United States Protection Agency (2010) 'Observational data for black carbon,' in *Report to Congress on black carbon*, EPA 450/R-12-001, pp. 115–37.

European Environment Agency (2013) *Status of black carbon Monitoring in Europe*, EEA Technical Report, No. 18/2013, European Environment Agency, Copenhagen.

Fischer, K. (1973) 'Mass absorption coefficients of natural aerosol particles in the 0.4 to 2.5 μm wavelength interval,' *Contributions to Atmospheric Physics*, **46**, 89–100.

Fleagle, R.G. and Businger, J.A. (1963) *An Introduction to Atmospheric Physics*, Academic Press, New York, Fig. 4.13.

Gast, P.R. (1960) 'Thermal radiation,' in *Handbook of Geophysics*, Macmillan, New York, pp. 16–21.

Gerber, H.E. (1979) 'Portable cell for simultaneously measuring the coefficients of light scattering and extinction for ambient aerosols,' *Applied Optics*, **18**, 1009–14.

Guzzi, R. and Rizzi, R. (1984) 'Water vapor absorption in the visible and near infrared: Results and field measurements,' *Applied Optics*, **23**, 1853–61.

Hänel, G. (1987) 'Radiation budget of the boundary layer: Simultaneous measurement of the mean solar volume absorption and extinction coefficients of particles,' *Contributions to Atmospheric Physics*, **60**, 241–7.

Hansen, A.D.A., Rosen, H. and Novakov, T. (1984) 'The aethalometer: An instrument for the real-time measurement of optical absorption by aerosol particles,' *Science of the Total Environment*, **36**, 191–6.

Hansen, J., Sato, M. and Ruedy, R. (1997) 'Radiative forcing and climate response,' *J. Geophysical Research*, **102**, 6831–64.

Harland, W.B. (2007) 'Origins and assessment of snowball Earth hypotheses,' *Geological Magazine*, **144**, 633–42.

Heal, M.R. and Quincey, P. (2012) 'The relationship between black carbon concentration and black smoke: A more general approach,' *Atmospheric Environment*, **54**, 538–44.

Hegglin, M.I., Krieger, U.K., Koop, T. and Peter, T. (2002) 'Organics induced fluorescence in Raman studies of sulfuric acid aerosols,' *Aerosol Sci. Tech.*, **36**, 510–12.

Henyey, L.G. and Greenstein, J.L. (1941) 'Diffuse radiation in the galaxy,' *Astrophysics Journal*, **93**, 70–83.

Hitzenberger, R., Dusek, U. and Berner, A. (1996) 'black carbon measurements using an integrating sphere,' *J. Geophysical Research*, **101**, 19601–6.

Hodkinson, R.J. (1966) 'Calculation of the color and visibility in urban atmospheres polluted by gaseous NO_2,' *International Journal of Air and Water Pollution*, **10**, 137–244.

Horvath, H. (1971) 'On the brown color of atmospheric haze,' *Atmospheric Environment*, **5**, 333–44.

Horvath, H. (1993) 'Atmospheric light absorption: A review,' *Atmospheric Environment*, **27A**, 293–317.

Horvath, H. (1997) 'Experimental calibration of aerosol light absorption measurements using the integrating plate method: Summary of the data,' *J. Aerosol Science*, **28**, 1149–61.

Horvath, H. (2009) 'Gustav Mie and the scattering and absorption of light by particles: Historic developments and basics,' *Journal of Quantitative Spectroscopy and Radiative Transfer*, **110**, 787–99.

Horvath, H. (2014) 'Photophoresis – a forgotten force?' *KONA Particle and Powder Journal*, **31**, 181–99.

Horvath, H., Catalan, L. and Trier, A. (1997) 'A study of the aerosol of Santiago de Chile III: Light absorption measurements,' *Atmospheric Environment*, **31**, 3737–44.

Horvath, H. and Gangl, M. (2003) 'A low-voltage spark generator for production of carbon particles,' *J. Aerosol Science*, **34**, 1581–8.

Horvath, H., Kreiner, I., Norek, C. and Georgi, B. (1988) 'Diesel emissions in Vienna,' *Atmospheric Environment*, **22**, 1255–69.

Horvath, H. and Metzig, G. (1990) 'Experimental determination of the accuracy of light absorption measurements with the integrating plate method,' *J. Aerosol Science*, **21**, s525–8.

Janzen, J. (1979) 'The refractive index of colloidal carbon,' *J. Colloid and Interface Science*, **69**, 436–47.

Kattawar, G.W. and Hood, D.A. (1976) 'Electromagnetic scattering from a spherical polydispersion of coated spheres,' *Applied Optics*, **15**, 1996–9.

Kiehl, J.T. and Yamanouchi, T. (1985) 'A parameterization for the absorption due to A, B, and γ oxygen bands,' *Tellus*, **37B**, 1–6.

Kim, J., Bauer, H., Dobrovicnik, T., Hitzenberger, R., Lottin, D., Ferry, D. and Petzold, A. (2014) 'Constraining optical properties and refractive index of soot through combined experimental and modelling studies,' submitted to *Aerosol Science and Technology.*

Kirchstetter, T.W., Novakov, T. and Hobbs, P.V. (2004) 'Evidence that the spectral dependence of light absorption by aerosols is affected by organic carbon,' *J. Geophysical Research*, **109**, D21208.

Knudsen, M. (1911) 'Die molekulare Wärmeleitfähigkeit der Gase und der Akkomodationskoeffizient,' *Annalen der Physik.*, **34**, 593–656.

Koch, D. and Del Genio, A.D. (2010) 'black carbon semi-direct effects on cloud cover: Review and synthesis,' *Atmospheric Chemistry and Physics*, **10**, 7685–96.

Kokhanovsky, A.A. and Zege, E.P. (1994) 'Extinction, absorption and light pressure cross sections of spherical particles in the modified geometrical optics approximation,' *Proc. SPIE*, **2222**, 356–66.

Kortüm, G.F. (1969) *Reflexionsspektrometrie*, Springer, Berlin.

Kumfer, B. and Kennedy, I. (2005) 'The role of soot in the health effects of inhaled airborne particles,' in Bockhorn, H., D'Anna, A., Sarofim, A.F. and Wang, H. (eds), *Combustion Generated Fine Carbonaceous Particles*, KIT Scientific Publishing, Karlsruhe, Germany, pp. 1–15.

Lack, D.A. and Cappa, C.D. (2010) 'Impact of brown and clear carbon on light absorption enhancement, single scattering albedo and absorption wavelength dependence of black carbon,' *Atmospheric Chemistry and Physics Discussions*, **10**, 785–819.

Lack, D.A. and Langridge (2013) 'On the attribution of black and black carbon light absorption using the Angström exponent,' *Atmospheric Chemistry and Physics*, **13**, 10535–43.

Laden, F., Neas, L.M., Dockery, D.W. and Schwartz, J. (2000) 'Association of fine particulate matter from different sources with daily mortality in six US cities,' *Environ. Health Perspective*, **108**, 941–7.

LaMer, V.K. and Sinclair, D. (1943) OSRD Report 1857, 29 September, Office of Publications Board, US Department of Commerce.

Lee, Y.H., Chen, K. and Adams, P.J. (2009) 'Development of a global model of mineral dust aerosol microphysics,' *Atmospheric Chemistry and Physics*, **9**, 241–58.

Limbeck, A., Kulmala, M. and Puxbaum, H. (2003) 'Secondary organic aerosol formation in the atmosphere via heterogeneous reaction of gaseous isoprene on acidic particles,' *Geophysical Research Letters*, **30**, doi:10.1029/2003GL017738.

Lin, C.I., Baker, M.B. and Charlson, R.J. (1973) 'Absorption coefficient of atmospheric aerosols: A method for measurement,' *Applied Optics*, **12**, 1356–63.

López-Iglesias, X., Schrader, P.E. and Michelson, H.A. (2014) 'Soot maturity and absorption cross sections,' *Journal of Aerosol Science*, **75**, 43–64.

Madelung, O., Rössler, U. and Schultz, M. (eds) (2000) 'Hematite (alpha-FE2O3): Optical properties, dielectric constants,' in Landolt-Börnstein – Group III Condensed Matter, *Numerical Data and Functional Relationships in Science and Technology. Volume 41D: Non Tetrahedally Bounded Binary Compounds II*, Springer-Verlag GmbH, Heidelberg, doi:10.1007/10681735_552.

Malissa, H., Puxbaum, H. and Pell, E. (1976) 'Simultaneous relative-conductometric determination of carbon and sulfur in dusts,' *Fresenius Zeitschrift für Analytische Chemie [Fresenius' Journal of Analytical Chemistry, now Analytical and Bioanalytical Chemistry]*, **282**(2), 109–13.

Mansfield, T., Hamilton, R., Ellis, B. and Newby P. (1991) 'Diesel particulate emissions and the implications for the soiling of buildings,' *Environmentalist*, **11**, 243–54.

Maricq, M.M. (2009) 'Electrical mobility based characterization of bimodal soot distributions in rich premixed flames,' in Bockhorn, H., D'Anna, A., Sarofim, A.F. and Wang, H. (eds), *Combustion Generated Fine Carbonaceous Particles*, KIT Scientific Publishing, Karlsruhe, Germany, pp. 348–66.

Maxwell Garnett, J.C. (1904) 'Colors in metal glasses and metallic films,' *Philosophic Transactions of the Royal Society*, **A203**, 385–420.

McCartney, J.T. and Ergun, S. (1967) *Optical Properties of Coals and Graphite*, Bulletin 641, US Department of the Interior, Bureau of Mines.

Michelson, H.A. (2003) 'Understanding and predicting the temporal response of laser induced incandescence form carbonaceous particles,' *Journal of Chemical Physics*, **118,** 7012–45.

Mie, G. (1908) 'Beiträge zur Optik trüber Medien speziell Kolloidaler Goldlösungen' ['Contributions to the optics of diffuse media, especially colloid metal solutions'], *Annalen der Physik*, **25**, 377–445.

Mogo, S., Cachorro, V.E., Sorribas, M., de Frutos, A. and Fernández, R. (2005) 'Measurements of continuous spectra of atmospheric absorption coefficients from UV to NIR via optical method,' *Geophysical Research Letters*, **32**, L13811.

Moosmüller, H., Chakrabarty, R.K. and Arnott, W.P. (2009) 'Aerosol light absorption and its measurement: A review,' *Journal of Quantitative Spectroscopy and Radiative Transfer*, **110**(11), 844–78.

Moosmüller, H., Varma, R. and Arnott W.P. (2005) 'Cavity ring down and cavity-enhanced detection techniques for the measurement of aerosol extinction,' *Aerosol Science and Technology*, **39**, 30–9.

Patterson, E.M. and Marshall, B.T. (1982) 'Diffuse reflectance and diffuse transmission measurements of aerosol absorption at the first international workshop on light absorption by aerosol particles,' *Applied Optics*, **21**, 387–93.

Petzold, A. and Schönlinner, M. (2004) 'Multi-angle absorption photometry: A new method for the measurement of aerosol light absorption and atmospheric black carbon,' *J. Aerosol Science*, **35**, 421–41.

Pickering, J.W., Moes, C.J.M., Sterenborg, H.J.C.M., Prahl, S.A. and van Gemert, M.J.C. (1992) 'Two integrating spheres with an intervening scattering sample,' *J. Opt. Soc. Am.*, **A9**, 621–31.

Pluchino, A.B., Goldberg, S.S., Dowling, J.N. and Randall, C.M. (1980) 'Refractive index measurements of single micron sized carbon particles,' *Applied Optics*, **19**, 3370–2.

Poss, G. and Metzig, G. (1987) 'extinction coefficient measurement of ambient aerosol particles using a "White Cell",' *Journal of Aerosol Science*, **18**(6), 285–8.

Prodi, V. (1972) 'A condensation aerosol generator for solid monodisperse particles,' in Mercer, T.T., Morrow, P.E. and Stoeber, W. (eds), *Assessment of Airborne Particles*, Charles C. Thomas Publisher, New York, pp. 169–81.

Reischl, G.P. (1991) 'Measurement of ambient aerosols by the differential mobility analyzer method: Concepts and realization criteria for the size range between 2 and 500 nm,' *Aerosol Science and Technology*, **14**, 5–24.

Rothman, L.S., Gordon, I.E., Babikov, Y., Barbe, A., Chris Benner, D., Bernath, P.F., Birk, M., Bizzocchi, L., Boudon, V., Brown, L.R., Campargue, A., Chance, K., Cohen, E.A., Coudert, L.H., Devi, V.M., Drouin, B.J., Fayt, A., Flaud, J.-M., Gamache, R.R., Harrison, J.J., Hartmann, J.-M., et al. (2013) 'The HITRAN2012 molecular spectroscopic database,' *Journal of Quantitative Spectroscopy & Radiative Transfer*, **130**, 4–50.

Rubinowitz, A. (1920) 'Radiometerkräfte und Ehrenhaftsche Photophorese (I and II),' *Annalen der Physik*, **62**, 691–37.

Saathof, H., Moehler, O., Schurath, U., Kamm, S., Dippel, B. and Mihelcic, D. (2003) 'The AIDA soot aerosol characterization campaign 1999,' *J. Aerosol Science*, **34**, 1277–96.

Sabbagh, N. (2005) 'Optische Eigenschaften von Aerosolpartikel: Absorptionserhöhung durch Beschichtung' ['Optical properties of aerosol particles: Absorption increase by coating'], MS thesis, University of Vienna, Vienna.

Schnaiter, M., Schmid, O., Petzold, A. Frischke, L., Klein, K.F., Andreae, M.O., Helas, G., Thielmann, A., Gimmler, M., Möhler, O.M., Linke, C. and Schurath, U. (2005) 'Measurement of wavelength resolved light absorption by aerosols utilizing a UV-VIS extinction cell,' *Aerosol Science and Technology*, **39**, 249–60.

Sedlacek, A. and Lee, J. (2007) 'Photothermal interferometric aerosol absorption spectrometry,' *Aerosol Science and Technology*, **41**, 1089–101.

Senftleben, H. and Benedict, E. (1918) 'Über die optischen Konstanten und die Strahlungsgesetze der Kohle' ['On the optical constants and the laws of radiation of coal'], *Annalen der Physik*, **54**, 65–78.

Shaddix, C.R. and Williams, T.C. (2009) 'Soot structure and dimensionless extinction coefficient in diffusion flames: Implications for index of refraction,' in Bockhorn, H., D'Anna, A., Sarofim, A.F. and Wang, H. (eds), *Combustion Generated Fine Carbonaceous Particles*, KIT Scientific Publishing, Karlsruhe, Germany, pp. 17–33.

Shettle, E.P. and Fenn, R.W. (1979) *Models for the Aerosols of the Lower Atmosphere and the Effects of Humidity Variations on Their Optical Properties*, Optical Physics Division Environmental Research Paper, Air Force Geophysics Lab., Hanscom AFB, MA.

Shifrin, K.S. (1951) *Scattering of Light in a Turbid Medium*, Gostekhizdat, Moscow (in Russian).

Smooke, M.D., Hall, R.J., Colket, M.B., Fielding, J., Long, M.B., McEnally, C.S. and Pfefferle, I.D. (2004) 'Investigation of the transition of lightly sooting towards heavy sooting co-flow ethylene diffusion flames,' *Combustion Theory and Modelling*, **8**, 593–606.

Smyth, K.C. and Shaddix, C.R. (1996) 'The elusive history of $m = 1.57 - 0.56i$ for the refractive index of soot,' *Combustion and Flame*, **107**, 314–20.

Sorensen, C.M. (2000) 'The optics of single particles and fractal aggregates,' *J. Aerosol Science*, **31**, 952–4.

Stagg, B.J. and Charalampopoulos, T.T. (1993) 'Refractive indices of pyrolytic graphite, amorphous carbon, and flame soot in the temperature range 25 to 600°C,' *Combustion and Flame*, **94**, 381–96.

Steipe, L. (1952) 'Untersuchungen über Magnetophotophorese' ['Investigations on magnetophotophoresis'], *Acta Physica Austriaca*, **6**, 1–6.

Stolarski, R.S. (1988) 'The arctic ozone hole,' *Scientific American*, **258**(1), 20–7.

Van de Hulst, H.C. (1980) *Multiple Light Scattering: Tables, Formulas, and Applications I, II*, Academic Press, New York.

Vecchi, R., Bernardoni, V., Paganelli, C., Valli, G., Vecchi, G., Bernardoni, G. and Valli, G. (2014) 'A filter-based light-absorption measurement with polar photometer: Effects of sampling artefacts from organic carbon,' *Journal of Aerosol Science*, **70**, 15–25.

Virkkula, A., Ahlquist, N.C., Covert, D., Sheridan, P., Arnott, W.P. and Ogren, J. (2005) 'A three-wavelength optical extinction cell for measuring aerosol light extinction and its application to determining light absorption coefficient,' *Aerosol Science And Technology*, **39**(1), 52–67.

Walter, R.S. (1980) 'An assessment of suspended particulates in relation to health,' *Atmospheric Environment*, **22**, 965–71.

Wentzel, M., Gorzawski, H., Naumann, K.H., Saathoff, H. and Weinbruch, S. (2003) 'Transmission electron microscopical and aerosol dynamical characterization of soot aerosols,' *Journal of Aerosol Science*, **34**, 1347–70.

White, J.U. (1976) 'Very long optical paths in air,' *J. Optical Society Am.*, **66**, 411–16.

7 Neural networks for particle parameter retrieval by multi-angle light scattering

Vladimir V. Berdnik and Valery A. Loiko

7.1 Introduction

The problem of retrieval of characteristics of dispersed particles and their concentration is one of the most important ones in the solution of many practical tasks of ecology, medicine, electronics, photonics, and other fields of science and technology (Masciangioli and Alper, 2012; Black et al., 1996; Ronliang, 2000). Many ways to solve this problem have been proposed. The choice of method depends on the characteristics of the particles and the measurement conditions. Among them, there is a significant place for methods of light-scattering optics that are fast and noninvasive.

Information on size, shape, and internal structure of particles can be extracted from the angular dependence of scattered light. This is a 'fingerprint' of the article. A problem is in the sensitivity of the retrieved parameters to variation of scattering characteristics. The retrieval of these characteristics is an ill-posed inverse problem.

The mathematical problems arising from retrieval of particle parameters are defined by methods of measurement of characteristics of the radiation interacting with particles. These can be divided into three classes, depending on what degree of separation of particles is used.

Methods of measurement in which the ensemble of polydisperse particles is considered as a whole (with no separation into a set of fractions) can be referred to as the first class. In this case, the problems of particle sizing are reduced to solution of the Fredholm equations of the first kind (Riefler and Wriedt, 2008; Riley and Agrawal, 1991; Capps et al., 1982; Schnablegger and Glatter, 1991; Mao et al., 1999). The method of neural networks is also applied (Ishimaru et al., 1990).

Measurements in which the ensemble of particles is divided into narrow fractions represent the second class. The problem of retrieval of particle size distribution thus essentially becomes simpler.

Methods of measurement in which the initial ensemble is divided into individual particles represent the third class (Spinrad and Brown, 1986; Ackleson and Spinrad, 1988; Bartholdi et al., 1980; Castagner and Bigio, 2006; Pluchino et al., 1980; Maltsev, 2000). The characteristics of each separate particle are measured in the regime 'one by one.' Despite the complexities of practical realization, this method is finding wider application. flow cytometry and particle counters concern this class, for example. Some methods to solve the retrieval problem for this case are elaborated on: the parametric method (Maltsev and Lopatin, 1997), the spectral decomposition method (Semyanov et al., 2004; Berdnik and Loiko, 2010), the two-angle method (Quist and Wyatt, 1985), the Fraunhofer diffraction method (Borovoi et al., 2000; Heffels et al., 1994), the fitting method (Ludlow and Everitt, 2000; Zakovic et al., 1998, 2005), and the neural network method (Berdnik and Loiko, 2010; Ulanowski et al., 1998; Wang et al., 1999; Berdnik et al., 2004b; 2004c; 2006; Berdnik and Loiko, 2009).

In this paper, we elucidate the problem of determination of characteristics of homogeneous particles by an artificial neural networks (NN) method. This is a versatile method to solve a wide range of inverse problems (Haykin, 1998; Gorban and Rossiev, 1996; Bishop, 1995). The term 'method of neural networks' refers to a set of methods of nonlinear approximation of multivariable functions, which are used for the inverse problem solution. The advantage of the method is a high speed. It admits determination of particle characteristics in a real-time scale.

In the majority of investigations devoted to this problem, spherical particles are considered. Several approaches are proposed to solve it.

Typically, in experiments, we deal with the limited number of angles available for measurement. That is why there is a set of publications in which neural network constructions are considered for different limited ranges of angles. The problem of particle size and refractive index determination by measurements of the intensity of the radiation scattered in the ranges of angles from 1.5° to 19° and 73° to 107° is treated in Spinrad and Brown (1986) and Ackleson and Spinrad (1988). In Bartholdi et al. (1980), a device for measuring the intensity of the radiation scattered by separate particles in the range of angles from 2.5° to 177.5° is described; in Castagner and Bigio (2006), a device for measuring the scattered radiation in the range of angles from 70° to 125° is described; in Pluchino et al. (1980), neural networks for the range of angles from 10° to 170° are constructed; in Maltsev (2000), a device for measuring the radiation scattered by an individual particle in the range of angles from 10° to 75° is described.

The solution of the retrieval problem by the angular dependence of the scattered radiation, when the range of angles from 0° to 180° is available for measurement, has been obtained in Ludlow and Everitt (2000). However, it is difficult to measure scattered radiation in such a wide range of angles in practice.

Angular dependence of scattered radiation usually can vary by several orders of magnitude. The degree of the angular dependence extension and the number of orders depend on the particle characteristics. To decrease the range of the processed data by the neural network, the logarithm of the difference between the measured intensities of the scattered radiation and calculated according to the Mie theory (Zakovic et al., 1998, 2005) is often used as an objective function (Zakovic et al., 2005). The experimentally measured angular dependence and the objective function are shown in Figs 7.1 and 7.2, respectively. Sometimes, the intensity of radiation is divided into a specially selected function $g(\theta)$ so that the result of dividing is weakly dependent on scattering angle θ.

The instability specific to inverse problems becomes apparent by the existence of many local minima in the objective function. Therefore, it is necessary to solve

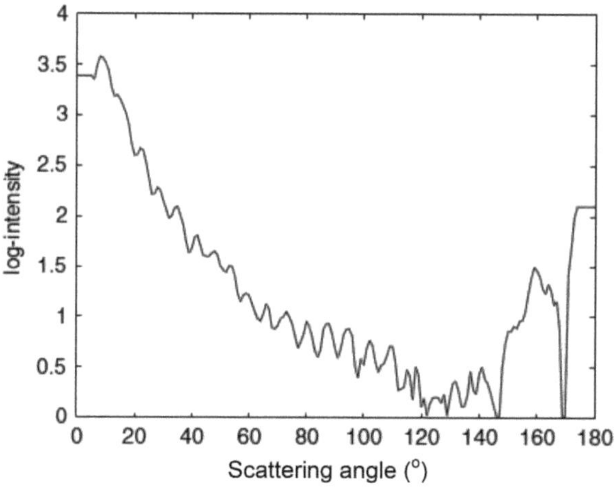

Fig. 7.1. Experimental scattering pattern. From Zakovic et al. (2005).

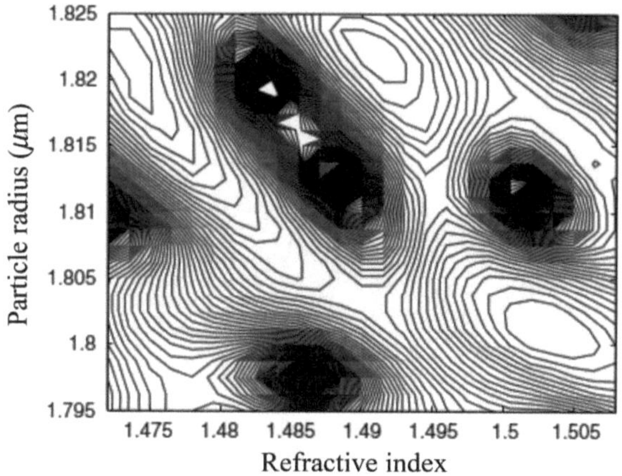

Fig. 7.2. Contour plot of the objective function. From Zakovic et al. (2005).

the problem of searching the global extremum of the multivariable function. That constitutes a complex problem, especially in the presence of the measurement errors. It is fair to say that the universal fitting method can be used to solve the problem. It consists of a multiple solution to the direct problem and selection of parameters of a particle to provide the calculations that most closely coincide with measured angular dependencies of scattered light. It uses the well-proven methods for the direct problem solution. An essential disadvantage of this method is the large time required for the extremum search.

Particle sizes have been determined by simple heuristically found approximate formulae. For example, the proportionality of the particle size and the characteristic frequency k found by the angular dependence of the scattered radiation (Semyanov et al., 2004; Berdnik and Loiko, 2010) is used. In Semyanov et al. (2004), this frequency has been found by the position of the maximum of the module of the Fourier transform from the product of the Hanning function and the angular dependence of the scattered radiation. In Berdnik and Loiko (2010), characteristic frequency has been found using the ratio of the intensity of the radiation scattered by a particle to the intensity of the radiation averaged with respect to particle parameters. That allows one to reduce the particle size determination error by 1.5–2 times.

Several types of neural networks are used. The Radial Basis Function Neural Networks (RBF NN) have been used to retrieve particle radius in the range of 0.5 μm to 1.5 μm and refractive index n in the range of 1.5–1.7 of spherical non-absorbing particles (Ulanowski et al., 1998; Wang et al., 1999). The relative refractive index in these papers varies in the range of 1.123 to 1.272 and the interval of scattering angles $0° = \theta = 180°$ is used. In these papers, the Sequences of Neural Networks have been used to increase the retrieval accuracy. The initial interval for radius R is divided into 10 overlapping sub-ranges. For learning global neural networks, 455 pairs of parameters from the initial range of parameters are generated. For training local neural networks, 144 pairs of each sub-range are generated. The errors in the retrieval of radius and refractive index are shown in Fig. 7.3. Table 7.1 summarizes the results of the evaluation of errors.

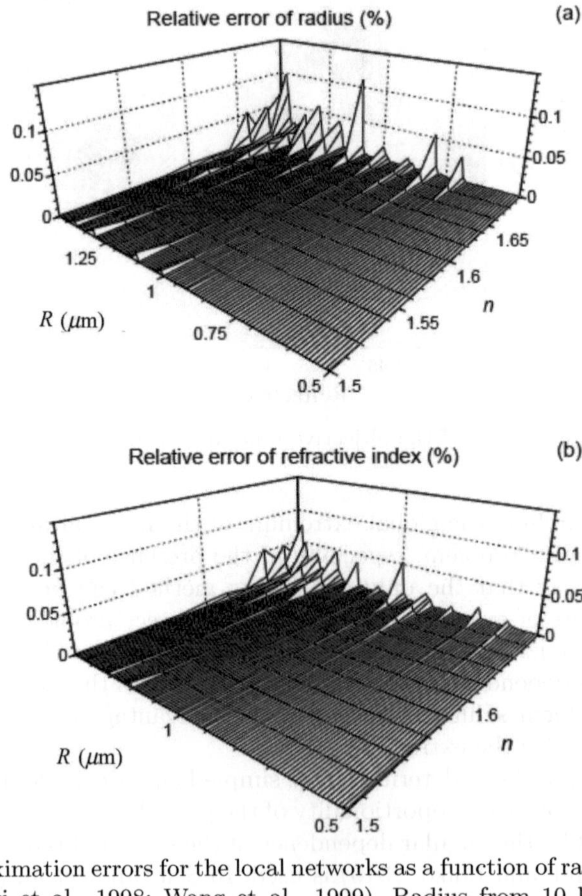

Fig. 7.3. Approximation errors for the local networks as a function of radius and refractive index (Ulanowski et al., 1998; Wang et al., 1999). Radius from 10 local networks are combined and relative errors are calculated separately for (a) R and (b) n.

Table 7.1. Relative approximation errors for local networks using five data scaling schemes.

Scheme	Scale and weighting	Radius error (%)		Refractive index error (%)	
		Mean	Max.	Mean	Max.
1	Log	0.0266	0.8243	0.0253	1.6833
2	Linear hypercube	0.1301	4.0570	0.0395	2.5768
3	Linear $(\sin\theta)^2$	0.0278	1.2762	0.0155	0.8739
4	Linear $(\sin\theta)^4$	0.0016	0.1106	0.0012	0.0765
5	Linear $(\sin\theta)^6$	0.0023	0.2170	0.0016	0.0752

Results are combined from 10 local networks. From Ulanowski et al. (1998) and Wang et al. (1999).

As one can see from the table, the minimum average relative retrieval error is 0.0016% and 0.0012% for radius and refractive index, respectively. They are extremely small, so a small initial interval of parameters and a small representative sample to train neural networks were used.

Since the forward- and backscattering regions are absent or distorted in experimental data, the local networks described in Ulanowski et al. (1998) and Wang et al. (1999) were also trained and tested with 'incomplete' input data. Data from 0° up to an angle, referred as a forward truncation angle, and/or between a backscattering truncation angle and 180° were removed. Approximation errors for such incomplete data are presented in Fig. 7.4.

In Berdnik et al. (2004b, 2004c, 2006), the High-Order Neural Networks (HO NN) have been used together with the Sequences of Neural Networks. In Berdnik and Loiko (2010) and Berdnik and Loiko (2009), the Multilayer Perceptron Neural Networks (MLP NN) have been used. They are created for particles with radii varying in the range from 0.6 μm to 13.6 μm, and relative refractive index varying in the range from 1.015 to 1.28.

The neural network method can be used for many types of particles. When compiling the training set, a large number of calculations should be performed. The number depends on the structure and characteristics of the particle. This stage, as well as the neural network training, takes a long time. The trained neural network solves the inverse problem fast. This is a great advantage of the method. It allows retrieving the particle parameters on a real-time scale. It is crucial, for example, for cytometry, where hundreds and thousands of cells per second need to be analyzed.

7.2 Formation and training of neural networks

In this paragraph, we consider the structure and capability of the neural networks (Haykin 1998; Gorban and Rossiev,1996; Bishop, 1995) to solve the problem of particle parameter retrieval by scattered radiation.

The neural network usually consists of a data-preprocessing system, a neural network calculator, and an interpreter. The latter is a data-processing system generated by the neural network calculator (Gorban and Rossiev, 1996).

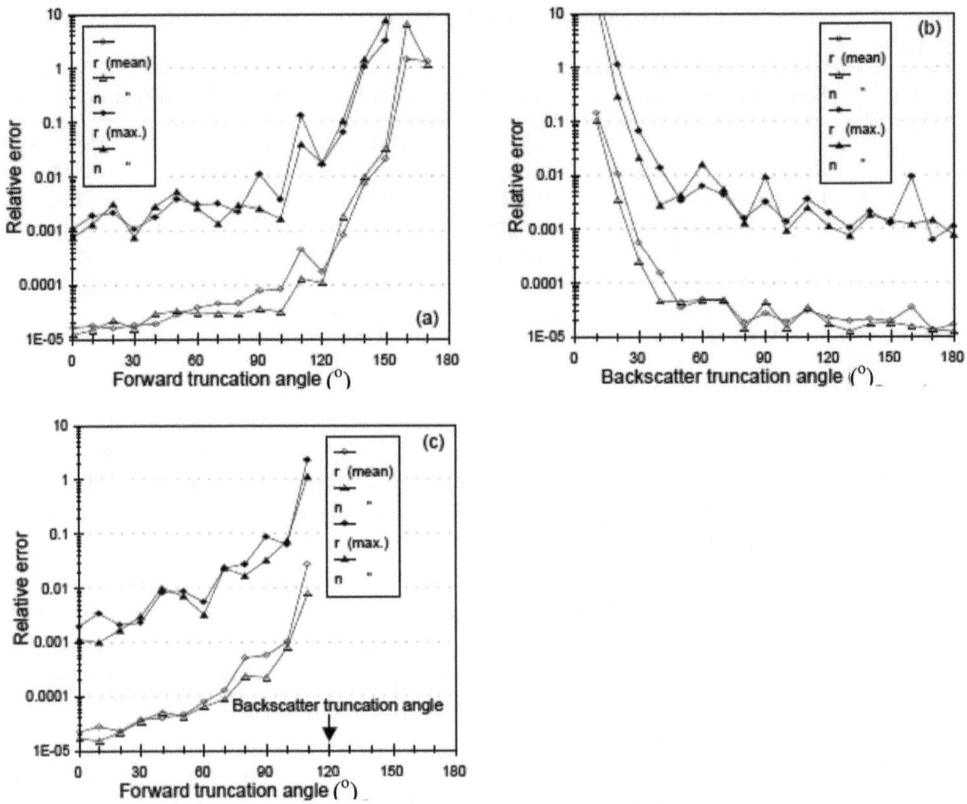

Fig. 7.4. Approximation errors for the local networks trained and tested with incomplete (truncated) data (Ulanowski et al., 1998; Wang et al., 1999). The errors are shown as a function of the truncation angle: the data from the interval (a) between the forward-scattering truncation angle and 180°; (b) between 0° and the backscattering truncation angle; and (c) between the forward truncation angle and 120° are used. Mean (open symbols) and maximum relative errors (filled symbols) for the radius (it is denoted here as r) and the refractive index (n) are given.

The data-preprocessing system is utilized for selection of the most informative parameters from a set of input data and for formation of the vector of input signals delivered to the input of the neural network calculator. The practice shows that the network is more effective if the pre-normalized data are used. Often, when using a linear transformation, each component of the vector of input signals Y_i, where i is the number of the vector coordinate, is normalized. Consequently, the minimum value of the components in the training set is equal to -1 and the maximum is equal to $+1$. Another way of normalization is a linear transformation to the vectors y_i. In this case, the average value of the normalized vectors is equal to zero and the mean square deviation is equal to unity:

$$y_i = \frac{Y_i - M(Y_i)}{\sigma(Y_i)} . \tag{7.1}$$

Here, y_i is the i-th component of the normalized vector; $M(Y_i) = \frac{1}{N_B}\sum_{\alpha} Y_{\alpha i}$ is the value of the vector Y_i averaged with respect to a training set; $\sigma(Y_i) = \left(\frac{1}{N_B}\sum_{\alpha}(Y_{\alpha i} - M(Y_i))^2\right)^{1/2}$ is the mean square deviation of the vector Y_i; α is the number of the example in the training set; and N_B is the total number of examples in the training set.

The neural network calculator is used directly for calculations. Several constructions are used for the nonlinear approximation of the multivariable functions, such as MLP NN, RBF NN, and HO NN.

The interpreter performs a function of transition from normalized values of the calculation result of the neural network calculator to the actual values of the unknown parameters. The function of the interpreter is to perform transition from normalized values of the calculation result of the neural network calculator to the actual values of the unknown parameters.

7.2.1 Multilayer perceptron neural networks

First, let us consider MLP NN. Schematic representation of the neural network calculator with three inner layers of neurons and the single neuron are shown in Fig. 7.5. The input signals are duplicated by the signal redistribution system, multiplied by the weight factors, and input to each neuron of the first hidden layer. All neurons perform the same actions. They summate signals coming to the neuron and converse this sum according to a nonlinear S-shaped function $F(x)$. It is called

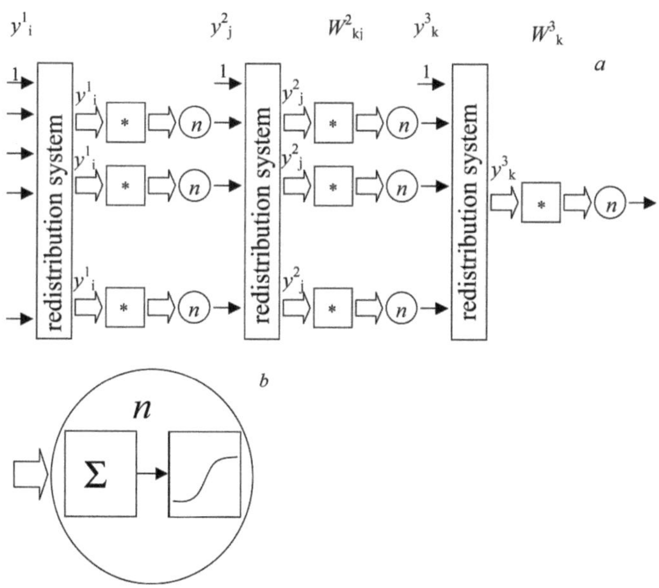

Fig. 7.5. Schematic diagram of the neural calculator of the Multilayer Perceptron Neural Network (MLP NN) with (a) three layers of neurons and (b) the single neuron.

a neuron activation function. Different functions are used for this aim. We use the hyperbolic tangent function (Haykin, 1998)

$$y = F(x) = \frac{1 - e^{-x}}{1 + e^{-x}} . \tag{7.2}$$

The output signals of the neurons of the first layer are the input signals for the neurons of the second layer. They are duplicated by the redistribution system, multiplied by weight factors of the second layer, and applied to neurons of the second layer. And so on.

To find the parameter (particle radius, refractive index, etc.), one has to form the proper neural network. Therefore, the last layer contains only one neuron. The sequence of output signals of each neuron layer of the neural network calculator with N hidden layers can be written as

$$y_i^1 = \begin{cases} 1 & i = 0 \\ Y_i^1 = y_i^{in} & i = 1, \dots, N^{in} \end{cases} , \tag{7.3}$$

$$y_j^2 = \begin{cases} 1 & j = 0 \\ Y_j^2 = F\left(\sum_{i=0}^{N^{in}} W_{ji}^1 y_i^1\right) & j = 1, \dots, N^1 \end{cases} , \tag{7.4}$$

$$y_k^3 = \begin{cases} 1 & k = 0 \\ Y_k^3 = F\left(\sum_{j=0}^{N^1} W_{kj}^2 y_j^2\right) & k = 1, \dots, N^2 \end{cases} , \tag{7.5}$$

$$\dots$$

$$y^{out} = Y_l^N = F\left(\sum_{k=0}^{N^N} W_l^N y_l^N\right) . \tag{7.6}$$

Here, y_i^{in} are the signals generated in the preprocessing system; N^{in} is the number of such signals; y_i^1, y_j^2, y_k^3, and y_l^N are the signals delivered to the first, second, third, and N-th hidden layer of neurons, respectively; y^{out} is the output signal of the neural network calculator; W_{jk}^1, W_{kj}^2, and W_l^N are the weight factors (weight coefficients) of neurons of the first, second, and N-th layers; and N^1 and N^2, N^N are the number of neurons in the first, second, ..., and N-th layers.

The factors $W_{jk}^1, W_{kj}^2, \dots, W_l^N$ completely determine the neural network calculator operation. Neural network training consists of finding these factors. A common method for finding the weight factors is to minimize the sum Φ of squared errors of the neural network on a set of data (training set)

$$\Phi = \frac{1}{2} \sum_{\alpha=1}^{N_B} \left(y^{out}\left(W_{ji}^1, \dots, W_l^N, y_{i\alpha}^{in}\right) - p_\alpha\right)^2 = \min , \tag{7.7}$$

where p is the unknown parameter, namely the particle radius R or the refractive index n; α is the number of the example in the training set; and N_B is the total number of examples in the training set. We use here the term 'objective function' for the Φ function.

To find the extremum of a multivariable function, several methods have been developed (Gill et al., 1982; Dennis and Schnabel, 1983). The most effective methods are the gradient methods that use partial derivatives of the objective function. The limited-memory Broyden–Fletcher–Goldfarb–Shanno (BFGS) method (Liu and Nocedal, 1989), implemented in the 'J' programming language, is used (*www.jsoftware.com*). The back propagation method is used to calculate the partial derivatives with respect to variables W^1_{jk}, $W^2_{kj}, \ldots, W^N_l$.

In the context of a forward trace, starting with the first layer, in accordance with Eqs (7.2)–(7.5) for all signals from the training set, matrices

$$y^1_{i\alpha} = \begin{cases} 1 & i = 0 \\ y^{in}_{\alpha i} & i = 1, \ldots, N^{in} \end{cases}, \tag{7.8}$$

$$y^2_{j\alpha} = \begin{cases} 1 & j = 0 \\ F\left(\sum\limits_{i=0}^{N^{in}} W^1_{ji} y^1_{i\alpha}\right) & j = 1, \ldots, N^1 \end{cases}, \tag{7.9}$$

$$y^3_{k\alpha} = \begin{cases} 1 & k = 0 \\ F\left(\sum\limits_{j=0}^{N^1} W^2_{kj} y^2_{j\alpha}\right) & k = 1, \ldots, N^2 \end{cases}, \tag{7.10}$$

$$\ldots$$

$$y^{out}_\alpha = F\left(\sum_{l=0}^{N^N} W^N_l y^N_{l\alpha}\right) \tag{7.11}$$

are calculated successively ($\alpha = 1, 2, \ldots, N_B$) ($N_B$ is the number of examples in the training set). The matrices $y^1_{i\alpha}$, $y^2_{j\alpha}, \ldots, y^N_{l\alpha}$ are stored.

Then, starting with the last layer and using the obtained values of the matrices $y^1_{i\alpha}$, $y^2_{j\alpha}, \ldots, y^N_{l\alpha}$, objective function derivatives with respect to the weight factors of the neural network are determined according to the equations

$$d^N_\alpha = \frac{\partial \Phi}{\partial y^{out}_\alpha} = y^{out}_\alpha - p_\alpha\,, \tag{7.12}$$

$$\frac{\partial \Phi}{\partial W^N_l} = \sum_\alpha^{N_B} d^N_\alpha F'\left(\sum_{l'}^{N^N} W^N_{l'} y^N_{l'\alpha}\right) y^N_{l\alpha}\,, \tag{7.13}$$

$$\ldots$$

$$d^3_{k\alpha} = \frac{\partial \Phi}{\partial y^3_{k\alpha}} = \sum_{s'}^{N_3} d^4_{s'\alpha} \cdot F'\left(\sum_{k'}^{N_2} W^3_{s'k'} y^3_{k'\alpha}\right) \cdot W^3_{s'k}\,, \tag{7.14}$$

$$\frac{\partial \Phi}{\partial W_{kj}^2} = \sum_{\alpha}^{N_B} d_{k\alpha}^3 F'\left(\sum_{j'}^{N_3} W_{kj'}^2 y_{j'\alpha}^2\right) y_{j\alpha}^2 , \tag{7.15}$$

$$d_{j\alpha}^2 = \frac{\partial \Phi}{\partial y_{j\alpha}^2} = \sum_{k'}^{N_2} d_{k'\alpha}^3 \cdot F'\left(\sum_{j'}^{N_2} W_{k'j'}^2 y_{j'\alpha}^2\right) \cdot W_{k'j}^2 , \tag{7.16}$$

$$\frac{\partial \Phi}{\partial W_{ji}^1} = \sum_{\alpha}^{N_B} d_{j\alpha}^2 F'\left(\sum_{i'}^{N_1} W_{ji'}^1 y_{i'\alpha}^1\right) y_{i\alpha}^1 . \tag{7.17}$$

Here, $F'(x)$ denotes the x-derivative of the function $F(x)$.

As a result, the values of all derivatives with respect to the internal parameters of the neural network are determined. This method to calculate derivatives takes significantly less time than a method based on calculations by the finite differences method. For example, for a network with one inner layer containing 30 neurons and a training set consisting of 10^4 examples, the gradient calculation time on the basis of the back propagation error is less by tens of times than that on the basis of the finite differences method and, if the layer contains 50 neurons, it is less by hundreds of times.

7.2.2 Radial basis function neural networks

A basic diagram showing a RBF NN is shown in Fig. 7.6. The network consists of two layers. Generally, Gaussian function is used as the activation function. Each neuron has its own activation function parameters.

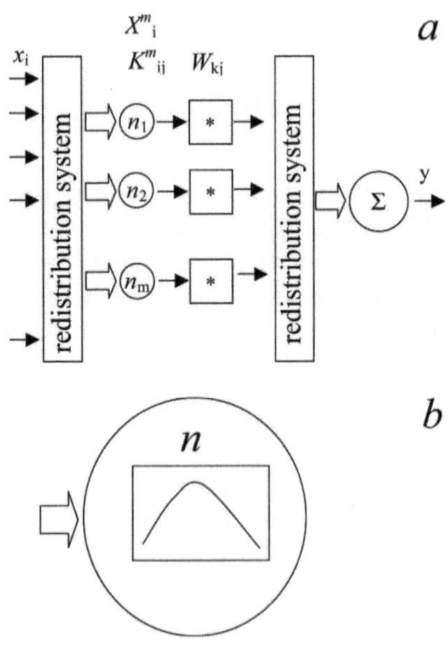

Fig. 7.6. Schematic diagram of the neural calculator of the (a) Radial Basis Function Neural Network (RBF NN) and (b) the single neuron.

The output signals of the neurons of the first layer are multiplied by the weight factors. The results are used as input signals for the summing neuron of the second layer. The output signal of the RBF NN is

$$y = \sum_m W_m \cdot \exp\left(- \sum_{i,j} K_{ij}^m \cdot \left(x_i - X_i^m\right) \left(x_j - X_j^m\right) \right) ! \tag{7.18}$$

Here, x_i is the vector of input signals; X_i^m is the matrix of shift factors of neurons; K_{ij}^m are the matrices of the form factors, characterizing the width of activation functions of neurons; W_m is the vector of weight factors of neurons; and W_m is the output signal of the network.

The neural network training means finding the characteristics of neurons X_i^m and K_i^m and the weight factors W_m such that the objective function reaches a minimum

$$\Phi\left(X_i^m, K_{ij}^m, W_m\right) = \sum_\alpha \left(\sum_m Q_\alpha^m \cdot W_m - y_\alpha \right)^2 = \min . \tag{7.19}$$

Here, $x_{\alpha i}$ and y_α are the sets of the training samples,

$$Q_\alpha^m = \exp - \sum_{i,j} K_{ij}^m \left(x_{\alpha i} - X_i^m\right) \left(x_{\alpha j} - X_j^m\right) . \tag{7.20}$$

If X_i^m and K_{ij}^m are specified, the weight factors W_m minimizing the objective function are found as a solution of the linear equation system

$$\sum_m Q_\alpha^m W_m = y_\alpha \tag{7.21}$$

by the least-squares method.

At the specified values X_i^m, the parameters K_{ij}^m are usually found using the gradient method by determination of the weight factors from Eq. (7.21). Partial derivatives of the objective function Φ with respect to the parameters K_{ij}^m are

$$\frac{\partial \Phi}{\partial K_{ij}^m} = -2W_m \sum_\alpha V_\alpha Q_\alpha^m \left(x_{\alpha i} - X_i^m\right) \left(x_{\alpha j} - X_j^m\right) . \tag{7.22}$$

Here, Q_α^m is determined by Eq. (7.20) as

$$V_\alpha = \sum_m Q_\alpha^m W_m - y_\alpha . \tag{7.23}$$

Equations (7.22) and (7.23) are found by differentiation of Eq. (7.18).

To solve this multidimensional optimization problem (Eq. (7.19)), the 'limited-memory BFGS' (LBFGS) method was used (Liu and Nocedal, 1989).

Various simplifications of the NN construction are possible. For example, if the shift factors X_i^m and the form factors K_{ij}^m are selected in advance (Ulanowski et al., 1998; Wang et al., 1999), the problem of neural network training is reduced to solution of the linear equation system (Eq. (7.21)). Another simplification consists of specification of the offset factors X_i^m on an ordered grid, and the factors K_{ij}^m in terms of $K_{ij}^m = K \cdot E_{ij}$, where E_{ij} is the unit matrix. In this case, the problem of training is reduced to searching for the extremum of a function of a single variable and solution of the linear algebraic equations at the intermediate stages.

7.2.3 High-order neural networks

A basic diagram showing a HO NN is presented in Fig. 7.7. The signal distributor groups the components of the vector of input signals in sets of n components and transfers them to the inputs of multiplying neurons. The last ones calculate the product of the input signals. The output signals y_l of the multiplying neurons are multiplied by the factors w_l and summed up by the summing neuron. The result of the network calculations is the p-value

$$p = \sum_{l=1}^{L} W_l y_l \,. \tag{7.24}$$

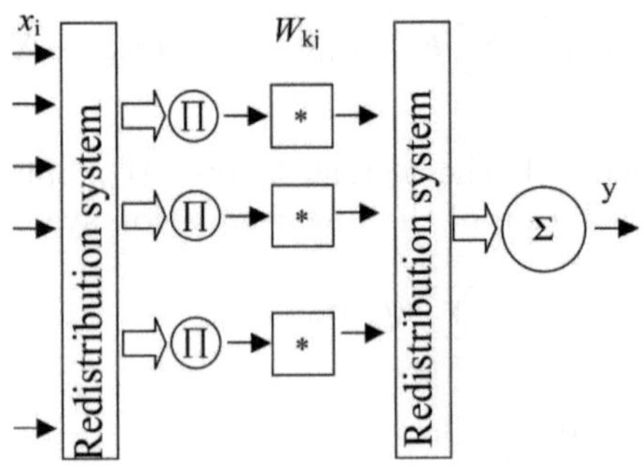

Fig. 7.7. Schematic diagram of the neural calculator of High-Order Neural Networks (HO NN).

Here, L is the number of unequal products of l components of the input signal vector x_i, $y_l = \prod_{q_m} x_{q_m}$. In this product, at m variation, the set of indexes q_m runs all unequal combinations of the numbers of components of the vector x_i in l elements.

The output signal p should estimate a value of the parameter to be determined. In our case, it is the particle radius R and the refractive index n. Training of the organized neural network consists of determination of the weight factors W_l. For this purpose, the sample of parameters R_α, n_α ($\alpha = 1, 2, \ldots, N_B$ is the number of the element and N_B is the total number of elements in the sample) is generated from the range of admissible values. Then, the intensity of the scattered radiation $I(\theta_j)$ is calculated for these parameters by the Mie theory. Next, in accordance with Eq. (7.1), the vectors $x_{\alpha l}$ are found for each α and the matrix y_{lk} is determined by multiplication of the vector components. After that, the solution of the equation system

$$R_\alpha = \sum_{l=1}^{L} y_{\alpha l} W_l^{(R)}, \tag{7.25}$$

$$n_\alpha = \sum_{l=1}^{L} y_{\alpha l} W_l^{(n)} \tag{7.26}$$

by the least square method provides the weight factors of neural networks $W_l^{(R)}$ and $W_l^{(n)}$ for retrieval of R and n.

Here, we consider HO NN of the second, third, fourth, and fifth orders.

7.2.4 Sequences of neural networks

The wider the range of variation of the particle parameters, the more difficult it is to provide a sufficient accuracy of parameter retrieval using one neural network. Therefore, it is necessary to divide the original variation range into some sub-ranges. To retrieve the parameters in each sub-range, the own local neural network has to be created.

Determination of the particle parameters in two steps was proposed in the papers of Ulanowski et al. (1998) and Wang et al. (1999). The principal scheme of the sequence of neural networks is shown in Fig. 7.8. The global neural network is used only to specify the local neural network to be used. The values of the unknown parameter p are obtained from the local network output. To reduce influence of a wrong choice of a local network, the sub-ranges used for formation of the local networks should be overlapped.

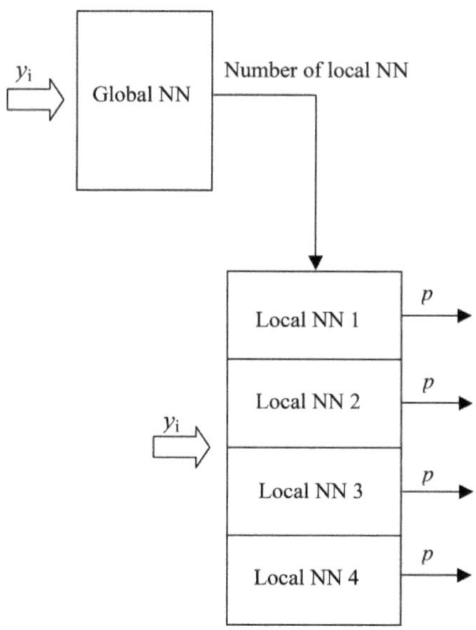

Fig. 7.8. Schematic diagram of Sequences of Neural Networks.

7.3 Formation of training data and input signals

The intensity of the scattered radiation $I_{s\alpha}$ is determined as follows (Newton, 1982; Deirmendjian, 1969; Bohren and Huffmen, 1983; Babenko et al., 2003; Kokhanovsky, 2004):

$$I_{s\alpha} = 2\pi F_0 \int_{\theta_l - \Delta\theta}^{\theta_l + \Delta\theta} \frac{d\sigma}{d\Omega} \sin\theta \, d\theta \,, \tag{7.27}$$

where F_0 is the flux density of the incident radiation; $\frac{d\sigma}{d\Omega}$ is the differential cross-section of scattering; θ_l is the scattering angle for photodetector with number l; and $\Delta\theta$ is a half of the field of view angle.

In the case of illumination of particles by natural radiation,

$$\frac{d\sigma}{d\Omega} = \frac{1}{k_0^2} (i_1 + i_2) \, . \tag{7.28}$$

Here, $k_0 = \frac{2\pi}{\lambda_0}$; λ_0 is the wavelength of the incident radiation; $i_1 = |S_1|^2$ and $i_2 = |S_2|^2$ are the dimensionless Mie intensities; and S_1 and S_2 are the dimensionless complex amplitudes of scattering (Deirmendjian, 1969).

Commonly, it is assumed that the photodetector signals are proportional to the intensities of the scattered radiation $I_l = I(\theta_l)$ and, therefore, are proportional to the integrals I_l of the Mie intensities:

$$I(\theta_l) = \int_{\theta_l - \Delta\theta}^{\theta_l + \Delta\theta} (i_1 + i_2) \sin\theta \, d\theta \, . \tag{7.29}$$

Here, l is number of the detector; and θ_l is the angle for the detector with number l.

Today, methods to calculate Mie intensities are well developed (Deirmendjian, 1969; Bohren and Huffmen, 1983; Babenko et al., 2003; Kokhanovsky, 2004).

We used logarithms of intensities $I(\theta_l)$ (determined by Eq. (7.29)) as a vector of input signals. As it follows from the numerical simulations, this procedure works well for particle radius retrieval. However, the refractive index retrieval errors are rather large (see section 7.4.1). This is due to the complex nature of the dependence of $I(\theta_l)$ on the refractive index. The dependence of the scattered radiation intensity is smoothed out when the polychromatic radiation is used (Fig. 7.9). It is reasonable to consider the logarithms of integrals of the intensities over the wavelength as the input signals as well

$$I^{int}(\theta_l) = \frac{1}{2\Delta\lambda} \int_{\lambda_0 - \Delta\lambda}^{\lambda_0 + \Delta\lambda} I(\theta_l, \lambda) \, d\lambda \, . \tag{7.30}$$

Here, $2\Delta\lambda$ is the width of the spectral range for the incident light. To calculate integral Eq. (7.30), the Gaussian quadrature rules (Abramowitz and Stegun, 1964) of order 100 should be used. This allows one to calculate the integrals of rapidly changing functions of λ with sufficient accuracy for the particles considered in the paper.

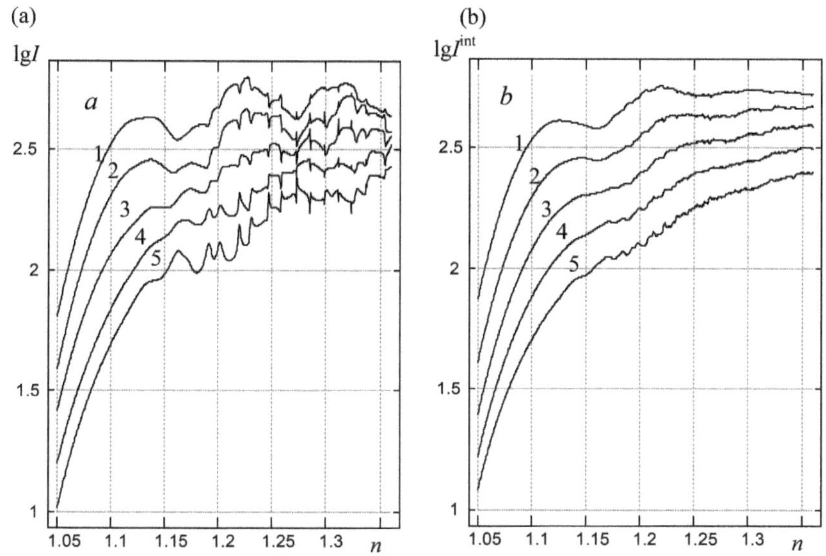

Fig. 7.9. Dependence of (a) $\lg I(\theta)$ and (b) $I^{int}(\theta)$ on particle refractive index n. $\lambda_0 = 0.55\ \mu$m; $\Delta\lambda = 0.1\mu$m; particle radius $R = 4\ \mu$m; $\theta(°) = 25$ (1); 30 (2); 35 (3); 40 (4); 45 (5).

7.4 Testing of neural networks

7.4.1 Parameter retrieval by intensity of scattered radiation

Let us test and compare the efficiency of neural networks by the example of the following problem.

Let the intensity of the scattered radiation $I(\theta_l)$ defined by Eq. (7.29) be known. It is required to retrieve the radius and the refractive index of particles. Figure 7.10a and 7.10b show the dependence of $\lg I(\theta = 30°)$ and $\lg I(\theta = 90°)$ on $\lg R$ and n, respectively. At $\theta = 30°$, the dependences of $\lg I(\theta)$ on $\lg R$ and n are close to the linear ones. The dependences of $\lg I(\theta)$ on $\lg R$ and n in the range of angles θ from $20°$ to $50°$ have similar behavior. Outside this range, the dependence of $\lg I(\theta)$ on $\lg R$ and n are more complex. The last one is illustrated by data displayed in Fig. 7.10b. The complexity of the dependence is shown in the accuracy of the retrieval. Therefore, it is reasonable to use a limited number of angles in the interval of $20°$ to $50°$. Here, the values of $\lg I(\theta_l)$ at $\theta_l = 25°, 30°, 35°, 40°$, and $45°$ at $\Delta\theta = 5°$ are used as a vector of input signals.

We formed a random sample of $N_B = 5 \times 10^3$ points from the range $\lg R\ (\mu\text{m}) \in [0, 1.5]$, $n \in [1.05, 1.35]$ and found the values for $\lg I(\theta_l)$ using the method described in the third section. MLP NN were trained according to these data (see section 7.2.1).

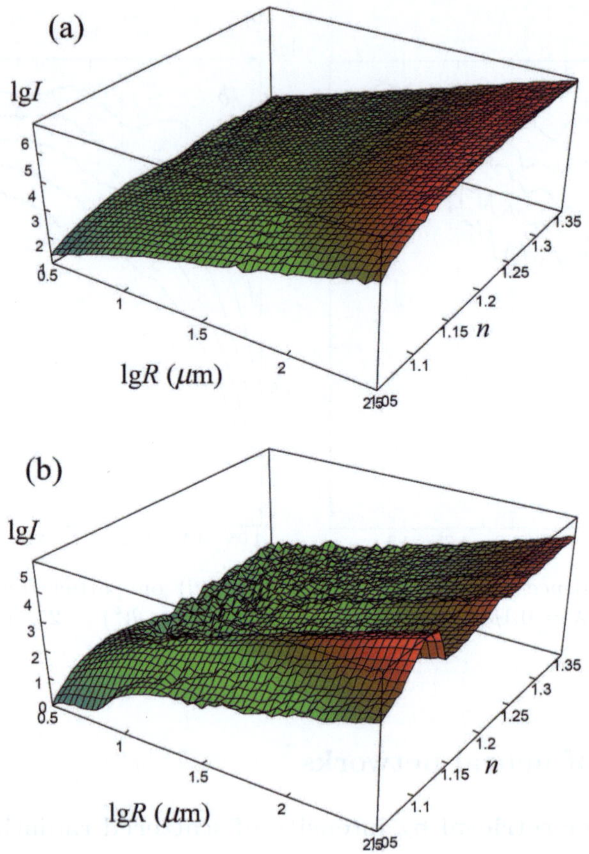

Fig. 7.10. Dependence of (a) $\lg I(\theta = 30°)$ and (b) $\lg I(\theta = 90°)$ on $\lg R$ and n.

The average relative errors of the retrieval of size and refractive index are used to assess the quality of the neural network operation:

$$\overline{\delta}R = \frac{1}{N_B} \sum_{\alpha=1}^{N_B} \delta R_\alpha \,, \tag{7.31}$$

$$\overline{\delta}n = \frac{1}{N_B} \sum_{\alpha=1}^{N_B} \delta n_\alpha \,. \tag{7.32}$$

Here, R_α and n_α are the values of particle radius and refractive index used for calculation of intensities in the example with the number α; $R_{e\alpha}$, $n_{e\alpha}$ are the retrieved parameter values in the example; $\delta R_\alpha = \left| 1 - \dfrac{R_{e\alpha}}{R_\alpha} \right|$ is the relative error of the retrieval of size; and $\delta n_\alpha = 1 - \dfrac{n_{e\alpha} - 1}{n_\alpha - 1}$ is the relative error of the refractive index retrieval.

Testing of neural networks is carried out on the samples that differ from the training ones. In the case of small training samples, the assessments of neural network functionality with teaching and testing samples can differ significantly. In our case ($N_B = 5 \times 10^3$), the relative error estimated by the testing sample is usually 1.02–1.04 times more than the relative error estimated by the training sample.

The dependences of the relative error of particle radius and refractive index retrieval by the neural network with one inner layer, having 10 neurons in the inner layer on the particle radius$R$, are shown in Fig. 7.11. The particle is illuminated by monochromatic radiation with the wavelength $\lambda_0 = 0.55\ \mu$m. The average relative errors of radius and refractive index retrieval are 0.058 and 0.084, respectively.

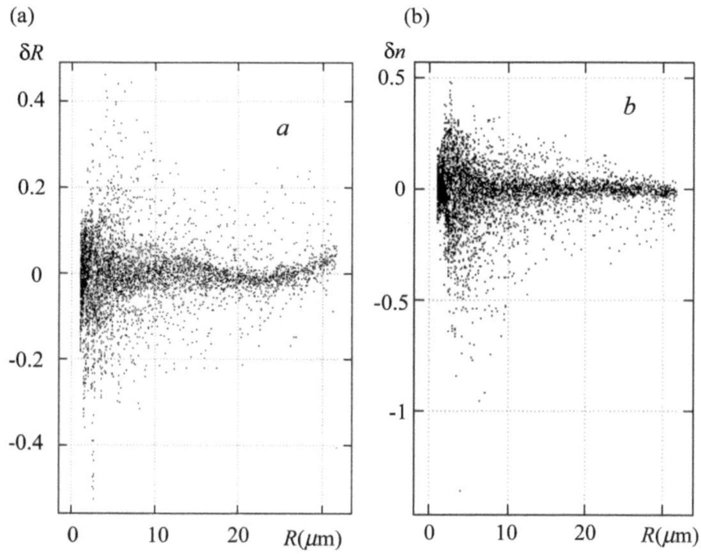

Fig. 7.11. Dependence of the relative error of retrieval of (a) the size and (b) the refractive index on the particle size. The network contains one inner layer with 10 neurons. Monochromatic illumination, $\lambda_0 = 0.55\ \mu$m.

The parameter retrieval error can be considerably reduced when the particles are illuminated by polychromatic radiation. This is due to the dependence of $\lg I^{int}(\theta_l)$ on $\lg R$ and n are much smoother than under the monochromatic radiation (see Figs. 7.12 and 7.13). The relative errors of retrieval of the radius and the refractive index of particles by the neural network with 10 neurons in the inner layer at polychromatic illumination are shown in Fig. 7.14. The average relative retrieval errors are 0.032 and 0.046, respectively.

It is possible to decrease the errors by increasing the number of neurons. The values of relative errors of parameter retrieval using neural networks with 30 neurons in the inner layer are shown in Fig. 7.15. The average relative errors are $\overline{\delta} R = 0.023$ and $\overline{\delta} n = 0.030$, respectively.

However, such a way significantly increases the time required for the neural networks training. This is illustrated by data given in Fig. 7.16. It displays the dependences of the objective function Φ (see Eq. (7.7)) on the number of training cycles N_C. Each training cycle contains 2,000 addresses to calculate the objective

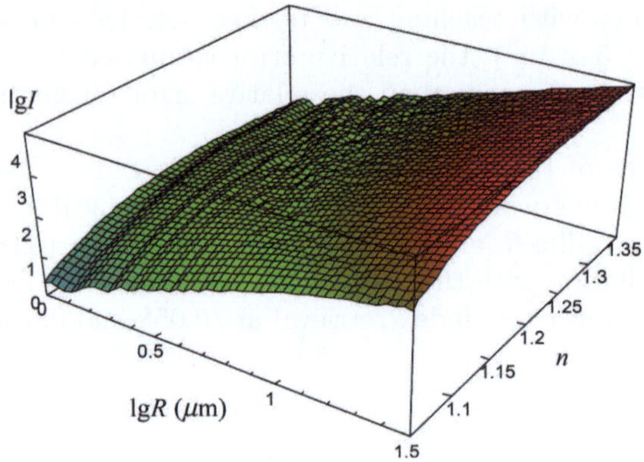

Fig. 7.12. Dependence of $\lg I(\theta = 30°)$ on $\lg R(\mu m)$ and n. Monochromatic illumination, $\lambda_0 = 0.55$ μm.

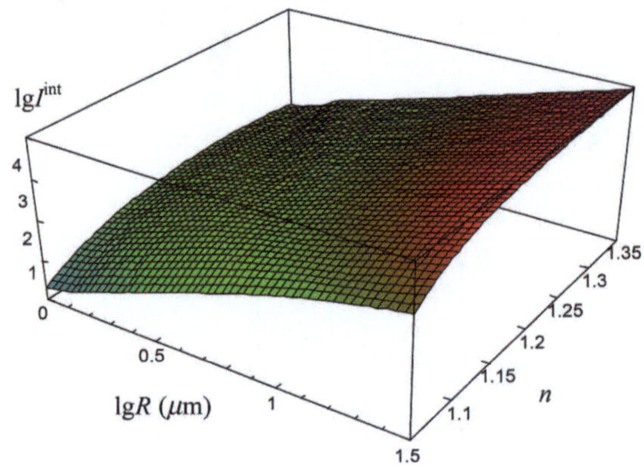

Fig. 7.13. Dependence of $\lg I(\theta = 30°)$ on $\lg R(\mu m)$ and n. Polychromatic illumination, $\lambda_0 = 0.55$ μm, $\Delta\lambda = 0.1$ μm.

function. The time required to complete one cycle depends on the number of neurons in a neural network, the number of layers, and the size of a training sample. For a three-layer neural network with one inner layer containing 30 neurons, it takes about 15 minutes on a computer with a 3.2 MHz central processing unit (CPU). As can be seen from the figure, training time for a neural network with 30 neurons is about three to four times longer that the training time for a network with 10 neurons.

So far, we have considered the functionality of neural networks on the assumption that the intensities of scattered radiation are known precisely. However, in practice, measurements are always carried out with errors. Therefore, it is important to estimate the resistance of neural networks to errors of input data. Rigorous assessments of the influence of measurement errors on particle parameter retrieval results should be based on an analysis of a specific measurement scheme. Generally, the influence of measurement errors is analyzed by addition of a noise to input signals and evaluation of parameter retrieval errors according to the noisy signals.

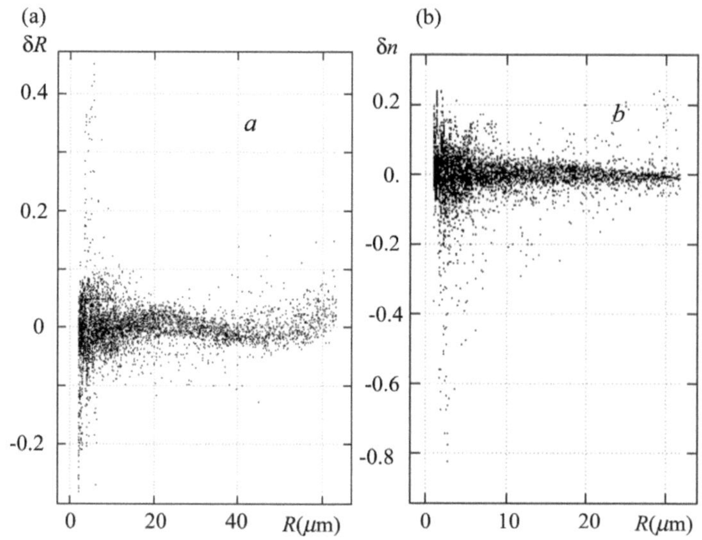

Fig. 7.14. Dependence of the relative error of retrieval of (a) the radius and (b) the refractive index on the particle radius. The neural network with one inner layer contains 10 neurons. The integrals over the wave length range from $\lambda_0 - \Delta\lambda$ to $\lambda_0 + \Delta\lambda$. $\lambda_0 = 0.55\ \mu$m, $\Delta\lambda = 0.1\ \mu$m.

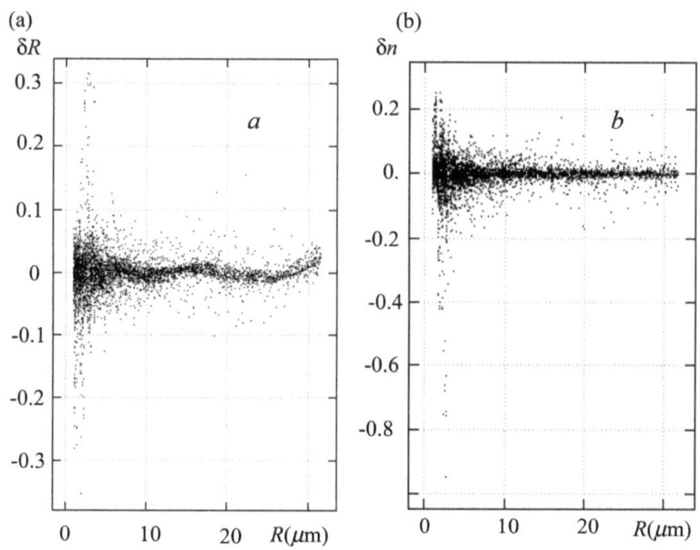

Fig. 7.15. Dependence of the relative error of retrieval of (a) the size and (b) the refractive index on particle size. The neural network with one inner layer contains 30 neurons. The integrals over the wavelength range from $\lambda_0 - \Delta\lambda$ to $\lambda_0 + \Delta\lambda$. $\lambda_0 = 0.55\ \mu$m, $\Delta\lambda = 0.1\ \mu$m.

At that, there is additional noise caused by dark noise of the photodetector and multiplicative noise mainly caused by errors of an amplification circuit. We assume that the intensity of a radiation source is rather large to neglect the dark noise of a

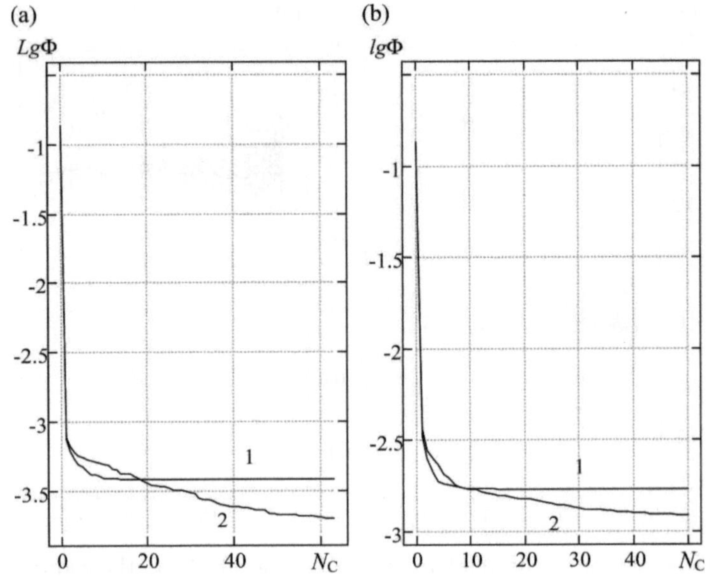

Fig. 7.16. Dependence of the logarithm of the objective function used in the neural networks training for retrieval of (a) the radius and (b) the refractive index on the number of training cycles, N_C. Neural networks contain 10 (curve 1) and 30 (curve 2) neurons in the inner layer.

photodetector. The multiplicative noise is simulated by multiplying the radiation intensity $I^{int}(\theta_l)$ by the value of $1+u$:

$$I^{int}(\theta_l) = I^{int}(\theta_l) \cdot (1+u)\,. \tag{7.33}$$

Here, u is the random variable with a Gaussian distribution, having zero mathematical expectation and standard deviation (SD) δ_e.

Figure 7.17 shows the dependences of the average relative error of parameter retrieval as a function of standard deviation δ_e. As can be seen from the figure, neural networks with 10 neurons in the inner layer for determining the radius are more resistant to measurement errors at high noises, when $\delta_e > 0.035$. However, in the range of low noise, the neural network with 30 neurons provides better accuracy in particle size retrieval.

Let us consider the RBF NN test results. The arrangement of nodes for networks with 10 and 30 neurons is displayed in Fig. 7.18. The values of relative errors of parameter retrieval using the network with 30 neurons are shown in Fig. 7.19. The average relative error of size retrieval is $\overline{\delta R} = 0.021$, with refractive index retrieval $\overline{\delta n} = 0.031$. Figure 7.20 shows the dependences of the average relative error of parameter retrieval using the RBF NN as a function of δ_e. The neural network with 30 neurons provides better parameter retrieval accuracy than the network with 10 neurons in the range of low noises. At $\delta_e < 0.015$, however, it has large noise sensitivity. At $\delta_e > 0.02$, its accuracy is lower than the accuracy of the neural network with 10 neurons.

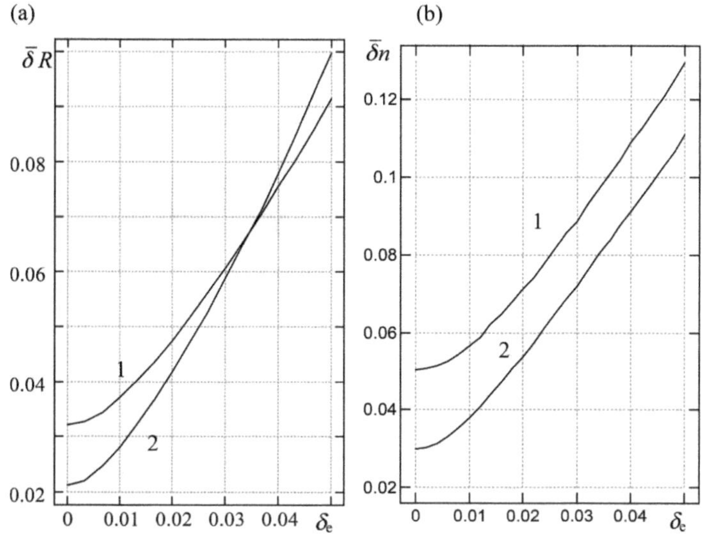

Fig. 7.17. Dependences of the average relative error of retrieval of 9a) the radius and (b) the refractive index using a three-layer neural network containing 10 neurons in the hidden layer (curve 1) and 30 neurons (curve 2) on standard deviation δ_e of the multiplicative noise.

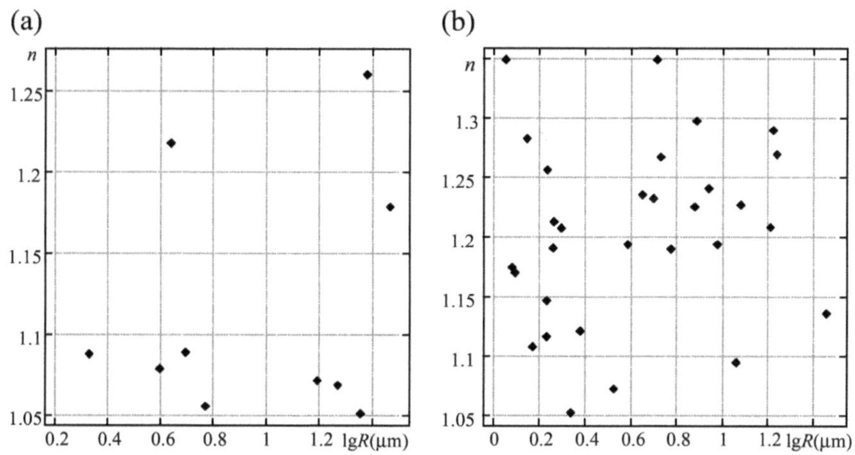

Fig. 7.18. Arrangement of the nodes in the plane (lg R, n) when forming a Radial Basis Function Neural Network (RBF NN) with (a) 10 and (b) 30 nodes.

Test results for the HO NN of orders 2, 3, and 4 are shown in Fig. 7.21. Table 7.2 provides the values $\bar{\delta}R$ and $\bar{\delta}n$ at zero noise. As seen from the given data, the methodological errors decrease as the order of the neural network increases. However, at that, the sensitivity of the network to multiplicative noise increases. When root-mean-square (r.m.s.) error exceeds two-hundredths, the parameter retrieval errors grow with the network order.

Now let us consider operation of Sequences of Neural Networks.

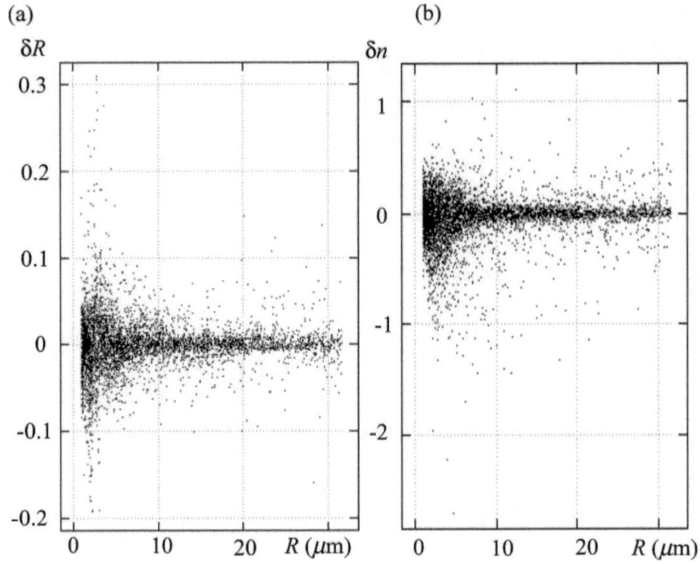

Fig. 7.19. Dependence of the relative error of retrieval of (a) the size and (b) the refractive index on the particle size. The Radial Basis Function Neural Network (RBF NN) with one inner layer contains 30 neurons. The integrals over the wavelength range from $\lambda_0 - \Delta\lambda$ to $\lambda_0 + \Delta\lambda$. $\lambda_0 = 0.55\ \mu$m, $\Delta\lambda = 0.1\ \mu$m.

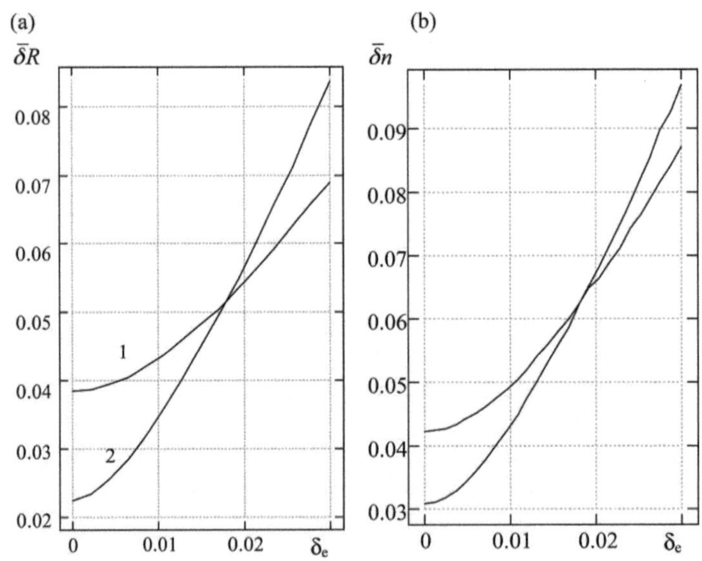

Fig. 7.20. Dependences of the average relative error of retrieval of (a) the radius and (b) the refractive index using a Radial Basis Function Neural Network (RBF NN) with 10 neurons in the hidden layer (curve 1) and 30 neurons (curve 2) on standard deviation δ_e of the multiplicative noise.

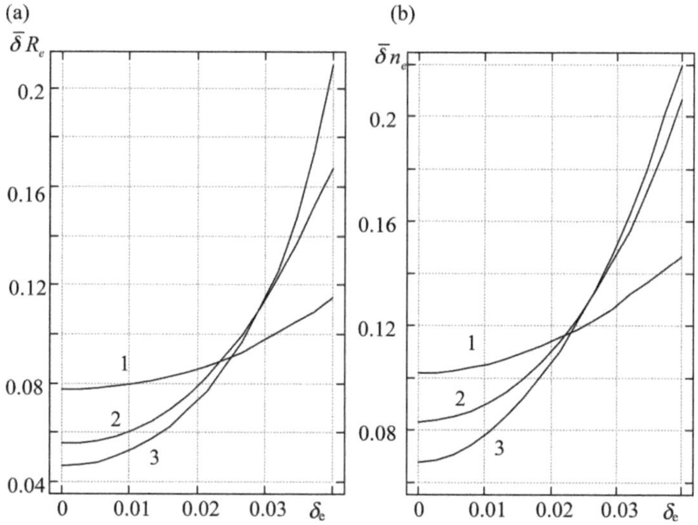

Fig. 7.21. Dependences of the average relative error of retrieval of (a) the radius and (b) the refractive index using High-Order Neural Networks (HO NN) of orders 2 (curve 1), 3 (curve 2), and 4 (curve 3) on standard deviation δ_e of the multiplicative noise.

Table 7.2. Retrieval errors at zero measurement noise.

Order of the HO NN	$\bar{\delta} R$	$\bar{\delta} n$
2	0.078	0.102
3	0.056	0.083
4	0.047	0.068

As we noted above, the sequence of neural networks contains a global neural network and a set of local neural networks. The global network operates on the original variation range of parameters ($0 \le \lg R \le 1.5$ (radius R is measured in μm), $1.05 \le n \le 1.35$). It is used to determine the local neural network to specify the particle parameters. We divided the initial variation range of the parameter $p_0 = \lg R$ into two parts ($0 \le \lg R < 0.75$ and $0.75 \le \lg R \le 1.5$). After that, we formed the MLP NN

$$d = y_{p0}^{out}(W, y) . \tag{7.34}$$

Here, $y = \lg I^{int}(\theta)$ is the vector of input signals applied to the input of the network; and W is the set of weight factors of the neural network: $W_{jk}^1, W_{kj}^2, \ldots, W_l^N$.

Then, we trained the network under the condition

$$\sum_{\alpha} \left(y_{p0}^{out} \left(W, y_\alpha \right) - V_\alpha \right)^2 = \min . \tag{7.35}$$

Here, $\alpha = 1, 2, \ldots, N_B$ is the number of the example from the training set for neural networks,

$$V_\alpha = \begin{cases} -0.25 & 0 \le \lg R_\alpha < 0.75 \\ 0.25 & 0.75 \le \lg R_\alpha \le 1.5 \end{cases} . \tag{7.36}$$

We assigned the particle to the first class if the discrimination function $y_{p0}^{out} = F_{p0}(W, y_\alpha) < 0$ (in this case, $0 \le \lg R < 0.75$) and to the second class if $y_{p0}^{out} = F_{p0}(W, y_\alpha) = 0$ (in this case, $0.75 \le \lg R \le 1.5$).

Similarly, to determine the refractive index, we divided the initial variation range of the parameter $p_1 = n$ into two parts ($1.05 \le n < 1.2$ and $1.2 \le n \le 1.35$) and formed the MLP NN

$$d = y_{p1}^{out}(W, y) . \tag{7.37}$$

We trained the network under the condition

$$\sum_\alpha \left(y_{p1}^{out}(W, y_\alpha) - V_\alpha\right)^2 = \min . \tag{7.38}$$

Here,

$$V_\alpha = \begin{cases} -0.25 & 1.05 \le n_\alpha < 1.2 \\ 0.25 & 1.2 \le n_\alpha \le 1.35 \end{cases} . \tag{7.39}$$

The particle relates to the first class if the discrimination function $y_{p1}^{out} = F_{p1}(W, y_\alpha) < 0$ (in this case, $1.05 \le n < 1.2$) and to the second class if $y_{p1}^{out} = F_{p1}(W, y_\alpha) = 0$ (in this case, $1.2 \le n \le 1.35$). These classes cannot be unambiguously divided because they are overlapped in the variablesy_α.

The histograms of distribution of the first and second classes, when determining the radius and refractive index, are shown in Fig. 7.22. The regions of parameter overlapping can be found from there while training the local neural networks.

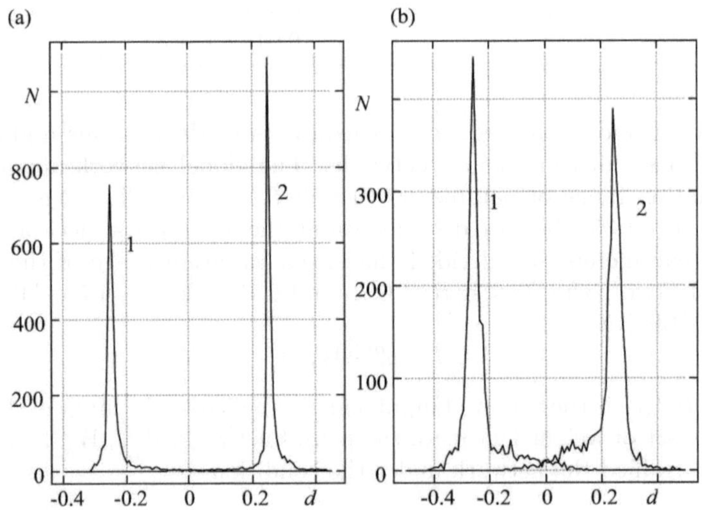

Fig. 7.22. Histograms of the distribution of the number of particles N for the first (curve 1) and the second (curve 2) classes of particles at discrimination functions with three-layer neural networks and 30 neurons in the inner layer for (a) radius and (b) refractive index determination.

Table 7.3 shows the boundaries of the initial variation range of parameters denoted as G and local ranges denoted as L0, L1, L2, and L3. These boundaries were used to generate samples for the training of global and local neural networks.

Table 7.3. Regions of particle parameters at the networks training.

Region	$\lg R_{\min}$ (μm)	$\lg R_{\max}$ (μm)	$n_{\min}$	$n_{\max}$
G	0	1.5	1.05	1.35
L0	0	0.86	1.05	1.23
L1	0	0.86	1.17	1.35
L2	0.51	1.5	1.05	1.23
L3	0.51	1.5	1.17	1.35

The test results of a sequence of neural networks and a neural network that is trained on the original variation range of parameters are shown in Fig. 7.23. At small noise, the use of local networks trained on ranges four times smaller than the original one allows reducing twice the particle parameter retrieval error. At significant noise, the error of parameter retrieval using the sequence of networks is less than the error of particle parameter retrieval using a single neuron network.

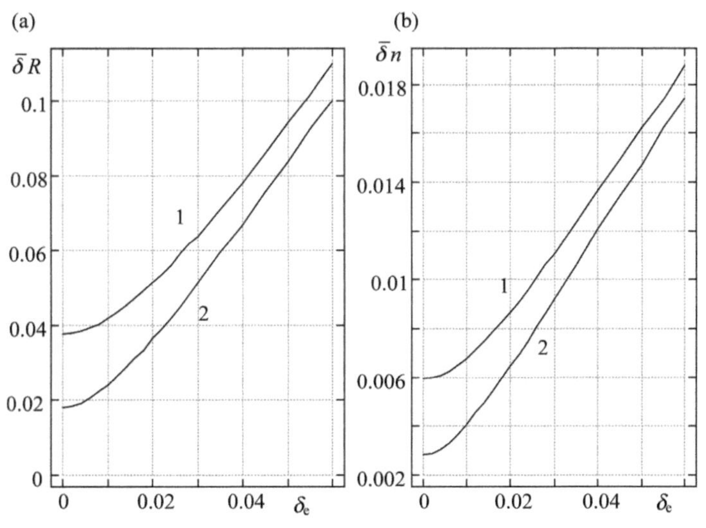

Fig. 7.23. Dependences of the average relative error of retrieval of (a) the radius and (b) the refractive index using neural networks containing 30 neurons in the hidden layer (curve 1) and the sequence of neural networks (curve 2) on standard deviation δ_e of the multiplicative noise.

7.4.2 Parameter retrieval by non-standardized intensities of scattered radiation

Above, we compared the operation of the neural networks for retrieval of particle parameters by measurements of the intensities $I(\theta_l)$ and $I^{int}(\theta_l)$ of the scattered radiation (see Eqs (7.29) and (7.30)).

In practice, the intensities $I(\theta_l)$ and $I^{int}(\theta_l)$ themselves are not measured, but some signals $U(\theta_l)$ and $U^{int}(\theta_l)$ are. Usually, they are proportional to intensities $I(\theta_l)$ and $I^{int}(\theta_l)$:

$$U(\theta_l) = aI(\theta_l)\,, \tag{7.40}$$

$$U^{int}(\theta_l) = bI^{int}(\theta_l)\,. \tag{7.41}$$

Here, a and b stand for the instrument constants.

The signal normalization process associated with the determination of instrument constants can be extremely difficult and time-consuming. Therefore, the investigation of possibilities of retrieving the particle parameters by the non-standardized signals is of particular interest.

If the field of view $(2\Delta\theta)$ and the particle size are not too large, the dependence of the intensity of the scattered radiation on the scattering angle has characteristic beatings. They are illustrated in Fig. 7.24. The frequency of these beatings strongly correlates with the size of the particles. This correlation is used to determine the size of particles (Semyanov et al., 2004; Berdnik and Loiko, 2010; Berdnik et al., 2006; Berdnik and Loiko, 2009). This frequency in Semyanov et al. (2004) is found by the position of the maximum of the Fourier transformation module from the product of the Hanning function and the angular dependence of the scattered radiation. For the angular dependence of the scattered radiation in the range of angles from 10° to 70°, the average relative error of radius retrieval at the unknown refractive index of particles is 3.6% (Semyanov et al., 2004).

The correlation between the characteristic frequency k and the particle radius R was used in (Berdnik and Loiko, 2010) as well. Here, the networks with particle radius and refractive index varied in the range of 0.6 μm $< R <$ 13.6 μm, $1.015 < n < 1.28$ are considered. However, the characteristic frequency k was determined otherwise. The signal $Y(\theta_l, R, n)$ was applied instead of the measured signal $U(\theta_l, R, n)$. It was organized as

$$Y(\theta_l, R, n) = \frac{U(\theta_l, R, n)}{\overline{I}(\theta_l) \cdot \dfrac{1}{N_L} \cdot \displaystyle\sum_{l=1}^{N_L} \dfrac{U(\theta_l, R, n)}{\overline{I}(\theta_l)}}\,. \tag{7.42}$$

Here, $\overline{I}(\theta_l) = \dfrac{1}{N_B} \cdot \displaystyle\sum_{\alpha=1}^{N_B} I(\theta_l, R_\alpha, n_\alpha)$ is the averaged (with respect to the parameters under consideration) angular dependence of the scattered radiation.

The signal determined by Eq. (7.42) is invariant with respect to the linear homogeneous transformations. Therefore, the intensities of the scattered radiation calculated according to the Mie theory are used to model function $Y(\theta_l, R, n)$. Note that the variation range of the signal $Y(\theta_l, R, n)$ at the angle variation is less than

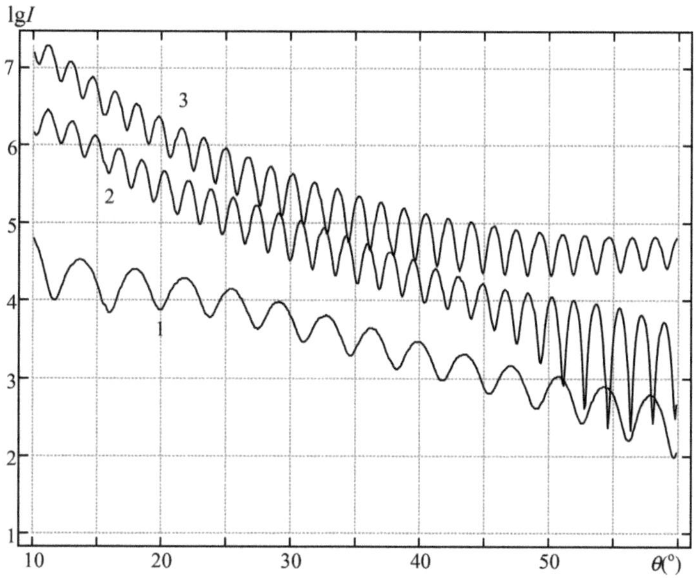

Fig. 7.24. Dependence of $\lg I$ on scattering angle θ for particles with $R = 4$ μm, $n = 1.2$ (curve 1), $1 + \lg I$ on θ for particles with $R = 8$ μm, $n = 1.1$ (curve 2), $2 + \lg I$ on θ for particles with $R = 8$ μm, $n = 1.05$ (curve 3).

the variation range of the intensity of the scattered radiation, as the division by $\overline{I}(\theta_l)$ excludes the greatly elongated middle component characteristic for optically soft particles.

The next step is to find the projections y_s of the signal $Y(\theta_l, R, n)$ on the vectors of the Karhunen–Loeve basis (Fukunaga, 1990):

$$y_s = \sum_l Y(\theta_l, R, n) \cdot H^Y_{ls} \,. \tag{7.43}$$

Here, H^Y_{ls} is the first K ($K = 4, 5, 6, 7$) vectors of the covariance matrix $C^Y_{ls} = \dfrac{1}{N_B} \cdot \sum\limits_{\alpha=1}^{N_B} Y(\theta_l, R_\alpha, n_\alpha) \cdot Y(\theta_s, R_\alpha, n_s)$ and to calculate the signal$Z(\theta_l)$:

$$Z(\theta_l) = Y(\theta_l, R, n) - \sum_{s=1}^{K} y_s H^Y_{ls} \,. \tag{7.44}$$

The characteristic frequency k was determined as a frequency whereby the modulus of the Fourier transform of Eq. (7.44) attains the maximum.

$$k = \arg\max \left| \sum_{l=1}^{N_L} Z(\theta_\alpha) \exp\left(-i\frac{2\pi k(l-1)}{N_L}\right) \right| \,. \tag{7.45}$$

As shown by the simulations, if the values of the intensity of radiation scattered in the range of [10°, 60°] are used, then at $K = 4$ and variation of parameters in the range of 0.6 μm $< R <$ 13.6 μm, $1.015 < n < 1.28$, the position of the global maximum of the modulus of the Fourier transform is an unambiguous and smooth function of R. Therefore, efficient algorithms for sorting in ascending order can be used to find the characteristic frequency.

A sequence of transformations of the measured (initial) signal and the Fourier transform in accordance with the above-described algorithm of the characteristic frequency k determination is illustrated by Figs 7.25–7.27. The initial signal (a) and its Fourier transform (b) are shown in Fig. 7.25. The result of the division of the initial signal into the averaged angular dependence of the scattered radiation $\overline{I}(\theta)$ and Fourier transform of the pre-processed signal are shown in Fig. 7.26a and 7.26b, respectively. The Fourier transform contains a considerable value for the low-frequency components. Such behavior is caused by the polynomial trend of the initial signal. Figure 7.27a and 7.26b show the processed signal (a) and its Fourier transform (b) after the polynomial trend exclusion. By this operation, the characteristic frequency is determined with sufficient accuracy.

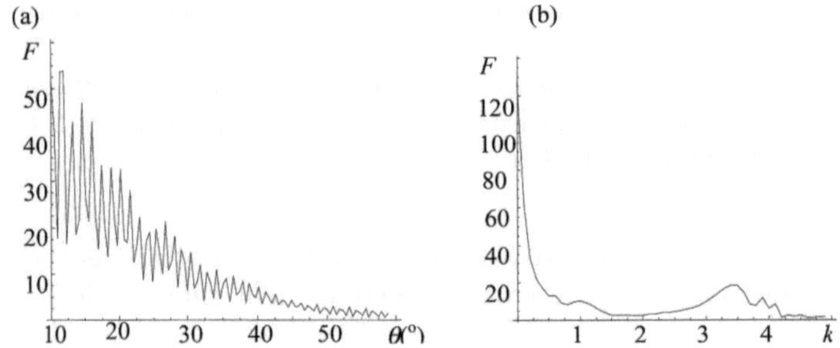

Fig. 7.25. (a) Initial signal and (b) its Fourier transform.

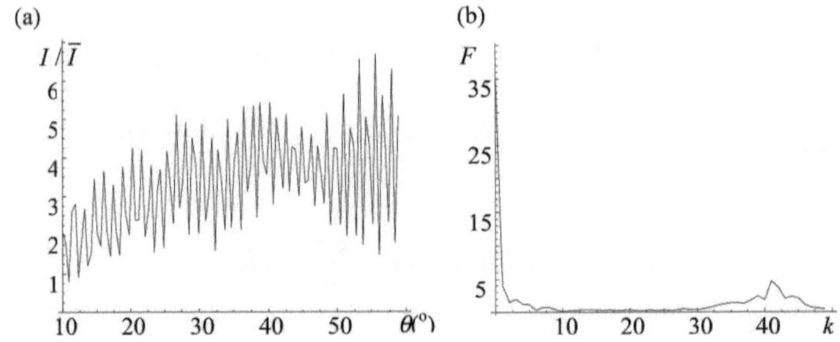

Fig. 7.26. (a) Pre-processed signal and (b) its Fourier transform after division by $\overline{I}(\theta)$.

Correlation dependence between the particle radius R and the characteristic frequency k found according to Eqs (7.42)–(7.45) is shown in Fig. 7.28a. As seen from this figure, generally the correlation is nearly linear. It can be used for particle radius determination. If a linear approximation of dependence of the radius on k is used, then the average relative error of radius retrieval is 0.023. That is 1.5 times less

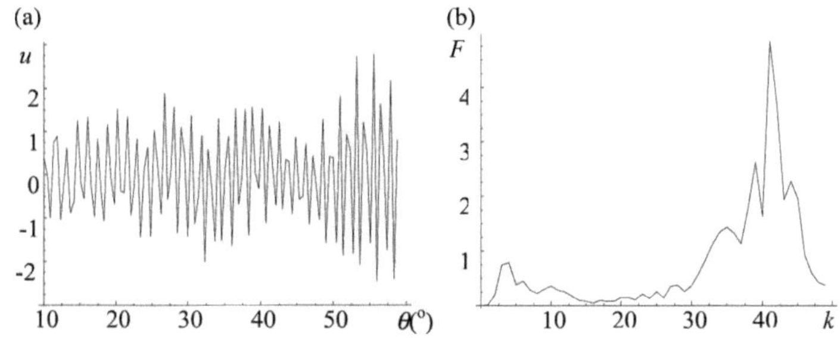

Fig. 7.27. (a) Processed signal and (b) its Fourier transform after exclusion of the polynomial trend.

than the error of 0.036 obtained in Semyanov et al. (2004) when using the Hanning function. In Fig. 7.28b, this dependence is enlarged for small sizes of particles. In this region, the linearity is broken. Therefore, to describe the correlation $R = f(k)$, in Berdnik and Loiko (2010), the approximation by quadratic spline (de Boor, 1978) was used on the grid: k_i = 1.828, 2.292, 2.755, 3.219, 3.682, 4.146, 4.61, 5.073, 5.536, 6, 49.168:

$$R = \varphi(k) \approx \sum_{m=1}^{12} B_m(k) S_m \,. \tag{7.46}$$

Here, m is the number of the basis spline (total number of quadratic basis splines on this grid is equal to 12); $B_m(k)$ are the basis splines; and S_l are spline factors that were defined by the least-squares method.

At such approximation of the function $\phi(k)$, the average relative error of particle radius retrieval is 0.018. It is reasonable to consider the function $\Psi(R, n) = R - \varphi(k)$. As can be seen from Fig. 7.29, the surface curvature is not strong. There-

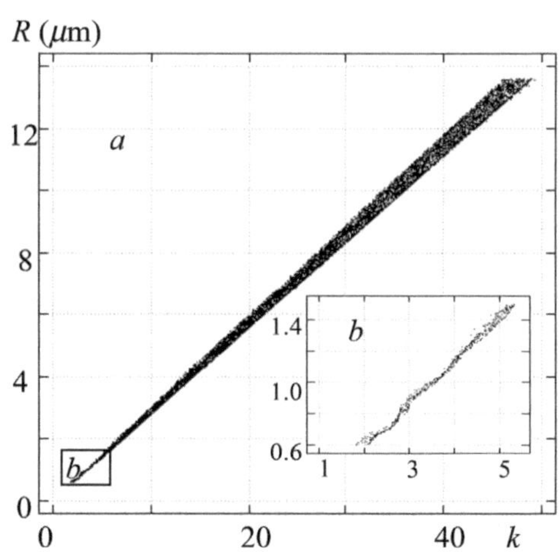

Fig. 7.28. Correlation dependence of the radius on the characteristic frequency k.

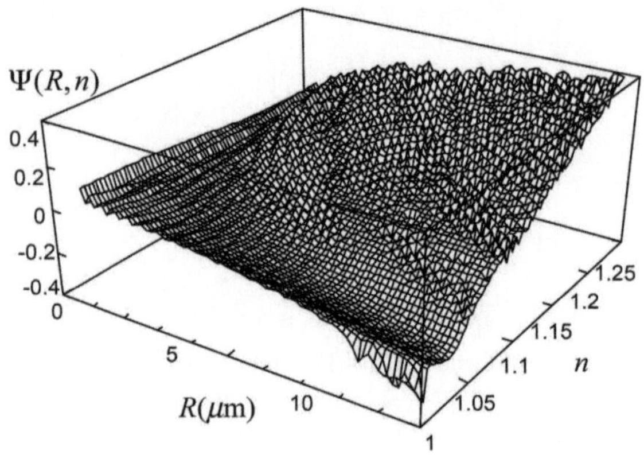

Fig. 7.29. Dependence of function $\Psi(R,n) = R - \varphi(k(R,n))$ on R and n.

fore, one can reduce the retrieval error if the function $\Psi(R,n)$ can be approximated by simply calculated functionals of the signal $Y(\theta_l, R, n)$.

The multilayer MLP NN are applied for this purpose (Berdnik and Loiko, 2010). To form the vector of input signals, the authors determine the characteristic frequency$k$; the parameters y_m defined by Eq. (7.44); and the projections z_s of the $|Z(\theta_l)|$ (see Eq. (7.45)) to the vectors $H_{ls}^{|Z|}$:

$$z_s = \sum_l |Z(\theta_l)| \cdot H_{ls}^{|Z|} . \tag{7.47}$$

Here, $|Z| = \begin{cases} Z & Z \geq 0 \\ -Z & Z < 0 \end{cases}$; and $H_{ls}^{|Z|}$ are the first K vectors of the covariance matrix: $C_{ls}^{|Z|} = \frac{1}{N_B} \cdot \sum_{\alpha=1}^{N_B} |Z(\theta_l, R_\alpha, n_\alpha)| \cdot |Z(\theta_s, R_\alpha, n_\alpha)|$.

The determined values of k (Eq. (7.45)), y_s (Eq. (7.43)), and z_s (Eq. (7.47)) are normalized as described in section 7.2. They are used as a vector of input parameters for a neural network. The weight factors of the network were found from the condition

$$\sum_{\alpha=1}^{N_B} \left(R_\alpha - \varphi(k_\alpha) - \Phi(W, y_\alpha^{in})\right) = \min . \tag{7.48}$$

When the factors of the neural network approximating the function $\Phi(y^{in}(R,n))$ are found, the particle radius is defined as

$$R = \phi(k) + \Phi(W, y^{in}) . \tag{7.49}$$

Let us present the data for the error of radius retrieval using such a neural network.

Figure 7.30 shows the dependence of the relative error of radius retrieval $\delta R = \left|1 - \frac{R_e}{R}\right|$, wherein R_e is the retrieved radius value, on the original values of the radius and the refractive index of particles. When finding the function$\Phi(W, y^{in})$,

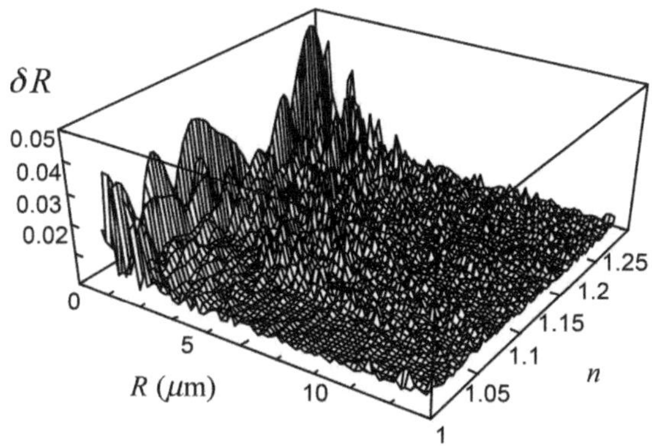

Fig. 7.30. Dependence of the relative error of the radius retrieval δR on R and n. No measurement errors.

the neural network with one inner layer containing 40 neurons is used. As can be seen in the figure, the maximum errors in the range of $R < 2$ μm attains the values $\delta R = 0.05$.

The average relative error $\overline{\delta R}$ of radius retrieval was estimated as the arithmetic average of relative errors with respect to 10^4 samples. Its value is equal to 0.003 if the multiplicative noise $\delta_e = 0$. Figure 7.31 shows the variation of the average relative error $\overline{\delta R}$ as a function of the multiplicative noise when using Eqs (7.46) and (7.49). At the approximation of the function $\Phi(W, y^{in})$ in accordance with Eq. (7.49), the data are given for the three-layer feed-forward neural network with 40 neurons in the inner layer. As can be seen from Fig. 7.31, the determination of radius by Eq. (7.49) reduces the error by five to six times compared with the approximation by Eq. (7.46) if the noise is small and $\delta_e < 0.1$. However, the sensitivity of the approximation by Eq. (7.46) to noise is significantly lower than by Eq. (7.49). At $\delta_e > 0.4$, approximation by Eq. (7.49) gives better results than by Eq. (7.9).

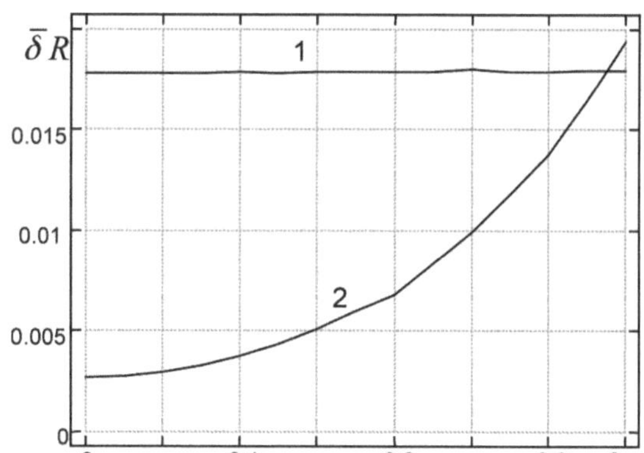

Fig. 7.31. Dependence of the average relative error of the radius retrieval $\overline{\delta R}$ on standard deviation of the multiplicative noise δ_e calculated by Eq. (7.46) (curve 1) and by Eq. (7.49) (curve 2).

7.4.3 Parameter retrieval by experimental data of scanning cytometry: comparison with the fitting method

Experimental verification using data of the flow cytometry has some features. It is a problem to select some of the measured particles and measure the size of the same particles by a standard microscopic method. Therefore, it is reasonable to compare the results of the processing of experimental data obtained by the neural network method and the fitting method (Berdnik et al., 2006; Berdnik and Loiko, 2009). The results of such an analysis carried out in Berdnik et al. (2006) with the HO NN are given in Table 7.4.

Table 7.4. Size and refractive index determined by the fitting method and by the High-Order Neural Networks (HO NN) with the experimental data on the scattered light intensity.

Particle type	Radius, R (μm)			Refractive index, n		
	Retrieval by neural network	Retrieval by fitting method	Relative difference (%)	Retrieval by neural network	Retrieval by fitting method	Relative difference (%)
Polystyrene bead	1.480	1.507	1.8	1.578	1.581	0.2
Polystyrene bead	2.220	2.208	0.5	1.686	1.720	2.0
Sphered erythrocyte	3.154	3.027	4.2	1.123	1.074	4.4

From Berdnik et al. (2006).

The experimental angular dependences of the intensity of scattered light obtained by the method of scanning flow cytometry (Maltsev and Semyanov, 2004) for two types of particles together with calculated angular dependences at parameters obtained with the MLP NN and the fitting method are shown in Figs 7.32 and 7.33. The results of the particle parameter retrieval by means of the neural networks and the fitting method are shown in Table 7.5 (Berdnik and Loiko, 2009) by the experimental data of Maltsev and Semyanov (2004). The following function was used as a criterion:

$$F = \sum_{l=1}^{N_L} \left(\lg \frac{U_{el}}{\frac{1}{L}\sum_{l=1}^{L} \frac{U_{el}}{\overline{U}_{el}}} - \lg \frac{U_l(p)}{\frac{1}{L}\sum_{l=1}^{L} \frac{U_l(p)}{\overline{U}_l}} \right)^2 . \tag{7.50}$$

Here, U_{el} is the experimental data; and $U_l(p)$ are the results of calculations by the Mie theory. The relative deviation of parameters found by the neural network and the fitting method does not exceed 4.3%.

7.5 Absorbing particles

Absorption significantly complicates the retrieval problem. Figure 7.34 illustrates the dependence of scattered radiation on the imaginary part of the refractive index

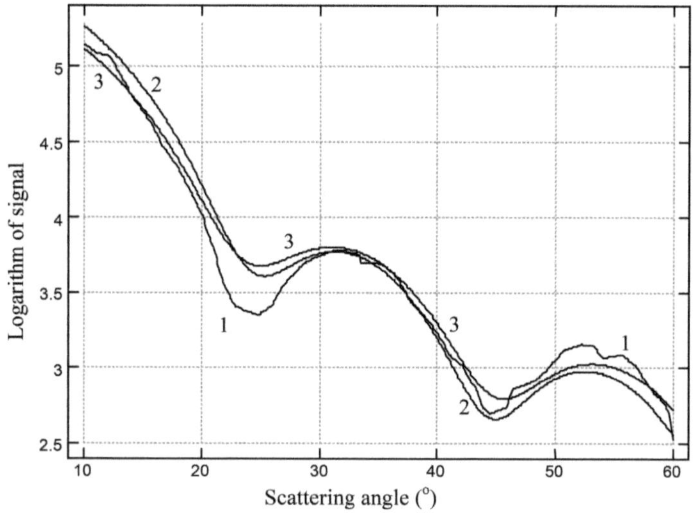

Fig. 7.32. Angular dependence of input signal for a particle of type 1: experimental data obtained by scanning flow cytometry (Dennis and Schnabel, 1983) (curve 1); results of calculation obtained under the Mie theory with the particle parameters retrieved by the neural network (curve 2); results of calculation obtained under the least-squares method (curve 3).

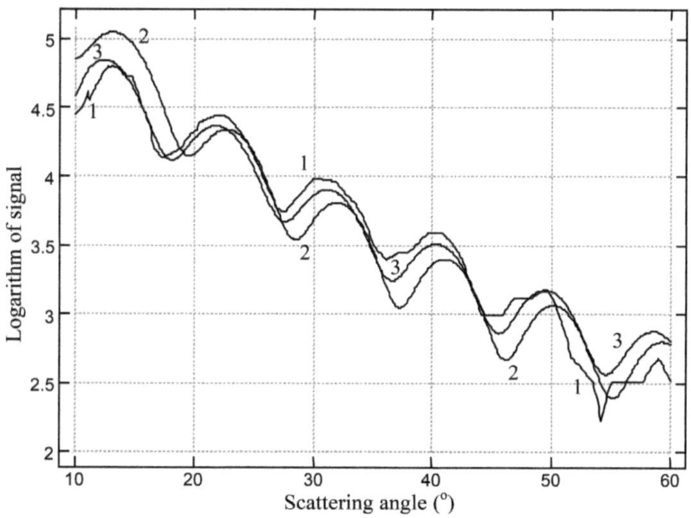

Fig. 7.33. Angular dependence of input signal (proportional to the scattered light intensity) for a particle of type 2: experimental data obtained by scanning flow cytometry (Dennis and Schnabel, 1983) (curve 1); results of calculation obtained under the Mie theory with the particle parameters retrieved by the neural network (curve 2); results of calculation obtained under the least-squares method (curve 3).

k. One can select a region of low absorption ($k < 0.03$), in which the intensity of the scattered radiation decreases with increasing absorption, and the region of strong

Table 7.5. Size and refractive index retrieved by the fitting method and by the Multilayer Perceptron Neural Networks (MLP NN).

Radius, R (μm)			Relative refractive index, n		
Retrieval by neural network	Retrieval by fitting method	Relative difference (%)	Retrieval by neural network	Retrieval by fitting method	Relative difference (%)
0.705	0.679	3.8	1.166	1.218	1.5
1.559	1.536	4.3	1.122	1.170	4.1

Experimental data from Maltsev and Semyanov (2004).

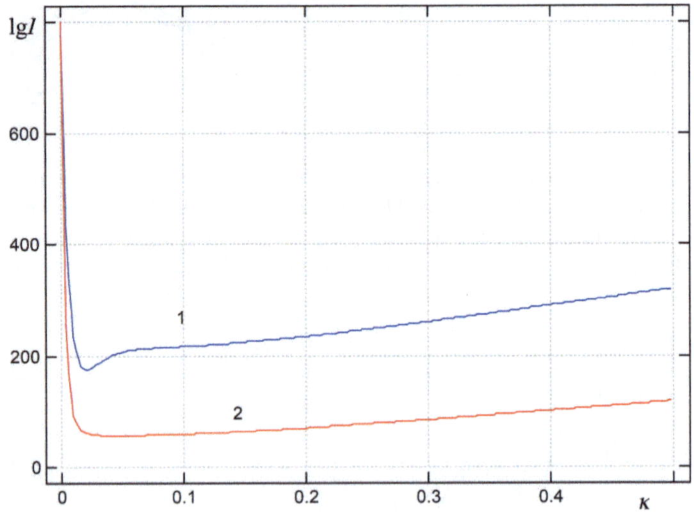

Fig. 7.34. The dependence of $\lg I(\theta = 45°)$ on the imaginary part of the refractive index k for particles with $R = 4$ μm, refractive index $n = 1.2$ (curve 1) and for particles with $R = 10$ μm, $n = 1.2$ (curve 2).

absorption ($k > 0.03$), in which the intensity of the scattered radiation increases with k.

Similarly to Berdnik et al. (2006), to unify the consideration, we show here the retrieval by data on the light intensity scattered in the range of angles from 10° to 60°. The ranges of the radius R, real n, and imaginary k parts of the refractive index are 0.6–10.6 μm, 1.02–1.38, and 0–0.03, respectively, which are the same as in Berdnik et al. (2006). It is worth noting that these intervals include range parameters of many types of biological cells.

To retrieve the particle parameters, the HO NN is used. Remember that the neural network consists of the preprocessor, signals distributor, layer of neurons-multiplicators, layer of synapses-amplifiers, and neuron-summator. The main advantages of the HO NN are the small time in training and the reliability. Training is reduced to the well-investigated problem of the solution of a system of linear algebraic equations.

As indicated above, the characteristic frequency of the angular dependence of scattered light correlates with size of the spherical particles. To select this frequency, the fast Fourier transform of the angular dependence of scattered light is used. However, for some particles, the intensity of components of the Fourier transform at low frequencies exceeds the intensity of the peak related to the size parameter of the particle. Thus, to isolate the characteristic frequency better, we found the ratio of intensity of scattered light I_j to intensity $\overline{I}_j$ averaged over all realizations of particle parameters (see Figs 7.25–7.27). Then, from the obtained dependence, we subtracted a polynomial trend. The coefficients of a polynomial trend contain information on size and refractive index of a particle. We found the first seven coefficients p_α of the expansion of the ratio $h_j = I_j/\overline{I}_j$ in the Legendre polynomials, number k_m, corresponding to maxima of the module of the discrete Fourier transform from function $\frac{I_j}{\overline{I}_j} - \sum_{\alpha=0}^{6} p_\alpha P_\alpha(\theta_j)$, and magnitude of this maxima F_m. The value of k_m correlates with particle size. Coefficients p_α and the value of F_m contain additional information about the size and refractive index of a particle. We used relative magnitudes p_α/p_0 and F_m/p_0 to exclude the dependence on the absolute value of intensity of scattered light.

To generate components of a vector of input signals, we used the generalized mean value (GMV) function $m(q)$ as well. This function is widely used in the theory of stochastic functions and applications. GMV function is defined for a system of N positive values h_j by the expression (Nigmatullin and Smith, 2005)

$$m(q) = \left(\frac{1}{N} \sum_{j=1}^{N} (h_j)^q \right)^{1/q} . \tag{7.51}$$

At $q = -1$, Eq. (7.51) determines the value of the harmonic mean; at $q = 1$, it determines the value of the arithmetic mean. Typical behavior of the GMV function $m(q)$ for some particles is illustrated by Fig. 7.35 (Berdnik et al., 2006).

As a vector of input signals, we used the following parameters: $\frac{p_\alpha}{p_0}, k_m, \frac{F_m}{p_0}, \frac{m(q)}{p_0}$ ($\alpha = 0, 1, 2$; $q = 0.1, 0.5, 5, 10$). Preprocessor forms a vector of input signals:

$$x_j = 1, \frac{p_1}{p_0}, \frac{p_2}{p_0}, k_m, \frac{F_m}{p_0}, \frac{m(q{=}0.1)}{p_0}, \frac{m(q{=}0.5)}{p_0}, \frac{m(q{=}5)}{p_0}, \frac{m(q{=}10)}{p_0}, \quad j = 0, \ldots, 8. \tag{7.52}$$

The signals distributor groups components of a vector of input signals on s ($s = \overline{1,9}$) elements and transfers them to the input of neurons, which find the product of the signals. Signals y_l from the outputs of neurons-multiplicators are multiplied by the weight coefficients w_l and summarized by a neuron-summator. The result of calculations is the magnitude p:

$$p = \sum_{l=1}^{L} W_l y_l , \tag{7.53}$$

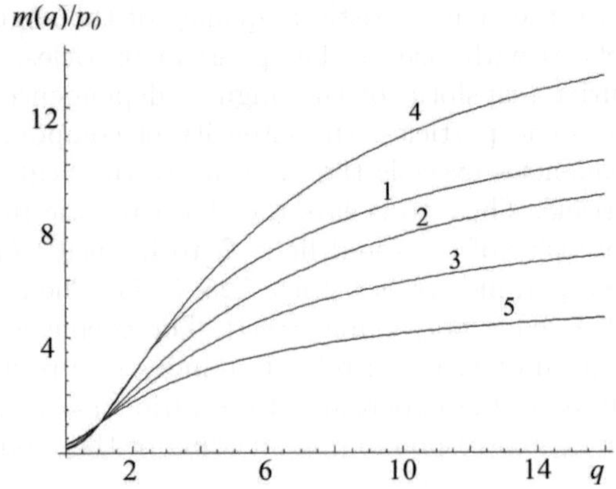

Fig. 7.35. The generalized mean value (GMV) function $m(q)/p_0$ for non-absorbing particles at $R = 0.7$ μm, $n = 1.1$ (curve 1); $R = 4$ μm, $n = 1.1$ (curve 2); $R = 9$ μm, $n = 1.1$ (curve 3); $R = 4$ μm, $n = 1.05$ (curve 4); and $R = 4$ μm, $n = 1.15$ (curve 5).

where L is the amount of unequal products of components of a vector of input signals x_j. Depending on the magnitude of s, we determine one, two, ..., and s-point neural networks.

The output signal p estimates the determined parameter. We deal with a particle radius R, real n, and imaginary k parts of the refractive index. The training of a neural network consists in determination of weight factors W_l. For this aim, from the interval of allowable values, we formed a sample of parameters R_i, n_i, k_i. The number of elements $i = 1, 2, \ldots, N_B$ (N_B is the dimension of the sample). The intensity of scattered light $I(\theta_j)$ was calculated using the Mie theory (Deirmendjian, 1969; Bohren and Huffmen, 1983; Babenko et al., 2003; Kokhanovsky, 2004). According to Eq. (7.53), for each i, we found vectors x_{ji} and determined matrix y_{li}, making products of the components of vector x_{ji}. Then, to retrieve values R, n, and k, we found weight coefficients of a neural network $W_l^{(R)}$, $W_l^{(n)}$, and $W_l^{(k)}$ by solving the systems of equations

$$R_i = \sum_{l=1}^{L} y_{il} W_l^{(R)}, \tag{7.54}$$

$$n_i = \sum_{l=1}^{L} y_{il} W_l^{(n)}, \tag{7.55}$$

$$k_i = \sum_{l=1}^{L} y_{il} W_l^{(k)} \tag{7.56}$$

using the least-squares method.

The matrix y_{il} is usually ill-posed. To solve the system (Eqs (7.54)–(7.56)), we used a pseudo-inverse matrix constructed by a method of singular expansion. In the construction of that matrix, we used first N singular numbers; the remaining ones we supposed as equal to zero.

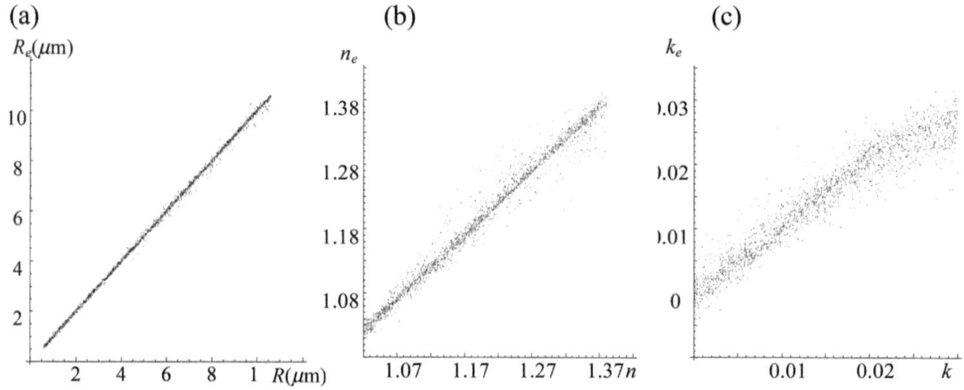

Fig. 7.36. Correlation dependencies of initial values of parameters R, n, and k, and values of (a) R_e, (b) n_e, and (c) k_e retrieved by a correlation High-Order Neural Network (HO NN).

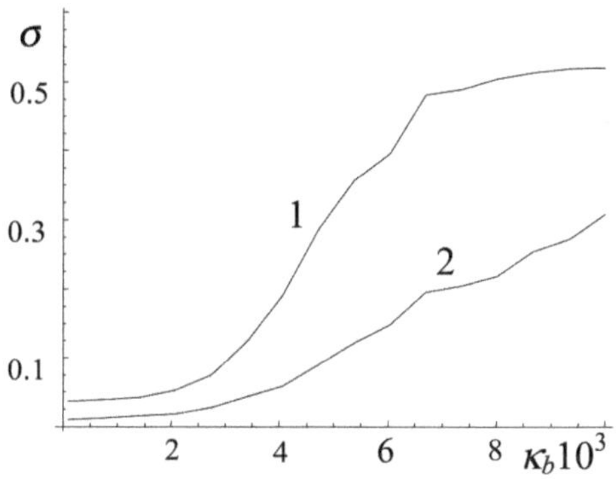

Fig. 7.37. Retrieval errors of radius σR_e and real part σn_e of the refractive index for absorbing particles as function k_b (k_b is an upper limit of the k-interval).

Correlation dependencies between the initial values of parameters R, n, and k, and values of R_e, n_e, and k_e retrieved by the correlation HO NN are shown in Fig. 7.36 (Berdnik et al., 2006).

The errors in the retrieval of radius R and the real part n of the refractive index of the absorbing particles are displayed in Fig. 7.37 (Berdnik et al., 2006). A thousand values of the intensity scattered by absorbing particles with the imaginary part of the refractive index uniformly distributed in the range $[0, k_b]$ (k_b is an upper limit of the k-interval) are given at the input of the net.

The errors in the retrieval of σR_e and σn_e shown in Fig. 7.37 are determined using the results of the retrieval of R and n by the neural network trained in non-absorbing particles. As seen from the displayed results, it is possible to use the neural network trained for non-absorbing particles to retrieve parameters R and n at $k_b < 0.002$.

7.6 Sizing of soft spheroidal particles

In this section, we consider the applicability of neural networks for sizing nonspherical and non-absorbing particles (Berdnik et al., 2004b, 2004c).

Let us have a spheroidal particle with refractive index n and semi-axes a (its direction coincides with the rotation axes of spheroid) and b illuminated by a plane wave with intensity I_0 with the wave vector parallel to the z-axis (Fig. 7.38). The symmetry axis of the particle is oriented in a direction determined by axial and azimuthal angles θ_0 and φ_0. These angles are not known beforehand and vary in intervals $[0, \pi/2]$ and $[0, 2\pi]$, respectively. The refractive index is not known beforehand also and is supposed to be in the range $[1.01, 1.02]$. The intensity of scattered light in directions θ_i $(i = 1, 2, \ldots, N_\theta$, where N_θ is the number of θ-angles) and φ_j $(i = 1, 2, \ldots, N_\varphi$. N_φ is the number of φ-angles) is measured. It is necessary to retrieve the lengths of semi-axes a and b from these data; the value a/b is the aspect ratio. Our calculations have shown that the best results will be obtained if one uses the radius of the equivolume sphere $R = (ab^2)^{1/3}$ and shape parameter $e = (a - b)/(a + b)$ while training a neural network.

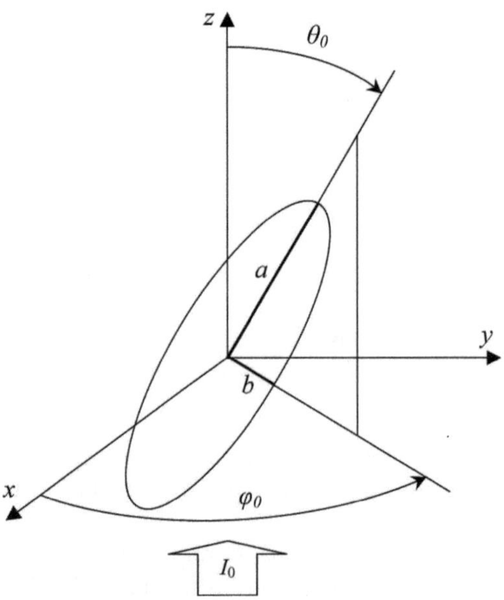

Fig. 7.38. Schematic drawing of a particle illumination. a is the major semi-axis; b is the minor semi-axis. The symmetry axis of the particle is oriented in a direction determined by axial and azimuthal angles θ_0 and φ_0, respectively. I0 is the incident light intensity.

To describe the scattered light intensity, the Rayleigh–Gans–Debye approach (RGD) (Ishimaru, 1978; Barber and Wang, 1978) is used. This approach is valid under the condition of $|n - 1| \cdot \max(a, b) \ll 1$. In this approach, only two terms of the amplitude-scattering matrix S_1 and S_2 $(S_2 = \cos\theta \cdot S_1)$ are distinct from zero. The simple formulae for calculation of intensity of light scattered by some types of nonspherical particles (spheroids, cylinders, etc.) are obtained. If the refractive index is close to unity, the particle sizes can vary in a wide interval to investigate the peculiarities of neural network construction.

The artificial neural network used to process the information contains two or more neural layers connected with each other (Berdnik et al., 2004c; Berdnik and Loiko, 2006). Neurons of the first layer execute a distributive function. They serve to transfer initial signal y_i to the inputs of the neurons of the next layer. Neurons of the second layer summarize input signals with their weights. Activation function F of this sum is an output signal and serves as an estimation of the desired parameters p_α. Usually, there is a sense adding a component, equal to unity, to a vector of input signals. It permits one to increase the accuracy of the approximation of the desired function.

Neural networks with a linear activation function can be used. In this case,

$$p_\alpha = \sum_{i=0}^{N_n} W_{\alpha i} y_i \,. \tag{7.57}$$

Here, $\alpha = 1, 2, \ldots, N_p$; N_n is the number of neurons of the first layer, equal to the dimensionality of an input vector; N_p is the number of neurons of the second layer, equal to the number of desired parameters; and $W_{\alpha i}$ is the matrix of weight factors.

The calculations (Berdnik et al., 2004b, 2004c) show that the best results in retrieval can be gained using multilevel neural networks. Its design is as follows.

While training a neural network of the first level, the training sample is created by using the whole initial range of parameters. As a result, the $W_{\alpha i}^{1,l_1}$-factors of the first-level network ($l_1 = 1$) are obtained. Then, the initial range of parameters is divided onto some intervals not overlapped with numbers $l_2 = 1, 2, \ldots, n_2$, where n_2 is the number of subintervals of the second level. For each of these intervals, the neural network with factors $W_{\alpha i}^{2,l_2}$ is trained. The set of these coefficients forms a neural network of the second level. As the variation of particle parameters in each interval is less than the initial range, the error of the parameter reconstruction by a neural network of the second level is decreased. The neural network of the first level serves to determine the set of $W_{\alpha i}^{2,l_2}$ factors.

Due to the errors of parameter determination by a neural network of the first level, the classification of particles cannot be carried out precisely. Therefore, one should train networks of the second level on the overlapped intervals. It has allowed us to reduce requirements to the accuracy of the classification of particles by a first-level network. One can continue this procedure and get networks of the third, the fourth, and the higher levels with factors $W_{\alpha i}^{3,l_3}$, $W_{\alpha i}^{3,l_3}$, etc.

A sketch of the processing by a three-level neural network is shown in Fig. 7.39. At the beginning, signal y is processed by a first-level neural network, which is trained on the initial range in which parameters change. The received value $p_\alpha = \sum_{i=0}^{N_n} W_{\alpha i}^{1,l_1} y_i$ is transferred to range l_2 of the second level to evaluate a second-level local neural network (it deals with $W_{\alpha i}^{2,l_2}$ factors). The input signal y is processed by this neural network. It is used to select a third-level local neural network (it deals with $W_{\alpha i}^{3,l_3}$ factors). The processing result from this network serves as an estimation of the p_α parameters.

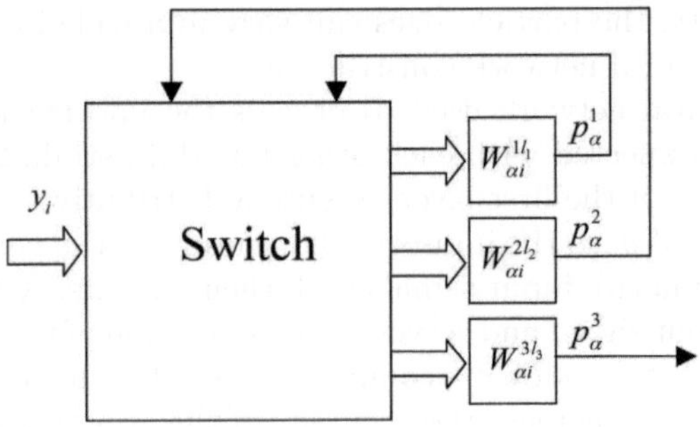

Fig. 7.39. Schematic representation of the three-level neural network.

A training sample of vectors y_n was formed by a random choice of particle parameters and orientations from the set of intervals. The module of the discrete Fourier transformation on coordinate φ from function $|S_1 = S_1(\theta_i, \varphi_i)|$ was used as a pattern vector y. It is invariant to the constant on the lattice shift. It has allowed one to reduce the disorder of pattern vectors while a particle is turned around the z-axis. To calculate $S_1 = S_1(\theta_i, \varphi_i)$ for spheroidal particles, we used analytical formulae (Ishimaru, 1978; Barber and Wang, 1978).

While training the global neural network (Nigmatullin and Smith, 2005), the equivolume radius of spheroids and aspect ratio a/b varied from 0.3 to 1.5 μm and 0 to 0.5, respectively; the ranges of θ_0 and φ_0 were from 0 to 0.5π and 0 to 2π, respectively. The refractive index of particle n varied from 1.01 to 1.02; $\lambda = 0.5$ μm. While training the local networks, the range of R values was divided into six intervals: 0.3–0.55, 0.45–0.75, 0.65–0.95, 0.85–1.15, 1.05–1.35, and 1.25–1.5 μm. While training the two-level local network, the range of aspect ratio a/b values was divided into intervals: 0.2–0.55, 0.25–0.75, 0.45–0.95, and 0.65–1.

The results of testing of the one- and three-level networks for determination of parameters R, e, and θ_0 for prolate spheroids are shown in Figs 7.40 and 7.41 (Berdnik et al., 2004c) (note that, in these figures and below, the angle is determined in radians). Values R_e, e_e, and θ_{0e} in these figures (as above) denote the proper parameters R, e, and θ_0 calculated by a neural network. Comparison of Figs 7.40 and 7.41 shows that the three-level neural network essentially increases the accuracy of the retrieval parameters.

The calculations show that the retrieval errors change, depending on particle parameters. To investigate this dependence, the process of the training of single-level neuron networks in a small range of particle parameters and the process of parameter retrieval by these networks should be simulated many times. An error of retrieval of prolate ellipsoids parameters (standard deviation, σ) as a function of e is shown in Fig. 7.42. The error increases dramatically if the shape of the particle is close to spherical. Especially, it strongly takes place at the retrieval of particle orientation.

To investigate the stability of the neural networks to the measuring errors, the random multiplicative error was added to the results of calculations of S_1 and the variance of the error of parameter retrieval by the results of multiple trials was determined. The calculations have shown that, when the measurement error

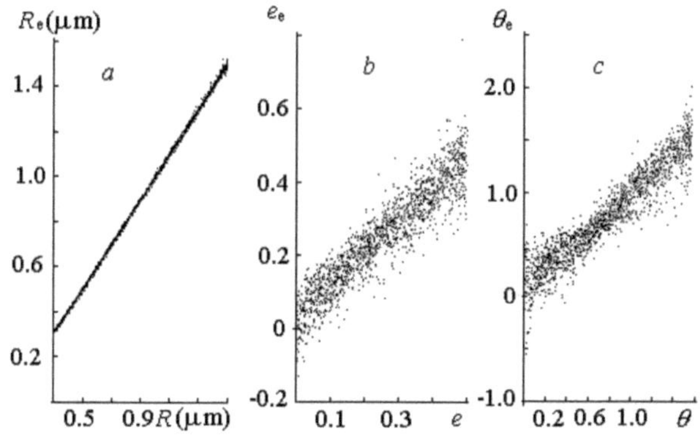

Fig. 7.40. Values of R_e, e_e, and θ_{0e} retrieved by the one-level neural network.

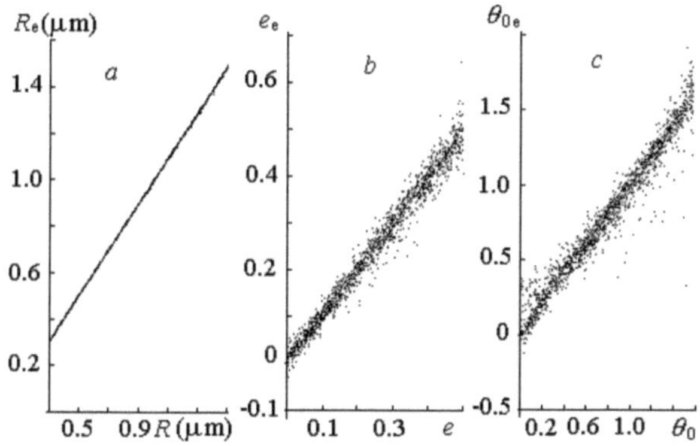

Fig. 7.41. Values of R_e, e_e, and θ_{0e} retrieved by the three-level neural network.

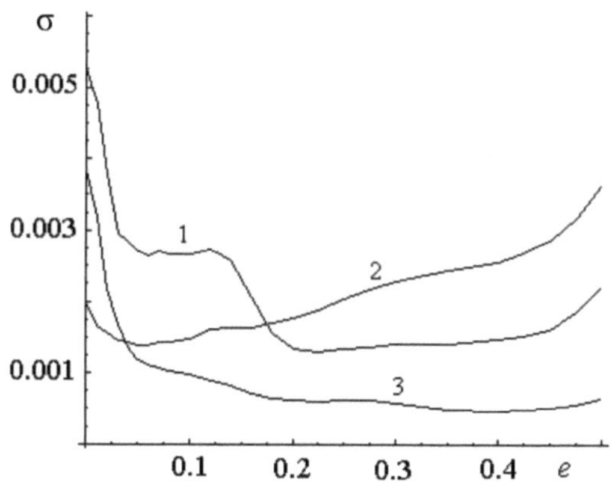

Fig. 7.42. Standard deviation σR_e of the retrieved value of the radius (curve 1), standard deviation σ_e of the retrieved value of shape parameter (divided into 10) (curve 2), standard deviation for the axial angle orientation θ_{0e} (divided into 100) (curve 3) versus shape parameter e. Radius of the equivolume sphere $R = (ab^2)^{1/3}$ is changed in the range from 0.95 to 1.0 μm.

is increasing, the retrieval errors of multilevel neural networks grow very quickly if the training of a neural network is carried out on the precise data (with no noise). However, if a small random error in measurement data is added to a training sample, the errors of the neural network retrieval increase at small noise, but decrease (to an acceptable quantity) at the large errors of measurement.

Figure 7.43 illustrates the dependencies of the errors of parameter retrieval for one-, two-, and three-level neural networks on the measurement error δ_e for oblate spheroids. The neural networks were trained at the training error $\delta_{\text{tr}} = 0.5\%$. The errors of the retrieval by a three-level neural network are the following: $\sigma_R = 0.003$ μm, $\sigma_e = 0.02$, $\sigma\theta_0 = 6°$. The coefficients of the influence (Berdnik et al., 2004c) of multiplicative measuring errors are $\eta R = 0.0015$ μm/%, $\eta R = 0.011$/%, $\eta\theta = 2°/\%$.

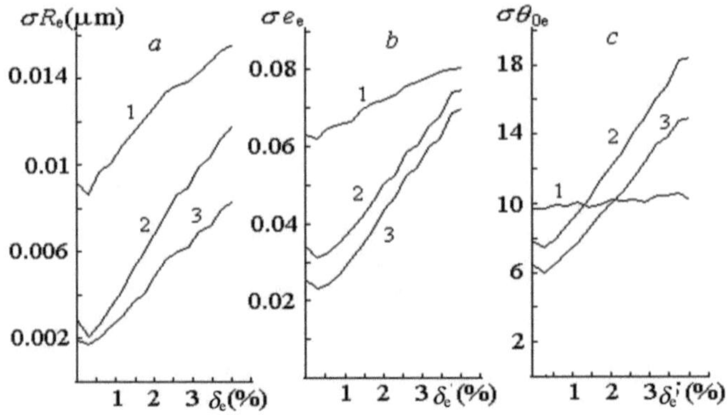

Fig. 7.43. Standard deviation in retrieved parameters R_e, e_e, and θ_{0e} of prolate spheroids versus the measurement error δ_e. The neural network was trained with noise in the input data. The training error $\delta_{\text{tr}} = 0.5\%$. One-level neural network (curve 1); two-level neural network (curve 2); three-level neural network (curve 3).

7.7 Sizing of spheroidal and cylindrical particles in a binary mixture

The problem of the retrieval of particle characteristics in a polydisperse ensemble is discussed in Berdnik et al. (2004a), and Berdnik and Loiko (2005). In Berdnik et al. (2004a), spherical particles are considered. Retrieval parameters of nonspherical particles in binary mixtures containing elongated and flattened spheroids and spheroids and cylindrical particles is considered in Berdnik and Loiko (2005). The method is based on the preliminary separation of the particles into classes using linear and quadrupole discrimination functions.

To retrieve the particle characteristics taken from a binary mixture, it is necessary to determine a particle kind, and then to use methods for the determination of particle sizes and orientation described in the previous section. To solve this problem, the discrimination function of an images vector $d(y)$ can be used. By its value, one can determine a kind of vector y. Two methods to construct the discrimination function are used in Berdnik and Loiko (2005).

The first one calculates the discrimination function as

$$d(y) = \sum_{i,j} C^1_{ij} \left(y_i - \overline{y}^1_i\right) \left(y_j - \overline{y}^1_j\right) - \sum_{i,j} C^2_{ij} \left(y_i - \overline{y}^2_i\right) \left(y_j - \overline{y}^2_j\right). \tag{7.58}$$

Here, $\overline{y}^1_i = (1/N_1) \sum_{n=1}^{N_1} y^{1n}_i$ is the mean value of an images vector for the first class; $\overline{y}^2_i = (1/N_2) \sum_{n=1}^{N_2} y^{2n}_i$ is the mean value of an images vector for the second class; and N_1 and N_2 are the number of elements of the first and the second class, respectively.

$$C^1_{ij} = \left(\frac{1}{N_1} \sum_{n=1}^{N_1} \left(y^{1n}_i - \overline{y}^1_i\right) \left(y^{1n}_j - \overline{y}^1_j\right) \right)^{-1} \tag{7.59}$$

is the covariance matrix of the first images;

$$C^2_{ij} = \left(\frac{1}{N_2} \sum_{n=1}^{N_2} \left(y^{2n}_i - \overline{y}^2_i\right) \left(y^{2n}_j - \overline{y}^2_j\right) \right)^{-1} \tag{7.60}$$

is the covariance matrix of the second images.

In the second method, reference vectors of the first $y^{1\alpha}_i$ and the second $y^{2\alpha}_i$ classes (α is the number of a vector) are chosen. The discrimination function is determined under the formula

$$d(y) = \sum_{\alpha} \frac{1}{\sum_{ij} C^1_{ij}(y_i - y^{1\alpha}_i)(y_j - y^{1\alpha}_j)} - \sum_{\beta} \frac{1}{\sum_{ij} C^2_{ij}(y_i - y^{2\beta}_i)(y_j - y^{2\beta}_j)}. \tag{7.61}$$

The distributions $N(d)$ of particle numbers N for prolate and oblate spheroids determined by the linear on C^1_{ij} (Eq. (7.59)) and C^2_{ij} (Eq. (7.60)) discrimination function Eq. (7.58) are shown in Fig. 7.44 (Berdnik et al., 2004a). The distributions determined by the quadrupole discrimination function (Eq. (7.61)) (Gorban and Rossiev, 1996) are shown in Fig. 7.45 (Berdnik and Loiko, 2005). In both cases, the discrimination functions for oblate and prolate spheroids are overlapped. It causes an error at particle classification.

The errors of parameter retrieval for a single particle taken from the mixture of particles are shown in Fig. 7.46. Here, the retrieval parameters of spheroidal particles with $e = (a - b)/(a + b)$ from -0.5 to 0.5 are shown. Values R_e, e_e, and θ_{0e} in this figure denote the proper parameters R, e, and θ_0 calculated by a neural network. The radius of the equivolume sphere is retrieved adequately, irrespective of the spheroid respect ratio. As seen from Fig. 7.46b, the wrong determination of a particle kind leads to errors in the retrieval of the shape parameter e. The errors in retrieval particle orientation are increased, too.

The retrieval parameters of a particle from the mixture of prolate spheroidal and cylindrical particles are shown in Fig. 7.47 (Berdnik and Loiko, 2005). As in the previous case, the radius of the equivolume sphere is retrieved with high accuracy. The errors in the retrieved shape parameter e and particle orientation increase in comparison with the case in which we deal with particles of only one kind.

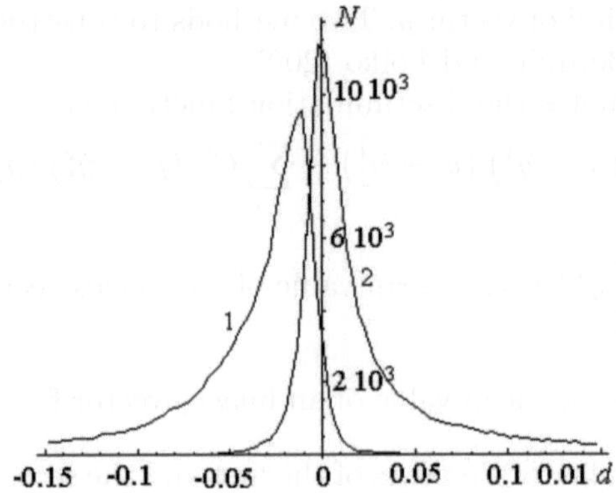

Fig. 7.44. Distribution functions of particle number – linear discrimination function. 1, oblate spheroids; 2, prolate spheroids.

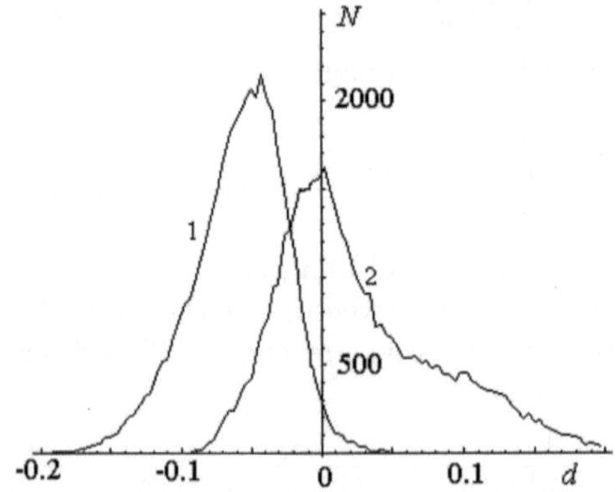

Fig. 7.45. Distribution functions of particle number – quadrupole discrimination function. Number of reference vectors is 50. 1, oblate spheroids; 2, prolate spheroids.

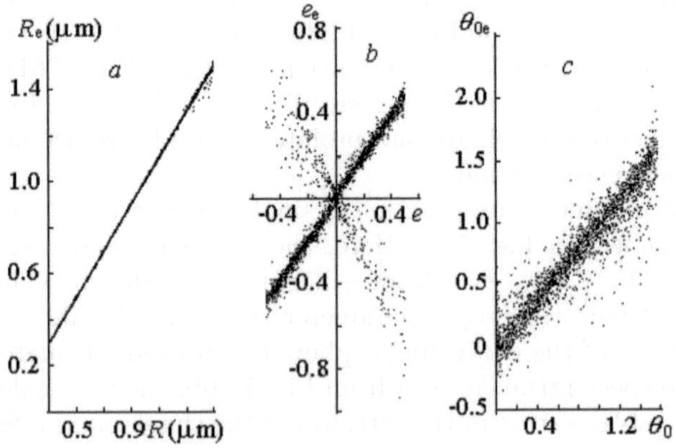

Fig. 7.46. Retrieved parameters R_e, e_e, and θ_{0e} of a particle from a mixture of oblate and prolate particles. Measuring error $\delta_e = 1\%$. Quadrupole discrimination function. Neural network was trained with noise in input data at $\delta_{\mathrm{tr}} = 0.5\%$.

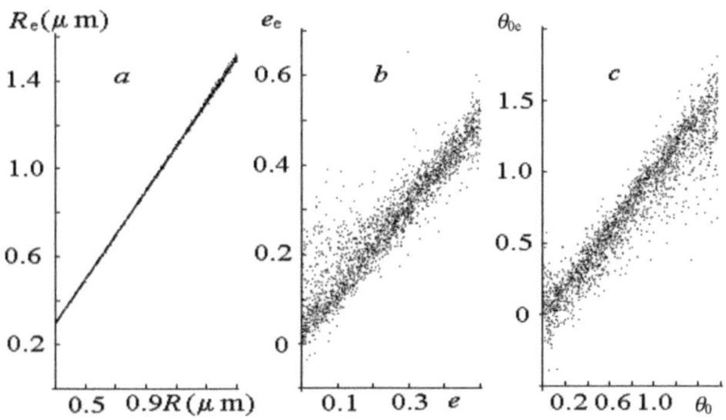

Fig. 7.47. Retrieved parameters R_e, e_e, and θ_{0e} of a particle from a mixture of prolate and cylindrical particles. Measuring error $\delta_e = 1\%$. Quadrupole discrimination function. Neural network was trained with noise in input data at $\delta_{\text{tr}} = 0.5\%$.

7.8 Conclusion

We described the principles of operation and training of some kinds of neural networks used as the inverse problem solution in scattering media optics. The MLP NN, RBF NN, HO NN, and Sequences of Neural Networks were discussed.

The results obtained for particle parameter retrieval by these networks for some ranges of parameters are outlined below shortly.

The errors of radius and refractive index retrieval using MLP NN with one inner layer containing 10 neurons were investigated as the example of the problem of retrieval of the size and the refractive index of spherical homogeneous particles with the radius of 1– 31.6 μm and the relative refractive index of 1.05–1.35 according to the data on the intensity of radiation scattered at angles $\theta_\alpha = 25°$, $30°$, $35°$, $40°$, and $45°$ at the field of view $10°$ and illumination by monochromatic radiation with the wavelength $\lambda_0 = 0.55$ μm. The average relative error (see Eq. (7.31)) of radius retrieval is equal to $\overline{\delta R} = 0.058$; the average relative error of refractive index retrieval (see Eq. (32)) is equal to $\overline{\delta n} = 0.084$ if there is no noise in the system. The errors of particle parameter retrieval typically increase with the particle size decreasing.

Substantial increase in accuracy of the parameter retrieval can be achieved if the particle is illuminated by polychromatic radiation. At the exposure of particles by uniformly distributed radiation in the range of wavelengths λ of 0.65–0.45 μm, the average relative errors of radius and refractive index retrieval by the three-layer MLP NN with 10 neurons in the inner layer are $\overline{\delta R} = 0.032$ and $\overline{\delta n} = 0.046$, respectively. They are less than on illumination by monochromatic light. This is due to the fact that the dependences of intensity of scattered radiation on the refractive index are smoothed out (see Figs 7.8 and 7.9).

At increase in the number of neurons in the network, the errors are reduced and tend to saturation, when the number of neurons in the hidden layer is 30–40. The errors in retrieval using a three-layer neural network with 30 neurons in the hidden layer are $\overline{\delta R} = 0.023$ and $\overline{\delta n} = 0.030$. At this, the network training time significantly increases. It is about five times more than the training time of a neural network with 10 neurons. Sensitivity to the measurement errors increases with the number of neurons.

RBF NN commonly have one inner layer with neurons. Gaussian function is usually used as the activation function. In contrast to the MLP NN, each neuron has properties that differ from those of other neurons. At this, the number of unknown internal network parameters sharply increases. Usually, to simplify the problem, a part of the internal parameters is specified in advance, and the other part is determined from the condition of the average error minimization. We set the matrix of shift factors in a random manner and use the limited-memory BFGS method (Gorban and Rossiev, 1996) to find the form factors K_{ij}^m. RBF NN with five inputs and 30 neurons has 465 unknown form factors, which require nonlinear optimization for their determination, while a three-layer MLP NN with the same number of inputs and 30 neurons in the hidden layer has only 211 unknown internal parameters. The time of training of the RBF NN significantly exceeds the time of training of the MLP NN. However, accuracy rates are nearly equal: $\overline{\delta R} = 0.021$ and $\overline{\delta n} = 0.031$ for the RBF NN and $\overline{\delta R} = 0.023$ and $\overline{\delta n} = 0.030$ for the MLP NN. At this, the noise resistance of the RBF NN is slightly lower.

For the HO NN, problem of neural network training is reduced to solution of a system of linear algebraic equations. The unique solution of this system is the advantage of the HO NN. However, the accuracy of parameter retrieval using the HO NN is significantly lower than that obtained by the MLP NN and RBF NN. The average relative errors of retrieval of the radius and refractive index by the HO NN of the fourth order are $\overline{\delta R} = 0.047$ and $\overline{\delta n} = 0.068$, respectively (see Table 7.1).

The sequence of neural networks consists of a global neural network trained on the initial variation range of parameters and several local neural networks trained on the local variation ranges of parameters. The use of local variation ranges of parameters with boundaries given in Table 7.1 allowed a reduction in the errors of $\overline{\delta R}$ and $\overline{\delta n}$ at zero noise to the values of 0.018 and 0.023, respectively.

The dependence of the intensity of scattered radiation on the scattering angle has characteristic beatings. Their frequency highly correlates with the size of the particles. This correlation allows one to determine the radius of the particle with the error $\overline{\delta R} = 0.018$ at the unknown refractive index of particles. The use of MLP NN allows a reduction in the average relative error at zero noise to the value of $\overline{\delta R} = 0.003$. The relative difference in the results for parameter retrieval performed by the neural network and the fitting method does not exceed 4% to 5%.

HO NN to determine the radius, real, and imaginary parts of the refractive index of homogeneous spherical particles by the intensity of light scattered in the range of angles 10° to 60° are considered. The results for particles with $0.6\ \mu\text{m} < R < 10.6\ \mu\text{m}$, $1.02 < n < 1.38$, $0 < k < 0.03$ are shown. The errors of retrieval of R, n, and k by a four-point single-level HO NN at $k < 0.03$ do not exceed 0.1 μm, 0.02, and 0.003, respectively. Maximum errors take place for small particles with a small refractive index. The errors decrease with increasing particle size.

The method for the construction of multilevel neuron networks to characterize arbitrary oriented optically soft non-absorbing spheroidal particles is discussed. We used two-layer neural networks with the linear activation function trained on the overlapped subintervals of the initial range of parameters, the discrete Fourier transform over the azimuthal angle, and the recursion procedure of parameter retrieval. The errors in retrieved particles sizes and orientation are investigated as a function of particle parameters and measuring errors. The errors of retrieval by a three-level neural network are the following: $\sigma R = 0.003$ μm, $\sigma e = 0.02$, $\sigma\theta_0 = 6°$. The coefficients of the influence of multiplicative measuring errors are $\eta R = 0.0015$ μm/%, $\eta R = 0.011/\%$, $\eta\theta = 2°/\%$. The equivolume radius of spheroids and parameter e varied from 0.3 to 1.5 μm and 0 to 0.5, respectively; the ranges of θ_0 and φ_0 were from 0 to 0.5π and 0 to 2π, respectively. The refractive index of particle n varied from 1.01 to 1.02; $\lambda = 0.5$ μm. Numerical modeling for optically soft spheroids has shown that the method is a perspective for characterizing nonspherical particle parameters from the measurements of scattered light intensity.

The problem in the retrieval of sizes of an individual optically soft particle taken from binary mixtures of either oblate and prolate spheroids or cylinders and oblate spheroids is considered (Berdnik et al., 2004a). The multilevel neural networks method with a linear activation function and the method of the discrimination functions are used. Neural networks to retrieve characteristics of cylinders and oblate and prolate spheroids are designed. The errors in retrieved particle characteristics are investigated for the radius of an equivolume sphere in the range of 0.3–1.5 μm, shape parameter $e = (a - b)/(a + b)$ of spheroidal and cylindrical particles from -0.5 to $+0.5$ and 0 to 0.5, respectively. The refractive index of particle n varied from 1.01 to 1.02; $\lambda = 0.5$ μm.

The results of numerical modeling show that the error in the retrieved radius of the equivolume sphere taken from the considered binary mixtures is the same as for an ensemble of particles belonging to only one kind. The errors in the retrieved shape parameter e and particle orientation direction θ_0 in mixtures increase significantly in comparison with the ensemble consisting of one kind of particle. The dominant reason is the errors in particle kind determination by the discrimination functions.

Acknowledgments

The investigations of the authors were supported by the Belarusian Republican Foundation for Fundamental Research (BRFFR), the scientific state program of Belarus 'Electronics and Photonics,' the scientific state program of Belarus 'Advanced Medical Technologies,' the Russian Foundation for Basic Research (RFBR): project 12-01-00333, and the NATO Science for Peace Program (SfP): project 977976.

References

Abramowitz, M. and Stegun, I.A. (eds) (1964) *Handbook of Mathematical Functions with Formulas, Graphs and Mathematical Tables*, Applied Mathematics Series-55 (AMS-55), National Bureau of Standards.

Ackleson, S.G. and Spinrad, R.W. (1988) 'Size and refractive index of individual marine particles: A flow cytometric approach,' *Appl. Opt.*, **27**(7), 1270–7.

Babenko, V.A., Astafyeva, L.A. and Kuzmin, V.N. (2003) *Electromagnetic Scattering in Disperse Media*, Springer, Praxis Publishing, Berlin.

Barber, P.W. and Wang, D.S. (1978) 'Rayleigh–Gans–Debye applicability to scattering by nonspherical particles,' *Appl. Opt.*, **17**(6), 797–803.

Bartholdi, M., Salzman, G.C., Hielbert, R.D. and Kerker, M. (1980) 'Differential Light Scattering photometer for rapid analysis of single particles in flow,' *Appl. Opt.*, **19**, 1573–81.

Berdnik, V.V. and Loiko, V.A. (2005) 'Sizing of spheroidal and cylindrical particles in a binary mixture by measurement of scattered light intensity: Application of neural networks,' *J. Quant. Spectrosc Radiat. Transfer.*, **91**, 1–10.

Berdnik, V.V. and Loiko, V.A. (2006) 'Particle sizing by multiangle light scattering data using the high order neural networks,' *J. Quant. Spectrosc Radiat. Transfer.*, **100**(1), 55–63.

Berdnik, V.V. and Loiko, V.A. (2009) 'Retrieval of size and refractive index of spherical particles by multiangle light scattering: Neural network method application,' *Appl. Opt.*, **48**(32), 6178–87.

Berdnik, V.V. and Loiko, V.A. (2010) 'Evaluation of the sizes of spherical particles on angular dependence of scattered radiation by a neural networks method,' Proceedings of the 13 All-Russia Scientifically Technical Conferences 'Neuroinformatika 2011', Moscow, Part 3, pp. 87–95 (in Russian).

Berdnik, V.V., Gilev, K., Shvalov, A., Maltsev, V.P. and Loiko, V.A. (2006) 'Characterization of spherical particles using high-order neural networks and scanning flow cytometry,' *J. Quantitative Spectrosk & Radiative Transfer.*, **102**(1), 62–72.

Berdnik, V.V., Mukhamedjarov, R.D. and Loiko, V.A. (2004a) 'Application of the neural network method for determining the characteristics of homogeneous spherical particles,' *Optics and Spectroscopy*, **92**, 285–91.

Berdnik, V.V., Mukhamedjarov, R.D. and Loiko, V.A. (2004b) 'Characterization of optically soft spheroidal particles by multiangle light-scattering data by use of the neural-networks method,' *Opt. Lett.*, **29**, 1019–21.

Berdnik, V.V., Mukhamedjarov, R.D. and Loiko, V.A. (2004c) 'Sizing of soft spheroidal particles by multiangle scattered light intensity data: Application of neural networks,' *Journal of Quant. Spectr. & Radiative Transfer.*, **89**, 279–89.

Bishop, C.M. (1995) *Neural Networks for Pattern Recognition*, Clarendon Press, Oxford.

Black, D., McQuay, M. and Bonin, M. (1996) 'Laser based techniques for particle size measurement: A review of sizing methods and their industrial applications,' *Prog. Energ. Combust.*, **22**, 267–306.

Bohren, C.F. and Huffmen, D.R. (1983) *Absorption and Scattering of Light by Small Particles*, John Wiley & Sons, New York.

Borovoi, A., Naats, E., Oppel, U. and Grishin, I. (2000) 'Shape characterization of a large nonspherical particle by use of its Fraunhofer diffraction pattern,' *Appl. Opt.*, **39**, 1989–97.

Capps, C.D., Henning, R.L. and Hess, G.M. (1982) 'Analytic inversion of remote-sensing data,' *Appl. Opt.*, **21**(19), 3581–7.

Castagner, J.-L. and Bigio, I.J. (2006) 'Polar nephelometer based on a rotational confocal imaging setup,' *Appl. Opt.*, **45**(10), 2232–9.

de Boor, C. (1978) *A Practical Guide to Splines*, Springer–Verlag, New York.

Deirmendjian, D. (1969) *Electromagnetic Scattering on Spherical Polydispersions*, The RAND Corporation, Santa Monica, CA.

Dennis, J.E. and Schnabel, R.B. (1983) *Numeric Methods for Unconstrained Optimization and Nonlinear Equations*, Prentice Hall, Englewood Cliffs, NJ.

Fukunaga, K. (1990) *Introduction to Statistical Pattern Recognition*, Academic, New York.

Gill, P.E., Murray, W. and Wright, M.H. (1982) *Practical Optimization*, Emerald Group Publishing Limited, West Yorkshire, UK.

Gorban, A.N. and Rossiev, D.A. (1996) *Neural Networks on Personal Computers*, Novosibirsk, Nauka (in Russian).
Haykin, S. (1998) *Neural Networks: A Comprehensive Foundation*, 2nd edn, Prentice Hall, Englewood Cliffs, NJ.
Heffels, C.M.G., Heitzman, D., Hirleman, E.D. and Scarlett, B. (1994) 'The use of azimuthal intensity variations in diffraction patterns for particle shape characterization,' *Part. Part. Syst. Charact.*, **11**, 194–9.
Ishimaru, A. (1978) *Wave Propagation and Scattering in Random Media*, Vol. 1: *Single Scattering and Transport Theory*, Academic Press, New York, San Francisco, London, Part 2.6.
Ishimaru, A., Marks, R.J., Tsang, L., Lam, C.M., Park, D.C. and Kitamura, S. (1990) 'Particle-size distribution determination using optical sensing and neural networks,' *Opt. Lett.*, **15**, 1221–3.
Kokhanovsky, A.A. (2004) *Light Scattering Media Optics: Problems and Solutions*, Springer, Berlin, Heidelberg, New York.
Liu, D. and Nocedal, J. (1989) 'On the limited memory BFGS method for large scale optimization,' *Mathematical Programming B.*, **45**, 503–28.
Ludlow, K. and Everitt, J. (2000) 'Inverse Mie problem,' *J. Opt. Soc. Am. A.*, **17**, 2229–35.
Maltsev, V.P. (2000) 'Scanning flow cytometry for individual particle analysis,' *Review of Scientific Instruments*, **71**(1), 243–55.
Maltsev, V.P. and Lopatin, V.N. (1997) 'A parametric solution of the inverse light-scattering problem for individual spherical particles,' *Appl. Opt.*, **37**(21), 6102–8.
Maltsev, V.P. and Semyanov, K.A. (2004) *Characterization of Bio-Particles from Light Scattering*, VSP, Utrecht.
Masciangioli, T. and Alper, J. (eds) (2012) *Challenges in Characterizing Small Particles: Exploring Particles from Nano to Macroscales*, NAS Press.
Newton, R.G. (1982) *Scattering Theory of Waves and Particles*, Couver Dover Publications Inc., Mineola, NY.
Nigmatullin, R.R. and Smith, G. (2005) 'The generalized mean value function approach: A new statistical tool for the detection of weak signals in spectroscopy,' *J. Phys. D: Appl. Phys.*, **38**, 328–37.
Pluchino, A.B., Goldberg, S.S., Dowling, J.M. and Randall, C.M. (1980) Refractive-index measurements of single micron-sized carbon particles,' *Appl. Opt.*, **19**(19), 3370–2.
Quist, G.M. and Wyatt, P.J. (1985) 'Empirical solution to the inverse scattering problem by the optical strip-map technique,' *J. Opt. Soc. Am. A.*, **2**, 1979–86.
Riefler, N. and Wriedt, T. (2008) 'Intercomparison of inversion algorithms for particle-sizing using Mie scattering,' *Part. Part. Syst. Charact.*, **25**, 216–30.
Riley, J.B. and Agrawal, Y.G. (1991) 'Sampling and inversion of data in diffraction particle sizing,' *Appl. Opt.*, **30**(33), 4800–17.
Ronliang, X. (2000) *Particle Characterization: Light Scattering Methods*, Kluwer Academic Publishers, The Netherlands.
Schnablegger, H. and Glatter, O. (1991) 'Optical sizing of small colloidal particles: An optimized regularization technique,' *Appl. Opt.*, **30**(33), 4889–96.
Semyanov, K.A., Tarasov, P.A., Zharinov, A.E., Chernyshev, A.V., Hoekstra, A.G. and Maltsev, V.P. (2004) 'Single-particle sizing from light scattering by spectral decomposition,' *Appl. Opt.*, **43**(26), 5110–15.
Spinrad, R.W. and Brown, J.F. (1986) 'Relative refraction index of marine microorganisms: A technique for flow cytometric estimation,' *Appl. Opt.*, **25**(12), 1930–4.
Ulanowski, Z., Wang, Z., Kaye, P.H. and Ludlow, I.K. (1998) 'Application of neural networks to the inverse light scattering problem for spheres,' *Appl. Opt.*, **37**, 4027–33.

Wang, Z., Ulanowski, Z. and Kaye, P.H. (1999) 'On solving the inverse scattering problem with RBF neural networks: Noise-free case,' *Neural Comput. & Applic.*, **8**, 177–86.

Ye, M., Wang, S., Lu, Y., Hu, T., Zhu, Z. and Xu, Y. (1999) 'Inversion of particle-size distribution from angular light-scattering data with genetic algorithms,' *Appl. Opt.*, **38**(12), 2677–85.

Zakovic, S., Ulanowski, Z.J. and Bartholomew-Biggs, M.C. (1998) 'Application of global optimization to particle identification problem using light scattering,' *Inverse Prob.*, **14**, 1053–67.

Zakovic, S., Ulanowski, Z.J. and Bartholomew-Biggs, M.C. (2005) 'Using global optimization for a microparticle identification problem with noise data,' *Journal of Global Optimization*, **32**(3), 325–47.

Index

MIX
Papier aus verantwortungsvollen Quellen
Paper from responsible sources
FSC® C105338

If you have any concerns about our products,
you can contact us on
ProductSafety@springernature.com

In case Publisher is established outside the EU,
the EU authorized representative is:
Springer Nature Customer Service Center GmbH
Europaplatz 3, 69115 Heidelberg, Germany

Printed by Libri Plureos GmbH
in Hamburg, Germany